INTRODUCTORY BIOTECHNOLOGY

INTRODUCTORY BIOTECHNOLOGY

Peter Kennedy
Santiago Salinas

Kruger Brentt
Publishers

2024

Kruger Brentt Publishers UK. LTD.
Company Number 9728962

Regd. Office: 68 St Margarets Road, Edgware, Middlesex HA8 9UU

ISBN: 978-1-78715-262-5

For information on all our publications visit our website at http://krugerbrentt.com/

PREFACE

Biotechnology, a groundbreaking field at the intersection of biology and technology, has emerged as a transformative force shaping the way we understand, manipulate, and harness living organisms for the betterment of society. The foundations of biotechnology lie in the seamless integration of biological principles with technological advancements, unlocking a realm of possibilities that spans from healthcare to agriculture and environmental sustainability. At its core, biotechnology leverages the intricate machinery of living systems, such as cells and genes, to develop innovative solutions that address complex challenges and improve the quality of human life. This interdisciplinary science draws from genetics, microbiology, biochemistry, and engineering, among other fields, weaving together a tapestry of knowledge that propels scientific discovery and drives the development of cutting-edge applications with profound implications for the future. As we delve into the foundations of biotechnology, we embark on a journey through the molecular intricacies of life, exploring the ways in which scientists and researchers manipulate biological entities to unlock unprecedented potential for innovation and progress.

The significance of the foundations of biotechnology reverberates across various facets of human existence, fundamentally altering the landscape of scientific inquiry and technological progress. At its core, biotechnology serves as a catalyst for innovation in medicine, agriculture, and industry, offering solutions to some of the most pressing challenges facing humanity. In the realm of healthcare, biotechnological advancements have paved the way for groundbreaking therapies, personalized medicine, and the development of vaccines that combat infectious diseases. In agriculture, the manipulation of crops at the genetic level has enhanced yields, improved nutritional content, and increased resistance to pests and environmental stressors. Moreover, biotechnology has become a cornerstone in the pursuit of sustainable practices, contributing to the development of biofuels, waste remediation, and environmentally friendly manufacturing processes. As we unravel the intricacies of biotechnological foundations, we uncover the immense potential to revolutionize how we approach global issues, emphasizing a future where science and technology work harmoniously to address the complex challenges of our time.

The book provides a holistic approach to biotechnology, covering essential topics from the basic principles of life to advanced laboratory techniques and applications in the field. Complex scientific concepts are explained in a clear and accessible manner, ensuring that readers with varying levels of familiarity with biotechnology can grasp the material. Emphasis is placed on the practical applications of biotechnological principles, providing readers with real-world insights into the industry. The book includes detailed discussions on laboratory instruments, measurements, and precision, offering practical insights into the tools and techniques used by biotechnicians. Chapters such as tructure and Function of Cells, Nucleic Acids- Structure & Functions and rotein- Structure and Functions provide an in-depth exploration of crucial biological components. The text is designed to reflect the latest advancements and trends in biotechnology, ensuring that readers are equipped with current and relevant information. This book is suitable for students pursuing undergraduate in biotechnology, biology, biochemistry, or related fields. It serves as a foundational text, offering a systematic and in-depth understanding of the essential concepts in biotechnology.

We are grateful to all those persons as well as various books, manuals, periodicals, newsletters, magazines, journals etc. that helped in the preparation of this book. In spite of the best efforts, it is possible that some errors may have occurred into the compilation and editing of the book. Further queries, constructive suggestions and criticisms for the further improvement of the book are always welcome and shall be thankfully acknowledged.

Peter Kennedy

Santiago Salinas

CONTENTS

CHAPTER-1

THE STUDY OF LIFE: AN INTRODUCTION

INTRODUCTION: AN OVERVIEW

Textbook introduction and summary: Given the broad definition of biotechnology applications and products, it is easy to see how there is enormous overlap within the fields of cellular biology, microbiology, chemistry, and biomedical engineering. It is the goal of this textbook to provide foundational knowledge to begin building your biotechnology tool kit and enter an exciting career of making a difference through biotechnology.

Figure 1.1 Some foods produced by microorganisms. Some of the products derived from the use of prokaryotes in early biotechnology include (a) cheese, (b) wine, (c) beer and bread, and (d) yogurt. (credit bread: modification of work by F. Rodrigo/Wikimedia Commons; credit wine: modification of work by Jon Sullivan; credit beer and bread: modification of work by Kris Miller; credit yogurt: modification of work by Jon Sullivan)

According to the United Nations Convention on *Biological Diversity, biotechnology* is *"any technological application that uses biological systems, living organisms, or derivatives thereof, to make or modify products or processes for specific use."* The concept of "specific use" typically involves a commercial application or benefit to humanity. Genetic engineering, artificial selection, antibiotic production, and cell culture are current topics of study in biotechnology and will be described in later chapters. However, humans were using microbes to create useful products long before Karl Ereky, a Hungarian engineer, coined the term biotechnology (https://en.wikipedia.org /wiki/Biotechnology). Some of the products of this early biotechnology are as familiar as cheese, bread, wine, beer, and yogurt, which employ both bacteria and other microbes, such as yeast, a fungus (Figure).

Ancient Biotechnological Practices: Cheese, Bread, Wine, Beer, and Yogurt

Cheese production began around 4,000 to 7,000 years ago when humans began to breed animals and process their milk. Fermentation, in this case, preserves nutrients: Milk will spoil relatively quickly, but when processed as cheese, it is more stable. As for ***beer***, the oldest records of brewing are about 6,000 years old and were an integral part of the Sumerian culture. Evidence indicates that the Sumerians discovered fermentation by chance. ***Wine*** has been produced for about 4,500 years, and evidence suggests that cultured milk products, like yogurt, have existed for at least 4,000 years.

In the early twentieth century, scientists gained a greater understanding of microbiology and explored ways of manufacturing specific products. In 1917, Chaim Weizmann first used a pure microbiological culture in an industrial process that of manufacturing corn starch using *Clostridium acetobutylicum*, to produce acetone, which was used to manufacture explosives during World War I. Shortly thereafter, in 1928, Alexander Fleming discovered the mold ***Penicillium***. His work led to the purification of the ***antibiotic*** compound formed by the mold by Howard Florey, Ernst Boris Chain, and Norman Heatley – to form what we today know as ***penicillin***. In 1940, penicillin became available for medicinal use to treat bacterial infections in humans.

The New Biotechnology

The field of modern biotechnology is generally thought of as having been born in 1971 when Paul Berg's (Stanford) experiments in ***gene splicing*** had early success. Herbert W. Boyer (Univ. Calif. at San Francisco) and Stanley N. Cohen (Stanford) significantly advanced the new technology in 1972 by transferring genetic material into a bacterium, such that the imported material would be reproduced, giving birth to the field of recombinant DNA technology. The commercial viability of a biotechnology industry was significantly expanded on June 16, 1980, when ***the***

United States Supreme Court ruled that a genetically modified microorganism could be patented. Technology breakthroughs since the 1980s, such as Polymerase Chain Reaction, Sanger Sequencing, Whole Genome Sequencing, and more recently, CRISPR have brought forth a new age of Biotechnology and products.

Given the broad definition of biotechnology applications and products, it is easy to see how there is enormous overlap within the fields of cellular biology, microbiology, chemistry, and biomedical engineering. It is the goal of this textbook to provide foundational knowledge to begin building your biotechnology toolkit and enter an exciting career of making a difference through biotechnology.

This textbook first explores the fundamentals of laboratory science and biotechnology. Unit 1 begins with a dive into the foundation of biology and chemistry by asking what is life and what are the molecules of life? The end of the unit offers a primer into laboratory science; preparing solutions and operating basic lab equipment. Unit 2 brings readers through a vital tool of biomanufacturing, understanding the growth and control of microbes. Many biotechnology products are made by harnessing the work of microbes, and this unit explores this invisible and intriguing world of microorganisms. The final unit, molecular biotechnology, offers a more in-depth look at how biological molecules such as DNA and protein are manipulated into creating useful products.

THEMES AND PRINCIPLES IN THE FIELD OF BIOLOGY

Biology is the science that studies life. What exactly is life? This may sound like a silly question with an obvious answer, but it is not easy to define life. For example, a branch of biology called virology studies viruses, which exhibit some of the characteristics of living entities but lack others. It turns out that although viruses can attack living organisms, cause diseases, and even reproduce, they do not meet the criteria that biologists use to define life.

From its earliest beginnings, biology has wrestled with four questions: What are the shared properties that make something "alive"? How do those various living things function? When faced with the remarkable diversity of life, how do we organize the different kinds of organisms so that we can better understand them? And, finally—what biologists ultimately seek to understand—how did this diversity arise and how is it continuing? As new organisms are discovered every day, biologists continue to seek answers to these and other questions.

Properties of Life

All groups of living organisms share several key characteristics or functions: order, sensitivity or response to stimuli, reproduction, adaptation, growth and development, regulation/homeostasis, and energy processing. When viewed together, these eight characteristics serve to define life.

Order

Organisms are highly organized structures that consist of one or more cells. Even very simple, single-celled organisms are remarkably complex. Inside each cell, atoms make up molecules. These in turn make up cell components or organelles. Multicellular organisms, which may consist of millions of individual cells, have an advantage over single-celled organisms in that their cells can be specialized to perform specific functions, and even sacrificed in certain situations for the good of the organism as a whole. How these specialized cells come together to form organs such as the heart, lung, or skin in organisms like the toad shown in Figure 1.2 will be discussed later.

Figure 1.2 A toad represents a highly organized structure consisting of cells, tissues, organs, and organ systems. (credit: "Ivengo(RUS)"/Wikimedia Commons)

Sensitivity or Response to Stimuli

Organisms respond to diverse stimuli. For example, plants can bend toward a source of light or respond to touch (Figure 1.3). Even tiny bacteria can move toward or away from chemicals (a process called chemotaxis) or light (phototaxis). Movement toward a stimulus is considered a positive response, while movement away from a stimulus is considered a negative response.

Reproduction

Single-celled organisms reproduce by first duplicating their DNA, which is the genetic material, and then dividing it equally as the cell prepares to divide to form two new cells. Many multi-cellular organisms (those made up of more than one cell) produce specialized reproductive cells that will form new individuals.

When reproduction occurs, DNA containing genes is passed along to an organism's offspring. These genes are the reason that the offspring will belong to the same species and will have characteristics similar to the parent, such as fur colour and blood type.

Figure 1.3 The leaves of this sensitive plant (Mimosa pudica) will instantly droop and fold when touched. After a few minutes, the plant returns to its normal state. (credit: Alex Lomas)

Adaptation

All living organisms exhibit a "fit" to their environment. Biologists refer to this fit as adaptation and it is a consequence of evolution by natural selection, which operates in every lineage of reproducing organisms. Examples of adaptations are diverse and unique, from heat-resistant Archaea that live in boiling hot springs to the tongue length of a nectar-feeding moth that matches the size of the flower from which it feeds. Adaptations enhance the reproductive potential of the individual exhibiting them, including their ability to survive to reproduce. Adaptations are not constant. As an environment changes, natural selection causes the characteristics of the individuals in a population to track those changes.

Growth and Development

Organisms grow and develop according to specific instructions coded for by their genes. These genes provide instructions that will direct cellular growth and development, ensuring that a species' young (Figure 1.4) will grow up to exhibit many of the same characteristics as its parents.

Figure 1.4 Although no two look alike, these kittens have inherited genes from both parents and share many of the same characteristics. (credit: Pieter & Renée Lanser)

Regulation/Homeostasis

Even the smallest organisms are complex and require multiple regulatory mechanisms to coordinate internal functions, such as the transport of nutrients, response to stimuli, and coping with environmental stresses. **Homeostasis** (literally, "steady state") refers to the relatively stable internal environment required to maintain life. For example, organ systems such as the digestive or circulatory systems perform specific functions like carrying oxygen throughout the body, removing wastes, delivering nutrients to every cell, and cooling the body.

Figure 1.5 Polar bears and other mammals living in ice-covered regions maintain their body temperature by generating heat and reducing heat loss through thick fur and a dense layer of fat under their skin. (credit: "longhorndave"/Flickr)

To function properly, cells require appropriate conditions such as proper temperature, pH, and concentrations of diverse chemicals. These conditions may, however, change from one moment to the next. Organisms are able to maintain homeostatic internal conditions within a narrow range almost constantly, despite environmental changes, by activation of regulatory mechanisms. For example, many organisms regulate their body temperature in a process known as thermo regulation. Organisms that live in cold climates, such as the polar bear (Figure 1.5), have body structures that help them withstand low temperatures and conserve body heat. In hot climates, organisms have methods (such as perspiration in humans or panting in dogs) that help them to shed excess body heat.

Energy Processing

All organisms (such as the California condor shown in Figure 1.6) use a source of energy for their metabolic activities. Some organisms capture energy from the Sun and convert it into chemical energy in food; others use chemical energy from molecules they take in.

Figure 1.6 A lot of energy is required for a California condor to fly. Chemical energy derived from food is used to power flight. California condors are an endangered species; scientists have strived to place a wing tag on each bird to help them identify and locate each individual bird. (credit: Pacific Southwest Region U.S. Fish and Wildlife)

Evolution

The diversity of life on Earth is a result of mutations, or random changes in hereditary material over time. These mutations allow the possibility for organisms to adapt to a changing environment. An organism that evolves characteristics fit for the environment will have greater reproductive success, subject to the forces of natural selection.

Levels of Organization of Living Things

Living things are highly organized and structured, following a hierarchy on a scale from small to large. The **atom** is the smallest and most fundamental unit of matter that retains the properties of an element. It consists of a nucleus surrounded by electrons. Atoms form molecules. A **molecule** is a chemical structure consisting of at least two atoms held together by a chemical bond. Many molecules that are biologically important are **macromolecules**, large molecules that are typically formed by combining smaller units called monomers. An example of a macromolecule is deoxyribonucleic acid (DNA) (Figure 1.7), which contains the instructions for the functioning of the organism that contains it.

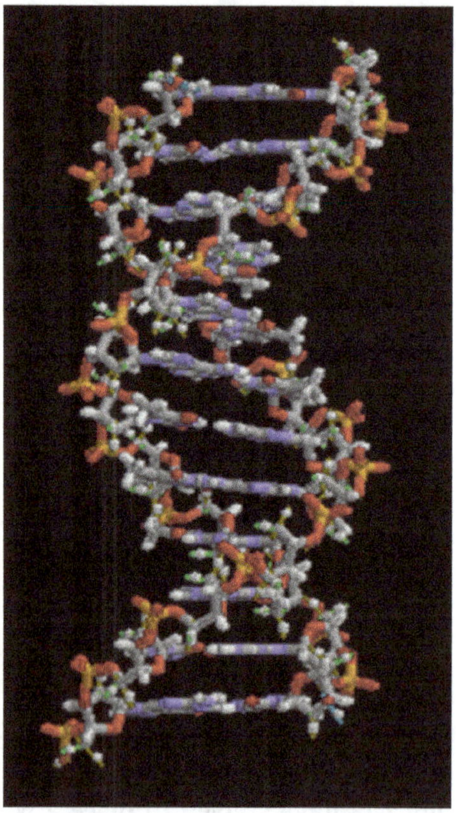

Figure 1.7 A molecule, like this large DNA molecule, is composed of atoms. (credit: "Brian0918"/Wikimedia Commons)

Concept in Action

Some cells contain aggregates of macromolecules surrounded by membranes; these are called **organelles.** Organelles are small structures that exist within cells and perform specialized functions. All living things are made of cells; the **cell** itself is the smallest fundamental unit of structure and function in living organisms. (This requirement is why viruses are not considered living: they are not made of cells. To make new viruses, they have to invade and hijack a living cell; only then can they obtain the materials they need to reproduce.) Some organisms consist of a single cell and others are multicellular. Cells are classified as prokaryotic or eukaryotic. **Prokaryotes** are single-celled organisms that lack organelles surrounded by a membrane and do not have nuclei surrounded by nuclear membranes; in contrast, the cells of **eukaryotes** do have membrane-bound organelles and nuclei.

In most multicellular organisms, cells combine to make **tissues**, which are groups of similar cells carrying out the same function. **Organs** are collections of tissues grouped together based on a common function. Organs are present not only in animals but also in plants. An **organ system** is a higher level of organization that consists of functionally related organs. For example vertebrate animals have many organ systems, such as the circulatory system that transports blood throughout the body and to and from the lungs; it includes organs such as the heart and blood vessels. **Organisms** are individual living entities. For example, each tree in a forest is an organism. Single-celled prokaryotes and single-celled eukaryotes are also considered organisms and are typically referred to as microorganisms.

Visual Connection

All the individuals of a species living within a specific area are collectively called a **population.** For example, a forest may include many white pine trees. All of these pine trees represent the population of white pine trees in this forest. Different populations may live in the same specific area. For example, the forest with the pine trees includes populations of flowering plants and also insects and microbial populations. A **community** is the set of populations inhabiting a particular area. For instance, all of the trees, flowers, insects, and other populations in a forest form the forest's community. The forest itself is an ecosystem. An **ecosystem** consists of all the living things in a particular area together with the abiotic, or non-living, parts of that environment such as nitrogen in the soil or rainwater. At the highest level of organization (Figure 1.8), the **biosphere** is the collection of all ecosystems, and it represents the zones of life on Earth. It includes land, water, and portions of the atmosphere.

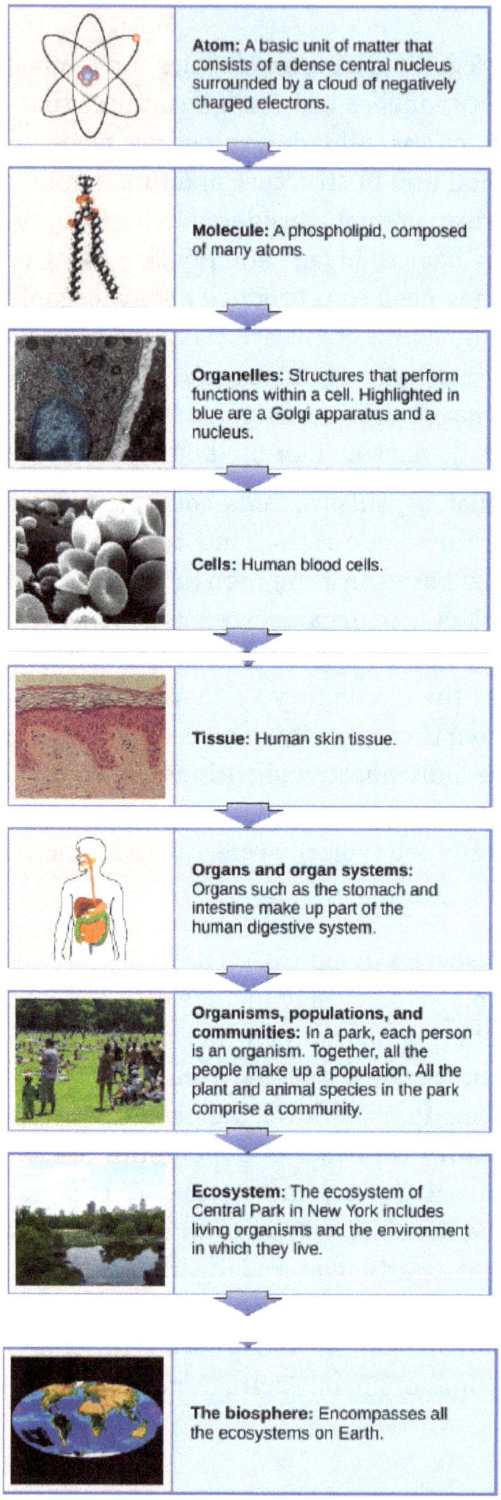

Figure 1.8 From an atom to the entire Earth, biology examines all aspects of life

The Diversity of Life

The science of biology is very broad in scope because there is a tremendous diversity of life on Earth. The source of this diversity is **evolution**, the process of gradual change during which new species arise from older species. Evolutionary biologists study the evolution of living things in everything from the microscopic world to ecosystems.

In the 18th century, a scientist named Carl Linnaeus first proposed organizing the known species of organisms into a hierarchical taxonomy. In this system, species that are most similar to each other are put together within a grouping known as a genus. Furthermore, similar genera (the plural of genus) are put together within a family. This grouping continues until all organisms are collected together into groups at the highest level. The current taxonomic system now has eight levels in its hierarchy, from lowest to highest, they are: species, genus, family, order, class, phylum, kingdom, domain. Thus species are grouped within genera, genera are grouped within families, families are grouped within orders, and so on (Figure 1.9).

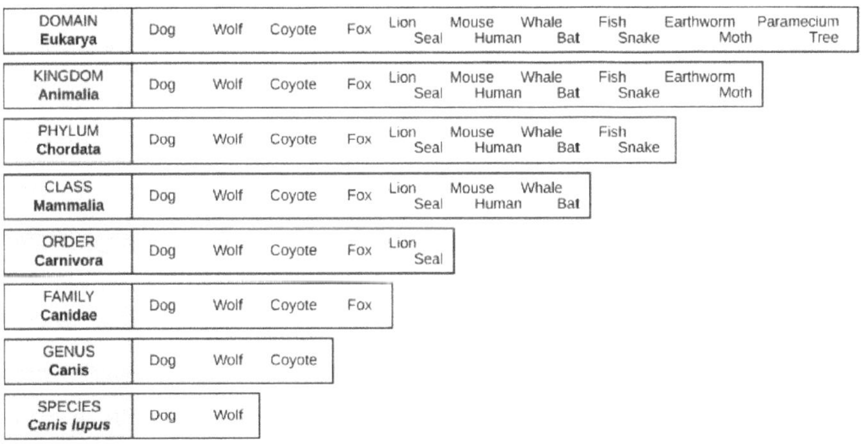

Figure 1.9 This diagram shows the levels of taxonomic hierarchy for a dog, from the broadest category—domain—to the most specific—species.

The highest level, domain, is a relatively new addition to the system since the 1970s. Scientists now recognize three domains of life, the Eukarya, the Archaea, and the Bacteria. The domain Eukarya contains organisms that have cells with nuclei. It includes the kingdoms of fungi, plants, animals, and several kingdoms of protists. The Archaea, are single-celled organisms without nuclei and include many extremophiles that live in harsh environments like hot springs. The Bacteria are another quite different group of single-celled organisms without nuclei (Figure 1.10). Both the Archaea and the Bacteria are prokaryotes, an informal name for cells without nuclei. The recognition in the 1970s that certain "bacteria," now known as the Archaea, were as different genetically and biochemically from other bacterial cells as they were from eukaryotes, motivated the recommendation to divide life into three domains. This dramatic change in our knowledge of the tree of life demonstrates that classifications are not permanent and will change when new information becomes available.

In addition to the hierarchical taxonomic system, Linnaeus was the first to name organisms using two unique names, now called the binomial naming system. Before Linnaeus, the use of common names to refer to organisms caused confusion because there were regional differences in these common names. Binomial names consist of the genus name (which is capitalized) and the species name (all lower-case). Both names are set in italics when they are printed. Every species is given a unique binomial which is recognized the world over, so that a scientist in any location can know which organism is being referred to. For example, the North American blue jay is known uniquely as *Cyanocitta cristata*. Our own species is *Homo sapiens*.

(a) (b) (c) (d)

Figure 1.10 These images represent different domains. The scanning electron micrograph shows (a) bacterial cells belong to the domain Bacteria, while the (b) extremophiles, seen all together as colored mats in this hot spring, belong to domain Archaea. Both the (c) sunflower and (d) lion are part of domain Eukarya.

Evolution In ACTION

Carl Woese and the Phylogenetic Tree

The evolutionary relationships of various life forms on Earth can be summarized in a phylogenetic tree. A **phylogenetic tree** is a diagram showing the evolutionary relationships among biological species based on similarities and differences in genetic or physical traits or both. A phylogenetic tree is composed of branch points, or nodes, and branches. The internal nodes represent ancestors and are points in evolution when, based on scientific evidence, an ancestor is thought to have diverged to form two new species. The length of each branch can be considered as estimates of relative time.

In the past, biologists grouped living organisms into five kingdoms: animals, plants, fungi, protists, and bacteria. The pioneering work of American microbiologist Carl Woese in the early 1970s has shown, however, that life on Earth has evolved along three lineages, now called domains—Bacteria, Archaea, and Eukarya. Woese proposed the domain as a new taxonomic level and Archaea as a new domain, to reflect the new phylogenetic tree (Figure 1.11). Many organisms belonging to the Archaea domain live under extreme conditions and are called extremophiles. To construct his tree, Woese used genetic relationships rather than similarities based on morphology (shape). Various genes were used in phylogenetic studies. Woese's tree was constructed from comparative sequencing of the genes that are universally distributed, found in some slightly altered form in every organism, conserved (meaning that these genes have remained only slightly changed throughout evolution), and of an appropriate length.

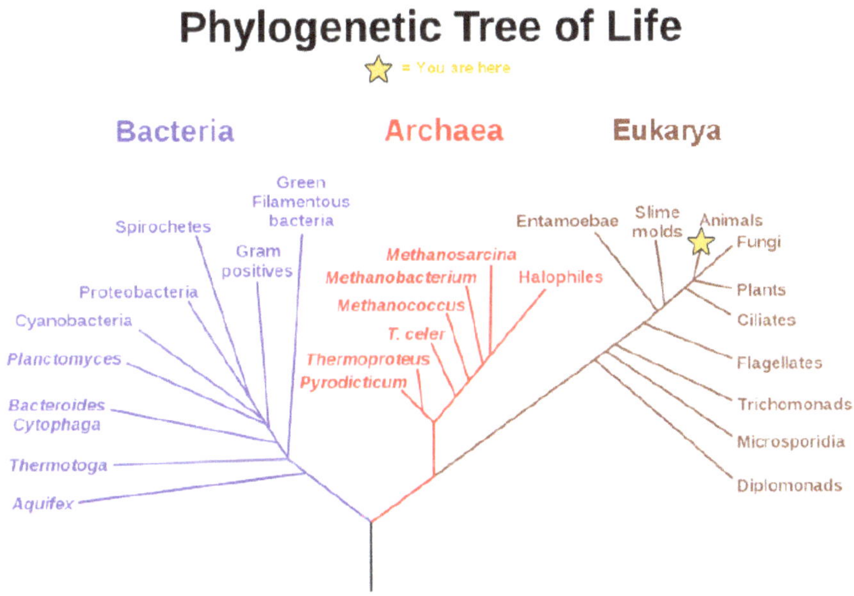

Figure 1.11 This phylogenetic tree was constructed by microbiologist Carl Woese using genetic relationships. The tree shows the separation of living organisms into three domains: Bacteria, Archaea, and Eukarya. Bacteria and Archaea are organisms without a nucleus or other organelles surrounded by a membrane and, therefore, are prokaryotes.

Branches of Biological Study

The scope of biology is broad and therefore contains many branches and sub disciplines. Biologists may pursue one of those sub disciplines and work in a more focused field. For instance, molecular biology studies biological processes at the molecular level, including interactions among molecules such as DNA, RNA, and proteins, as well as the way they are regulated. Microbiology is the study of the structure and function of microorganisms. It is quite a broad branch itself, and depending on the subject of study, there are also microbial physiologists, ecologists, and geneticists, among others.

Another field of biological study, neurobiology, studies the biology of the nervous system, and although it is considered a branch of biology, it is also recognized as an interdisciplinary field of study known as neuroscience. Because of its interdisciplinary nature, this sub discipline studies different functions of the nervous system using molecular, cellular, developmental, medical, and computational approaches.

Figure 1.12 Researchers work on excavating dinosaur fossils at a site in Castellón, Spain.

Paleontology, another branch of biology, uses fossils to study life's history (Figure 1.12). Zoology and botany are the study of animals and plants, respectively. Biologists can also specialize as biotechnologists, ecologists, or physiologists, to name just a few areas. Biotechnologists apply the knowledge of biology to create useful products. Ecologists study the interactions of organisms in their environments. Physiologists study the workings of cells, tissues and organs. This is just a small sample of the many fields that biologists can pursue. From our own bodies to the world we live in, discoveries in biology can affect us in very direct and important ways. We depend on these discoveries for our health, our food sources, and the benefits provided by our ecosystem. Because of this, knowledge of biology can benefit us in making decisions in our day-to-day lives.

The development of technology in the twentieth century that continues today, particularly the technology to describe and manipulate the genetic material, DNA, has transformed biology. This transformation will allow biologists to continue to understand the history of life in greater detail, how the human body works, our human origins, and how humans can survive as a species on this planet despite the stresses caused by our increasing numbers. Biologists continue to decipher huge mysteries about life suggesting that we have only begun to understand life on the planet, its history, and our relationship to it. For this and other reasons, the knowledge of biology gained through this textbook and other printed and electronic media should be a benefit in whichever field you enter.

Careers in ACTION

Forensic Scientist

Forensic science is the application of science to answer questions related to the law. Biologists as well as chemists and biochemists can be forensic scientists. Forensic scientists provide scientific evidence for\ use in courts, and their job involves examining trace material associated with crimes. Interest in forensic science has increased in the last few years, possibly because of popular television shows that feature forensic scientists on the job. Also, the development of molecular techniques and the establishment of DNA databases have updated the types of work that forensic scientists can do. Their job activities are primarily related to crimes against people such as murder, rape, and assault. Their work involves analyzing samples such as hair, blood, and other body fluids and also processing DNA (Figure 1.13) found in many different environments and materials. Forensic scientists also analyze other biological evidence left at crime scenes, such as insect parts or pollen grains. Students who want to pursue careers in forensic science will most likely be required to take chemistry and biology courses as well as some intensive math courses.

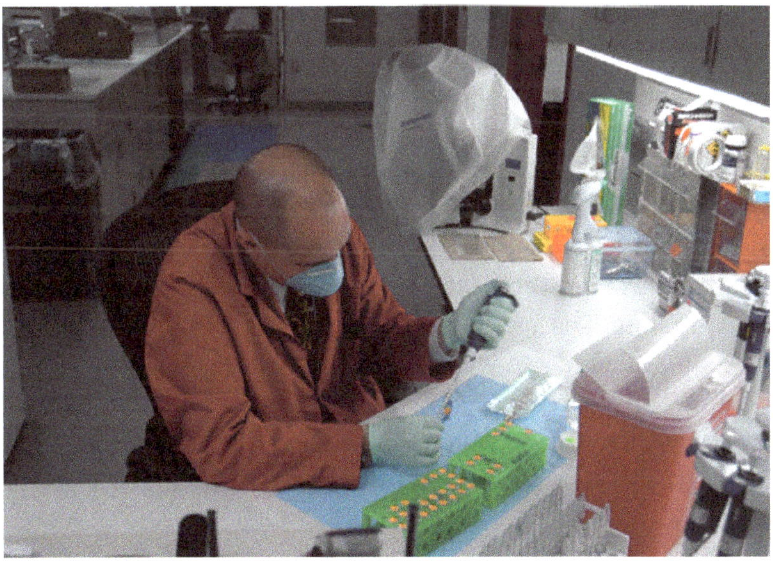

Figure 1.13 This forensic scientist works in a DNA extraction room at the U.S. Army Criminal Investigation Laboratory.)

Scientific Ethics

Scientists must ensure that their efforts do not cause undue damage to humans, animals, or the environment. They also must ensure that their research and communications are free of bias and that they properly balance financial, legal, safety, replicability, and other considerations. Bioethics is an important and continually evolving field, in which researchers collaborate with other thinkers and organizations. They work to define guidelines for current practice, and also continually consider new developments and emerging technologies in order to form answers for the years and decades to come.

Unfortunately, the emergence of bioethics as a field came after a number of clearly unethical practices, where biologists did not treat research subjects with dignity and in some cases did them harm. In the 1932 Tuskegee syphilis study, 399 African American men were diagnosed with syphilis but were never informed that they had the disease, leaving them to live with and pass on the illness to others. Doctors even withheld proven medications because the goal of the study was to understand the impact of untreated syphilis on Black men.

While the decisions made in the Tuskegee study are unjustifiable, some decisions are genuinely difficult to make. For example, bioethicists may examine the implications of gene editing technologies, including the ability to create organisms that may displace others in the environment, as well as the ability to "design" human beings. In that effort, ethicists will likely seek to balance the positive outcomes - such as improved therapies or prevention of certain illnesses - with negative outcomes.

Bioethics are not simple, and often leave scientists balancing benefits with harm. In this text and course, you will discuss medical discoveries that, at their core, have what many consider an ethical lapse. In 1951, Henrietta Lacks, a 30-year-old African American woman, was diagnosed with cervical cancer at Johns Hopkins Hospital. Unique characteristics of her illnesses gave her cells the ability to divide continuously, essentially making them "immortal." Without her knowledge or permission, researchers took samples of her cells and with them created the immortal HeLa cell line. These cells have contributed to major medical discoveries, including the polio vaccine and work related to cancer, AIDS, cell aging, and even very recently in COVID-19 research. For the most part, Lacks has not been credited for her role in those discoveries, and her family has not benefited from the billions of dollars in pharmaceutical profits obtained partly through the use of her cells.

Today, harvesting tissue or organs from a dying patient without consent is not only considered unethical but also illegal, regardless of whether such an act could save other patients' lives. Part of the role of ethics in scientific research is to examine similar issues before, during, and after research or practice takes place, as well as to adhere to established professional principles and consider the dignity and safety of all organisms involved or affected by the work.

THE PROCESS OF SCIENCE

By the end of this section, you will be able to:

- Identify the shared characteristics of the natural sciences
- Understand the process of scientific inquiry
- Compare inductive reasoning with deductive reasoning
- Describe the goals of basic science and applied science

Figure 1.14 Formerly called blue-green algae, the (a) cyanobacteria seen through a light microscope are some of Earth's oldest life forms. These (b) stromatolites along the shores of Lake Thetis in Western Australia are ancient structures formed by the layering of cyanobacteria in shallow waters. (credit a: modification of work by NASA; scalebardata from Matt Russell; credit b: modification of work by Ruth Ellison)

Like geology, physics, and chemistry, biology is a science that gathers knowledge about the natural world. Specifically, biology is the study of life. The discoveries of biology are made by a community of researchers who work individually and together using agreed-on methods. In this sense, biology, like all sciences is a social enterprise like politics or the arts. The methods of science include careful observation, record keeping, logical and mathematical reasoning, experimentation, and submitting conclusions to the scrutiny of others. Science also requires considerable imagination and creativity; a well-designed experiment is commonly described as elegant, or beautiful. Like politics, science has considerable practical implications and some science is dedicated to practical applications, such as the prevention of disease (see Figure 1.15). Other science proceeds largely motivated by curiosity. Whatever its goal, there is no doubt that science, including biology, has transformed human existence and will continue to do so.

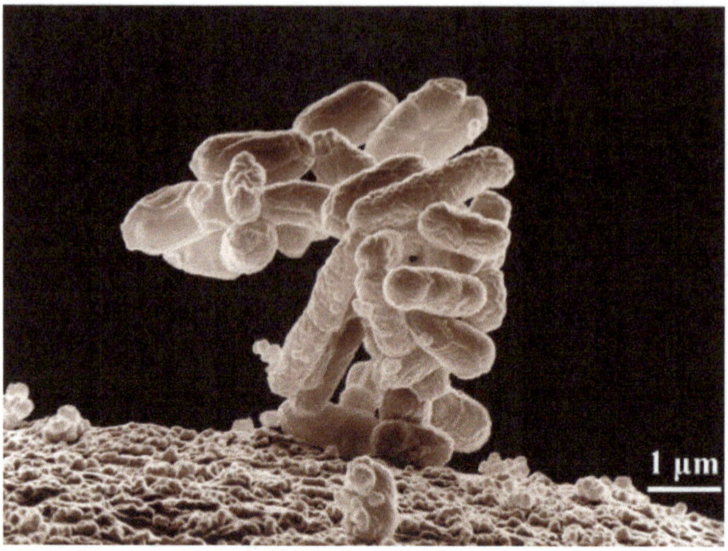

Figure 1.15 Biologists may choose to study Escherichia coli (E. coli), a bacterium that is a normal resident of our digestive tracts but which is also sometimes responsible for disease outbreaks. In this micrograph, the bacterium is visualized using a scanning electron microscope and digital colorization. (credit: Eric Erbe; digital colorization by Christopher Pooley, USDA-ARS)

The Nature of Science

Biology is a science, but what exactly is science? What does the study of biology share with other scientific disciplines? **Science** (from the Latin *scientia*, meaning "knowledge") can be defined as knowledge about the natural world.

Science is a very specific way of learning, or knowing, about the world. The history of the past 500 years demonstrates that science is a very powerful way of knowing about the world; it is largely responsible for the technological revolutions that have taken place during this time. There are however, areas of knowledge and human experience that the methods of science cannot be applied to. These include such things as answering purely moral questions, aesthetic questions, or what can be generally categorized as spiritual questions. Science cannot investigate these areas because they are outside the realm of material phenomena, the phenomena of matter and energy, and cannot be observed and measured.

The **scientific method** is a method of research with defined steps that include experiments and careful observation. The steps of the scientific method will be examined in detail later, but one of the most important aspects of this method is the testing of hypotheses. A **hypothesis** is a suggested explanation for an event, which can be tested. Hypotheses, or tentative explanations, are generally produced within the context of a **scientific theory**. A scientific theory is a generally accepted, thoroughly tested and confirmed explanation for a set of observations or phenomena. Scientific theory is the foundation of scientific knowledge. In addition, in many scientific disciplines (less so in biology) there are **scientific laws,** often expressed in mathematical formulas, which describe how elements of nature will behave under certain specific conditions. There is not an evolution of hypotheses through theories to laws as if they represented some increase in certainty about the world. Hypotheses are the day-to-day material that scientists work with and they are developed within the context of theories. Laws are concise descriptions of parts of the world that are amenable to formulaic or mathematical description.

Natural Sciences

What would you expect to see in a museum of natural sciences? Frogs? Plants? Dinosaur skeletons? Exhibits about how the brain functions? A planetarium? Gems and minerals? Or maybe all of the above? Science includes such diverse fields as astronomy, biology, computer sciences, geology, logic, physics, chemistry, and mathematics (Figure 1.16). However, those fields of science related to the physical world and its phenomena and processes are considered natural sciences. Thus, a museum of natural **sciences** might contain any of the items listed above.

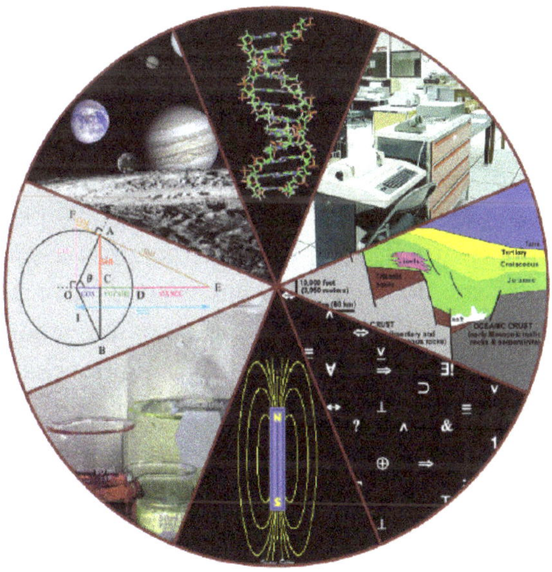

Figure 1.16 Some fields of science include astronomy, biology, computer science, geology, logic, physics, chemistry, and mathematics. (credit: "Image Editor"/Flickr) There is no complete agreement when it comes to defining what the natural sciences include. For some experts, the natural sciences are astronomy, biology, chemistry, earth science, and physics. Other scholars choose to divide natural sciences into life sciences, which study living things and include biology, and physical sciences, which study nonliving matter and include astronomy, physics, and chemistry. Some disciplines such as biophysics and biochemistry build on two sciences and are interdisciplinary.

Scientific Inquiry

One thing is common to all forms of science: an ultimate goal "to know." Curiosity and inquiry are the driving forces for the development of science. Scientists seek to understand the world and the way it operates. Two methods of logical thinking are used: inductive reasoning and deductive reasoning.

Inductive reasoning is a form of logical thinking that uses related observations to arrive at a general conclusion. This type of reasoning is common in descriptive science. A life scientist such as a biologist makes observations and records them. These data can be qualitative (descriptive) or quantitative (consisting of numbers), and the raw data can be supplemented with drawings, pictures, photos, or videos. From many observations, the scientist can infer conclusions (inductions) based on evidence. Inductive reasoning involves formulating generalizations inferred from careful observation and the analysis of a large amount of data. Brain studies often work this way. Many brains are observed while people are doing a task. The part of the brain that lights up, indicating activity, is then demonstrated to be the part controlling the response to that task.

Deductive reasoning or deduction is the type of logic used in hypothesis-based science. In deductive reasoning, the pattern of thinking moves in the opposite direction as compared to inductive reasoning. **Deductive reasoning** is a form of logical thinking that uses a general principle or law to predict specific results. From those general principles, a scientist can deduce and predict the specific results that would be valid as long as the general principles are valid. For example, a prediction would be that if the climate is becoming warmer in a region, the distribution of plants and animals should change. Comparisons have been made between distributions in the past and the present, and the many changes that have been found are consistent with a warming climate. Finding the change in distribution is evidence that the climate change conclusion is a valid one.

Both types of logical thinking are related to the two main pathways of scientific study: descriptive science and hypothesis-based science. **Descriptive** (or discovery) **science** aims to observe, explore, and discover, while hypothesis-based science begins with a specific question or problem and a potential answer or solution that can be tested. The boundary between these two forms of study is often blurred, because most scientific endeavors combine both approaches. Observations lead to questions, questions lead to forming a hypothesis as a possible answer to those questions, and then the hypothesis is tested. Thus, descriptive science and **hypothesis-based science** are in continuous dialogue.

Hypothesis Testing

Biologists study the living world by posing questions about it and seeking science-based responses. This approach is common to other sciences as well and is often referred to as the scientific method. The scientific method was used even in ancient times, but it was first documented by England's Sir Francis Bacon (1561–1626) (Figure 1.17), who set up inductive methods for scientific inquiry. The scientific method is not exclusively used by biologists but can be applied to almost anything as a logical problem-solving method.

Figure 1.17 Sir Francis Bacon is credited with being the first to document the scientific method.

The scientific process typically starts with an observation (often a problem to be solved) that leads to a question. Let's think about a simple problem that starts with an observation and apply the scientific method to solve the problem. One Monday morning, a student arrives at class and quickly discovers that the classroom is too warm.

That is an observation that also describes a problem: the classroom is too warm. The student then asks a question: "Why is the classroom so warm?" Recall that a hypothesis is a suggested explanation that can be tested. To solve a problem, several hypotheses may be proposed. For example, one hypothesis might be, "The classroom is warm because no one turned on the air conditioning." But there could be other responses to the question, and therefore other hypotheses may be proposed. A second hypothesis might be, "The classroom is warm because there is a power failure, and so the air conditioning doesn't work."

Once a hypothesis has been selected, a prediction may be made. A prediction is similar to a hypothesis but it typically has the format "If . . . then . . . ." For example, the prediction for the first hypothesis might be, "If the student turns on the air conditioning, then the classroom will no longer be too warm."

A hypothesis must be testable to ensure that it is valid. For example, a hypothesis that depends on what a bear thinks is not testable, because it can never be known what a bear thinks. It should also be **falsifiable,** meaning that it can be disproven by experimental results. An example of an unfalsifiable hypothesis is "Botticelli's Birth of Venus is beautiful." There is no experiment that might show this statement to be false. To test a hypothesis, a researcher will conduct one or more experiments designed to eliminate one or more of the hypotheses. This is important. A hypothesis can be disproven, or eliminated, but it can never be proven. Science does not deal in proofs like mathematics. If an experiment fails to disprove a hypothesis, then we find support for that explanation, but this is not to say that down the road a better explanation will not be found, or a more carefully designed experiment will be found to falsify the hypothesis.

Each experiment will have one or more variables and one or more controls. A **variable** is any part of the experiment that can vary or change during the experiment. A **control** is a part of the experiment that does not change. Look for the variables and controls in the example that follows. As a simple example, an experiment might be conducted to test the hypothesis that phosphate limits the growth of algae in freshwater ponds. A series of artificial ponds are filled with water and half of them are treated by adding phosphate each week, while the other half are treated by adding a salt that is known not to be used by algae. The variable here is the phosphate (or lack of phosphate), the experimental or treatment cases are the ponds with added phosphate and the control ponds are those with something inert added, such as the salt. Just adding something is also a control against the possibility that adding extra matter to the pond has an effect. If the treated ponds show lesser growth of algae, then we have found support for our hypothesis. If they do not, then we reject our hypothesis. Be aware that rejecting one hypothesis does not determine whether or not the other hypotheses can be accepted; it simply eliminates one hypothesis that is not valid (Figure 1.18). Using the scientific method, the hypotheses that are inconsistent with experimental data are rejected.

In recent years a new approach of testing hypotheses has developed as a result of an exponential rowth of data deposited in various databases. Using computer algorithms and statistical analyses of data in databases, a new field of so-called "data research" (also referred to as "in silico" research) provides new methods of data analyses and their interpretation. This will increase the demand for specialists in both biology and computer science, a promising career opportunity.

Visual Connection

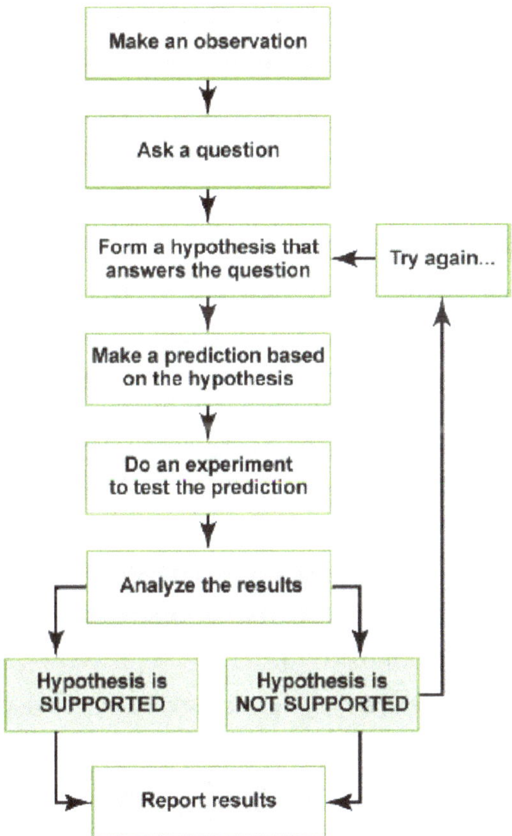

Figure 1.18 The scientific method is a series of defined steps that include experiments and careful observation. If a hypothesis is not supported by data, a new hypothesis can be proposed.

In the example below, the scientific method is used to solve an everyday problem. Which part in the example below is the hypothesis? Which is the prediction? Based on the results of the experiment, is the hypothesis supported? If it is not supported, propose some alternative hypotheses.

- My toaster doesn't toast my bread.
- Why doesn't my toaster work?
- There is something wrong with the electrical outlet.
- If something is wrong with the outlet, my coffeemaker also won't work when plugged into it.
- I plug my coffeemaker into the outlet.
- My coffeemaker works.

In practice, the scientific method is not as rigid and structured as it might at first appear. Sometimes an experiment leads to conclusions that favor a change in approach; often, an experiment brings entirely new scientific questions to the puzzle. Many times, science does not operate in a linear fashion; instead, scientists continually draw inferences and make generalizations, finding patterns as their research proceeds. Scientific reasoning is more complex than the scientific method alone suggests.

Basic and Applied Science

The scientific community has been debating for the last few decades about the value of different types of science. Is it valuable to pursue science for the sake of simply gaining knowledge, or does scientific knowledge only have worth if we can apply it to solving a specific problem or bettering our lives? This question focuses on the differences between two types of science: basic science and applied science.

Basic science or "pure" science seeks to expand knowledge regardless of the short-term application of that knowledge. It is not focused on developing a product or a service of immediate public or commercial value. The immediate goal of basic science is knowledge for knowledge's sake, though this does not mean that in the end it may not result in an application.

In contrast, **applied science** or "technology," aims to use science to solve real-world problems, making it possible, for example, to improve a crop yield, find a cure for a particular disease, or save animals threatened by a natural disaster. In applied science, the problem is usually defined for the researcher.

Some individuals may perceive applied science as "useful" and basic science as "useless." A question these people might pose to a scientist advocating knowledge acquisition would be, "What for?" A careful look at the history of science, however, reveals that basic knowledge has resulted in many remarkable applications of great value. Many scientists think that a basic understanding of science is necessary before an application is developed; therefore, applied science relies on the results generated through basic science. Other scientists think that it is time to move on from basic science and instead to find solutions to actual problems. Both approaches are valid. It is true that there are problems that demand immediate attention; however, few solutions would be found without the help of the knowledge generated through basic science.

One example of how basic and applied science can work together to solve practical problems occurred after the discovery of DNA structure led to an understanding of the molecular mechanisms governing DNA replication. Strands of DNA, unique in every human, are found in our cells, where they provide the instructions necessary for life. During DNA replication, new copies of DNA are made, shortly before a cell divides to form new cells. Understanding the mechanisms of DNA replication enabled scientists to develop laboratory techniques that are now used to identify genetic diseases, pinpoint individuals who were at a crime scene, and determine paternity. Without basic science, it is unlikely that applied science could exist.

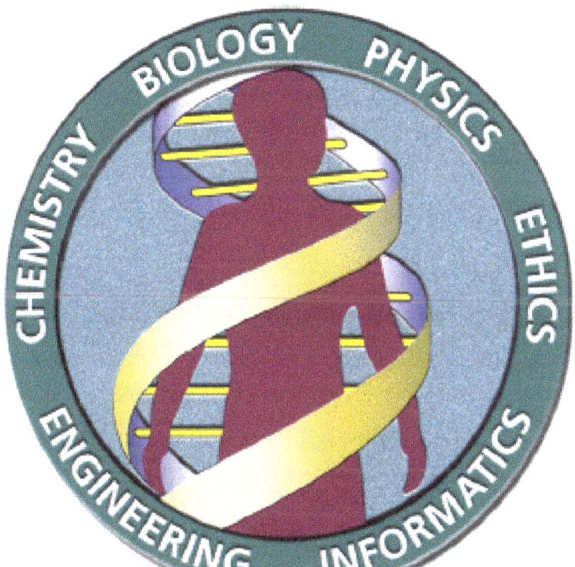

Figure 1.19 The Human Genome Project was a 13-year collaborative effort among researchers working in several different fields of science. The project was completed in 2003. (credit: the U.S. Department of Energy Genome Programs)

Another example of the link between basic and applied research is the Human Genome Project, a study in which each human chromosome was analyzed and mapped to determine the precise sequence of DNA subunits and the exact location of each gene. (The gene is the basic unit of heredity represented by a specific DNA segment that codes for a functional molecule.) Other organisms have also been studied as part of this project to gain a better understanding of human chromosomes. The Human Genome Project (Figure 1.19) relied on basic research carried out with non-human organisms and, later, with the human genome. An important end goal eventually became using the data for applied research seeking cures for genetically related diseases.

While research efforts in both basic science and applied science are usually carefully planned, it is important to note that some discoveries are made by serendipity, that is, by means of a fortunate accident or a lucky surprise. Penicillin was discovered when biologist Alexander Fleming accidentally left a petri dish of Staphylococcus bacteria open. An unwanted mold grew, killing the bacteria. The mold turned out to be Penicillium, and a new critically important antibiotic was discovered. In a similar manner, Percy Lavon Julian was an established medicinal chemist working on a way to mass produce compounds with which to manufacture important drugs. He was focused on using soybean oil in the production of progesterone (a hormone important in the menstrual cycle and pregnancy), but it wasn't until water accidentally leaked into a large soybean oil storage tank thathe found his method. Immediately recognizing the resulting substance as stigmasterol, a primary ingredient in progesterone and similar drugs, he began the process of replicating and industrializing the process in a manner that has helped millions of people. Even in the highly organized world of science, luck—when combined with an observant, curious mind focused on the types of reasoning discussed above—can lead to unexpected breakthroughs.

Reporting Scientific Work

Whether scientific research is basic science or applied science, scientists must share their findings for other researchers to expand and build upon their discoveries. Communication and collaboration within and between sub disciplines of science are key to the advancement of knowledge in science. For this reason, an important aspect of a scientist's work is disseminating results and communicating with peers. Scientists can share results by presenting them at a scientific meeting or conference, but this approach can reach only the limited few who are present. Instead, most scientists present their results in peer-reviewed articles that are published in scientific journals. Peer-reviewed articles are scientific papers that are reviewed, usually anonymously by a scientist's colleagues, or peers. These colleagues are qualified individuals, often experts in the same research area, who judge whether or not the scientist's work is suitable for publication. The process of peer review helps to ensure

that the research described in a scientific paper or grant proposal is original, significant, logical, and thorough. Grant proposals, which are requests for research funding, are also subject to peer review. Scientists publish their work so other scientists can reproduce their experiments under similar or different conditions to expand on the findings.

There are many journals and the popular press that do not use a peer-review system. A large number of online open-access journals, journals with articles available without cost, are now available many of which use rigorous peer-review systems, but some of which do not. Results of any studies published in these forums without peer review are not reliable and should not form the basis for other scientific work. In one exception, journals may allow a researcher to cite a personal communication from another researcher about unpublished results with the cited author's permission.

SUMMARY

Biology is the science that studies living organisms and their interactions with one another and their environments. Science attempts to describe and understand the nature of the universe in whole or in part. Science has many fields; those fields related to the physical world and its phenomena are considered natural sciences.

A hypothesis is a tentative explanation for an observation. A scientific theory is a well-tested and consistently verified explanation for a set of observations or phenomena. A scientific law is a description, often in the form of a mathematical formula, of the behaviour of an aspect of nature under certain circumstances. Two types of logical reasoning are used in science. Inductive reasoning uses results to produce general scientific principles. Deductive reasoning is a form of logical thinking that predicts results by applying general principles. The common thread throughout scientific research is the use of the scientific method. Scientists present their results in peer- reviewed scientific papers published in scientific journals.

Science can be basic or applied. The main goal of basic science is to expand knowledge without any expectation of short-term practical application of that knowledge. The primary goal of applied research, however, is to solve practical problems.

CHAPTER-2

THE KEY BUILDING BLOCKS OF LIFE

INTRODUCTION: AN OVERVIEW

Life, in all its complexity and diversity, is intricately woven from a set of fundamental elements often referred to as the key building blocks. These foundational components form the essence of biological existence, shaping the structures and processes that define living organisms. From the microscopic world of cells to the vast ecosystems that blanket the planet, understanding these key building blocks is essential for unravelling the mysteries of life and exploring the interconnected web of biological systems. In this exploration, we delve into the core components that serve as the basis for the remarkable tapestry of life on Earth.

The elements carbon, hydrogen, nitrogen, oxygen, sulphur, and phosphorus are the key building blocks of the chemicals found in living things. They form the carbohydrates, nucleic acids, proteins, and lipids (all of which will be defined later in this chapter) that are the fundamental molecular components of all organisms. In this chapter, we will discuss these important building blocks and learn how the unique properties of the atoms of different elements affect their interactions with other atoms to form the molecules of life.

Food provides an organism with nutrients—the matter it needs to survive. Many of these critical nutrients come in the form of biological macromolecules, or large molecules necessary for life. These macromolecules are built from different combinations of smaller organic molecules. What specific types of biological macromolecules do living things require? How are these molecules formed? What functions do they serve? In this chapter, we will explore these questions.

THE BUILDING BLOCKS OF MOLECULES

At its most fundamental level, life is made up of matter. **Matter** occupies space and has mass. All matter is composed of **elements**, substances that cannot be broken down or transformed chemically into other substances. Each element is made of atoms, each with a constant number of protons and unique properties. A total of 118 elements have been defined; however, only 92 occur naturally, and fewer than 30 are found in living cells. The remaining 26 elements are unstable and, therefore, do not exist for very long or are theoretical and have yet to be detected.

Each element is designated by its chemical symbol (such as H, N, O, C, and Na), and possesses unique properties. These unique properties allow elements to combine and to bond with each other in specific ways.

Atoms

An atom is the smallest component of an element that retains all of the chemical properties of that element. For example, one hydrogen atom has all of the properties of the element hydrogen, such as it exists as a gas at room temperature, and it bonds with oxygen to create a water molecule. Hydrogen atoms cannot be broken down into anything smaller while still retaining the properties of hydrogen. If a hydrogen atom were broken down into subatomic particles, it would no longer have the properties of hydrogen.

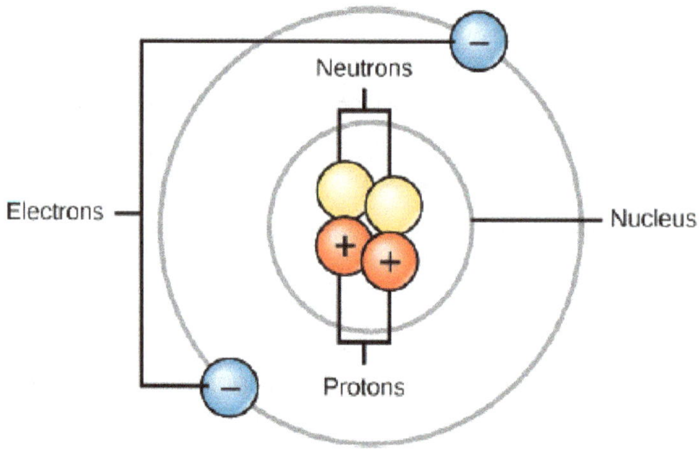

Figure 2.1 Atoms are made up of protons and neutrons located within the nucleus, and electrons surrounding the nucleus.

At the most basic level, all organisms are made of a combination of elements. They contain atoms that combine together to form molecules. In multicellular organisms, such as animals, molecules can interact to form cells that combine to form tissues, which make up organs. These combinations continue until entire multicellular organisms are formed.

All atoms contain protons, electrons, and neutrons (Figure 2.1). The most common isotope of hydrogen (H) is the only exception and is made of one proton and one electron with no neutrons. A **proton** is a positively charged particle that resides in the **nucleus** (the core of the atom) of an atom and has a mass of 1 and a charge of +1. An **electron** is a negatively charged particle that travels in the space around the nucleus. In other words, it resides outside of the nucleus. It has a negligible mass and has a charge of −1.

Neutrons, like protons, reside in the nucleus of an atom. They have a mass of 1 and no charge. The positive (protons) and negative (electrons) charges balance each other in a neutral atom, which has a net zero charge.

Because protons and neutrons each have a mass of 1, the mass of an atom is equal to the number of protons and neutrons of that atom. The number of electrons does not factor into the overall mass, because their mass is so small.

As stated earlier, each element has its own unique properties. Each contains a different number of protons and neutrons, giving it its own atomic number and mass number. The **atomic number** of an element is equal to the number of protons that element contains. The **mass number**, or atomic mass, is the number of protons plus the number of neutrons of that element. Therefore, it is possible to determine the number of neutrons by subtracting the atomic number from the mass number.

These numbers provide information about the elements and how they will react when combined. Different elements have different melting and boiling points, and are in different states (liquid, solid, or gas) at room temperature. They also combine in different ways. Some form specific types of bonds, whereas others do not. How they combine is based on the number of electrons present. Because of these characteristics, the elements are arranged into the **periodic table of elements**, a chart of the elements that includes the atomic number and relative atomic mass of each element. The periodic table also provides key information about the properties of elements (Figure 2.2)—often indicated by colour-coding. The arrangement of the table also shows how the electrons in each element are organized and provides important details about how atoms will react with each other to form molecules.

Isotopes are different forms of the same element that have the same number of protons, but a different number of neutrons. Some elements, such as carbon, potassium, and uranium, have naturally occurring isotopes. Carbon-12, the most common isotope of carbon, contains six protons and six neutrons. Therefore, it has a mass number of 12 (six protons and six neutrons) and an atomic number of 6 (which makes it carbon). Carbon-14 contains six protons and eight neutrons. Therefore, it has a mass number of 14 (six protons and eight neutrons) and an atomic number of 6, meaning it is still the element carbon. These two alternate forms of carbon are isotopes. Some isotopes are unstable and will lose protons, other subatomic particles, or energy to form more stable elements. These are called **radioactive isotopes** or radioisotopes.

Visual Connection

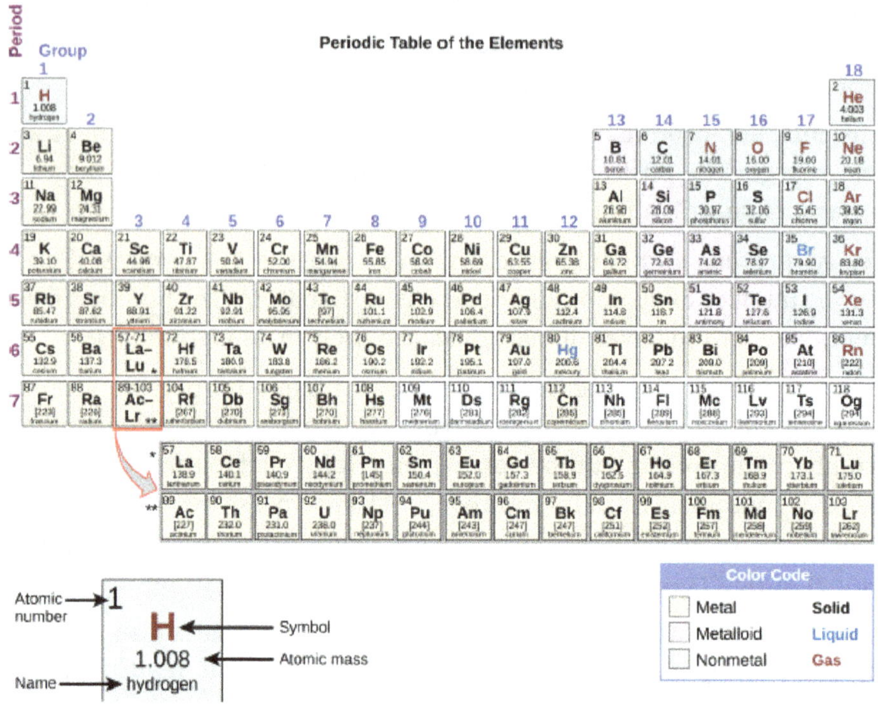

Figure 2.2 Arranged in columns and rows based on the characteristics of the elements, the periodic table provides key information about the elements and how they might interact with each other to form molecules. Most periodic tables provide a key or legend to the information they contain.

How many neutrons do (K) potassium-39 and potassium-40 have, respectively?

Carbon Dating

Carbon-14 (^{14}C) is a naturally occurring radioisotope that is created in the atmosphere by cosmic rays. This is a continuous process, so more ^{14}C is always being created. As a living organism develops, the relative level of ^{14}C in its body is equal to the concentration of ^{14}C in the atmosphere. When an organism dies, it is no longer ingesting ^{14}C, so the ratio will decline. ^{14}C decays to ^{14}N by a process called beta decay; it gives off energy in this slow process.

After approximately 5,730 years, only one-half of the starting concentration of ^{14}C will have been converted to ^{14}N. The time it takes for half of the original concentration of an isotope to decay to its more stable form is called its half-life. Because the half-life of ^{14}C is long, it is used to age formerly living objects, such as fossils. Using the ratio of the ^{14}C concentration found in an object to the amount of ^{14}C detected in the atmosphere, the amount of the isotope that has not yet decayed can be determined. Based on this amount, the age of the fossil can be calculated to about 50,000 years (Figure 2.3). Isotopes with longer half-lives, such as potassium-40, are used to calculate the ages of older fossils. Through the use of carbon dating, scientists can reconstruct the ecology and biogeography of organisms living within the past 50,000 years.

Figure 2.3 The age of remains that contain carbon and are less than about 50,000 years old, such as this pygmy mammoth, can be determined using carbon dating. (credit: Bill Faulkner/NPS)

Chemical Bonds

How elements interact with one another depends on how their electrons are arranged and how many openings for electrons exist at the outermost region where electrons are present in an atom. Electrons exist at energy levels that form shells around the nucleus. The closest shell can hold up to two electrons. The closest shell to the nucleus is always filled first, before any other shell can be filled. Hydrogen has one electron; therefore, it has only one spot occupied within the lowest shell. Helium has two electrons; therefore, it can completely fill the lowest shell with its two electrons. If you look at the periodic table, you will see that hydrogen and helium are the only two elements in the first row. This is because they only have electrons in their first shell. Hydrogen and helium are the only two elements that have the lowest shell and no other shells.

The second and third energy levels can hold up to eight electrons. The eight electrons are arranged in four pairs and one position in each pair is filled with an electron before any pairs are completed.

Looking at the periodic table again (Figure 2.2), you will notice that there are seven rows. These rows correspond to the number of shells that the elements within that row have. The elements within a particular row have increasing numbers of electrons as the columns proceed from left to right. Although each element has the same number of shells, not all of the shells are completely filled with electrons. If you look at the second row of the periodic table, you will find lithium (Li), beryllium (Be), boron (B), carbon (C), nitrogen (N), oxygen (O), fluorine (F), and neon (Ne). These all have electrons that occupy only the first and second shells. Lithium has only one electron in its outermost shell, beryllium has two electrons, boron has three, and so on, until the entire shell is filled with eight electrons, as is the case with neon.

Not all elements have enough electrons to fill their outermost shells, but an atom is at its most stable when all of the electron positions in the outermost shell are filled. Because of these vacancies in the outermost shells, we see the formation of **chemical bonds**, or interactions between two or more of the same or different elements that result in the formation of molecules. To achieve greater stability, atoms will tend to completely fill their outer shells and will bond with other elements to accomplish this goal by sharing electrons, accepting electrons from another atom, or donating electrons to another atom. Because the outermost shells of the elements with low atomic numbers (up to calcium, with atomic number 20) can hold eight electrons, this is referred to as the **octet rule.** An element can donate, accept, or share electrons with other elements to fill its outer shell and satisfy the octet rule.

When an atom does not contain equal numbers of protons and electrons, it is called an **ion.** Because the number of electrons does not equal the number of protons, each ion has a net charge. Positive ions are formed by losing electrons and are called **cations.** Negative ions are formed by gaining electrons and are called **anions.**

For example, sodium only has one electron in its outermost shell. It takes less energy for sodium to donate that one electron than it does to accept seven more electrons to fill the outer shell. If sodium loses an electron, it now has 11 protons and only 10 electrons, leaving it with an overall charge of +1. It is now called a sodium ion.

The chlorine atom has seven electrons in its outer shell. Again, it is more energy-efficient for chlorine to gain one electron than to lose seven. Therefore, it tends to gain an electron to create an ion with 17 protons and 18 electrons, giving it a net negative (–1) charge. It is now called a chloride ion. This movement of electrons from one element to another is referred to as electron transfer. As Figure 2.4 illustrates, a sodium atom (Na) only has one electron in its outermost shell, whereas a chlorine atom (Cl) has seven electrons in its outermost shell. A sodium atom will donate its one electron to empty its shell, and a chlorine atom will accept that electron to fill its shell, becoming chloride. Both ions now satisfy the octet rule and have complete outermost shells. Because the number of electrons is no longer equal to the number of protons, each is now an ion and has a +1 (sodium) or −1 (chloride) charge.

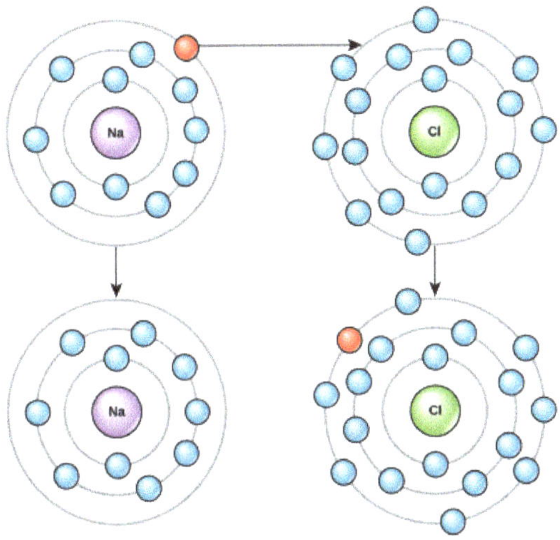

Figure 2.4 Elements tend to fill their outermost shells with electrons. To do this, they can either donate or accept electrons from other elements.

Ionic Bonds

There are four types of bonds or interactions: ionic, covalent, hydrogen bonds, and van der Waals interactions. Ionic and covalent bonds are strong interactions that require a larger energy input to break apart. When an element donates an electron from its outer shell, as in the sodium atom example above, a positive ion is formed. The element accepting the electron is now negatively charged. Because positive and negative charges attract, these ions stay together and form an ionic bond, or a bond between ions. The elements bond together with the electron from one element staying predominantly with the other element. When Na^+ and Cl^- ions combine to produce NaCl, an electron from a sodium atom stays with the other seven from the chlorine atom, and the sodium and chloride ions attract each other in a lattice of ions with a net zero charge.

Covalent Bonds

Another type of strong chemical bond between two or more atoms is a **covalent bond**. These bonds form when an electron is shared between two elements and are the strongest and most common form of chemical bond in living organisms. Covalent bonds form between the elements that make up the biological molecules in our cells. Unlike ionic bonds, covalent bonds do not dissociate in water.

The hydrogen and oxygen atoms that combine to form water molecules are bound together by covalent bonds. The electron from the hydrogen atom divides its time between the outer shell of the hydrogen atom and the incomplete outer shell of the oxygen atom. To completely fill the outer shell of an oxygen atom, two electrons from two hydrogen atoms are needed, hence the subscript "2" in H_2O. The electrons are shared between the atoms, dividing their time between them to "fill" the outer shell of each. This sharing is a lower energy state for all of the atoms involved than if they existed without their outer shells filled.

There are two types of covalent bonds: polar and nonpolar. **Nonpolar covalent bonds** form between two atoms of the same element or between different elements that share the electrons equally. For example, an oxygen atom can bond with another oxygen atom to fill their outer shells. This association is nonpolar because the electrons will be equally distributed between each oxygen atom. Two covalent bonds form between the two oxygen atoms because oxygen requires two shared electrons to fill its outermost shell. Nitrogen atoms will form three covalent bonds (also called triple covalent) between two atoms of nitrogen because each nitrogen atom needs three electrons to fill its outermost shell. Another example of a nonpolar covalent bond is found in the methane (CH_4) molecule. The carbon atom has four electrons in its outermost shell and needs four more to fill it. It gets these four from four hydrogen atoms, each atom providing one. These elements all share the electrons equally, creating four nonpolar covalent bonds (Figure 2.5).

In a **polar covalent bond**, the electrons shared by the atoms spend more time closer to one nucleus than to the other nucleus. Because of the unequal distribution of electrons between the different nuclei, a slightly positive ($\delta+$) or slightly negative ($\delta-$) charge develops. The covalent bonds between hydrogen and oxygen atoms in water are polar covalent bonds. The shared electrons spend more time near the oxygen nucleus, giving it a small negative charge, than they spend near the hydrogen nuclei, giving these molecules a small positive charge.

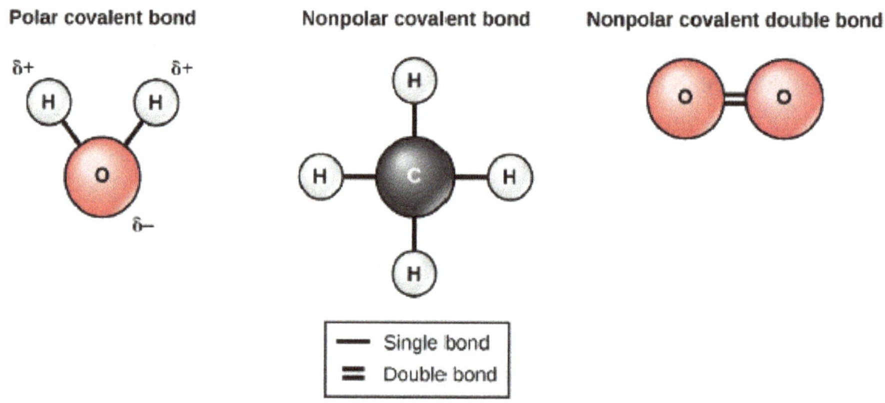

Figure 2.5 The water molecule (left) depicts a polar bond with a slightly positive charge on the hydrogen atoms and a slightly negative charge on the oxygen. Examples of nonpolar bonds include methane (middle) and oxygen (right).

Hydrogen Bonds

Ionic and covalent bonds are strong bonds that require considerable energy to break. However, not all bonds between elements are ionic or covalent bonds. Weaker bonds can also form. These are attractions that occur between positive and negative charges that do not require much energy to break. Two weak bonds that occur frequently are hydrogen bonds and van der Waals interactions. These bonds give rise to the unique properties of water and the unique structures of DNA and proteins.

When polar covalent bonds containing a hydrogen atom form, the hydrogen atom in that bond has a slightly positive charge. This is because the shared electron is pulled more strongly toward the other element and away from the hydrogen nucleus. Because the hydrogen atom is slightly positive (δ+), it will be attracted to neighbouring negative partial charges (δ–). When this happens, a weak interaction occurs between the δ+ charge of the hydrogen atom of one molecule and the δ– charge of the other molecule. This interaction is called a **hydrogen bond**. This type of bond is common; for example, the liquid nature of water is caused by the hydrogen bonds between water molecules (Figure 2.6). Hydrogen bonds give water the unique properties that sustain life. If it were not for hydrogen bonding, water would be a gas rather than a liquid at room temperature.

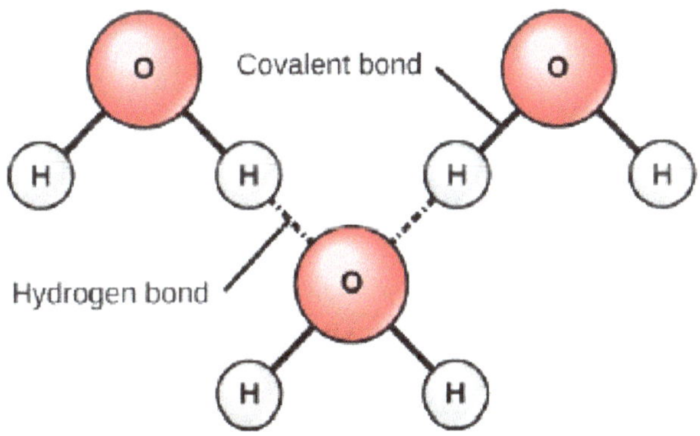

Figure 2.6 Hydrogen bonds form between slightly positive (δ+) and slightly negative (δ–) charges of polar covalent molecules, such as water.

Hydrogen bonds can form between different molecules and they do not always have to include a water molecule. Hydrogen atoms in polar bonds within any molecule can form bonds with other adjacent molecules. For example, hydrogen bonds hold together two long strands of DNA to give the DNA molecule its characteristic double- stranded structure. Hydrogen bonds are also responsible for some of the three-dimensional structure of proteins.

Van Der Waals Interactions

Like hydrogen bonds, **van der Waals interactions** are weak attractions or interactions between molecules. They occur between polar, covalently bound, atoms in different molecules. Some of these weak attractions are caused by temporary partial charges formed when electrons move around a nucleus. These weak interactions between molecules are important in biological systems.

PROPERTIES OF WATER

Do you ever wonder why scientists spend time looking for water on other planets? It is because water is essential to life; even minute traces of it on another planet can indicate that life could or did exist on that planet. Water is one of the more abundant molecules in living cells and the one most critical to life as we know it. Approximately 60–70 percent of your body is made up of water. Without it, life simply would not exist.

Water Is Polar

The hydrogen and oxygen atoms within water molecules form polar covalent bonds. The shared electrons spend more time associated with the oxygen atom than they do with hydrogen atoms. There is no overall charge to a water molecule, but there is a slight positive charge on each hydrogen atom and a slight negative charge on the oxygen atom.

Figure 2.7 As this macroscopic image of oil and water show, oil is a nonpolar compound and, hence, will not dissolve in water. Oil and water do not mix. (credit: Gautam Dogra)

Because of these charges, the slightly positive hydrogen atoms repel each other and form the unique shape seen in Figure 2.6. Each water molecule attracts other water molecules because of the positive and negative charges in the different parts of the molecule. Water also attracts other polar molecules (such as sugars), forming hydrogen bonds. When a substance readily forms hydrogen bonds with water, it can dissolve in water and is referred to as **hydrophilic** ("water-loving").

Hydrogen bonds are not readily formed with nonpolar substances like oils and fats (Figure 2.7). These nonpolar compounds are **hydrophobic** ("water-fearing") and will not dissolve in water.

Water Stabilizes Temperature

The hydrogen bonds in water allow it to absorb and release heat energy more slowly than many other substances. **Temperature** is a measure of the motion (kinetic energy) of molecules. As the motion increases, energy is higher and thus temperature is higher. Water absorbs a great deal of energy before its temperature rises. Increased energy disrupts the hydrogen bonds between water molecules. Because these bonds can be created and disrupted rapidly, water absorbs an increase in energy and temperature changes only minimally. This means that water moderates temperature changes within organisms and in their environments. As energy input continues, the balance between hydrogen-bond formation and destruction swings toward the destruction side. More bonds are broken than are formed. This process results in the release of individual water molecules at the surface of the liquid (such as a body of water, the leaves of a plant, or the skin of an organism) in a process called **evaporation**. Evaporation of sweat, which is 90 percent water, allows for cooling of an organism, because breaking hydrogen bonds requires an input of energy and takes heat away from the body.

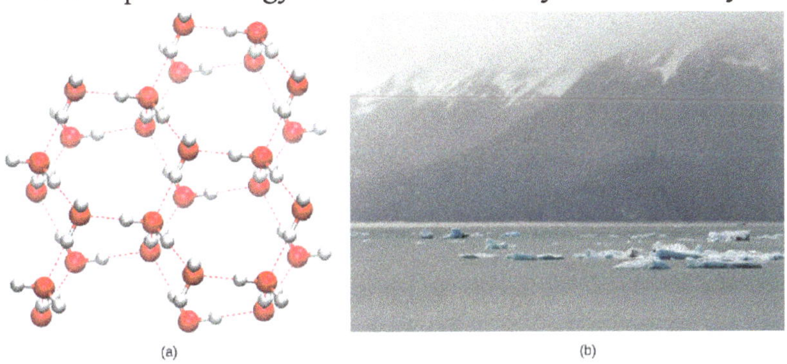

Figure 2.8 (a) The lattice structure of ice makes it less dense than the freely flowing molecules of liquid water. Ice's lower density enables it to (b) float on water. (credit a: modification of work by Jane Whitney; credit b: modification of work by Carlos Ponte)

Conversely, as molecular motion decreases and temperatures drop, less energy is present to break the hydrogen bonds between water molecules. These bonds remain intact and begin to form a rigid, lattice-like structure (e.g., ice) (Figure 2.8a). When frozen, ice is less dense than liquid water (the molecules are farther apart). This means that ice floats on the surface of a body of water (Figure 2.8b). In lakes, ponds, and oceans, ice will form on the surface of the water, creating an insulating barrier to protect the animal and plant life beneath from freezing in the water. If this did not happen, plants and animals living in water would freeze in a block of ice and could not move freely, making life in cold temperatures difficult or impossible.

Excellent Solvent

Because water is polar, with slight positive and negative charges, ionic compounds and polar molecules can readily dissolve in it. Water is, therefore, what is referred to as a **solvent**—a substance capable of dissolving another substance. The charged particles will form hydrogen bonds with a surrounding layer of water molecules. This is referred to as a sphere of hydration and serves to keep the particles separated or dispersed in the water. In the case of table salt (NaCl) mixed in water (Figure 2.9), the sodium and chloride ions separate, or dissociate, in the water, and spheres of hydration are formed around the ions. A positively charged sodium ion is surrounded by the partially negative charges of oxygen atoms in water molecules. A negatively charged chloride ion is surrounded by the partially positive charges of hydrogen atoms in water molecules. These spheres of hydration are also referred to as hydration shells. The polarity of the water molecule makes it an effective solvent and is important in its many roles in living systems.

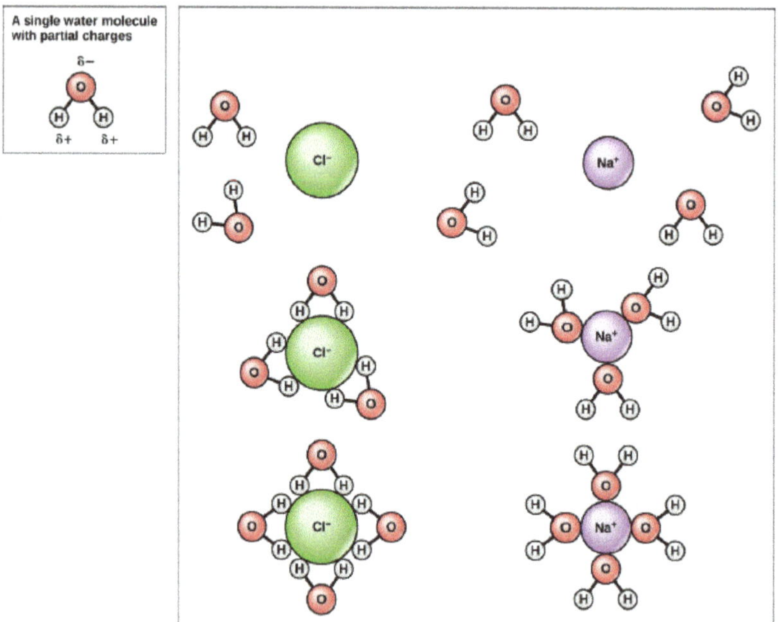

Figure 2.9 When table salt (NaCl) is mixed in water, spheres of hydration form around the ions.

Cohesive

Have you ever filled up a glass of water to the very top and then slowly added a few more drops? Before it overflows, the water actually forms a dome-like shape above the rim of the glass. This water can stay above the glass because of the property of **cohesion.** In cohesion, water molecules are attracted to each other (because of hydrogen bonding), keeping the molecules together at the liquid-air (gas) interface, although there is no more room in the glass. Cohesion gives rise to **surface tension**, the capacity of a substance to withstand rupture when placed under tension or stress. When you drop a small scrap of paper onto a droplet of water, the paper floats on top of the water droplet, although the object is denser (heavier) than the water. This occurs because of the surface tension that is created by the water molecules. Cohesion and surface tension keep the water molecules intact and the item floating on the top. It is even possible to "float" a steel needle on top of a glass of water if you place it gently, without breaking the surface tension (Figure 2.10).

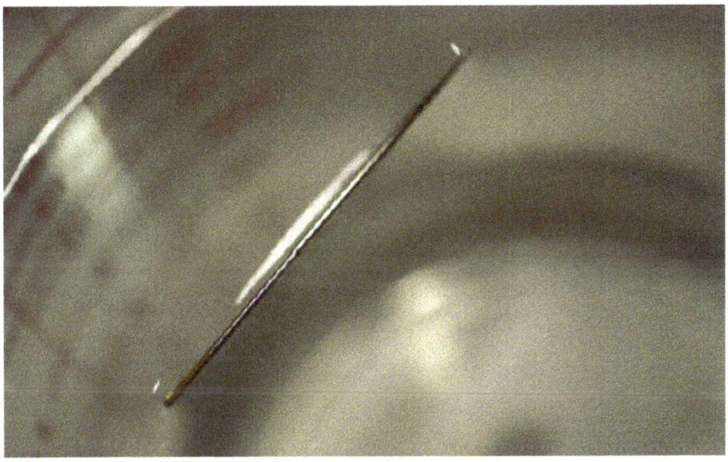

Figure 2.10 The weight of a needle on top of water pulls the surface tension downward; at the same time, the surface tension of the water is pulling it up, suspending the needle on the surface of the water and keeping it from sinking. Notice the indentation in the water around the needle. (credit: Cory Zanker)

These cohesive forces are also related to the water's property of **adhesion**, or the attraction between water molecules and other molecules. This is observed when water "climbs" up a straw placed in a glass of water. You will notice that the water appears to be higher on the sides of the straw than in the middle. This is because the water molecules are attracted to the straw and therefore adhere to it.

Cohesive and adhesive forces are important for sustaining life. For example, because of these forces, water can flow up from the roots to the tops of plants to feed the plant.

Buffers, pH, Acids, and Bases

The pH of a solution is a measure of its acidity or bascicity. You have probably used **litmus paper**, paper that has been treated with a natural water-soluble dye so it can be used as a pH indicator, to test how much acid or base (basicity) exists in a solution. You might have even used some to make sure the water in an outdoor swimming pool is properly treated. In both cases, this pH test measures the amount of hydrogen ions that exists in a given solution. High concentrations of hydrogen ions yield a low pH, whereas low levels of hydrogen ions result in a high pH. The overall concentration of hydrogen ions is inversely related to its pH and can be measured on the **pH scale** (Figure 2.11). Therefore, the more hydrogen ions present, the lower the pH; conversely, the fewer hydrogen ions, the higher the pH.

The pH scale ranges from 0 to 14. A change of one unit on the pH scale represents a change in the concentration of hydrogen ions by a factor of 10, a change in two units represents a change in the concentration of hydrogen ions by a factor of 100. Thus, small changes in pH represent large changes in the concentrations of hydrogen ions. Pure water is neutral. It is neither acidic nor basic, and has a pH of 7.0. Anything below 7.0 (ranging from 0.0 to 6.9) is acidic, and anything above 7.0 (from 7.1 to 14.0) is alkaline. The blood in your veins is slightly alkaline (pH = 7.4). The environment in your stomach is highly acidic (pH = 1 to 2). Orange juice is mildly acidic (pH = approximately 3.5), whereas baking soda is basic (pH = 9.0).

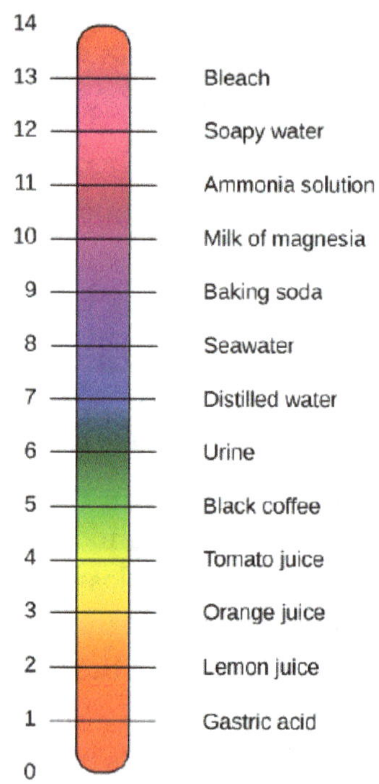

Figure 2.11 The pH scale measures the amount of hydrogen ions (H^+) in a substance. (credit: modification of work by Edward Stevens)

Acids are substances that provide hydrogen ions (H^+) and lower pH, whereas **bases** provide hydroxide ions (OH^-) and raise pH. The stronger the acid, the more readily it donates H^+. For example, hydrochloric acid and lemon juice are very acidic and readily give up H^+ when added to water. Conversely, bases are those substances that readily donate OH^-. The OH^- ions combine with H^+ to produce water, which raises a substance's pH. Sodium hydroxide and many household cleaners are very alkaline and give up OH^- rapidly when placed in water, thereby raising the pH.

Most cells in our bodies operate within a very narrow window of the pH scale, typically ranging only from 7.2 to 7.6. If the pH of the body is outside of this range, the respiratory system malfunctions, as do other organs in the body. Cells no longer function properly, and proteins will break down. Deviation outside of the pH range can induce coma or even cause death.

So how is it that we can ingest or inhale acidic or basic substances and not die? Buffers are the key. **Buffers** readily absorb excess H^+ or OH^-, keeping the pH of the body carefully maintained in the aforementioned narrow range. Carbon dioxide is part of a prominent buffer system in the human body; it keeps the pH within the proper range. This buffer system involves carbonic acid (H_2CO_3) and bicarbonate (HCO^-) anion. If too much H^+ enters the body, bicarbonate will combine with the H^+ to create carbonic acid and limit the decrease in pH. Likewise, if too much OH^- is introduced into the system, carbonic acid will rapidly dissociate into bicarbonate and H^+ ions. The H^+ ions can combine with the OH^- ions, limiting the increase in pH. While carbonic acid is an important product in this reaction, its presence is fleeting because the carbonic acid is released from the body as carbon dioxide gas each time we breathe. Without this buffer system, the pH in our bodies would fluctuate too much and we would fail to survive.

BUFFERING SYSTEMS: HUMAN BLOOD

Buffering System in Action: Bicarbonate

There are many critical buffering systems in the human body responsible for maintaining homeostasis. In the following example, you will explore the transport of oxygen in the blood and factors that affect oxygen binding to hemoglobin and how the bicarbonate buffer system carries CO2 to keep your blood pH in balance.

Once the oxygen diffuses across the alveoli, it enters the bloodstream and is transported to the tissues where it is unloaded, and carbon dioxide diffuses out of the blood and into the alveoli to be expelled from the body. Although gas exchange is a continuous process, the oxygen and carbon dioxide are transported by different mechanisms.

Transport of Oxygen in the Blood

Although oxygen dissolves in blood, only a small amount of oxygen is transported this way. Only 1.5 percent of oxygen in the blood is dissolved directly into the blood itself. Most oxygen—98.5 percent—is bound to a protein called hemoglobin and carried to the tissues.

Hemoglobin

Hemoglobin, or Hb, is a protein molecule found in red blood cells (erythrocytes) made of four subunits: two alpha subunits and two beta subunits (Figure 2.12). Each subunit surrounds a central heme group that contains iron and binds one oxygen molecule, allowing each hemoglobin molecule to bind four oxygen molecules. Molecules with more oxygen bound to the heme groups are brighter red. As a result, oxygenated arterial blood where the HB is carrying four oxygen molecules is bright red, while venous blood that is deoxygenated is darker red.

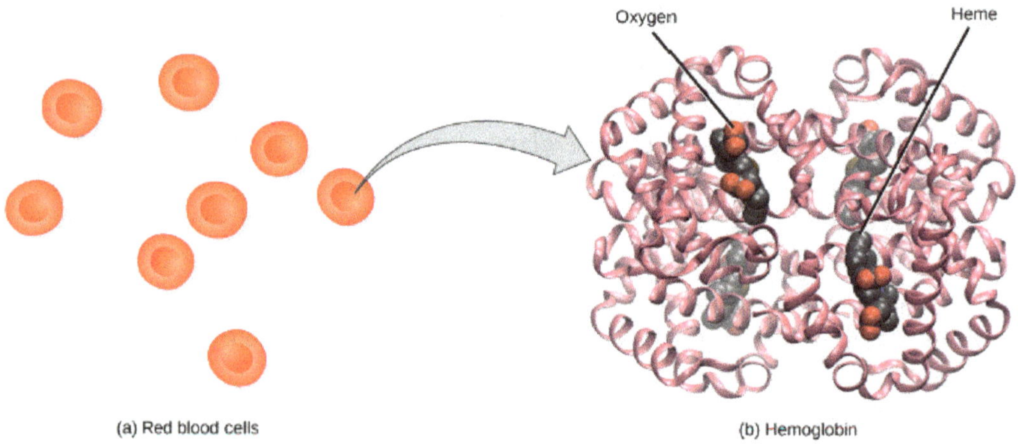

Figure 2.12 The protein inside (a) red blood cells that carries oxygen to cells and carbon dioxide to the lungs is (b) hemoglobin. Hemoglobin is made up of four symmetrical subunits and four heme groups. Iron associated with the heme binds oxygen. It is the iron in hemoglobin that gives blood its red colour.

It is easier to bind a second and third oxygen molecule to Hb than the first molecule. This is because the hemoglobin molecule changes its shape, or conformation, as oxygen binds. The fourth oxygen is then more difficult to bind. The binding of oxygen to hemoglobin can be plotted as a function of the partial pressure of oxygen in the blood (x-axis) versus the relative Hb-oxygen saturation (y-axis). The resulting graph—an **oxygen dissociation curve**—is sigmoidal, or S-shaped (Figure 2.13). As the partial pressure of oxygen increases, the hemoglobin becomes increasingly saturated with oxygen.

Visual Connection

The kidneys are responsible for removing excess H+ ions from the blood. If the kidneys fail, what would happen to blood pH and to hemoglobin affinity for oxygen?

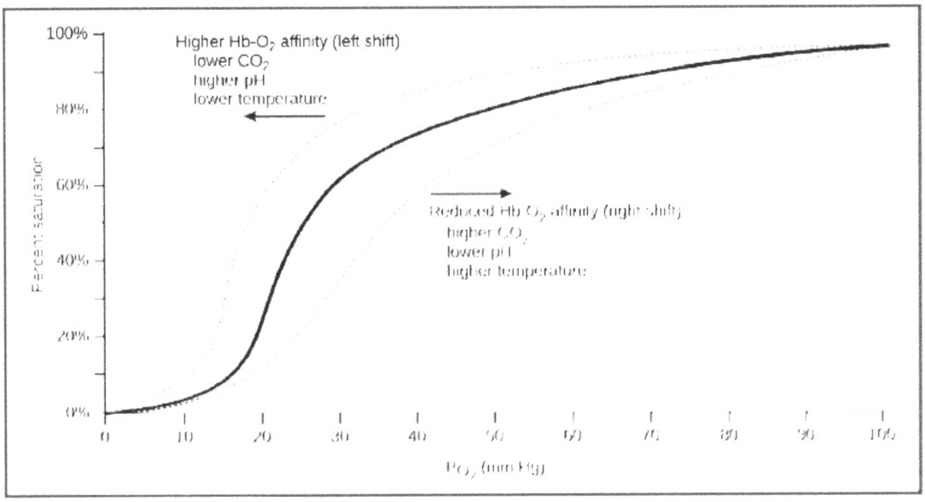

Figure 2.13 The oxygen dissociation curve demonstrates that, as the partial pressure of oxygen increases, more oxygen binds hemoglobin. However, the affinity of hemoglobin for oxygen may shift to the left or the right depending on environmental conditions.

Factors Affecting Oxygen Binding

The oxygen-carrying capacity of hemoglobin determines how much oxygen is carried in the blood. In addition to PO_2, other environmental factors and diseases can affect oxygen carrying capacity and delivery.

Carbon dioxide levels, blood pH, and body temperature affect oxygen-carrying capacity (Figure 2.13). When carbon dioxide is in the blood, it reacts with water to form bicarbonate (HCO^-_3) and hydrogen ions (H^+). As the level of carbon dioxide in the blood increases, more H^+ is produced and the pH decreases. This increase in carbon dioxide and subsequent decrease in pH reduce the affinity of hemoglobin for oxygen. The oxygen dissociates from the Hb molecule, shifting the oxygen dissociation curve to the right. Therefore, more oxygen is needed to reach the same hemoglobin saturation level as when the pH was higher. A similar shift in the curve also results from an increase in body temperature. Increased temperature, such as from increased activity of skeletal muscle, causes the affinity of hemoglobin for oxygen to be reduced.

Diseases like sickle cell anemia and thalassemia decrease the blood's ability to deliver oxygen to tissues and its oxygen-carrying capacity. In **sickle cell anemia**, the shape of the red blood cell is crescent-shaped, elongated, and stiffened, reducing its ability to deliver oxygen (Figure 2.14). In this form, red blood cells cannot pass through the capillaries. This is painful when it occurs. **Thalassemia** is a rare genetic disease caused by a defect in either the alpha or the beta subunit of Hb. Patients with thalassemia produce a high number of red blood cells, but these cells have lower-than-normal levels of hemoglobin. Therefore, the oxygen-carrying capacity is diminished.

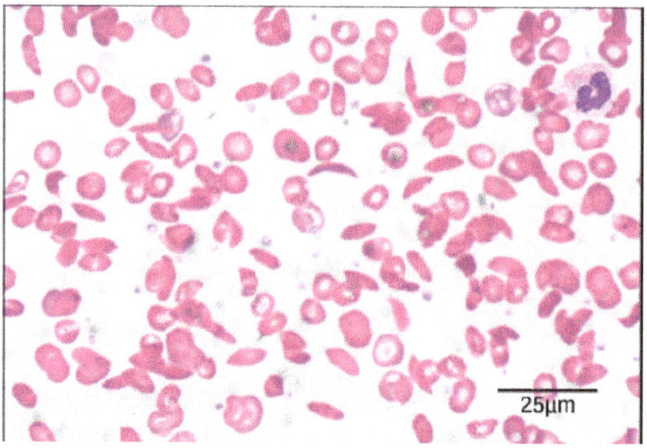

Figure 2.14 Individuals with sickle cell anemia have crescent-shaped red blood cells. (credit: modification of work by Ed Uthman; scale-bar data from Matt Russell)

Transport of Carbon Dioxide in the Blood

Carbon dioxide molecules are transported in the blood from body tissues to the lungs by one of three methods: dissolution directly into the blood, binding to hemoglobin, or carried as a bicarbonate ion. Several properties of carbon dioxide in the blood affect its transport. First, carbon dioxide is more soluble in blood than oxygen. About 5 to 7 percent of all carbon dioxide is dissolved in the plasma. Second, carbon dioxide can bind to plasma proteins or can enter red blood cells and bind to hemoglobin. This form transports about 10 percent of the carbon dioxide. When carbon dioxide binds to hemoglobin, a molecule called **carbamino-hemoglobin** is formed. Binding of carbon dioxide to hemoglobin is reversible. Therefore, when it reaches the lungs, the carbon dioxide can freely dissociate from the hemoglobin and be expelled from the body.

Third, the majority of carbon dioxide molecules (85 percent) are carried as part of the **bicarbonate buffer system**. In this system, carbon dioxide diffuses into the red blood cells. **Carbonic anhydrase (CA)** within the red blood cells quickly converts the carbon dioxide into carbonic acid (H_2CO_3). Carbonic acid is an unstable intermediate molecule that immediately dissociates into **bicarbonate ions (HCO_3^-)** and hydrogen (H+) ions.

Since carbon dioxide is quickly converted into bicarbonate ions, this reaction allows for the continued uptake of carbon dioxide into the blood down its concentration gradient. It also results in the production of H^+ ions. If too much H^+ is produced, it can alter blood pH. However, hemoglobin binds to the free H^+ ions and thus limits shifts in pH. The newly synthesized bicarbonate ion is transported out of the red blood cell into the liquid component of the blood in exchange for a chloride ion (Cl^-); this is called the **chloride shift**. When the blood reaches the lungs, the bicarbonate ion is transported back into the red blood cell in exchange for the chloride ion. The H^+ ion dissociates from the hemoglobin and binds to the bicarbonate ion. This produces the carbonic acid intermediate, which is converted back into carbon dioxide through the enzymatic action of CA. The carbon dioxide produced is expelled through the lungs during exhalation.

$$+ H_2O \leftrightarrow \underset{\text{(carbonic acid)}}{H_2CO_3} \leftrightarrow \underset{\text{(bicarbo}}{HCO_3^{\ -}}$$

The benefit of the bicarbonate buffer system is that carbon dioxide is "soaked up" into the blood with little change to the pH of the system. This is important because it takes only a small change in the overall pH of the body for severe injury or death to result. The presence of this bicarbonate buffer system also allows for people to travel and live at high altitudes: When the partial pressure of oxygen and carbon dioxide change at high altitudes, the bicarbonate buffer system adjusts to regulate carbon dioxide while maintaining the correct pH in the body.

Carbon Monoxide Poisoning

While carbon dioxide can readily associate and dissociate from hemoglobin, other molecules such as carbon monoxide (CO) cannot. Carbon monoxide has a greater affinity for hemoglobin than oxygen. Therefore, when carbon monoxide is present, it binds to hemoglobin preferentially over oxygen. As a result, oxygen cannot bind to hemoglobin, so very little oxygen is transported through the body (Figure 2.15). Carbon monoxide is a colourless, odourless gas and is therefore difficult to detect. It is produced by gas-powered vehicles and tools. Carbon monoxide can cause headaches, confusion, and nausea; long-term exposure can cause brain damage or death. Administering 100 percent (pure) oxygen is the usual treatment for carbon monoxide poisoning. Administration of pure oxygen speeds up the separation of carbon monoxide from hemoglobin.

BIOLOGICAL MOLECULES AND ITS TYPES

The large molecules necessary for life that are built from smaller organic molecules are called biological **macromolecules**. There are four major classes of biological macromolecules (carbohydrates, lipids, proteins, and nucleic acids), and each is an important component of the cell and performs a wide array of functions. Combined, these molecules make up the majority of a cell's dry mass. Biological macromolecules are organic, meaning that they contain carbon (with some exceptions, like carbon dioxide). In addition, they may contain hydrogen, oxygen, nitrogen, phosphorus, sulfur, and additional minor elements.

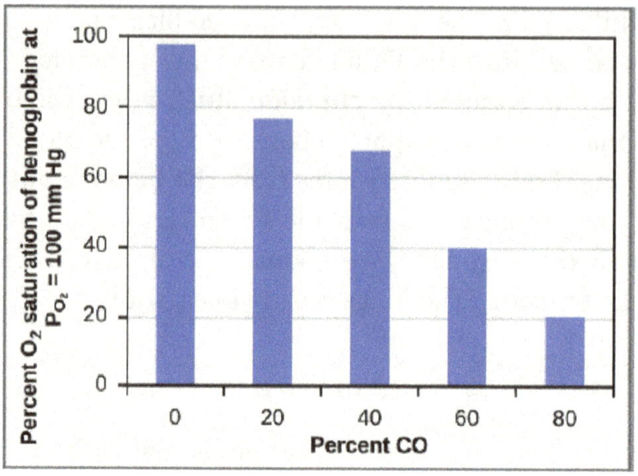

Figure 2.15 As percent CO increases, the oxygen saturation of hemoglobin decreases.

Carbon

It is often said that life is "carbon-based." This means that carbon atoms, bonded to other carbon atoms or other elements, form the fundamental components of many, if not most, of the molecules found uniquely in living things. Other elements play important roles in biological molecules, but carbon certainly qualifies as the "foundation" element for molecules in living things. It is the bonding properties of carbon atoms that are responsible for its important role.

Carbon Bonding

Carbon contains four electrons in its outer shell. Therefore, it can form four covalent bonds with other atoms or molecules. The simplest organic carbon molecule is methane (CH_4), in which four hydrogen atoms bind to a carbon atom (Figure 2.16).

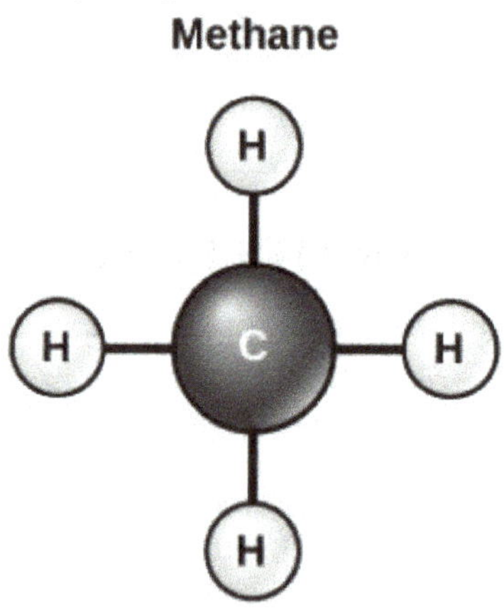

Figure 2.16 Carbon can form four covalent bonds to create an organic molecule. The simplest carbon molecule is methane (CH_4), depicted here.

However, structures that are more complex are made using carbon. Any of the hydrogen atoms can be replaced with another carbon atom covalently bonded to the first carbon atom. In this way, long and branching chains of carbon compounds can be made (Figure 2.17a). The carbon atoms may bond with atoms of other elements, such as nitrogen, oxygen, and phosphorus (Figure 2.17b). The molecules may also form rings, which themselves can link with other rings (Figure 2.17c). This diversity of molecular forms accounts for the diversity of functions of the biological macromolecules and is based to a large degree on the ability of carbon to form multiple bonds with itself and other atoms.

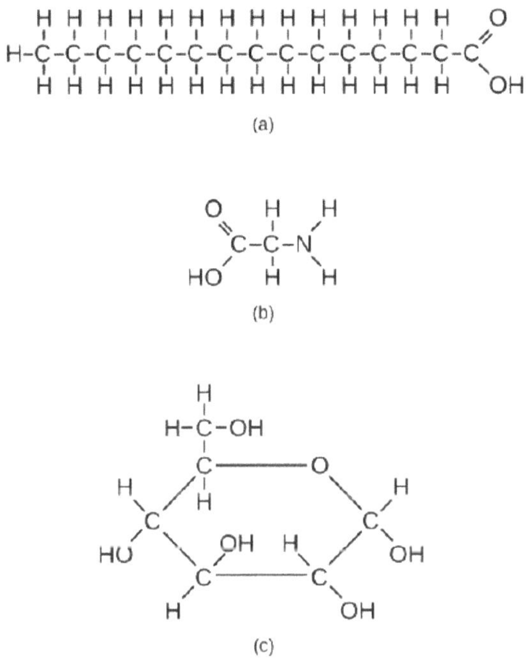

Figure 2.17 These examples show three molecules (found in living organisms) that contain carbon atoms bonded in various ways to other carbon atoms and the atoms of other elements. (a) This molecule of stearic acid has a long chain of carbon atoms. (b) Glycine, a component of proteins, contains carbon, nitrogen, oxygen, and hydrogen atoms. (c) Glucose, a sugar, has a ring of carbon atoms and one oxygen atom.

Carbohydrates

Carbohydrates are macromolecules with which most consumers are somewhat familiar. To lose weight, some individuals adhere to "low-carb" diets. Athletes, in contrast, often "carb-load" before important competitions to ensure that they have sufficient energy to compete at a high level. Carbohydrates are, in fact, an essential part of our diet; grains, fruits, and vegetables are all natural sources of carbohydrates. Carbohydrates provide energy to the body, particularly through glucose, a simple sugar. Carbohydrates also have other important functions in humans, animals, and plants.

Carbohydrates can be represented by the formula $(CH_2O)_n$, where *n* is the number of carbon atoms in the molecule. In other words, the ratio of carbon to hydrogen to oxygen is 1:2:1 in carbohydrate molecules. Carbohydrates are classified into three subtypes: monosaccharides, disaccharides, and polysaccharides.

Monosaccharides (mono- = "one"; sacchar- = "sweet") are simple sugars, the most common of which is glucose. In monosaccharides, the number of carbon atoms usually ranges from three to six. Most monosaccharide names end with the suffixose. Depending on the number of carbon atoms in the sugar, they may be known as trioses (three carbon atoms), pentoses (five carbon atoms), and hexoses (six carbon atoms).

Monosaccharides may exist as a linear chain or as ring-shaped molecules; in aqueous solutions, they are usually found in the ring form.

The chemical formula for glucose is $C_6H_{12}O_6$. In most living species, glucose is an important source of energy. During cellular respiration, energy is released from glucose, and that energy is used to help make adenosine triphosphate (ATP). Plants synthesize glucose using carbon dioxide and water by the process of photosynthesis, and the glucose, in turn, is used for the energy requirements of the plant. The excess synthesized glucose is often stored as starch that is broken down by other organisms that feed on plants.

Galactose (part of lactose, or milk sugar) and fructose (found in fruit) are other common monosaccharides. Although glucose, galactose, and fructose all have the same chemical formula ($C_6H_{12}O_6$), they differ structurally and chemically (and are known as isomers) because of differing arrangements of atoms in the carbon chain (Figure 2.18).

Disaccharides (di- = "two") form when two monosaccharides undergo a dehydration reaction (a reaction in which the removal of a water molecule occurs). During this process, the hydroxyl group (–OH) of one monosaccharide combines with a hydrogen atom of another monosaccharide, releasing a molecule of water (H_2O) and forming a covalent bond between atoms in the two sugar molecules.

Common disaccharides include lactose, maltose, and sucrose. Lactose is a disaccharide consisting of the monomers glucose and galactose. It is found naturally in milk. Maltose, or malt sugar, is a disaccharide formed from a dehydration reaction between two glucose molecules. The most common disaccharide is sucrose, or table sugar, which is composed of the monomers glucose and fructose.

A long chain of monosaccharides linked by covalent bonds is known as a **polysaccharide** (poly- = "many"). The chain may be branched or unbranched, and it may contain different types of monosaccharides. Polysaccharides may be very large molecules. Starch, glycogen, cellulose, and chitin are examples of polysaccharides.

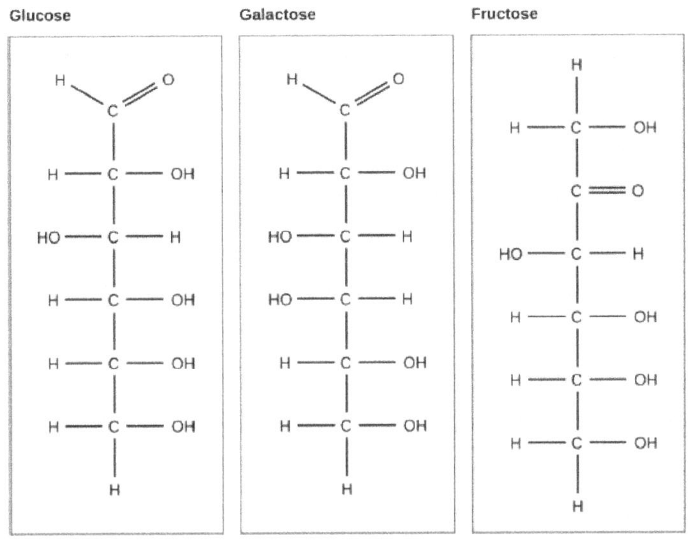

Figure 2.18 Glucose, galactose, and fructose are isomeric monosaccharides, meaning that they have the same chemical formula but slightly different structures.

Starch is the stored form of sugars in plants and is made up of amylose and amylopectin (both polymers of glucose). Plants are able to synthesize glucose, and the excess glucose is stored as starch in different plant parts, including roots and seeds. The starch that is consumed by animals is broken down into smaller molecules, such as glucose. The cells can then absorb the glucose.

Glycogen is the storage form of glucose in humans and other vertebrates, and is made up of monomers of glucose. Glycogen is the animal equivalent of starch and is a highly branched molecule usually stored in liver and muscle cells. Whenever glucose levels decrease, glycogen is broken down to release glucose.

Cellulose is one of the most abundant natural biopolymers. The cell walls of plants are mostly made of cellulose, which provides structural support to the cell. Wood and paper are mostly cellulosic in nature. Cellulose is made up of glucose monomers that are linked by bonds between particular carbon atoms in the glucose molecule.

Every other glucose monomer in cellulose is flipped over and packed tightly as extended long chains. This gives cellulose its rigidity and high tensile strength—which is so important to plant cells. Cellulose passing through our digestive system is called dietary fiber. While the glucose-glucose bonds in cellulose cannot be broken down by human digestive enzymes, herbivores such as cows, buffalos, and horses are able to digest grass that is rich in cellulose and use it as a food source. In these animals, certain species of bacteria reside in the digestive system of herbivores and secrete the enzyme cellulase. The appendix also contains bacteria that break down cellulose, giving it an important role in the digestive systems of some ruminants. Cellulases can break down cellulose into glucose monomers that can be used as an energy source by the animal.

Carbohydrates serve other functions in different animals. Arthropods, such as insects, spiders, and crabs, have an outer skeleton, called the exoskeleton, which protects their internal body parts. This exoskeleton is made of the biological macromolecule **chitin,** which is a nitrogenous carbohydrate. It is made of repeating units of a modified sugar containing nitrogen.

Thus, through differences in molecular structure, carbohydrates are able to serve the very different functions of energy storage (starch and glycogen) and structural support and protection (cellulose and chitin) (Figure 2.19).

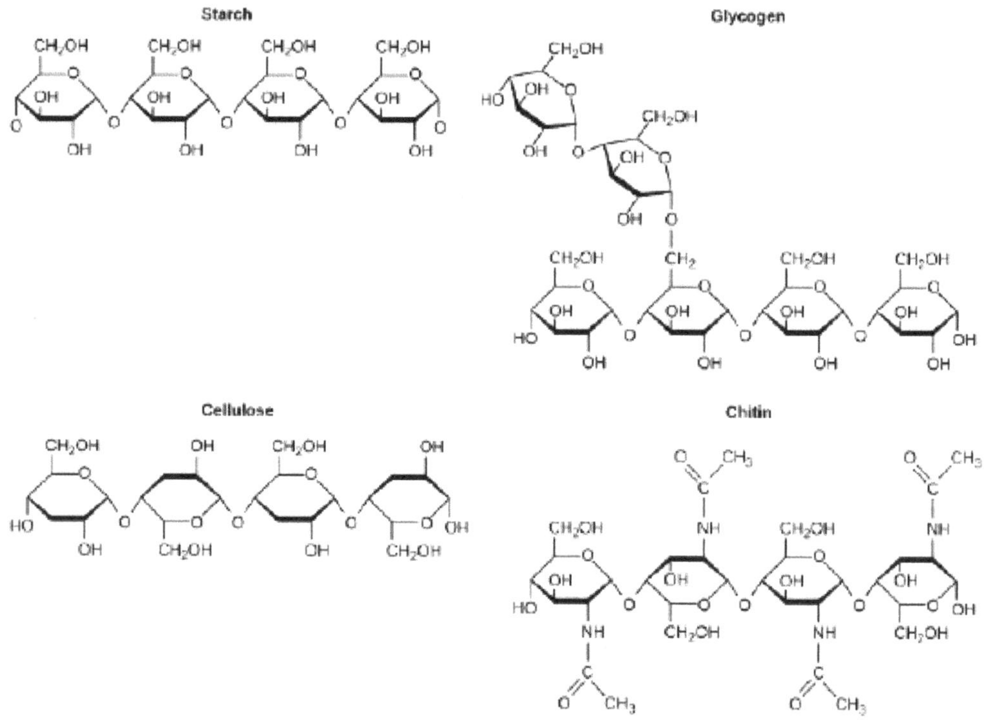

Figure 2.19 Although their structures and functions differ, all polysaccharide carbohydrates are made up of monosaccharides and have the chemical formula (CH_2O)*n*.

Registered Dietitian

Obesity is a worldwide health concern, and many diseases, such as diabetes and heart disease, are becoming more prevalent because of obesity. This is one of the reasons why registered dietitians are increasingly sought after for advice. Registered dietitians help plan food and nutrition programs for individuals in various settings. They often work with patients in health-care facilities, designing nutrition plans to prevent and treat diseases. For example, dietitians may teach a patient with diabetes how to manage blood-sugar levels by eating the correct types and amounts of carbohydrates. Dietitians may also work in nursing homes, schools, and private practices.

To become a registered dietitian, one needs to earn at least a bachelor's degree in dietetics, nutrition, food technology, or a related field. In addition, registered dietitians must complete a supervised internship program and pass a national exam. Those who pursue careers in dietetics take courses in nutrition, chemistry, biochemistry, biology, microbiology, and human physiology. Dietitians must become experts in the chemistry and functions of food (proteins, carbohydrates, and fats).

Lipids

Lipids include a diverse group of compounds that are united by a common feature. Lipids are hydrophobic ("water-fearing"), or insoluble in water, because they are nonpolar molecules. This is because they are hydrocarbons that include only nonpolar carbon-carbon or carbon-hydrogen bonds. Lipids perform many different functions in a cell. Cells store energy for long-term use in the form of lipids called fats. Lipids also provide insulation from the environment for plants and animals (Figure 2.20). For example, they help keep aquatic birds and mammals dry because of their water-repelling nature. Lipids are also the building blocks of many hormones and are an important constituent of the plasma membrane. Lipids include fats, oils, waxes, phospholipids, and steroids.

Figure 2.20 Hydrophobic lipids in the fur of aquatic mammals, such as this river otter, protect them from the elements. (credit: Ken Bosma)

A **fat** molecule, such as a triglyceride, consists of two main components—glycerol and fatty acids. Glycerol is an organic compound with three carbon atoms, five hydrogen atoms, and three hydroxyl (-OH) groups. Fatty acids have a long chain of hydrocarbons to which an acidic carboxyl group is attached, hence the name "fatty acid." The number of carbons in the fatty acid may range from 4 to 36; most common are those containing 12–18 carbons. In a fat molecule, a fatty acid is attached to each of the three oxygen atoms in the -OH groups of the glycerol molecule with a covalent bond (Figure 2.21).

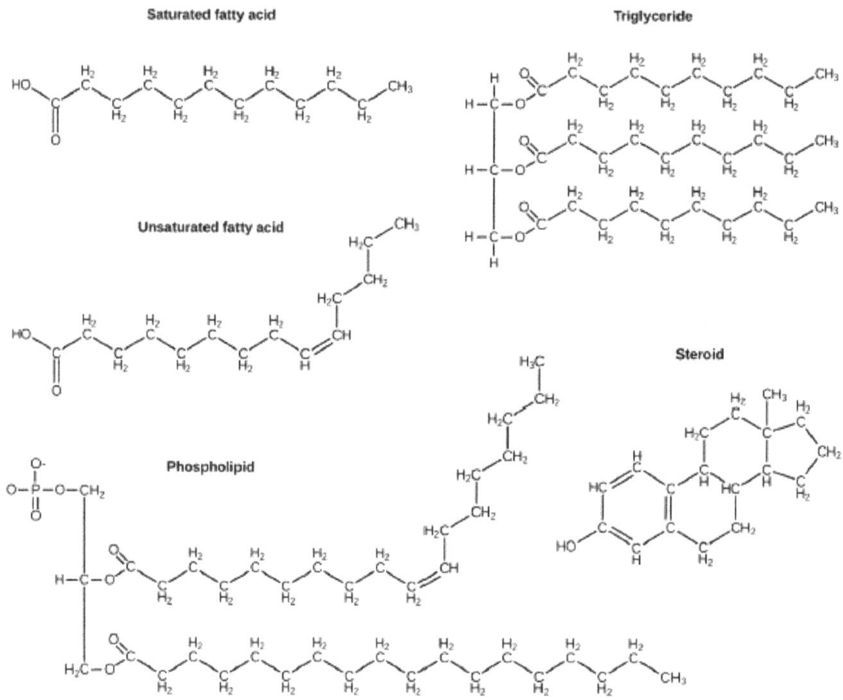

Figure 2.21 Lipids include fats, such as triglycerides, which are made up of fatty acids and glycerol, phospholipids, and steroids.

During this covalent bond formation, three water molecules are released. The three fatty acids in the fat may be similar or dissimilar. These fats are also called **triglycerides** because they have three fatty acids. Some fatty acids have common names that specify their origin. For example, palmitic acid, a saturated fatty acid, is derived from the palm tree. Arachidic acid is derived from *Arachis hypogaea*, the scientific name for peanuts.

Fatty acids may be saturated or unsaturated. In a fatty acid chain, if there are only single bonds between neighbouring carbons in the hydrocarbon chain, the fatty acid is saturated. **Saturated fatty acids** are saturated with hydrogen; in other words, the number of hydrogen atoms attached to the carbon skeleton is maximized.

When the hydrocarbon chain contains a double bond, the fatty acid is an **unsaturated fatty acid.**

Most unsaturated fats are liquid at room temperature and are called **oils.** If there is one double bond in the molecule, then it is known as a monounsaturated fat (e.g., olive oil), and if there is more than one double bond, then it is known as a polyunsaturated fat (e.g., canola oil).

Saturated fats tend to get packed tightly and are solid at room temperature. Animal fats with stearic acid and palmitic acid contained in meat, and the fat with butyric acid contained in butter, are examples of saturated fats. Mammals store fats in specialized cells called adipocytes, where globules of fat occupy most of the cell. In plants, fat or oil is stored in seeds and is used as a source of energy during embryonic development.

Unsaturated fats or oils are usually of plant origin and contain unsaturated fatty acids. The double bond causes a bend or a "kink" that prevents the fatty acids from packing tightly, keeping them liquid at room temperature. Olive oil, corn oil, canola oil, and cod liver oil are examples of unsaturated fats. Unsaturated fats help to improve blood cholesterol levels, whereas saturated fats contribute to plaque formation in the arteries, which increases the risk of a heart attack.

In the food industry, oils are artificially hydrogenated to make them semi-solid, leading to less spoilage and increased shelf life. Simply speaking, hydrogen gas is bubbled through oils to solidify them. During this hydrogenation process, double bonds of the *cis*-conformation in the hydrocarbon chain may be converted to double bonds in the *trans*-conformation. This forms a ***trans*-fat** from a *cis*-fat. The orientation of the double bonds affects the chemical properties of the fat (Figure 2.22).

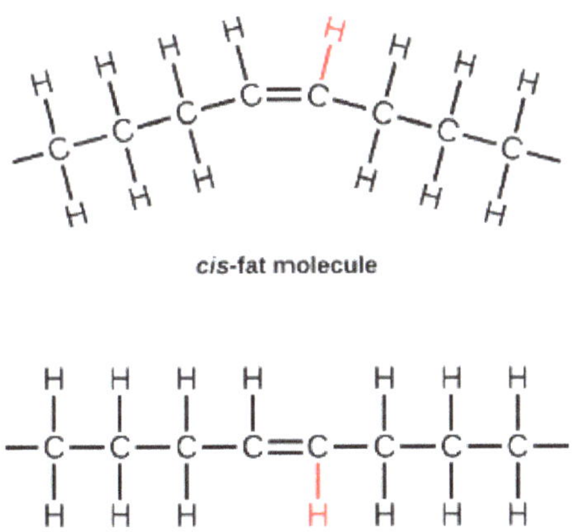

Figure 2.22 During the hydrogenation process, the orientation around the double bonds is changed, making a *trans*-fat from a *cis*-fat. This changes the chemical properties of the molecule.

Margarine, some types of peanut butter, and shortening are examples of artificially hydrogenated *trans*-fats. Recent studies have shown that an increase in *trans*-fats in the human diet may lead to an increase in levels of low-density lipoprotein (LDL), or "bad" cholesterol, which, in turn, may lead to plaque deposition in the arteries, resulting in heart disease. Many fast food restaurants have recently eliminated the use of *trans*-fats, and U.S. food labels are now required to list their *trans*-fat content.

Essential fatty acids are fatty acids that are required but not synthesized by the human body. Consequently, they must be supplemented through the diet. Omega-3 fatty acids fall into this category and are one of only two known essential fatty acids for humans (the other being omega-6 fatty acids). They are a type of polyunsaturated fat and are called omega-3 fatty acids because the third carbon from the end of the fatty acid participates in a double bond.

Salmon, trout, and tuna are good sources of omega-3 fatty acids. Omega-3 fatty acids are important in brain function and normal growth and development. They may also prevent heart disease and reduce the risk of cancer.

Like carbohydrates, fats have received a lot of bad publicity. It is true that eating an excess of fried foods and other "fatty" foods leads to weight gain. However, fats do have important functions. Fats serve as long-term energy storage. They also provide insulation for the body. Therefore, "healthy" unsaturated fats in moderate amounts should be consumed on a regular basis.

Phospholipids are the major constituent of the plasma membrane. Like fats, they are composed of fatty acid chains attached to a glycerol or similar backbone. Instead of three fatty acids attached, however, there are two fatty acids and the third carbon of the glycerol backbone is bound to a phosphate group. The phosphate group is modified by the addition of an alcohol.

A phospholipid has both hydrophobic and hydrophilic regions. The fatty acid chains are hydrophobic and exclude themselves from water, whereas the phosphate is hydrophilic and interacts with water.

Cells are surrounded by a membrane, which has a bilayer of phospholipids. The fatty acids of phospholipids face inside, away from water, whereas the phosphate group can face either the outside environment or the inside of the cell, which are both aqueous.

Steroids and Waxes

Unlike the phospholipids and fats discussed earlier, **steroids** have a ring structure. Although they do not resemble other lipids, they are grouped with them because they are also hydrophobic. All steroids have four, linked carbon rings and several of them, like cholesterol, have a short tail.

Cholesterol is a steroid. Cholesterol is mainly synthesized in the liver and is the precursor of many steroid hormones, such as testosterone and estradiol. It is also the precursor of vitamins E and K. Cholesterol is the precursor of bile salts, which help in the breakdown of fats and their subsequent absorption by cells. Although cholesterol is often spoken of in negative terms, it is necessary for the proper functioning of the body. It is a key component of the plasma membranes of animal cells.

Waxes are made up of a hydrocarbon chain with an alcohol (–OH) group and a fatty acid. Examples of animal waxes include beeswax and lanolin. Plants also have waxes, such as the coating on their leaves, that helps prevent them from drying out.

Proteins

Proteins are one of the most abundant organic molecules in living systems and have the most diverse range of functions of all macromolecules. Proteins may be structural, regulatory, contractile, or protective; they may serve in transport, storage, or membranes; or they may be toxins or enzymes. Each cell in a living system may contain thousands of different proteins, each with a unique function. Their structures, like their functions, vary greatly. They are all, however, polymers of amino acids, arranged in a linear sequence.

The functions of proteins are very diverse because there are 20 different chemically distinct amino acids that form long chains, and the amino acids can be in any order. For example, proteins can function as enzymes or hormones. **Enzymes**, which are produced by living cells, are catalysts in biochemical reactions (like digestion) and are usually proteins. Each enzyme is specific for the substrate (a reactant that binds to an enzyme) upon which it acts. Enzymes can function to break molecular bonds, to rearrange bonds, or to form new bonds. An example of an enzyme is salivary amylase, which breaks down amylose, a component of starch.

Hormones are chemical signalling molecules, usually proteins or steroids, secreted by an endocrine gland or group of endocrine cells that act to control or regulate specific physiological processes, including growth, development, metabolism, and reproduction. For example, insulin is a protein hormone that maintains blood glucose levels.

Proteins have different shapes and molecular weights; some proteins are globular in shape whereas others are fibrous in nature. For example, hemoglobin is a globular protein, but collagen, found in our skin, is a fibrous protein. Protein shape is critical to its function. Changes in temperature, pH, and exposure to chemicals may lead to permanent changes in the shape of the protein, leading to a loss of function or **denaturation** (to be discussed in more detail later). All proteins are made up of different arrangements of the same 20 kinds of amino acids.

Amino acids are the monomers that make up proteins. Each amino acid has the same fundamental structure, which consists of a central carbon atom bonded to an amino group (-NH_2), a carboxyl group (–COOH), and a hydrogen atom. Every amino acid also has another variable atom or group of atoms bonded to the central carbon atom known as the R group. The R group is the only difference in structure between the 20 amino acids; otherwise, the amino acids are identical (Figure 2.23).

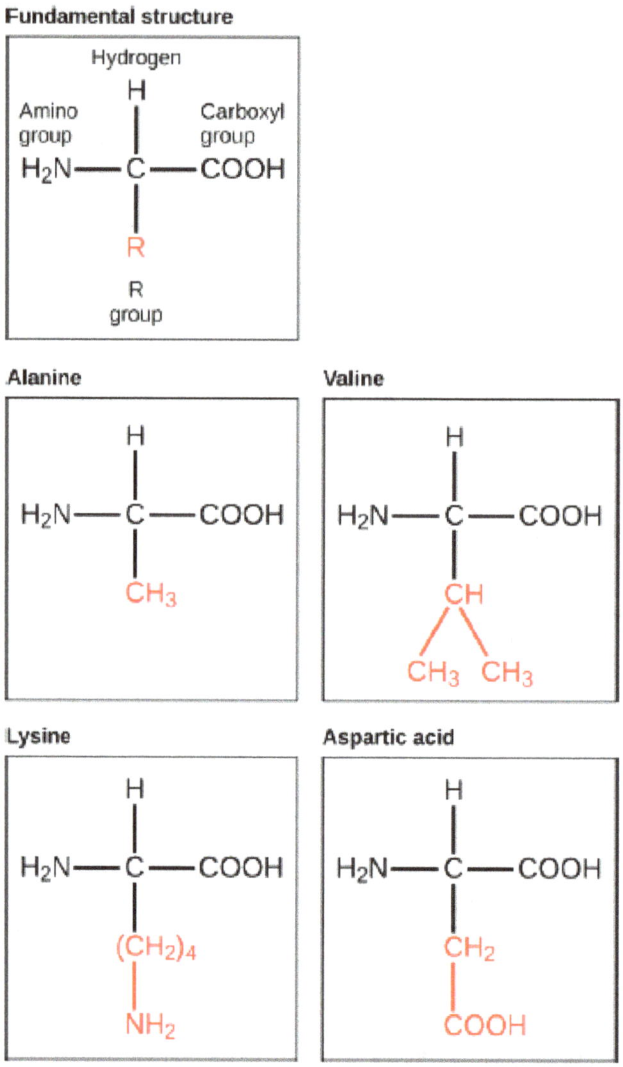

Figure 2.23 Amino acids are made up of a central carbon bonded to an amino group (–NH2), a carboxyl group (–COOH), and a hydrogen atom. The central carbon's fourth bond varies among the different amino acids, as seen in these examples of alanine, valine, lysine, and aspartic acid.

The chemical nature of the R group determines the chemical nature of the amino acid within its protein (that is, whether it is acidic, basic, polar, or nonpolar).

The sequence and number of amino acids ultimately determine a protein's shape, size, and function. Each amino acid is attached to another amino acid by a covalent bond, known as a peptide bond, which is formed by a dehydration reaction. The carboxyl group of one amino acid and the amino group of a second amino acid combine, releasing a water molecule. The resulting bond is the peptide bond.

The products formed by such a linkage are called polypeptides. While the terms polypeptide and protein are sometimes used interchangeably, a **polypeptide** is technically a polymer of amino acids, whereas the term protein is used for a polypeptide or polypeptides that have combined together, have a distinct shape, and have a unique function.

Significance of Cytochrome

Cytochrome c is an important component of the molecular machinery that harvests energy from glucose. Because this protein's role in producing cellular energy is crucial, it has changed very little over millions of years. Protein sequencing has shown that there is a considerable amount of sequence similarity among cytochrome c molecules of different species; evolutionary relationships can be assessed by measuring the similarities or differences among various species' protein sequences.

For example, scientists have determined that human cytochrome c contains 104 amino acids. For each cytochrome c molecule that has been sequenced to date from different organisms, 37 of these amino acids appear in the same position in each cytochrome c. This indicates that all of these organisms are descended from a common ancestor. On comparing the human and chimpanzee protein sequences, no sequence difference was found. When human and rhesus monkey sequences were compared, a single difference was found in one amino acid. In contrast, human-to-yeast comparisons show a difference in 44 amino acids, suggesting that humans and chimpanzees have a more recent common ancestor than humans and the rhesus monkey, or humans and yeast.

Structure of Protein

As discussed earlier, the shape of a protein is critical to its function. To understand how the protein gets its final shape or conformation, we need to understand the four levels of protein structure: primary, secondary, tertiary, and quaternary .

The unique sequence and number of amino acids in a polypeptide chain is its primary structure. The unique sequence for every protein is ultimately determined by the gene that encodes the protein. Any change in the gene sequence may lead to a different amino acid being added to the polypeptide chain, causing a change in protein structure and function. William Warrick Cardozo showed that sickle-cell anemia is caused by a change in protein structure as a result of gene encoding, meaning that it is an inherited disorder. In sickle cell anemia, the hemoglobin β chain has a single amino acid substitution, causing a change in both the structure and function of the protein. What is most remarkable to consider is that a hemoglobin molecule is made up of two alpha chains and two beta chains that each consist of about 150 amino acids. The molecule, therefore, has about 600 amino acids. The structural difference between a normal hemoglobin molecule and a sickle cell molecule—that dramatically decreases life expectancy in the affected individuals—is a single amino acid of the 600.

Because of this change of one amino acid in the chain, the normally biconcave, or disc-shaped, red blood cells assume a crescent or "sickle" shape, which clogs arteries. This can lead to a myriad of serious health problems, such as breathlessness, dizziness, headaches, and abdominal pain for those who have this disease.

Folding patterns resulting from interactions between the non-R group portions of amino acids give rise to the secondary structure of the protein. The most common are the alpha (α)-helix and beta (β)-pleated sheet structures. Both structures are held in shape by hydrogen bonds. In the alpha helix, the bonds form between every fourth amino acid and cause a twist in the amino acid chain.

In the β-pleated sheet, the "pleats" are formed by hydrogen bonding between atoms on the backbone of the polypeptide chain. The R groups are attached to the carbons, and extend above and below the folds of the pleat. The pleated segments align parallel to each other, and hydrogen bonds form between the same pairs of atoms on each of the aligned amino acids. The α-helix and β-pleated sheet structures are found in many globular and fibrous proteins.

The unique three-dimensional structure of a polypeptide is known as its tertiary structure. This structure is caused by chemical interactions between various amino acids and regions of the polypeptide. Primarily, the interactions among R groups create the complex three-dimensional tertiary structure of a protein. There may be ionic bonds formed between R groups on different amino acids, or hydrogen bonding beyond that involved in the secondary structure. When protein folding takes place, the hydrophobic R groups of nonpolar amino acids lay in the interior of the protein, whereas the hydrophilic R groups lay on the outside. The former types of interactions are also known as hydrophobic interactions.

In nature, some proteins are formed from several polypeptides, also known as subunits, and the interaction of these subunits forms the quaternary structure. Weak interactions between the subunits help to stabilize the overall structure. For example, hemoglobin is a combination of four polypeptide subunits.

Each protein has its own unique sequence and shape held together by chemical interactions. If the protein is subject to changes in temperature, pH, or exposure to chemicals, the protein structure may change, losing its shape in what is known as denaturation as discussed earlier. Denaturation is often reversible because the primary structure is preserved if the denaturing agent is removed, allowing the protein to resume its function. Sometimes denaturation is irreversible, leading to a loss of function. One example of protein denaturation can be seen when an egg is fried or boiled. The albumin protein in the liquid egg white is denatured when placed in a hot pan, changing from a clear substance to an opaque white substance. Not all proteins are denatured at high temperatures; for instance, bacteria that survive in hot springs have proteins that are adapted to function at those temperatures.

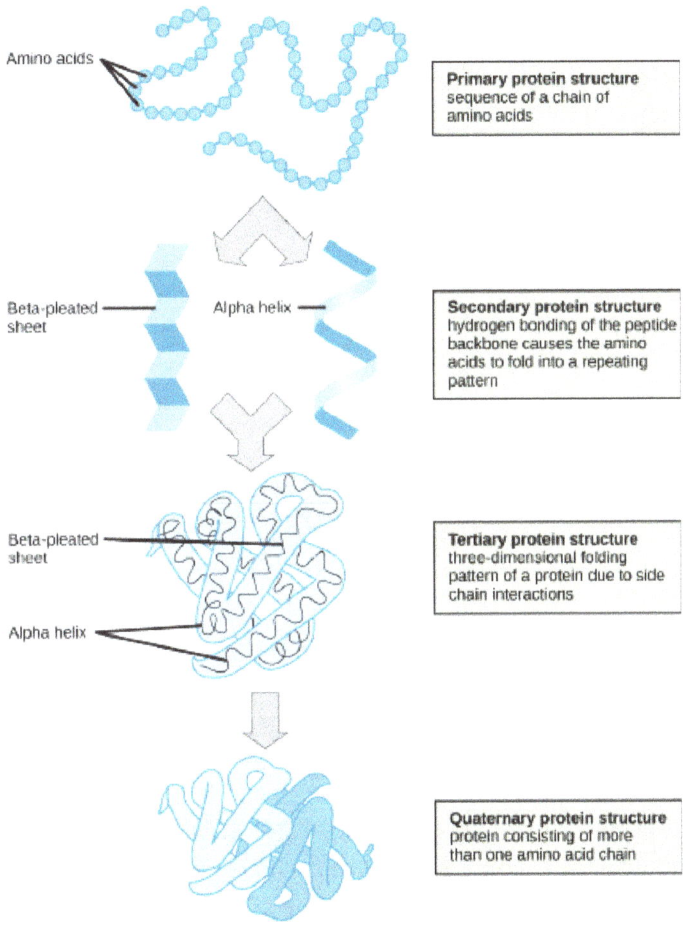

Figure 2.24 The four levels of protein structure can be observed in these illustrations. (credit: modification of work by National Human Genome Research Institute)

Nucleic Acids

Nucleic acids are key macromolecules in the continuity of life. They carry the genetic blueprint of a cell and carry instructions for the functioning of the cell.

The two main types of **nucleic acids** are **deoxyribonucleic acid (DNA)** and **ribonucleic acid (RNA).** DNA is the genetic material found in all living organisms, ranging from single-celled bacteria to multicellular mammals.

The other type of nucleic acid, RNA, is mostly involved in protein synthesis. The DNA molecules never leave the nucleus, but instead use an RNA intermediary to communicate with the rest of the cell. Other types of RNA are also involved in protein synthesis and its regulation.

DNA and RNA are made up of monomers known as **nucleotides.** The nucleotides combine with each other to form a polynucleotide, DNA or RNA. Each nucleotide is made up of three components: a nitrogenous base, a pentose (five-carbon) sugar, and a phosphate group (Figure 2.25). Each nitrogenous base in a nucleotide is attached to a sugar molecule, which is attached to a phosphate group.

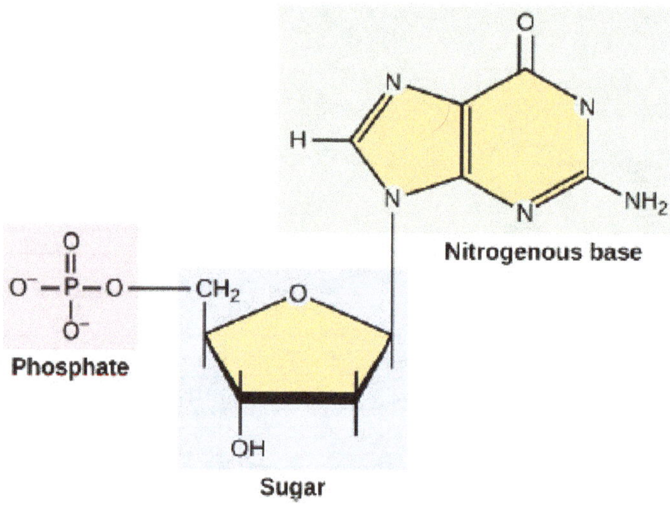

Figure 2.25 A nucleotide is made up of three components: a nitrogenous base, a pentose sugar, and a phosphate group.

Double-Helical Structure of DNA

DNA has a double-helical structure (Figure 2.26). It is composed of two strands, or polymers, of nucleotides. The strands are formed with bonds between phosphate and sugar groups of adjacent nucleotides. The strands are bonded to each other at their bases with hydrogen bonds, and the strands coil about each other along their length, hence the "double helix" description, which means a double spiral.

The alternating sugar and phosphate groups lie on the outside of each strand, forming the backbone of the DNA. The nitrogenous bases are stacked in the interior, like the steps of a staircase, and these bases pair; the pairs are bound to each other by hydrogen bonds. The bases pair in such a way that the distance between the backbones of the two strands is the same all along the molecule.

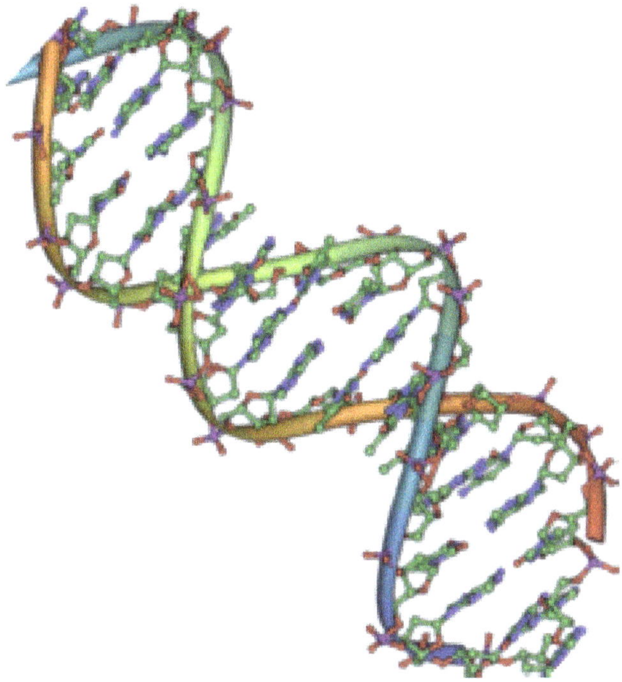

Figure 2.26 The double-helix model shows DNA as two parallel strands of intertwining molecules. (credit: Jerome Walker, Dennis Myts)

FRIEDRICH MIESCHER & THE STRUCTURE OF NUCLEIC ACIDS

Johannes Friedrich Miescher (13 August 1844 – 26 August 1895) was a Swiss physician and biologist, and the **first researcher to isolate nucleic acid (https://en.wikipedia.org/wiki/Nucleic_acid).** Miescher isolated various phosphate-rich chemicals, which he called nuclein (now known as nucleic acids!), from the nuclei of white blood cells in 1869 in Felix Hoppe-Seyler's laboratory at the University of Tübingen, Germany, paving the way for the identification of DNA as the carrier of inheritance. The significance of the discovery, first published in 1871, was not at first apparent, and it was Albrecht Kossel who made the initial inquiries into its chemical structure. Later, **Friedrich Miescher raised the idea that the nucleic acids could be involved in heredity.**

After a typhus infection, Miescher lost hearing in one ear. Miescher felt that his partial deafness would be challenging as a doctor, so he turned to physiological chemistry. He originally wanted to study lymphocytes but was encouraged by Felix Hoppe-Seyler to study neutrophils. He was interested in studying the chemistry of the nucleus. Lymphocytes were difficult to obtain in enough numbers to study, while neutrophils were known to be one of the main and first components in pus and could be obtained from bandages at the nearby hospital. The problem was, however, washing the cells off the bandages without damaging them.

Figure 2.27 Image: Johannes Friedrich Miescher (13 August 1844 – 26 August 1895). Image credit http://www.pbs.org/ wgbh/nova/photo51/images/befo-miescher.jpg (http://www.pbs.org/wgbh/nova/photo51/images/befo- miescher.jpg)

Miescher devised different salt solutions, eventually producing one with sodium sulfate. The cells were filtered. Since centrifuges were not available at the time, the cells were left to settle to the bottom of a beaker. He then tried to isolate the nuclei free of cytoplasm. He subjected the purified nuclei to an alkaline extraction followed by acidification, resulting in the formation of a precipitate that Miescher called nuclein. He found that this contained phosphorus and nitrogen, but not sulfur. The discovery was so unlike anything else at the time that Hoppe-Seyler repeated all of Miescher's research himself before publishing it in his journal. Miescher then went on to study physiology at Leipzig in the laboratory of Carl Ludwig for a year before being appointed a professor of physiology.

Although Miescher and his students aggressively researched nucleic acid chemistry, its function remained unknown. However, his discovery played an important part in the identification of nucleic acids as the carriers of inheritance. The importance of Miescher's discovery was not apparent until Albrecht Kossel, a German physiologist specializing in the physiological chemistry of the cell and its nucleus and of proteins researched its' chemical structure.

CHAPTER-3

INTRODUCTION TO BIOTECHNOLOGY

INTRODUCTION: AN OVERVIEW

Biotechnology, a dynamic and multidisciplinary field at the intersection of biology, chemistry, genetics, and technology, represents the harnessing of living organisms, cells, and biological systems to develop innovative solutions and products that benefit society. With roots dating back to ancient practices such as fermentation and selective breeding, modern biotechnology has evolved into a sophisticated science that leverages our understanding of the molecular and cellular processes of living organisms. This transformative discipline encompasses a wide range of applications, from the development of novel drugs and therapies to the engineering of crops for improved agricultural yield and sustainability. As we delve into the complexities of manipulating biological systems at the molecular level, biotechnology not only holds the promise of addressing global challenges in health, agriculture, and the environment but also sparks ethical discussions and considerations surrounding the responsible use of this powerful technology. In this era of rapid scientific advancements, the study and application of biotechnology play a pivotal role in shaping the future of medicine, agriculture, and industry.

The latter half of the twentieth century began with the discovery of the structure of DNA, then progressed to the development of the basic tools used to study and manipulate DNA. These advances, as well as advances in our understanding of and ability to manipulate cells, have led some to refer to the twenty-first century as the biotechnology century. The rate of discovery and of the development of new applications in medicine, agriculture, and energy is expected to accelerate, bringing huge benefits to humankind and perhaps also significant risks. Many of these developments are expected to raise significant ethical and social questions that human societies have not yet had to consider.

Biotechnology is the use of artificial methods to modify the genetic material of living organisms or cells to produce novel compounds or to perform new functions.

Biotechnology has been used for improving livestock and crops since the beginning of agriculture through selective breeding. Since the discovery of the structure of DNA in 1953, and particularly since the development of tools and methods to manipulate DNA in the 1970s, biotechnology has become synonymous with the manipulation of organisms' DNA at the molecular level. The primary applications of this technology are in medicine (for the production of vaccines and antibiotics) and in agriculture (for the genetic modification of crops). Biotechnology also has many industrial applications, such as fermentation, the treatment of oil spills, and the production of biofuel, as well as many household applications such as the use of enzymes in laundry detergent.

Manipulating Genetic Material

To accomplish the applications described above, biotechnologists must be able to extract, manipulate, and analyze nucleic acids.

Review of Nucleic Acid Structure

To understand the basic techniques used to work with nucleic acids, remember that nucleic acids are macromolecules made of nucleotides (a sugar, a phosphate, and a nitrogenous base). The phosphate groups on these molecules each have a net negative charge. An entire set of DNA molecules in the nucleus of eukaryotic organisms is called the genome. DNA has two complementary strands linked by hydrogen bonds between the paired bases.

Unlike DNA in eukaryotic cells, RNA molecules leave the nucleus. Messenger RNA (mRNA) is analyzed most frequently because it represents the protein-coding genes that are being expressed in the cell.

Isolation of Nucleic Acids

To study or manipulate nucleic acids, the DNA must first be extracted from cells. Various techniques are used to extract different types of DNA (Figure 3.1). Most nucleic acid extraction techniques involve steps to break open the cell, and then the use of enzymatic reactions to destroy all undesired macromolecules. Cells are broken open using a detergent solution containing buffering compounds. To prevent degradation and contamination, macromolecules such as proteins and RNA are inactivated using enzymes. The DNA is then brought out of solution using alcohol. The resulting DNA, because it is made up of long polymers, forms a gelatinous mass.

RNA is studied to understand gene expression patterns in cells. RNA is naturally very unstable because enzymes that break down RNA are commonly present in nature. Some are even secreted by our own skin and are very difficult to inactivate. Similar to DNA extraction, RNA extraction involves the use of various buffers and enzymes to inactivate other macromolecules and preserve only the RNA.

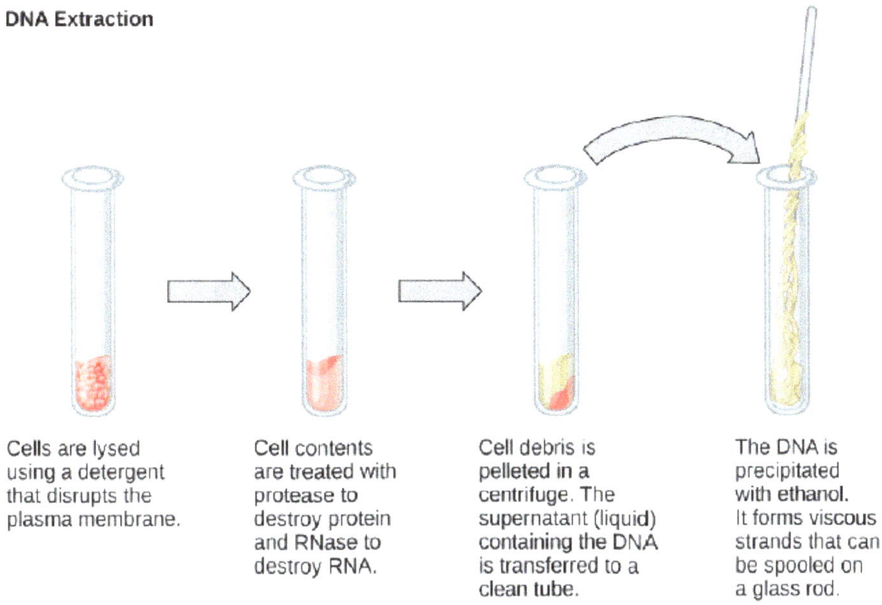

Figure 3.1 The basic method used for the extraction of DNA.

Gel Electrophoresis

Because nucleic acids are negatively charged ions at neutral or alkaline pH in an aqueous environment, they can be moved by an electric field. **Gel electrophoresis** is a technique used to separate charged molecules on the basis of size and charge. The nucleic acids can be separated as whole chromosomes or as fragments. The nucleic acids are loaded into a slot at one end of a gel matrix, an electric current is applied, and negatively charged molecules are pulled toward the opposite end of the gel (the end with the positive electrode). Smaller molecules move through the pores in the gel faster than larger molecules; this difference in the rate of migration separates the fragments on the basis of size. The nucleic acids in a gel matrix are invisible until they are stained with a compound that allows them to be seen, such as a dye. Distinct fragments of nucleic acids appear as bands at specific distances from the top of the gel (the negative electrode end) that are based on their size (Figure 3.2). A mixture of many fragments of varying sizes appear as a long smear, whereas uncut genomic DNA is usually too large to run through the gel and forms a single large band at the top of the gel.

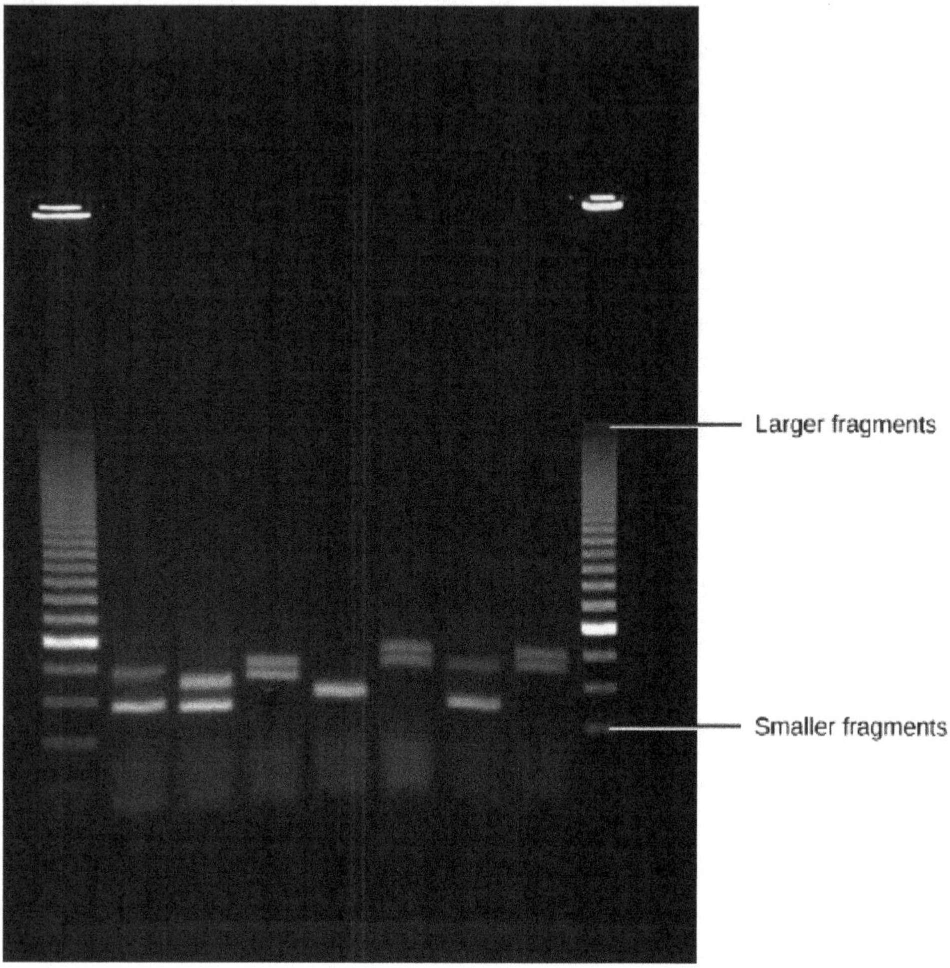

Figure 3.2 DNA fragments from six samples run on a gel, stained with a fluorescent dye and viewed under UV light. (credit: modification of work by James Jacob, Tompkins Cortland Community College)

Polymerase Chain Reaction(PCR)

DNA analysis often requires focusing on one or more specific regions of the genome. It also frequently involves situations in which only one or a few copies of a DNA molecule are available for further analysis. These amounts are insufficient for most procedures, such as gel electrophoresis. Polymerase chain reaction (PCR) is a technique used to rapidly increase the number of copies of specific regions of DNA for further analyses (Figure 3.3). PCR uses a special form of DNA polymerase, the enzyme that replicates DNA, and other short nucleotide sequences called primers that base pair to a specific portion of the DNA being replicated. PCR is used for many purposes in laboratories. These include: 1) the identification of the owner of a DNA sample left at a crime scene; 2) paternity analysis; 3) the comparison of small amounts of ancient DNA with modern organisms; and 4) determining the sequence of nucleotides in a specific region.

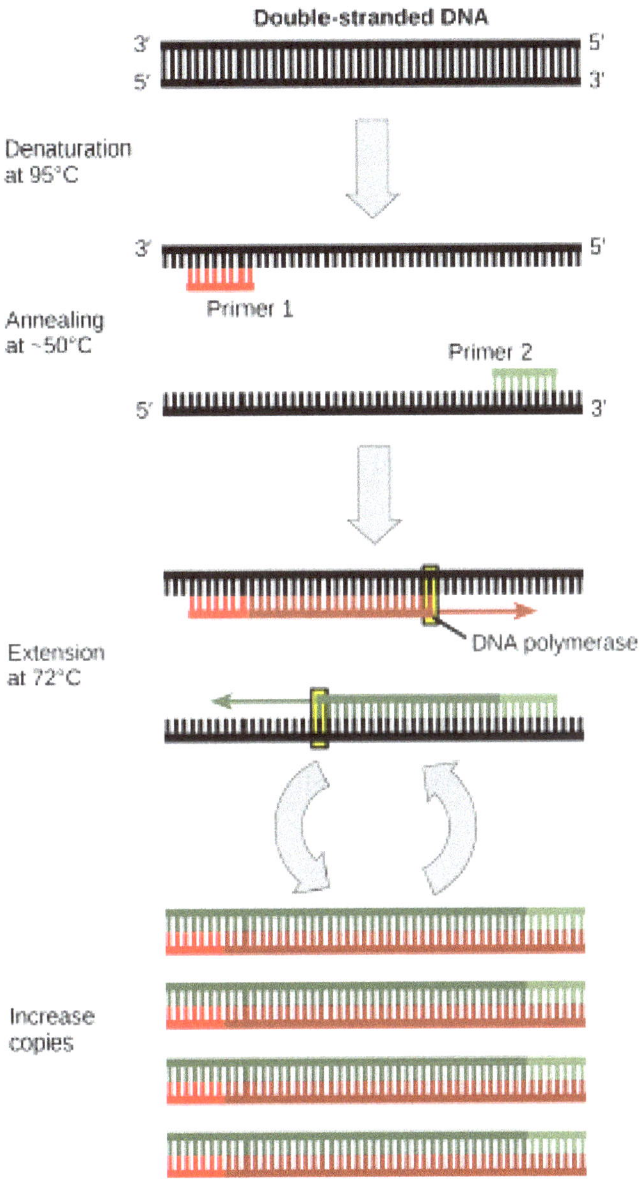

Figure 3.3 Polymerase chain reaction, or PCR, is used to produce many copies of a specific sequence of DNA using a special form of DNA polymerase.

CLONING

In general, **cloning** means the creation of a perfect replica. Typically, the word is used to describe the creation of a genetically identical copy. In biology, the re-creation of a whole organism is referred to as "reproductive cloning." Long before attempts were made to clone an entire organism, researchers learned how to copy short stretches of DNA—a process that is referred to as molecular cloning. The technique offered methods to create new medicines and to overcome difficulties with existing ones. When Lydia Villa-Komaroff, working in the Gilbert Lab at Harvard, published the first paper outlining the technique for producing synthetic insulin, diabetes researchers and patients received new hope in fighting the disease. Insulin at that time was only produced using pig and cow pancreases, and the life-saving substance was often in short supply. Synthetic insulin, once mass produced, would solve that problem for many patients. These early discoveries led to the "BioTech Boom," and spurred continued research and funding for newer and better ways to improve health.

Molecular Cloning

Cloning allows for the creation of multiple copies of genes, expression of genes, and study of specific genes. To get the DNA fragment into a bacterial cell in a form that will be copied or expressed, the fragment is first inserted into a plasmid. A **plasmid** (also called a vector in this context) is a small circular DNA molecule that replicates independently of the chromosomal DNA in bacteria. In cloning, the plasmid molecules can be used to provide a "vehicle" in which to insert a desired DNA fragment. Modified plasmids are usually reintroduced into a bacterial host for replication. As the bacteria divide, they copy their own DNA (including the plasmids). The inserted DNA fragment is copied along with the rest of the bacterial DNA. In a bacterial cell, the fragment of DNA from the human genome (or another organism that is being studied) is referred to as foreign DNA to differentiate it from the DNA of the bacterium (the host DNA).

Plasmids occur naturally in bacterial populations (such as *Escherichia coli*) and have genes that can contribute favourable traits to the organism, such as antibiotic resistance (the ability to be unaffected by antibiotics). Plasmids have been highly engineered as vectors for molecular cloning and for the subsequent large-scale production of important molecules, such as insulin. A valuable characteristic of plasmid vectors is the ease with which a foreign DNA fragment can be introduced. These plasmid vectors contain many short DNA sequences that can be cut with different commonly available **restriction enzymes.** Restriction enzymes (also called restriction endonucleases) recognize specific DNA sequences and cut them in a predictable manner; they are naturally produced by bacteria as a defense mechanism against foreign DNA. Many restriction enzymes make staggered cuts in the two strands of DNA, such that the cut ends have a 2- to 4-nucleotide single-stranded overhang. The sequence that is recognized by the restriction enzyme is a four- to eight-nucleotide sequence that is a palindrome.

Like with a word palindrome, this means the sequence reads the same forward and backward. In most cases, the sequence reads the same forward on one strand and backward on the complementary strand. When a staggered cut is made in a sequence like this, the overhangs are complementary (Figure 3.4).

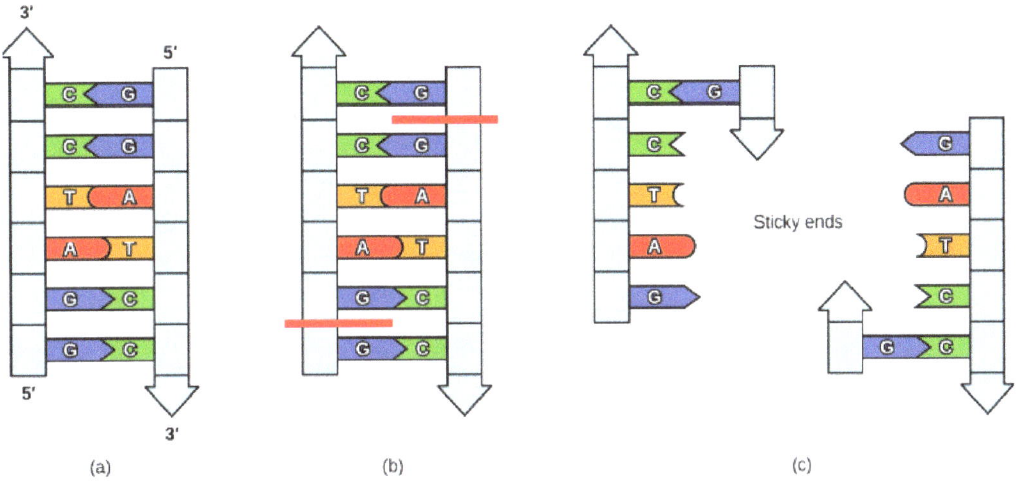

Figure 3.4 In this (a) six-nucleotide restriction enzyme recognition site, notice that the sequence of six nucleotides reads the same in the 5' to 3' direction on one strand as it does in the 5' to 3' direction on the complementary strand. This is known as a palindrome. (b) The restriction enzyme makes breaks in the DNA strands, and (c) the cut in the DNA results in "sticky ends". Another piece of DNA cut on either end by the same restriction enzyme could attach to these sticky ends and be inserted into the gap made by this cut.

Because these overhangs are capable of coming back together by hydrogen bonding with complementary overhangs on a piece of DNA cut with the same restriction enzyme, these are called "sticky ends." The process of forming hydrogen bonds between complementary sequences on single strands to form double-stranded DNA is called **annealing.** Addition of an enzyme called DNA ligase, which takes part in DNA replication in cells, permanently joins the DNA fragments when the sticky ends come together. In this way, any DNA fragment can be spliced between the two ends of a plasmid DNA that has been cut with the same restriction enzyme (Figure 3.5).

Plasmids with foreign DNA inserted into them are called **recombinant DNA** molecules because they contain new combinations of genetic material. Proteins that are produced from recombinant DNA molecules are called **recombinant proteins.** Not all recombinant plasmids are capable of expressing genes. Plasmids may also be engineered to express proteins only when stimulated by certain environmental factors, so that scientists can control the expression of the recombinant proteins.

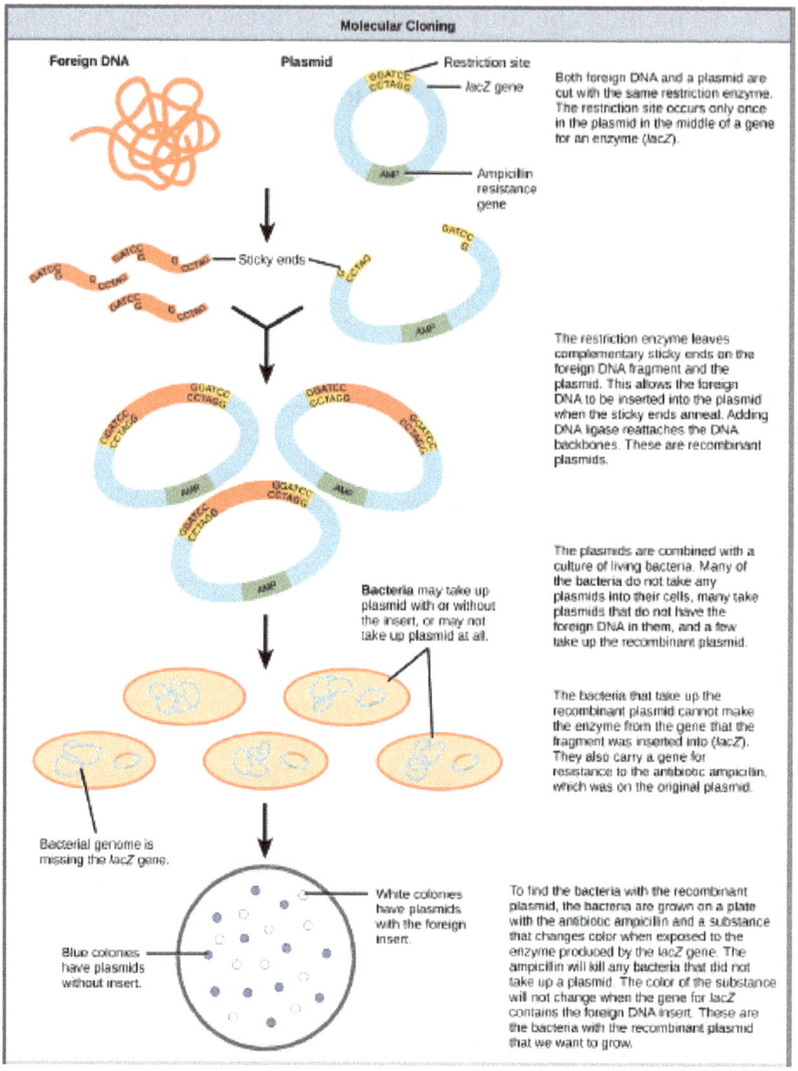

Figure 3.5 The steps involved in molecular cloning.

Reproductive Cloning

Reproductive cloning is a method used to make a clone or an identical copy of an entire multicellular organism. Most multicellular organisms undergo reproduction by sexual means, which involves the contribution of DNA from two individuals (parents), making it impossible to generate an identical copy or a clone of either parent. Recent advances in biotechnology have made it possible to reproductively clone mammals in the laboratory.

Natural sexual reproduction involves the union, during fertilization, of a sperm and an egg. Each of these gametes is haploid, meaning they contain one set of chromosomes in their nuclei. The resulting cell, or zygote, is then diploid and contains two sets of chromosomes. This cell divides mitotically to produce a multicellular organism. However, the union of just any two cells cannot produce a viable zygote; there are components in the cytoplasm of the egg cell that are essential for the early development of the embryo during its first few cell divisions. Without these provisions, there would be no subsequent development. Therefore, to produce a new individual, both a diploid genetic complement and an egg cytoplasm are required. The approach to producing an artificially cloned individual is to take the egg cell of one individual and to remove the haploid nucleus. Then a diploid nucleus from a body cell of a second individual, the donor, is put into the egg cell. The egg is then stimulated to divide so that development proceeds. This sounds simple, but in fact it takes many attempts before each of the steps is completed successfully.

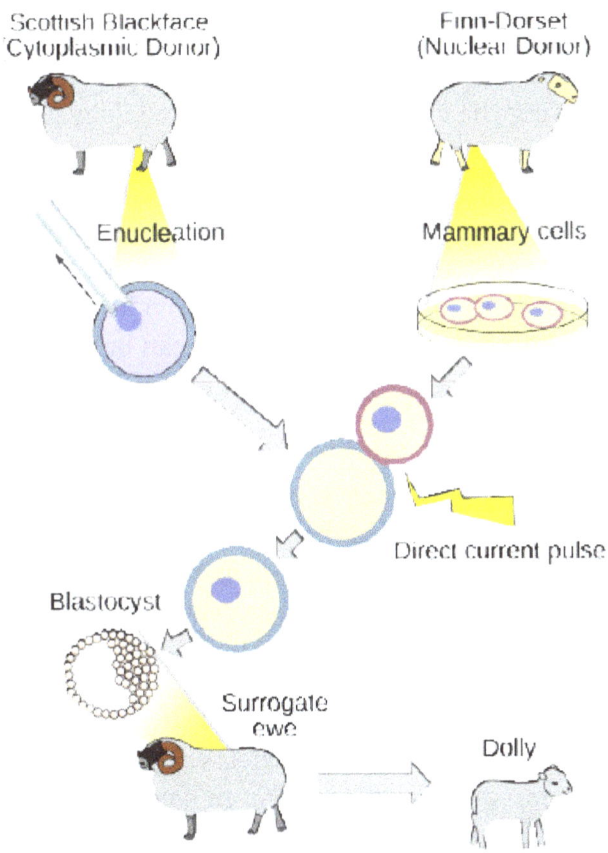

Figure 3.6 Dolly the sheep was the first agricultural animal to be cloned. To create Dolly, the nucleus was removed from a donor egg cell. The enucleated egg was placed next to the other cell, then they were shocked to fuse. They were shocked again to start division. The cells were allowed to divide for several days until an early embryonic stage was reached, before being implanted in a surrogate mother.

Genetic Engineering

Using recombinant DNA technology to modify an organism's DNA to achieve desirable traits is called **genetic engineering**. Addition of foreign DNA in the form of recombinant DNA vectors that are generated by molecular cloning is the most common method of genetic engineering. An organism that receives the recombinant DNA is called a **genetically modified organism** (GMO). If the foreign DNA that is introduced comes from a different species, the host organism is called **transgenic**. Bacteria, plants, and animals have been genetically modified since the early 1970s for academic, medical, agricultural, and industrial purposes. These applications will be examined in more detail in the next module.

Although the classic methods of studying the function of genes began with a given phenotype and determined the genetic basis of that phenotype, modern techniques allow researchers to start at the DNA sequence level and ask: "What does this gene or DNA element do?" This technique, called **reverse genetics**, has resulted in reversing the classical genetic methodology. One example of this method is analogous to damaging a body part to determine its function. An insect that loses a wing cannot fly, which means that the wing's function is flight. The classic genetic method compares insects that cannot fly with insects that can fly, and observes that the non-flying insects have lost wings. Similarly in a reverse genetics approach, mutating or deleting genes provides researchers with clues about gene function. Alternately, reverse genetics can be used to cause a gene to overexpress itself to determine what phenotypic effects may occur.

ROLE OF BIOTECHNOLOGY IN MEDICINE AND AGRICULTURE

It is easy to see how biotechnology can be used for medicinal purposes. Knowledge of the genetic makeup of our species, the genetic basis of heritable diseases, and the invention of technology to manipulate and fix mutant genes provides methods to treat diseases. Biotechnology in agriculture can enhance resistance to disease, pests, and environmental stress to improve both crop yield and quality.

Genetic Diagnosis and Gene Therapy

The process of testing for suspected genetic defects before administering treatment is called genetic diagnosis by genetic testing. In some cases in which a genetic disease is present in an individual's family, family members may be advised to undergo genetic testing. For example, mutations in the *BRCA* genes may increase the likelihood of developing breast, ovarian, and some other cancers. A person with breast cancer can be screened for these mutations. If one of the high-risk mutations is found, relatives may also wish to be screened for that particular mutation, or simply be more vigilant for the occurrence of cancers. Genetic testing is also offered for fetuses (or embryos with in vitro fertilization) to determine the presence or absence of disease-causing genes in families with specific debilitating diseases.

Gene therapy is a genetic engineering technique that may one day be used to cure certain genetic diseases. In its simplest form, it involves the introduction of a non-mutated gene at a random location in the genome to cure a disease by replacing a protein that may be absent in these individuals because of a genetic mutation. The non-mutated gene is usually introduced into diseased cells as part of a vector transmitted by a virus, such as an adenovirus, that can infect the host cell and deliver the foreign DNA into the genome of the targeted cell (Figure 3.7). To date, gene therapies have been primarily experimental procedures in humans. A few of these experimental treatments have been successful, but the methods may be important in the future as the factors limiting its success are resolved.

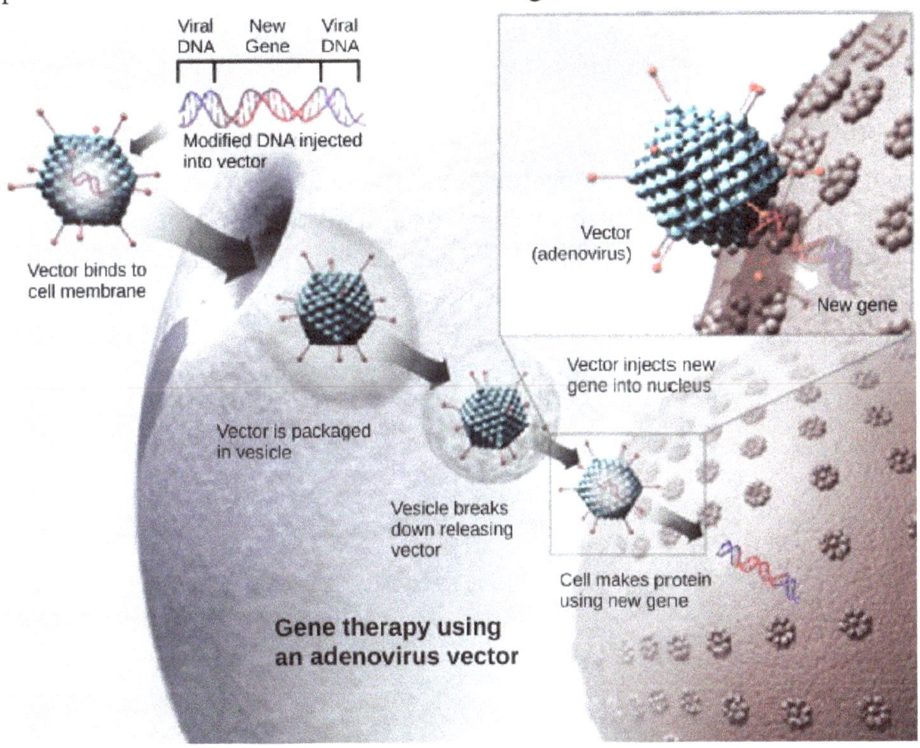

Figure 3.7 This diagram shows the steps involved in curing disease with gene therapy using an adenovirus vector. (credit: modification of work by NIH)

The Manufacture of Vaccines, Antibiotics, and Hormones

Traditional vaccination strategies use weakened or inactive forms of microorganisms or viruses to stimulate the immune system. Modern techniques use specific genes of microorganisms cloned into vectors and mass- produced in bacteria to make large quantities of specific substances to stimulate the immune system. The substance is then used as a vaccine. In some cases, such as the H1N1 flu vaccine, genes cloned from the virus have been used to combat the constantly changing strains of this virus.

Antibiotics kill bacteria and are naturally produced by microorganisms such as fungi; penicillin is perhaps the most well-known example. Antibiotics are produced on a large scale by cultivating and manipulating fungal cells. The fungal cells have typically been genetically modified to improve the yields of the antibiotic compound.

Recombinant DNA technology was used to produce large-scale quantities of the human hormone insulin in *E. coli* as early as 1978. Previously, it was only possible to treat diabetes with pig insulin, which caused allergic reactions in many humans because of differences in the insulin molecule. In addition, human growth hormone (HGH) is used to treat growth disorders in children. The HGH gene was cloned from a cDNA (complementary DNA) library and inserted into *E. coli* cells by cloning it into a bacterial vector.

Transgenic Animals

Although several recombinant proteins used in medicine are successfully produced in bacteria, some proteins need a eukaryotic animal host for proper processing. For this reason, genes have been cloned and expressed in animals such as sheep, goats, chickens, and mice. Animals that have been modified to express recombinant DNA are called transgenic animals (Figure 3.8).

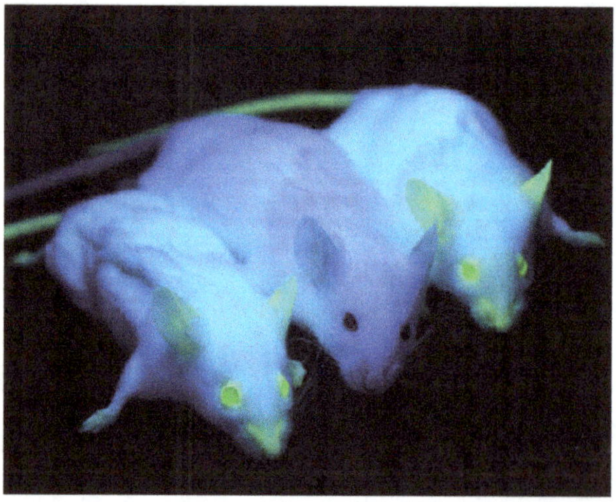

Figure 3.8 It can be seen that two of these mice are transgenic because they have a gene that causes them to fluoresce under a UV light. The non-transgenic mouse does not have the gene that causes fluorescence. (credit: Ingrid Moen et al.)

Several human proteins are expressed in the milk of transgenic sheep and goats. In one commercial example, the FDA has approved a blood anticoagulant protein that is produced in the milk of transgenic goats for use in humans. Mice have been used extensively for expressing and studying the effects of recombinant genes and mutations.

Transgenic Plants

Manipulating the DNA of plants (creating genetically modified organisms, or GMOs) has helped to create desirable traits such as disease resistance, herbicide, and pest resistance, better nutritional value, and better shelf life (Figure 3.9). Plants are the most important source of food for the human population. Farmers developed ways to select for plant varieties with desirable traits long before modern-day biotechnology practices were established.

Figure 3.9 Corn, a major agricultural crop used to create products for a variety of industries, is often modified through plant biotechnology. (credit: Keith Weller, USDA)

Transgenic plants have received DNA from other species. Because they contain unique combinations of genes and are not restricted to the laboratory, transgenic plants and other GMOs are closely monitored by government agencies to ensure that they are fit for human consumption and do not endanger other plant and animal life. Because foreign genes can spread to other species in the environment, particularly in the pollen and seeds of plants, extensive testing is required to ensure ecological stability. Staples like corn, potatoes, and tomatoes were the first crop plants to be genetically engineered.

Transformation of Plants Using Agrobacterium Tumefaciens

In plants, tumors caused by the bacterium *Agrobacterium tumefaciens* occur by transfer of DNA from the bacterium to the plant. The artificial introduction of DNA into plant cells is more challenging than in animal cells because of the thick plant cell wall. Researchers used the natural transfer of DNA from *Agrobacterium* to a plant host to introduce DNA fragments of their choice into plant hosts. In nature, the disease-causing *A. tumefaciens* have a set of plasmids that contain genes that integrate into the infected plant cell's genome. Researchers manipulate the plasmids to carry the desired DNA fragment and insert it into the plant genome.

The Organic Insecticide Bacillus Thuringiensis

Bacillus thuringiensis (Bt) is a bacterium that produces protein crystals that are toxic to many insect species that feed on plants. Insects that have eaten Bt toxin stop feeding on the plants within a few hours. After the toxin is activated in the intestines of the insects, death occurs within a couple of days. The crystal toxin genes have been cloned from the bacterium and introduced into plants, therefore allowing plants to produce their own crystal Bt toxin that acts against insects. Bt toxin is safe for the environment and non-toxic to mammals (including humans). As a result, it has been approved for use by organic farmers as a natural insecticide. There is some concern, however, that insects may evolve resistance to the Bt toxin in the same way that bacteria evolve resistance to antibiotics.

FlavrSavr Tomato

The first GM crop to be introduced into the market was the FlavrSavr Tomato produced in 1994. Molecular genetics technology was used to slow down the process of softening and rotting caused by fungal infections, which led to increased shelf life of the GM tomatoes. Additional genetic modification improved the flavour of this tomato. The FlavrSavr tomato did not successfully stay in the market because of problems maintaining and shipping the crop.

THE HISTORY OF PHARMACEUTICAL REGULATIONS

As you just learned in this module, the exposure to a virus can sometimes provide immunity to future infections and is the basis for vaccination. Edward Jenner created the first vaccine in the 1770s, in an unusually unethical way. Small pox was a deadly disease in this time, and he had a theory about protecting people from infection. To test his theory, he gave a small box cowpox, then infected him with small pox. He was right – exposure to the cowpox had protected him from the small pox virus.

The artwork in the figure is titled *The Cow-Pock-or-the Wonderful Effects of the New Inoculation!* By James Gillray in 1802. This cartoon is a caricature of a Smallpox and Inoculation Hospital at St. Pancras, showing cowpox vaccine administered to frightened young women, and cows emerging from different parts of people's bodies. The cartoon inspiration was the controversy over inoculating against the dreaded disease, smallpox. Opponents of vaccination had depicted cases of vaccines developing bovine features, and this is picked up and exaggerated by Gillray. Thanks to an aggressive vaccine campaign, the deadly smallpox virus was eradicated in the 80s. Unfortunately, the resurgence of anti-vaccination sentiment (https://www.ncbi.nlm.nih.gov/pmc/ articles /PMC6122668/) in the U.S. (and worldwide) has returned resulting in unusual and preventable viral outbreaks causing grave concern of scientists globally.

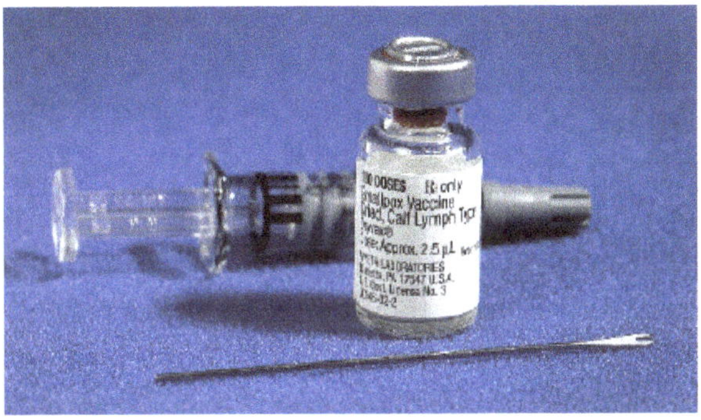

Figure 3.10 small pox vaccine and needle. Image credit: https://commons. wikimedia.org/wiki/ File:Smallpox_vaccine.jpg (https://commons.wikimedia. org/wiki/File:Smallpox_vaccine.jpg)

Figure 3.11 1802 art work, The Cow-Pock, James Gillray. Images and content credit: https://simple.wikipedia.org/ wiki/Vaccine (https://simple.wikipedia. org/wiki/Vaccine), Artwork, By James Gillray - Library of Congress, Prints & Photographs Division, LC-USZC4-3147, Public Domain, https://simple.wikipedia. org/wiki/Vaccine#/media/ File:The_cow_pock.jpg (https://simple.wikipedia. org/wiki/Vaccine#/media/File:The_cow_pock.jpg)

Consumer Safety and Health

The idea of protecting the safety and health of consumers is not new. There were many societies, such as ancient Rome, where the penalties for producing an unsafe product were quite harsh. In contrast, our modern- day society's view of regulations and protections is that of prevention. Of course, there are still "after-the-fact" remedies available to the consumer. However, ***regulations in modern-day are designed to build quality and safety into the production process***, thereby limiting if not eliminating health and safety concerns before the product reaches the consumer. In this module, we will examine some of the most significant regulations, explore how regulations come to pass, and discuss the role of the Food and Drug Administration (FDA) in regulations of consumer products, especially those in health and wellness.

THE FOOD & DRUG ADMINISTRATION

The FDA (http://www.fda.gov) is an administrative agency created to regulate food and drug supplies in the United States for the safety and health of its citizens. FDA is an agency within the Department of Health and Human (http://legacy.cnx.org/content/m71330/1.1/) Services (http://legacy.cnx.org/content/m71330/ 1.1/) . Although the FDA has traditionally focused on the US markets, however, with the global market growth (of imports and exports of products and raw materials) the FDA now manages over 2 trillion dollars of goods manufactured in over 150 nations worldwide (FDA Global engagement report, 2016).

FDA Organization

The FDA, like most organizations, change with the economy, political appointments, world harmonization, and emerging technologies and products. The current FDA organizational hierarchy (https://www.fda.gov/ aboutfda/centersoffices/organizationcharts/ucm393155.htm) consists of the Office of the Commissioner overseeing five offices and directorates. Those offices oversee eight centers. Like the federal government, the FDA possesses the following powers:

- **Legislative:** The FDA has the authority to create and issue rules ***(not laws!).***
- **Executive:** The FDA has the power to conduct investigations.
- **Judicial:** The FDA has the jurisdiction to review evidence and make judgments on a product.

Field Offices & Centers

The field offices are located throughout the United States and are *the eyes and ears* of the FDA, and it is from these offices that operational personnel enforce the law. Under the Offices, the FDA has Product Centers headquartered largely in the Washington, D.C. area.

Office of the Commissioner:

The leadership of the agency's scientific activities, communication, legislative liaison, policy and planning, women's and minority health. The current FDA Commissioner is Dr. Scott Gottlieb, was appointed by President Trump, and sworn in May 2017.

National Center for Toxicological Research (NCTR): FDA's research center conducts peer-reviewed research and develops new scientific tools for FDA to improve public health. This research produces innovative tools to assist in solving complex health issues and anticipated toxicological problems and enhances the science of regulatory decision-making at the FDA.

Office of Medical Products and Tobacco:

Provides advice and counsel to the Commissioner on all medical product and tobacco-related programs and issues and oversees the following centers: CDER, CBER, CDRH, and CTP.

Center for Drug Evaluation and Research (CDER): The CDER oversees the regulation of drugs (https://www.fda.gov/Drugs/default.htm). The official definition of a drug is "an article intended for use in the diagnosis, cure, mitigation, treatment, or prevention of disease in man or other animals; articles designed to affect the structure or any function of the body of man or other animals." (FDA, 2018) . CDER regulates over- the-counter and prescription drugs, including biological therapeutics, which act like drugs, and generic drugs.

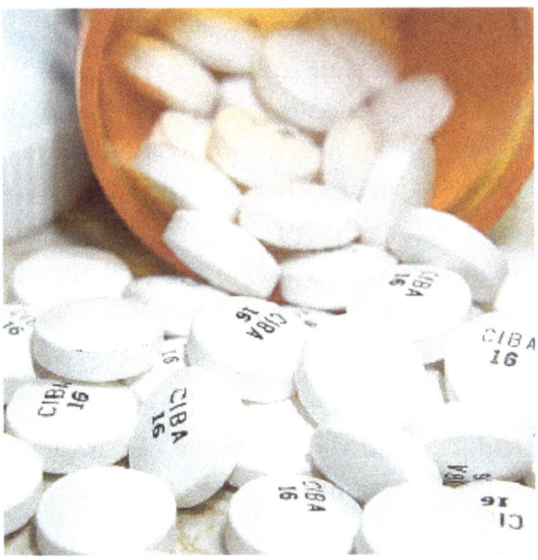

Figure 3.12 The FDA product centers oversee many different types of regulated products. Ritalin, a drug approved by the FDA in 1955. Image Credit: https://fr.m.wikipedia.org/wiki/Fichier:Ritalin-SR-20mg-1000x1000.jpg (https://fr.m.wikipedia.org/wiki/Fichier:Ritalin-SR-20mg-1000x1000.jpg)

Ritalin (Figure 3.12) is one of many drugs that CDER oversees. In 2018, CDER approved a staggering 59 novel drugs, far surpassing the previous five years. In totality, the FDA has approved over 1600 drugs to date. However, CDER's reach covers more than just traditional medicines. For example, the FDA classifies all fluoride toothpaste, antiperspirants, and dandruff shampoos as over-the-counter drugs.

Figure 3.13 dandruff shampoo, the FDA classifies as an over-the-counter drug. Image Credit: Mike Mozart, https://www.flickr.com/photos/jeepersmedia/16162915122 (https://www.flickr.com/photos/jeepersmedia/ 16162915122)

Center for Biologics Evaluation and Research (CBER): The CBER oversees the regulation of vaccines, blood, and biologics (https://www.fda.gov/BiologicsBloodVaccines/default.htm). The official definition of a biological product is "**any virus, therapeutic serum, toxin, antitoxin, vaccine, blood, blood component or derivative, allergenic product,** or analogous product" (FDA, 2018). Note, therapeutic biologics, biologics that act like drugs, are overseen by CDER instead. CDER oversees all drugs, even biologic (think, recombinant proteins!)!

Center for Devices and Radiological Health (CDRH): Oversees the regulation of medical devices (https://www.fda.gov/MedicalDevices/default.htm) and radiation-emitting products, and includes biotechnology products used in diagnostics, such as HIV or pregnancy tests. You will be surprised to learn about some unusual products that are considered medical devices.

Center for Tobacco Products (CTP): The Center for Tobacco Products (https://www.fda.gov/ TobaccoProducts/default.htm) (CTP) oversees the implementation of the Family Smoking Prevention and Tobacco Control Act. Some of the Agency's responsibilities under the law include setting performance standards, reviewing premarket applications for new and modified-risk tobacco products, requiring new warning labels, and establishing and enforcing advertising and promotion restrictions.

Office of Foods and Veterinary Medicine:

Leads a functionally unified FDA Foods Program that addresses food and feed safety, nutrition, and other critical areas to achieve public health goals. The Office of Food & Veterinary Medicine oversees two centers, CFSAN and CVM.

Center for Food Safety and Applied Nutrition (CFSAN): CFSAN oversees food (https://www.fda.gov/Food/ default.htm) safety and purity. It has the power to regulate all domestic and imported food *except for meat, poultry, and eggs* (USDA regulates those). They oversee the safety of food ingredients developed through biotechnology, dietary supplements, food additives, and proper labeling of food. CFSAN is also concerned with food contamination, such as biological pathogens and naturally occurring toxins.

Center for Veterinary Medicine, and (CVM): The CVM https://www.fda.gov/ AnimalVeterinary/default.htm) is a product center which oversees regulation of pet food and pet food additives, as well as drugs and biologics for animals, such as oral heart-worm medication and annual vaccines (Figure 3.14). The CVM does conduct research that helps FDA ensure the safety of animal drugs, food for animals, and food products made from animals, however, they do not oversee pre-clinical animal studies.

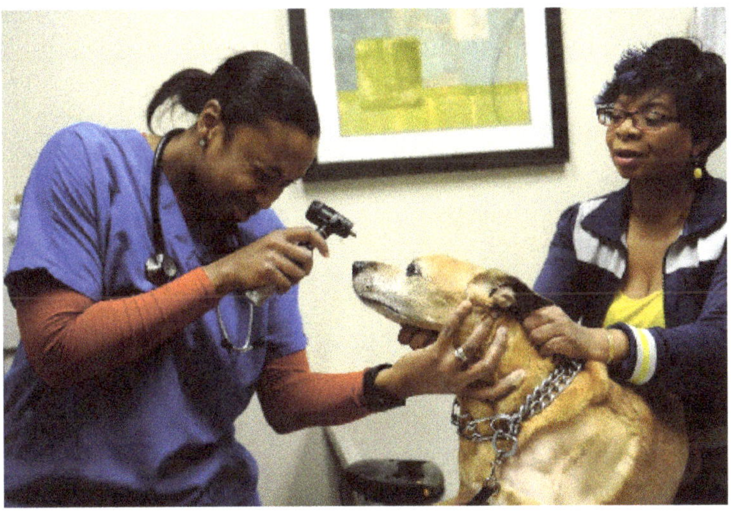

Figure 3.14 Dr. Nina Griffin is giving a dog his annual exam and vaccinations. Image credit: https://www.jble.af.mil/ News/Article-Display/Article/843261/ langley-veterinary-clinic-keeping-our-furry-friends-healthy/ (https://www.jble.af.mil/News/Article-Display/Article/843261/langley-veterinary-clinic-keeping-our-furry- friends-healthy/)

Office of Global Regulatory Operations and Policy:

Provides leadership for FDA's domestic and international product quality and safety efforts.

The FDA's **Office of Regulatory Affairs (ORA)** leads all FDA field activities and provides FDA leadership on imports, inspections, and enforcement policy. ORA supports the agency in the following ways: By inspecting regulated products and manufacturers. ? Performing sample analysis on regulated products, reviewing imported goods, developing policies on compliance and enforcement, and executing the FDA's Import Strategy and Food Protection Plans.

The **Office of International Programs** leads the FDA's international effort is divided into Office of the Director, International Agreement Staff, International Relations Staff, International Scientific Activities and Standards Staff, International Planning, and Resources Management Staff. Conflicting regulatory standards made international trade of FDA-regulated products a challenge until an effort began in the 1990s to harmonize international standards so that important medical products could be bought and sold with less regulatory 'red tape.' The FDA assists both foreign and domestic manufacturers in compliance with health and safety regulations.

CHAPTER-4

TOOLS FOR BIOTECHNICIANS: MEASUREMENTS AND PRECISION

INTRODUCTION: AN OVERVIEW

In the realm of biotechnology, the effectiveness and reliability of scientific endeavours hinge upon the precision of measurements, underscoring the crucial role played by specialized tools. From pipettes and spectrophotometers to advanced imaging devices, the toolkit available to biotechnicians is diverse and tailored for the meticulous quantification and analysis essential in this field. Accurate measurements not only form the foundation of experimental outcomes but also contribute significantly to the development of groundbreaking solutions in areas such as healthcare, agriculture, and environmental science. This exploration into the tools employed by biotechnicians for measurements and precision unveils the meticulous craftsmanship and technological sophistication integral to advancing our understanding of the biological world and addressing complex challenges in various industries.

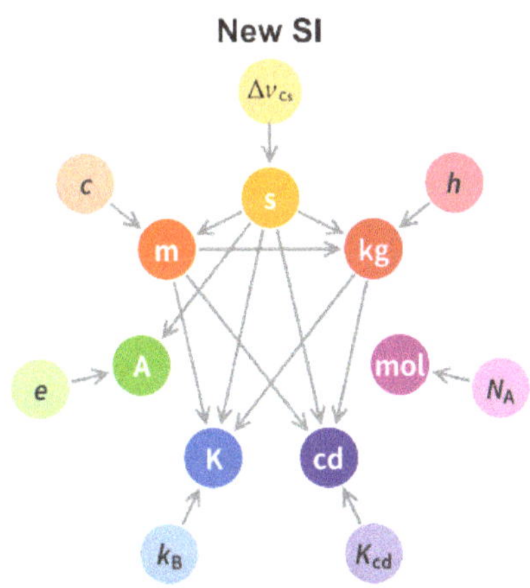

Figure 4.1 The SI system after the 2019 redefinition: Dependence of base unit definitions on physical constants with fixed numerical values and other base units. Image credit: By Emilio Pisanty - Own work, CC BY-SA 4.0, https://commons.wikimedia.org/w/index.php?curid=50713156(https://commons.wikimedia.org/w/ index.php?curid=50713156)

In November 2018, the 26th General Conference on Weights and Measures (CGPM) voted unanimously in favour of revised definitions of the SI base units, which the International Committee for Weights and Measures (CIPM) had proposed earlier that year. The new definitions will come into force on May 20th, 2019. The kilogram, ampere, Kelvin, and mole will then be defined by setting exact numerical values for the Planck constant (h), the elementary electric charge (e), the Boltzmann constant (k), and the Avogadro constant (NA), respectively. The meter and candela are already defined by physical constants, subject to correction to their present definitions. The new definitions aim to improve the SI without changing the size of any units, thus ensuring continuity with existing measurements.

MEASUREMENTS

Measurements provide much of the information that informs the hypotheses, theories, and laws describing the behaviour of matter and energy in both the macroscopic and microscopic domains of chemistry. Every measurement provides three kinds of information: the size or magnitude of the measurement (a number); a standard of comparison for the measurement (a unit); and an indication of the uncertainty of the measurement. While the number and unit are explicitly represented when a quantity is written, the uncertainty is an aspect of the measurement result that is more implicitly represented and will be discussed later.

The number in the measurement can be represented in different ways, including decimal form and scientific notation. (Scientific notation is also known as exponential notation; a review of this topic can be found in Appendix B (https://legacy.cnx.org/content/m68860/latest/) .) For example, the maximum takeoff weight of a Boeing 777-200ER airliner is 298,000 kilograms, which can also be written as 2.98×10^5 kg. The mass of the average mosquito is about 0.0000025 kilograms, which can be written as 2.5×10^{-6} kg.

Units, such as liters, pounds, and centimeters, are standards of comparison for measurements. A 2-liter bottle of a soft drink contains a volume of beverage that is twice that of the accepted volume of 1 liter. The meat used to prepare a 0.25-pound hamburger weighs one-fourth as much as the accepted weight of 1 pound. Without units, a number can be meaningless, confusing, or possibly life threatening. Suppose a doctor prescribes phenobarbital to control a patient's seizures and states a dosage of "100" without specifying units. Not only will this be confusing to the medical professional giving the dose, but the consequences can be dire: 100 mg given three times per day can be effective as an anticonvulsant, but a single dose of 100 g is more than 10 times the lethal amount.

The measurement units for seven fundamental properties ("base units") are listed in Table 4.1. The standards for these units are fixed by international agreement, and they are called the International System of Units or SI Units (from the French, *Le Système International d'Unités*). SI units have been used by the United States National Institute of Standards and Technology (NIST) since 1964. Units for other properties may be derived from these seven base units.

Table 4.1 Base Units of the SI System

Property Measured	Name of Unit	Symbol of Unit
length	meter	m
mass	kilogram	kg
time	second	s
temperature	kelvin	K
electric current	ampere	A
amount of substance	mole	mol
luminous intensity	candela	cd

Everyday measurement units are often defined as fractions or multiples of other units. Milk is commonly packaged in containers of 1 gallon (4 quarts), 1 quart (0.25 gallon), and one pint (0.5 quart). This same approach is used with SI units, but these fractions or multiples are always powers of 10. Fractional or multiple SI units are named using a prefix and the name of the base unit. For example, a length of 1000 meters is also called a kilometer because the prefix *kilo* means "one thousand," which in scientific notation is 10^3 (1 kilometer = 1000 m = 10^3 m). The prefixes used and the powers to which 10 are raised are listed in Table 4.2.

Table 4.2 Common Unit Prefixes

Prefix	Symbol	Factor	Example
femto	f	10^{-15}	1 femtosecond (fs) = 1×10^{-15} s (0.000000000000001 s)
pico	p	10^{-12}	1 picometer (pm) = 1×10^{-12} m (0.000000000001 m)
nano	n	10^{-9}	4 nanograms (ng) = 4×10^{-9} g (0.000000004 g)
micro	µ	10^{-6}	1 microliter (µL) = 1×10^{-6} L (0.000001 L)
milli	m	10-3	2 millimoles (mmol) = 2×10^{-3} mol (0.002 mol)
centi	c	10^{-2}	7 centimeters (cm) = 7×10^{-2} m (0.07 m)
deci	d	10^{-1}	1 deciliter (dL) = 1×10^{-1} L (0.1 L)
kilo	k	10^{3}	1 kilometer (km) = 1×10^{3} m (1000 m)
mega	M	10^{6}	3 megahertz (MHz) = 3×10^{6} Hz (3,000,000 Hz)
giga	G	10^{9}	8 gigayears (Gyr) = 8×10^{9} yr (8,000,000,000 yr)
tera	T	10^{12}	5 terawatts (TW) = 5×10^{12} W (5,000,000,000,000 W)

Need a refresher or more practice with scientific notation? Visit this site (http://openstax.org/l/16notation) to go over the basics of scientific notation.

SI Base Units

The initial units of the metric system, which eventually evolved into the SI system, were established in France during the French Revolution. The original standards for the meter and the kilogram were adopted there in 1799 and eventually by other countries. This section introduces four of the SI base units commonly used in chemistry. Other SI units will be introduced in subsequent chapters.

Length

The standard unit of length in both the SI and original metric systems is the meter (m). A meter was originally specified as 1/10,000,000 of the distance from the North Pole to the equator. It is now defined as the distance light in a vacuum travels in 1/299,792,458 of a second. A meter is about 3 inches longer than a yard (Figure 4.2); one meter is about 39.37 inches or 1.094 yards. Longer distances are often reported in kilometers (1 km = 1000 m = 10^3 m), whereas shorter distances can be reported in centimeters (1 cm = 0.01 m = 10^{-2} m) or millimeters (1 mm = 0.001 m = 10^{-3} m).

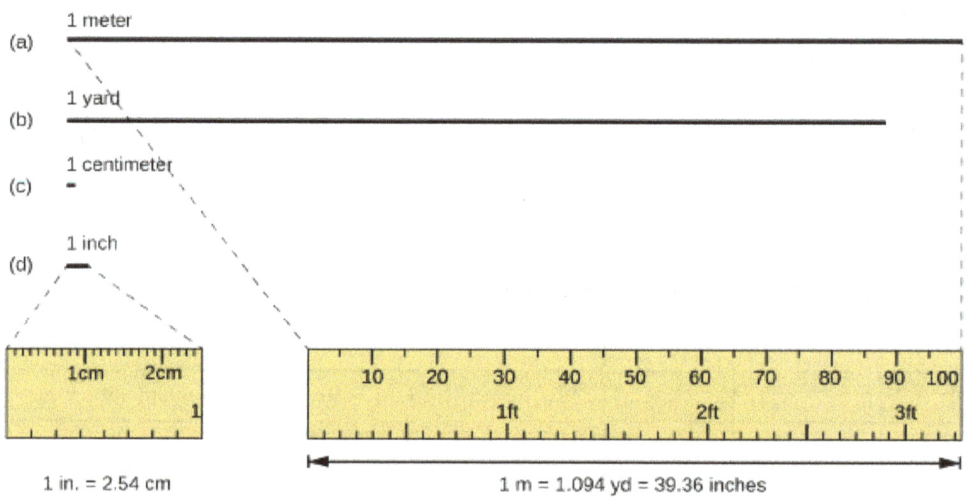

Figure 4.2 The relative lengths of 1 m, 1 yd, 1 cm, and 1 in. are shown (not actual size), as well as comparisons of 2.54 cm and 1 in., and of 1 m and 1.094 yd.

Mass

The standard unit of mass in the SI system is the **kilogram (kg).** The kilogram was previously defined by the International Union of Pure and Applied Chemistry (IUPAC) as the mass of a specific reference object. This object was originally one liter of pure water, and more recently it was a metal cylinder made from a platinum-iridium alloy with a height and diameter of 39 mm (Figure 4.3). In May 2019, this definition was changed to one that is based instead on precisely measured values of several fundamental physical constants. One kilogram is about 2.2 pounds. The gram (g) is exactly equal to 1/1000 of the mass of the kilogram (10-3 kg).

Figure 4.3 This replica prototype kilogram as previously defined is housed at the National Institute of Standards and Technology (NIST) in Maryland. (credit: National Institutes of Standards and Technology)

Temperature

Temperature is an intensive property. The SI unit of temperature is the **kelvin (K).** The IUPAC convention is to use kelvin (all lowercase) for the word, K (uppercase) for the unit symbol, and neither the word "degree" nor the degree symbol (°). The degree **Celsius (°C)** is also allowed in the SI system, with both the word "degree" and the degree symbol used for Celsius measurements. Celsius degrees are the same magnitude as those of kelvin, but the two scales place their zeros in different places. Water freezes at 273.15 K (0°C) and boils at 373.15 K (100°C) by definition, and normal human body temperature is approximately 310 K (37°C). The conversion between these two units and the Fahrenheit scale will be discussed later in this chapter.

Time

The SI base unit of time is the **second (s).** Small and large time intervals can be expressed with the appropriate prefixes; for example, 3 microseconds = 0.000003 s

= 3 × 10^{-6} and 5 megaseconds = 5,000,000 s = 5 × 10^{6} s. Alternatively, hours, days, and years can be used.

Derived SI Units

We can derive many units from the seven SI base units. For example, we can use the base unit of length to define a unit of volume, and the base units of mass and length to define a unit of density.

Volume

Volume is the measure of the amount of space occupied by an object. The standard SI unit of volume is defined by the base unit of length (Figure 4.4). The standard volume is a cubic meter (m^3), a cube with an edge length of exactly one meter. To dispense a cubic meter of water, we could build a cubic box with edge lengths of exactly one meter. This box would hold a cubic meter of water or any other substance.

A more commonly used unit of volume is derived from the decimeter (0.1 m, or 10 cm). A cube with edge lengths of exactly one decimeter contains a volume of one cubic decimeter (dm^3). A **liter (L)** is the more common name for the cubic decimeter. One liter is about 1.06 quarts.

A cubic centimeter (cm^3) is the volume of a cube with an edge length of exactly one centimeter. The abbreviation cc (for cubic centimeter) is often used by health professionals. A cubic centimeter is equivalent to a milliliter (mL) and is 1/1000 of a liter.

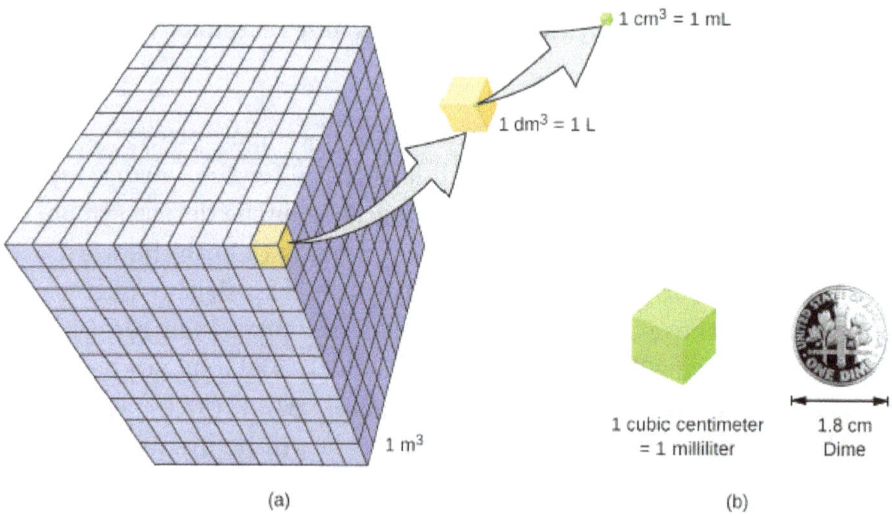

Figure 4.4 (a) The relative volumes are shown for cubes of 1 m3, 1 dm3 (1 L), and 1 cm3 (1 mL) (not to scale). (b) The diameter of a dime is compared relative to the edge length of a 1-cm3 (1-mL) cube.

Density

We use the mass and volume of a substance to determine its density. Thus, the units of density are defined by the base units of mass and length.

The **density** of a substance is the ratio of the mass of a sample of the substance to its volume. The SI unit for density is the kilogram per cubic meter (kg/m^3). For many situations, however, this is an inconvenient unit, and we often use grams per cubic centimeter (g/cm^3) for the densities of solids and liquids, and grams per liter (g/L) for gases. Although there are exceptions, most liquids and solids have densities that range from about 0.7 g/cm^3 (the density of gasoline) to 19 g/cm^3 (the density of gold). The density of air is about 1.2 g/L. Table 4.3 shows the densities of some common substances.

Table 4.3 Densities of Common Substances

Solids	Liquids	Gases (at 25 °C and 1 atm)
ice (at 0 °C) 0.92 g/cm^3	water 1.0 g/cm^3	dry air 1.20 g/L
oak (wood) 0.60–0.90 g/cm^3	ethanol 0.79 g/cm^3	oxygen 1.31 g/L
iron 7.9 g/cm^3	acetone 0.79 g/cm^3	nitrogen 1.14 g/L
copper 9.0 g/cm^3	glycerin 1.26 g/cm^3	carbon dioxide 1.80 g/L
lead 11.3 g/cm^3	olive oil 0.92 g/cm^3	helium 0.16 g/L
silver 10.5 g/cm^3	gasoline 0.70–0.77 g/cm^3	neon 0.83 g/L
gold 19.3 g/cm^3	mercury 13.6 g/cm^3	radon 9.1 g/L

While there are many ways to determine the density of an object, perhaps the most straightforward method involves separately finding the mass and volume of the object, and then dividing the mass of the sample by its volume. In the following example, the mass is found directly by weighing, but the volume is found indirectly through length measurements.

$$\text{density} = \frac{\text{mass}}{\text{volume}}$$

= 202.09459... dm^3 (value from calculator)

= 202 dm^3, or 202 L (answer rounded to three significant figures)

ACCURACY AND PRECISION

Scientists typically make repeated measurements of a quantity to ensure the quality of their findings and to evaluate both the **precision** and the **accuracy** of their results. Measurements are said to be precise if they yield very similar results when repeated in the same manner. A measurement is considered accurate if it yields a result that is very close to the true or accepted value. Precise values agree with each other; accurate values agree with a true value. These characterizations can be extended to other contexts, such as the results of an archery competition (Figure 4.6).

Figure 4.6 (a) These arrows are close to both the bull's eye and one another, so they are both accurate and precise.

These arrows are close to one another but not on target, so they are precise but not accurate. (c) These arrows are neither on target nor close to one another, so they are neither accurate nor precise.

Suppose a quality control chemist at a pharmaceutical company is tasked with checking the accuracy and precision of three different machines that are meant to dispense 10 ounces (296 mL) of cough syrup into storage bottles. She proceeds to use each machine to fill five bottles and then carefully determines the actual volume dispensed, obtaining the results tabulated in Table 4.4.

Table 4.4 Volume (mL) of Cough Medicine Delivered by 10-oz (296 mL) Dispensers

Dispenser #1	Dispenser #2	Dispenser #3
283.3	298.3	296.1
284.1	294.2	295.9
283.9	296.0	296.1
284.0	297.8	296.0
284.1	293.9	296.1

Considering these results, she will report that dispenser #1 is precise (values all close to one another, within a few tenths of a milliliter) but not accurate (none of the values are close to the target value of 296 mL, each being more than 10 mL too low). Results for dispenser #2 represent improved accuracy (each volume is less than 3 mL away from 296 mL) but worse precision (volumes vary by more than 4 mL). Finally, she can report that dispenser #3 is working well, dispensing cough syrup both accurately (all volumes within 0.1 mL of the target volume) and precisely (volumes differing from each other by no more than 0.2 mL).

METHOD VALIDATION PROCESS

Method validation is the process used to determine the conditions to obtain a result reliably, and the limits of that procedure. Formal validation of test methods is required in GMP compliant laboratories. However, the aspects of validation are appropriate for any research or testing laboratory.

Accuracy

Accuracy is the closeness of a test result to the true or accepted value. Accuracy can be tested by using a reference standard value is known. If it is a method being tested, results are compared to those of a standardized assay. Tests for accuracy are also used to validate equipment. Accuracy is calculated by determining the percent error of the mean (PEM).

$$\text{PEM} = (\text{calculated mean} - \text{true value}) \div \text{true value} \times 100\%$$

Precision

Precision is the degree of agreement between individual test results when the procedure is applied repeatedly. Repeatability is the precision of measurements made under uniform conditions and reproducibility (ruggedness) is the precision of measurements made under non-uniform conditions such as in two different laboratories. When precision is high, there is a high degree of agreement in data. Precision is calculated using standard deviation.

Limit of Detection (LOD)

LOD is the lowest concentration of the analyte which can be detected above background by a method. There is always some error associated with any analytical measurement. An important determination that must be made is how significant a signal must be before it can be distinguishable from background. The accepted rule in analytical chemistry is that the signal must be at least three times greater than the background noise. It is important to note that an analytical method can never prove that a substance is not present; rather it can show only the limit which the sample cannot be detected. Interpolate from linear regression to determine the concentration value of LOD.

$$\text{LOD} = \text{Instrument Output of the blank} + 3 \times \text{SD of the blank}$$

Limit of Quantitation (LOQ)

LOQ is the lowest concentration of the analyte that the method can quantify with acceptable accuracy and precision. Depending on the application, this is 5-10X the standard deviation of the blank. Interpolate from linear regression to determine the concentration value of LOQ.

$$\text{LOQ} = \text{Instrument Output of the blank} + 10 \times \text{SD of the blank}$$

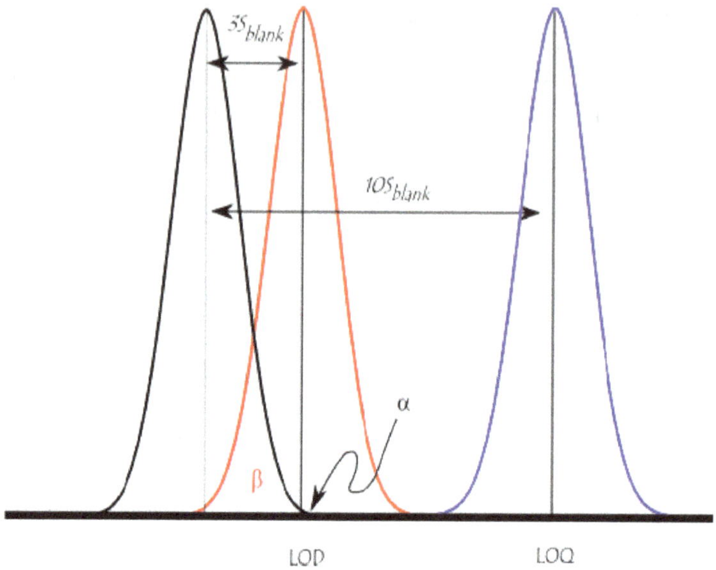

Figure 4.7 Illustration of the concept of detection limit and quantitation limit by showing the theoretical normal distributions associated with blank, detection limit, and quantification limit level samples. Image credit: https://commons.wikimedia.org/wiki/File:LOD.png(https://commons.wikimedia.org/wiki/File:LOD.png)

Specificity

Specificity is a measure of the extent to which a method can identify the presence of a compound in a sample without interference from other materials that are present. This is sometimes referred to as selectivity. A very selective test will only give a positive result for the compound of interest.

Linearity

Linearity is the ability of a method to give test results that are directly proportional to the concentration of the material of interest within a given concentration range. The range is defined as the limits of concentrations, from the lowest to the highest, that a method can measure with acceptable results. Tests and Assays have a range in which they exhibit linearity. For example, many assays for proteins display a standard curve that is linear at lower concentrations but flattens out at higher concentrations of the protein. Therefore, a test for protein will be accurate only in the range of lower concentrations when the curve is linear.

Robustness

Robustness is a measure of the capacity of a method to remain unaffected when there are small, deliberate variations in method parameters. It indicates the method's reliability during normal use.

Ruggedness

Ruggedness is the degree of reproducibility of the results obtained by the analysis of the same samples under a variety of conditions. Ruggedness provides a measure of reproducibility across variations of conditions normally expected, such as, from lab to lab and analyst to analyst.

The Clinical Laboratory Improvement Amendments

The Clinical Laboratory Improvement Amendments (CLIA) of 1988 are United States federal regulatory standards that apply to all clinical laboratory testing performed on humans in the United States, except clinical trials and basic research. CLIA "defines quality standards for all laboratory testing to ensure accuracy, reliability, and timeliness of patient results regardless of where the test was performed" (CMS. gov).

Table 4.5 Three federal agencies are responsible for CLIA, each having a unique role in assuring quality standards: The Food and Drug Administration (FDA), Center for Medicaid Services (CMS) and the Center for Disease Control (CDC).

FDA	Categorizes tests based on the complexity Reviews requests for Waiver by Application Develops rules/guidance for CLIA complexity categorization
CMS (http://www.cms.gov/Regulations-and-Guidance/ Legislation/CLIA/index.html?redirect=/clia/)	Issues laboratory certificates Collects user fees Conducts inspections and enforces regulatory compliance Monitors laboratory performance on Proficiency Testing Publishes CLIA rules and regulations
CDC (http://wwwn.cdc.gov/CLIA/Default.aspx)	Provides analysis, research, and technical assistance Develops technical standards and laboratory practice guidelines Conducts laboratory quality improvement studies Monitors proficiency testing practices Develops and distributes educational resources

CHAPTER-5

BIOTECHNICIAN INSTRUMENTS: SOLUTION PREPARATION

INTRODUCTION: AN OVERVIEW

In the realm of biotechnology, the proficiency of a biotechnician is not only measured by their expertise in analytical techniques but also by their adeptness in the precise preparation of solutions. The accurate formulation of solutions is a fundamental aspect of experimental success, influencing outcomes in diverse areas such as genetic engineering, pharmaceutical development, and molecular biology. This exploration delves into the essential instruments employed by biotechnicians in the art and science of solution preparation, unraveling the meticulous processes that underpin advancements in biotechnological research and applications. From meticulous measurements to controlled mixing, the tools in a biotechnician's arsenal for solution preparation form the backbone of innovative strides in the dynamic landscape of biotechnology.

Your alarm goes off and, after hitting "snooze" once or twice, you pry yourself out of bed. You make a cup of coffee to help you get going, and then you shower, get dressed, eat breakfast, and check your phone for messages. On your way to school, you stop to fill your car's gas tank, almost making you late for the first day of chemistry class. As you find a seat in the classroom, you read the question projected on the screen: "Welcome to class! Why should we study chemistry?"

Do you have an answer? You may be studying chemistry because it fulfills an academic requirement, but if you consider your daily activities, you might find chemistry interesting for other reasons. Most everything you do and encounter during your day involves chemistry. Making coffee, cooking eggs, and toasting bread involve chemistry. The products you use—like soap and shampoo, the fabrics you wear, the electronics that keep you connected to your world, the gasoline that propels your car—all of these and more involve chemical substances and processes. Whether you are aware or not, chemistry is part of your everyday world. In this course, you will learn many of the essential principles underlying the chemistry of modern-day life.

MASS FORMULAS AND THE CONCEPT OF MOLES

Many argue that modern chemical science began when scientists started exploring the quantitative as well as the qualitative aspects of chemistry. For example, Dalton's atomic theory was an attempt to explain the results of measurements that allowed him to calculate the relative masses of elements combined in various compounds. Understanding the relationship between the masses of atoms and the chemical formulas of compounds allows us to quantitatively describe the composition of substances.

Formula Mass

An earlier chapter of this text described the development of the atomic mass unit, the concept of average atomic masses, and the use of chemical formulas to represent the elemental makeup of substances. These ideas can be extended to calculate the **formula mass** of a substance by summing the average atomic masses of all the atoms represented in the substance's formula.

Formula Mass for Covalent Substances

For covalent substances, the formula represents the numbers and types of atoms composing a single molecule of the substance; therefore, the formula mass may be correctly referred to as a molecular mass. Consider chloroform ($CHCl_3$), a covalent compound once used as a surgical anesthetic and now primarily used in the production of tetrafluoroethylene, the building block for the "anti-stick" polymer, Teflon. The molecular formula of chloroform indicates that a single molecule contains one carbon atom, one hydrogen atom, and three chlorine atoms. The average molecular mass of a chloroform molecule is therefore equal to the sum of the average atomic masses of these atoms. Figure 5.1 outlines the calculations used to derive the molecular mass of chloroform, which is 119.37 amu.

Element	Quantity		Average atomic mass (amu)		Subtotal (amu)
C	1	×	12.01	=	12.01
H	1	×	1.008	=	1.008
Cl	3	×	35.45	=	106.35
			Molecular mass		119.37

Figure 5.1 The average mass of a chloroform molecule, $CHCl_3$, is 119.37 amu, which is the sum of the average atomic masses of each of its constituent atoms. The model shows the molecular structure of chloroform.

Likewise, the molecular mass of an aspirin molecule, $C_9H_8O_{4,}$ is the sum of the atomic masses of nine carbon atoms, eight hydrogen atoms, and four oxygen atoms, which amounts to 180.15 amu (Figure 5.2).

Element	Quantity		Average atomic mass (amu)		Subtotal (amu)
C	9	×	12.01	=	108.09
H	8	×	1.008	=	8.064
O	4	×	16.00	=	64.00
			Molecular mass		180.15

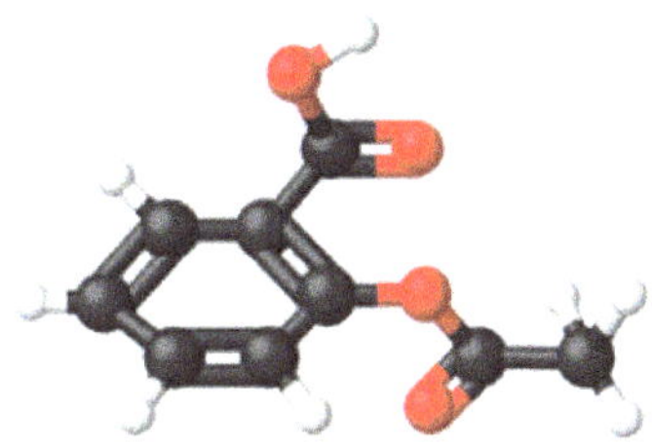

Figure 5.2 The average mass of an aspirin molecule is 180.15 amu. The model shows the molecular structure of aspirin, $C^9H^8O^4$.

Formula Mass for Ionic Compounds

Ionic compounds are composed of discrete cations and anions combined in ratios to yield electrically neutral bulk matter. The formula mass for an ionic compound is calculated in the same way as the formula mass for covalent compounds: by summing the average atomic masses of all the atoms in the compound's formula. Keep in mind, however, that the formula for an ionic compound does not represent the composition of a discrete molecule, so it may not correctly be referred to as the "molecular mass."

As an example, consider sodium chloride, NaCl, the chemical name for common table salt. Sodium chloride is an ionic compound composed of sodium cations, Na^+, and chloride anions, Cl^-, combined in a 1:1 ratio. The formula mass for this compound is computed as 58.44 amu (see Figure 5.3).

Element	Quantity		Average atomic mass (amu)		Subtotal
Na	1	×	22.99	=	22.99
Cl	1	×	35.45	=	35.45
			Formula mass		58.44

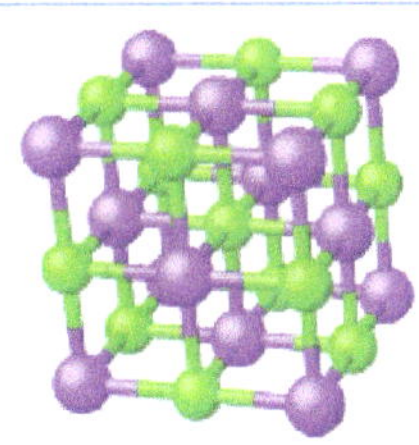

Figure 5.3 Table salt, NaCl, contains an array of sodium and chloride ions combined in a 1:1 ratio. Its formula mass is 58.44 amu.

Note that the average masses of neutral sodium and chlorine atoms were used in this computation, rather than the masses for sodium cations and chlorine anions. This approach is perfectly acceptable when computing the formula mass of an ionic compound. Even though a sodium cation has a slightly smaller mass than a sodium atom (since it is missing an electron), this difference will be offset by the fact that a chloride anion is slightly more massive than a chloride atom (due to the extra electron). Moreover, the mass of an electron is negligibly small with respect to the mass of a typical atom. Even when calculating the mass of an isolated ion, the missing or additional electrons can generally be ignored, since their contribution to the overall mass is negligible, reflected only in the non-significant digits that will be lost when the computed mass is properly rounded. The few exceptions to this guideline are very light ions derived from elements with precisely known atomic masses.

The Mole

The identity of a substance is defined not only by the types of atoms or ions it contains, but by the quantity of each type of atom or ion. For example, water, H_2O, and hydrogen peroxide, H_2O_2, are alike in that their respective molecules are composed of hydrogen and oxygen atoms. However, because a hydrogen peroxide molecule contains two oxygen atoms, as opposed to the water molecule, which has only one, the two substances exhibit very different properties. Today, sophisticated instruments allow the direct measurement of these defining microscopic traits; however, the same traits were originally derived from the measurement of macroscopic properties (the masses and volumes of bulk quantities of matter) using relatively simple tools (balances and volumetric glassware). This experimental approach required the introduction of a new unit for amount of substances, the *mole*, which remains indispensable in modern chemical science.

The *mole* is an amount unit similar to familiar units like pair, dozen, gross, etc. It provides a specific measure of *the number* of atoms or molecules in a sample of matter. One Latin connotation for the word "mole" is "large mass" or "bulk," which is consistent with its use as the name for this unit. The mole provides a link between an easily measured macroscopic property, bulk mass, and an extremely important fundamental property, number of atoms, molecules, and so forth. A **mole** of substance is that amount in which there are $6.02214076 \times 10^{23}$ discrete entities (atoms or molecules). This large number is a fundamental constant known as **Avogadro's number (N_A)** or the Avogadro constant in honor of Italian scientist Amedeo Avogadro. This constant is properly reported with an explicit unit of "per mole," a conveniently rounded version being 6.022×10^{23}/mol.

Consistent with its definition as an amount unit, 1 mole of any element contains the same number of atoms as 1 mole of any other element. The masses of 1 mole of different elements, however, are different, since the masses of the individual atoms are drastically different. The molar mass of an element (or compound) is the mass in grams of 1 mole of that substance, a property expressed in units of grams per mole (g/mol) (see Figure 5.4).

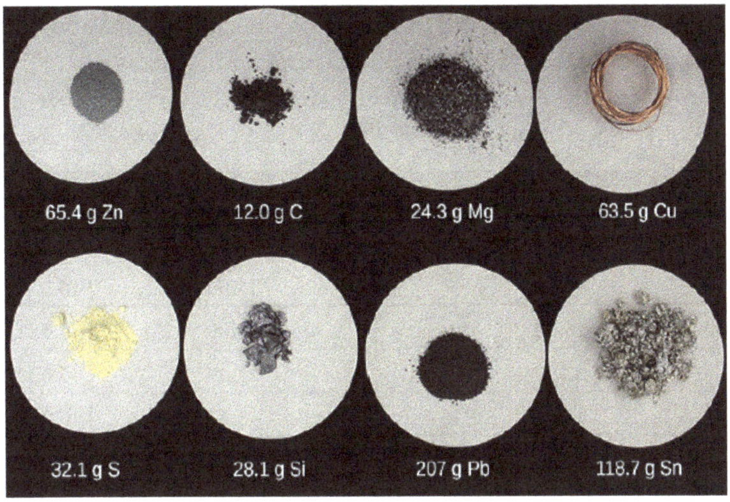

Figure 5.4 Each sample contains 6.022 × 10^{23} atoms —1.00 mol of atoms. From left to right (top row): 65.4 g zinc, 12.0 g carbon, 24.3 g magnesium, and 63.5 g copper. From left to right (bottom row): 32.1 g sulfur, 28.1 g silicon, 207g lead, and 118.7 g tin. (credit: modification of work by Mark Ott)

The molar mass of any substance is numerically equivalent to its atomic or formula weight in amu. Per the amu definition, a single ^{12}C atom weighs 12 amu (its atomic mass is 12 amu). A mole of ^{12}C weighs 12 g (its molar mass is 12 g/mol). This relationship holds for all elements, since their atomic masses are measured relative to that of the amu-reference substance, ^{12}C. Extending this principle, the molar mass of a compound in grams is likewise numerically equivalent to its formula mass in amu (Figure 5.5).

Figure 5.5 Each sample contains 6.02 × 10^{23} molecules or formula units—1.00 mol of the compound or element.

Clock-wise from the upper left: 130.2 g of $C_8H_{17}OH$ (1-octanol, formula mass 130.2 amu), 454.4 g of HgI_2 (mercury(II) iodide, formula mass 454.4 amu), 32.0 g of CH_3OH (methanol, formula mass 32.0 amu) and 256.5 g of S_8 (sulfur, formula mass 256.5 amu). (credit: Sahar Atwa)

Element	Average Atomic Mass (amu)	Molar Mass (g/mol)	Atoms/Mole		
C	12.01	12.01	6.022	×	10^{23}
H	1.008	1.008	6.022	×	10^{23}
O	16.00	16.00	6.022	×	10^{23}
Na	22.99	22.99	6.022	×	10^{23}
Cl	35.45	35.45	6.022	×	10^{23}

While atomic mass and molar mass are numerically equivalent, keep in mind that they are vastly different in terms of scale, as represented by the vast difference in the magnitudes of their respective units (amu versus g). To appreciate the enormity of the mole, consider a small drop of water weighing about 0.03 g (see Figure 5.6). Although this represents just a tiny fraction of 1 mole of water (~18 g), it contains more water molecules than can be clearly imagined. If the molecules were distributed equally among the roughly seven billion people on earth, each person would receive more than 100 billion molecules.

Figure 5.6 The number of molecules in a single droplet of water is roughly 100 billion times greater than the number of people on earth. (credit: "tanakawho"/ Wikimedia commons)

The mole is used in chemistry to represent 6.022 × 1023 of something, but it can be difficult to conceptualize such a large number. Watch this video (http://openstax.org/l/16molevideo) and then complete the "Think" questions that follow. Explore more about the mole by reviewing the information under "Dig Deeper."

The relationships between formula mass, the mole, and Avogadro's number can be applied to compute various quantities that describe the composition of substances and compounds, as demonstrated in the next several example problems.

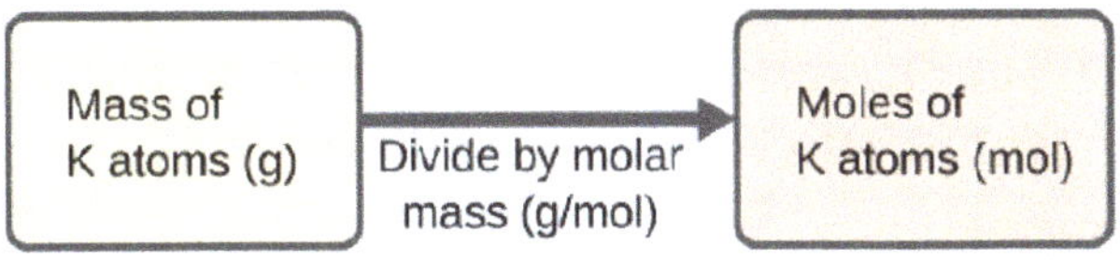

In this case, logic dictates (and the factor-label method supports) multiplying the provided amount (mol) by the molar mass (g/mol):

$$9.2 \times 10^{-4} \cancel{\text{mol Ar}} \left(\frac{39.95 \text{ g Ar}}{\cancel{\text{mol Ar}}} \right) = 0.037 \text{ g Ar}$$

The result is in agreement with our expectations, around 0.04 g Ar.

Copper is commonly used to fabricate electrical wire (**Figure 5.8**). How many copper atoms are in 5.00 g of copper wire?

Figure 5.7 Copper wire is composed of many, many atoms of Cu. (credit: Emilian Robert Vicol)

Counting Neurotransmitter Molecules in the Brain

The brain is the control center of the central nervous system (Figure 5.8). It sends and receives signals to and from muscles and other internal organs to monitor and control their functions; it processes stimuli detected by sensory organs to guide interactions with the external world; and it houses the complex physiological processes that give rise to our intellect and emotions. The broad field of neuroscience spans all aspects of the structure and function of the central nervous system, including research on the anatomy and the BRAIN Initiative, a federal initiative announced in 2013, aims to accelerate and capitalize on these advances through the concerted efforts of various industrial, academic, and government agencies.

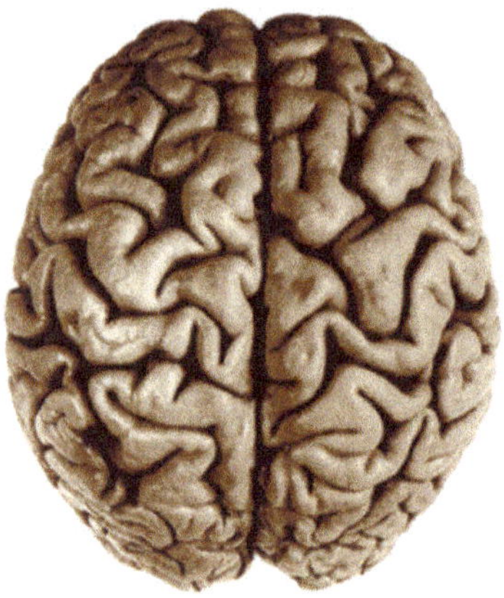

(a)

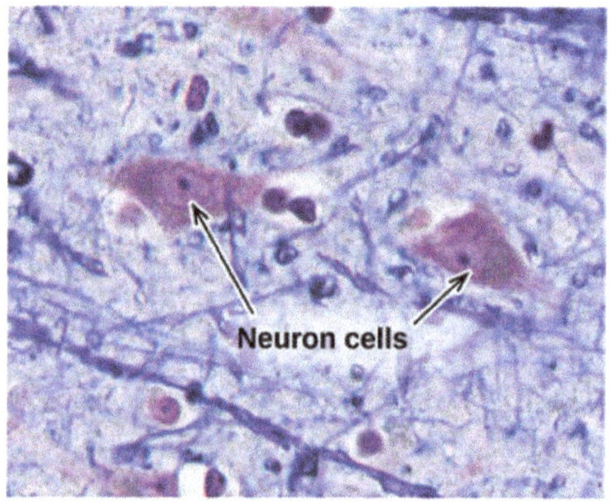

(b)

Figure 5.8 (a) A typical human brain weighs about 1.5 kg and occupies a volume of roughly 1.1 L. (b) Information is transmitted in brain tissue and throughout the central nervous system by specialized cells called neurons (micrograph shows cells at 1600× magnification).

Specialized cells called neurons transmit information between different parts of the central nervous system by way of electrical and chemical signals. Chemical signaling occurs at the interface between different neurons when one of the cells releases molecules (called neurotransmitters) that diffuse across the small gap between the cells (called the synapse) and bind to the surface of the other cell. These neurotransmitter molecules are stored in small intracellular structures called vesicles that fuse to the cell membrane and then break open to release their contents when the neuron is appropriately stimulated. This process is called exocytosis (see Figure 5.9). One neurotransmitter that has been very extensively studied is dopamine, $C_8H_{11}NO_2$. Dopamine is involved in various neurological processes that impact a wide variety of human behaviors. Dysfunctions in the dopamine systems of the brain underlie serious neurological diseases such as Parkinsons and schizophrenia.

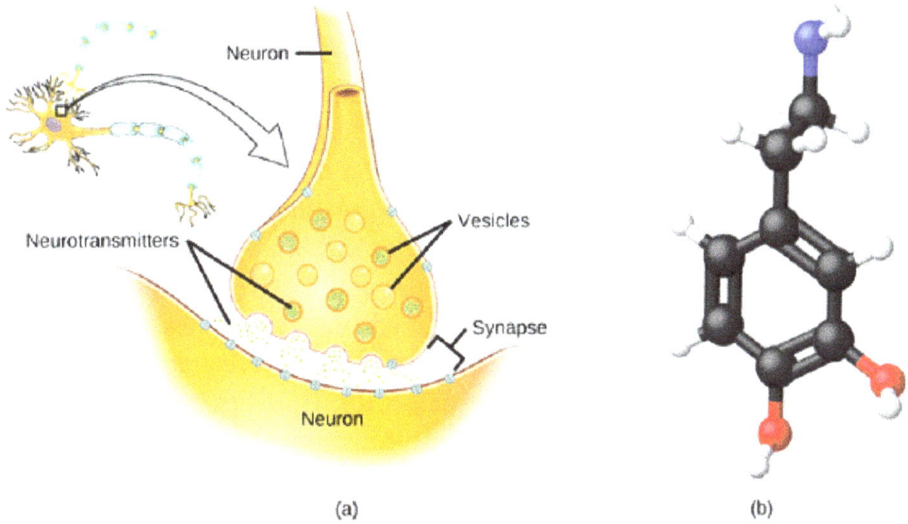

Figure 5.9 (a) Chemical signals are transmitted from neurons to other cells by the release of neurotransmitter molecules into the small gaps (synapses) between the cells. (b) Dopamine, $C_8H_{11}NO_2$, is a neurotransmitter involved in a number of neurological processes.

One important aspect of the complex processes related to dopamine signaling is the number of neurotransmitter molecules released during exocytosis. Since this number is a central factor in determining neurological response (and subsequent human thought and action), it is important to know how this number changes with certain controlled stimulations, such as the administration of drugs. It is also important to understand the mechanism responsible for any changes in the number of neurotransmitter molecules released—for example, some dysfunction in exocytosis, a change in the number of vesicles in the neuron, or a change in the number of neurotransmitter molecules in each vesicle.

Significant progress has been made recently in directly measuring the number of dopamine molecules stored in individual vesicles and the amount actually released when the vesicle undergoes exocytosis. Using miniaturized probes that can selectively detect dopamine molecules in very small amounts, scientists have determined that the vesicles of a certain type of mouse brain neuron contain an average of 30,000 dopamine molecules per vesicle (about 5×10^{-20} mol or 50 zmol). Analysis of these neurons from mice subjected to various drug therapies shows significant changes in the average number of dopamine molecules contained in individual vesicles, increasing or decreasing by up to three-fold, depending on the specific drug used. These studies also indicate that not all of the dopamine in a given vesicle is released during exocytosis, suggesting that it may be possible to regulate the fraction released using pharmaceutical therapies.[1]

MOLARITY

Mixtures—samples of matter containing two or more substances physically combined—are more commonly encountered in nature than are pure substances. Similar to a pure substance, the relative composition of a mixture plays an important role in determining its properties. The relative amount of oxygen in a planet's atmosphere determines its ability to sustain aerobic life. The relative amounts of iron, carbon, nickel, and other elements in steel (a mixture known as an "alloy") determine its physical strength and resistance to corrosion. The relative amount of the active ingredient in a medicine determines its effectiveness in achieving the desired pharmacological effect. The relative amount of sugar in a beverage determines its sweetness (see Figure 5.10). This section will describe one of the most common ways in which the relative compositions of mixtures may be quantified.

Figure 5.10 Sugar is one of many components in the complex mixture known as coffee. The amount of sugar in a given amount of coffee is an important determinant of the beverage's sweetness. (credit: Jane Whitney)

Solutions

Solutions have previously been defined as homogeneous mixtures, meaning that the composition of the mixture (and therefore its properties) is uniform throughout its entire volume. Solutions occur frequently in nature and have also been implemented in many forms of manmade technology. A more thorough treatment of solution properties is provided in the chapter on solutions and colloids, but provided here is an introduction to some of the basic properties of solutions.

The relative amount of a given solution component is known as its **concentration.** Often, though not always, a solution contains one component with a concentration that is significantly greater than that of all other components. This component is called the **solvent** and may be viewed as the medium in which the other components are dispersed, or **dissolved.** Solutions in which water is the solvent are, of course, very common on our planet. A solution in which water is the solvent is called an **aqueous solution.**

A **solute** is a component of a solution that is typically present at a much lower concentration than the solvent. Solute concentrations are often described with qualitative terms such as **dilute** (of relatively low concentration) and **concentrated** (of relatively high concentration).

Concentrations may be quantitatively assessed using a wide variety of measurement units, each convenient for particular applications. **Molarity (*M*)** is a useful concentration unit for many applications in chemistry. Molarity is defined as the number of moles of solute in exactly 1 liter (1 L) of the solution:

$$M = \frac{\text{mol solute}}{\text{L solution}}$$

Calculating Molar Concentrations from the Mass of Solute

Distilled white vinegar (Figure 5.11) is a solution of acetic acid, CH_3CO_2H, in water. A 0.500-L vinegar solution contains 25.2 g of acetic acid. What is the concentration of the acetic acid solution in units of molarity?

Figure 5.11 Distilled white vinegar is a solution of acetic acid in water.

Dilution of Solutions

Dilution is the process whereby the concentration of a solution is lessened by the addition of solvent. For example, a glass of iced tea becomes increasingly diluted as the ice melts. The water from the melting ice increases the volume of the solvent (water) and the overall volume of the solution (iced tea), thereby reducing the relative concentrations of the solutes that give the beverage its taste (Figure 5.12).

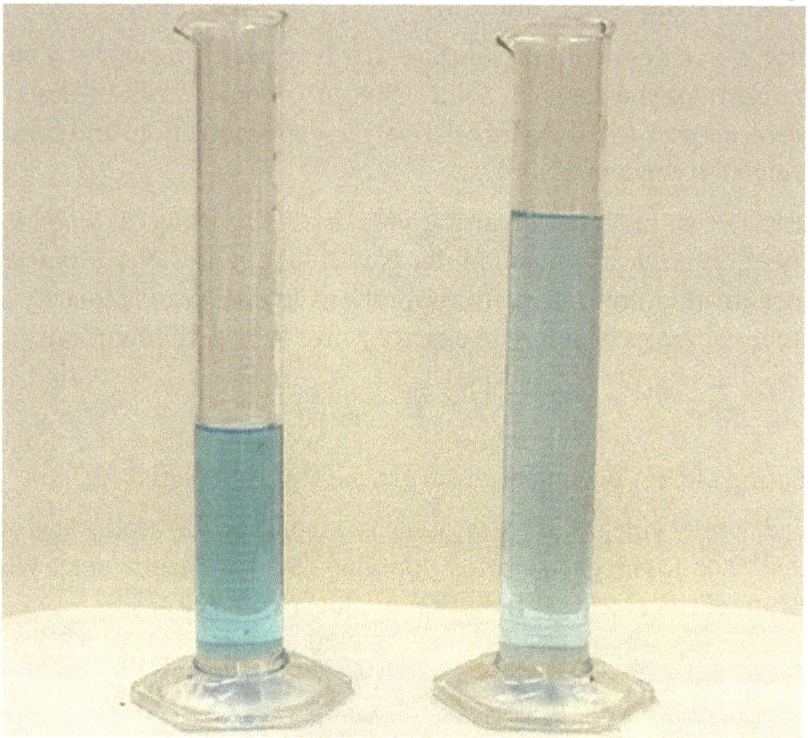

Figure 5.12 Both solutions contain the same mass of copper nitrate. The solution on the right is more dilute because the copper nitrate is dissolved in more solvent. (credit: Mark Ott).

Dilution is also a common means of preparing solutions of a desired concentration. By adding solvent to a measured portion of a more concentrated *stock solution*, a solution of lesser concentration may be prepared. For example, commercial pesticides are typically sold as solutions in which the active ingredients are far more concentrated than is appropriate for their application. Before they can be used on crops, the pesticides must be diluted. This is also a very common practice for the preparation of a number of common laboratory reagents.

A simple mathematical relationship can be used to relate the volumes and concentrations of a solution before and after the dilution process. According to the definition of molarity, the number of moles of solute in a solution (n) is equal to the product of the solution's molarity (M) and its volume in liters (L):

$$n = ML$$

Expressions like these may be written for a solution before and after it is diluted:

$$n_1 = M_1L_1$$

$$n_2 = M_2L_2$$

where the subscripts "1" and "2" refer to the solution before and after the dilution, respectively. Since the dilution process *does not change the amount of solute in the solution,* $n_1 = n_2$. Thus, these two equations may be set equal to one another:

$$M_1L_1 = M_2L_2$$

This relation is commonly referred to as the dilution equation. Although this equation uses molarity as the unit of concentration and liters as the unit of volume, other units of concentration and volume may be used as long as the units properly cancel per the factor-label method. Reflecting this versatility, the dilution equation is often written in the more general form:

$$C_1V_1 = C_2V_2$$

where *C* and *V* are concentration and volume, respectively.

OTHER UNITS FOR SOLUTION CONCENTRATIONS

- Define the concentration units of mass percentage, volume percentage, mass-volume percentage,
- parts-per-million (ppm), and parts-per-billion (ppb) Perform computations relating a solution's concentration and its components' volumes and/or masses using these units

The previous section introduced molarity, a very useful measurement unit for evaluating the concentration of solutions. However, molarity is only one measure of concentration. This section will describe some other units of concentration that are commonly used in various applications, either for convenience or by convention.

Mass Percentage

Earlier in this chapter, percent composition was introduced as a measure of the relative amount of a given element in a compound. Percentages are also commonly used to express the composition of mixtures, including solutions. The **mass percentage** of a solution component is defined as the ratio of the component's mass to the solution's mass, expressed as a percentage:

Mass percentage is also referred to by similar names such as *percent mass, percent weight, weight/weight percent*, and other variations on this theme. The most common symbol for mass percentage is simply the percent sign, %, although more detailed symbols are often used including %mass, %weight, and (w/w)%. Use of these more detailed symbols can prevent confusion of mass percentages with other types of percentages, such as volume percentages (to be discussed later in this section).

Mass percentages are popular concentration units for consumer products. The label of a typical liquid bleach bottle (Figure 5.13) cites the concentration of its active ingredient, sodium hypochlorite (NaOCl), as being 7.4%. A 100.0-g sample of bleach would therefore contain 7.4 g of NaOCl.

Figure 5.13 Liquid bleach is an aqueous solution of sodium hypochlorite (NaOCl). This brand has a concentration of 7.4% NaOCl by mass.

Mass-Volume Percentage

Mixed" percentage units, derived from the mass of solute and the volume of solution, are popular for certain biochemical and medical applications. A **mass-volume percent** is a ratio of a solutes mass to the solutions volume expressed as a percentage. The specific units used for solute mass and solution volume may vary, depending on the solution. For example, physiological saline solution, used to prepare intravenous fluids, has a concentration of 0.9% mass/volume (m/v), indicating that the composition is 0.9 g of solute per 100 mL of solution. The concentration of glucose in blood (commonly referred to as "blood sugar") is also typically expressed in terms of a mass-volume ratio. Though not expressed explicitly as a percentage, its concentration is usually given in milligrams of glucose per deciliter (100 mL) of blood (Figure 5.14).

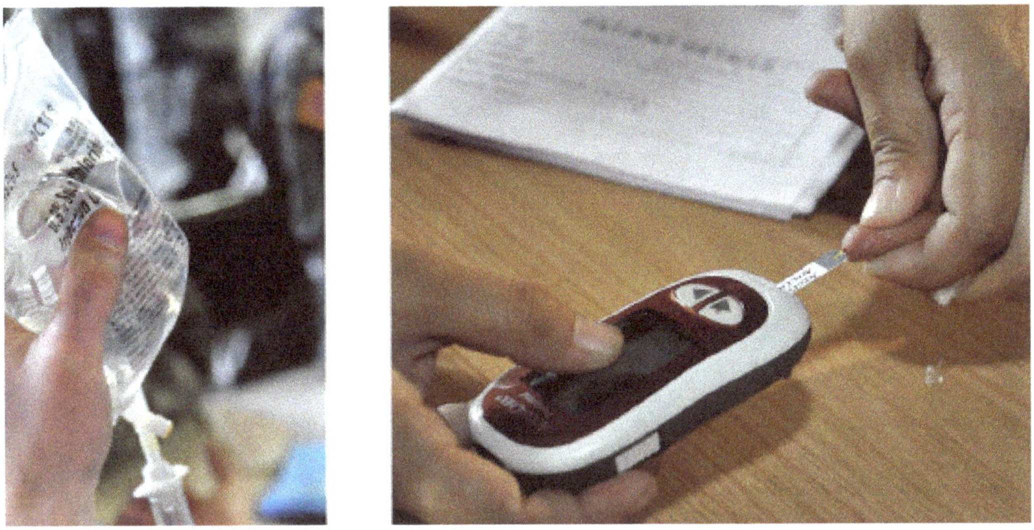

Figure 5.14 "Mixed" mass-volume units are commonly encountered in medical settings. (a) The NaCl concentration of physiological saline is 0.9% (m/v). (b) This device measures glucose levels in a sample of blood. The normal range for glucose concentration in blood (fasting) is around 70–100 mg/dL. (credit a: modification of work by "The National Guard"/Flickr; credit b: modification of work by Biswarup Ganguly)

Parts per Million and Parts per Billion

Very low solute concentrations are often expressed using appropriately small units such as **parts per million (ppm)** or **parts per billion (ppb)**. Like percentage ("part per hundred") units, ppm and ppb may be defined in terms of masses, volumes, or mixed mass-volume units. There are also ppm and ppb units defined with respect to numbers of atoms and molecules.

The mass-based definitions of ppm and ppb are given here:

$$\text{ppm} = \frac{\text{mass solute}}{\text{mass solution}} \times 10^6 \text{ ppm}$$

$$\text{ppb} = \frac{\text{mass solute}}{\text{mass solution}} \times 10^9 \text{ ppb}$$

Both ppm and ppb are convenient units for reporting the concentrations of pollutants and other trace contaminants in water. Concentrations of these contaminants are typically very low in treated and natural waters, and their levels cannot exceed relatively low concentration thresholds without causing adverse effects on health and wildlife. For example, the EPA has identified the maximum safe level of fluoride ion in tap water to be 4 ppm. Inline water filters are designed to reduce the concentration of fluoride and several other trace-level contaminants in tap water (Figure 5.15).

(a) (b)

Figure 5.15 (a) In some areas, trace-level concentrations of contaminants can render unfiltered tap water unsafe for drinking and cooking. (b) Inline water filters reduce the concentration of solutes in tap water. (credit a: modification of work by Jenn Durfey; credit b: modification of work by "vastateparkstaff"/ Wikimedia commons)

BUFFERS

A buffer solution is an aqueous solution consisting of a mixture of a weak acid and its conjugate base, or vice versa. Within a buffer solution, pH changes very little when adding a small amount of strong acid or base. In nature, many systems use buffering for pH regulation. For example, the bicarbonate buffering system is used to regulate the pH of blood. Buffer solutions help maintain a biological system at its proper pH.

It is essential when working with protein (and other biomolecular) solutions that you pay attention to pH. The pH determines not only the overall native conformation (and proper functioning) of the protein it can also determine how the protein interacts with other biomolecules and how the protein interacts with non-biological surfaces such as an ion exchange column. When a solution is buffered, it resists a change in pH, even when H+ ions are added or lost from the system. Many chemicals can act as a buffer and can work at a wide range of pH.

Factors Impacting Buffer Selection:

- The pKa of a buffer is the pH at which the buffer experiences little change in pH with the addition of acids and bases. When choosing a buffer, select one with a pKa one unit above or below the pH of the solution. Buffers are effective in resisting a change in pH within a range of approximately one pH unit. For example, an acetate buffer has a pKa of 4.8 and is, therefore, effective in a range of 3.8-5.8 pH units.

- Not all buffering systems are compatible with all biochemistry procedures, and technicians may need to research the appropriate buffer to use in a biochemical application. For example, although Tris is the most common buffer used in a biotechnology lab, it is inappropriate for some protein assays because it tends to react with the assay components resulting in erroneous data.
- Two factors that influence the pH of a buffer: concentration and temperature. Although it is convenient to prepare a buffer as a concentrated stock, buffer pH can vary when diluted. Always check the final pH of a buffered solution if a concentrated stock was used to prepare it. Temperature also affects the pH of a buffer. Always pH a buffer at the temperature it will be used.

Buffer Preparation by Titration:

Sodium phosphate buffers are commonly used for physiological applications since it mimics certain components of extracellular fluids and is non-toxic to cells. Unlike Tris buffer, it is not sensitive to temperature changes but is sensitive to dilution; therefore, it is best to prepare stock solutions at or close to the final concentration. It is also highly susceptible to microorganism contamination and therefore, stored at 4°C.

A sodium phosphate buffer is a useful solution in a biosciences lab because the phosphoric acid has multiple dissociation constants, permitting the preparation of phosphate buffers near any of the three pKa values (pH values); 2.15, 6.86 and 12.32. This buffer is most commonly prepared at pH 7.0 using monosodium phosphate and its conjugate base, disodium phosphate. Phosphate buffers are prepared by mixing stock solutions of the desired concentration of dibasic and monobasic in known ratios, or through titration. To prepare a sodium phosphate buffer, first determine the concentration and working pH of the buffer. Using Henderson-Hassel Bach equation, calculate how much acid and base is required. Alternatively, use one of the many published tables available, such as the one below.

To prepare 1 L of 0.1 M sodium phosphate buffer of the desired pH, dilute the following mixtures should to 1 L (final volume) with H2O. Reference: Cold Springs Harbor Protocols, http://cshprotocols.cshlp.org (http://cshprotocols.cshlp.org)

pH	Volume (mL) of 1 M Na 2 HPO4	Volume (mL) of 1M NaH2 PO4
5.8	7.9	92.1
6.0	12.0	88.0
6.2	17.8	82.2

6.4	25.5	74.5
6.6	35.2	64.8
6.8	46.3	53.7
7.0	57.7	42.3
7.2	68.4	31.6
7.4	77.4	22.6
7.6	84.5	15.5
7.8	89.6	10.4
8.0	93.2	6.8

WATER STANDARDS

Water is the primary solvent used in a biosciences lab. In a pharmaceutical company, water can be used for cleaning, raw material, formulation, and as a solvent. Therefore it is critical to use the highest quality water and comply with water standards. However, what does 'high quality' mean? How do we qualify water? Each application requires different quality standards, and therefore, water quality is one of the most important considerations for a biotechnology lab. It is also one of the most costly expenditures!

TYPES OF WATER

The FDA & USP recognize eight types of water:

1. Non-potable
2. Potable (drinkable) water
3. USP purified water
4. USP water for injection (WFI)
5. USP sterile water for injection
6. LUSP sterile water for inhalation
7. USP bacteriostatic water for injection
8. USP sterile water for irrigation

Source Water:

Source water, such as municipal water, can only be used for non-contact with product uses, such as cooling systems. Water for pharmaceutical manufacturing and testing purposes must be highly purified. Tap water can have high levels of chlorine and other ions, particulate matter and non-indicator microorganisms and viruses not tested for, by drinking quality standards.

Ultra-Pure Water:

The United States Pharmacopoeia (USP) distinguishes between water for injections (WFI) and purified water. WFI water must be produced by distillation or reverse osmosis and must have extremely low endotoxin levels. It is also the water of choice for maintenance of mammalian and insect cell lines, as they are sensitive to the presence of endotoxin. Endotoxins are pyrogens, which induce fever in humans. They are part of the cell wall in gram-negative bacteria and are lipopolysaccharides—molecules containing lipids and polysaccharides. A person with a systemic gram-negative infection runs the risk of dying from septic shock when treated with an antibiotic that kills the bacterial infection. The killed bacteria release endotoxin triggering the person's immune system resulting in septic shock. Tap water is contaminated with endotoxin and therefore NOT appropriate for injection. The concentration of endotoxin is endotoxin units per milliliter (EU/mL). A common water quality test performed on biopharmaceuticals is Limulus Amebocyte Lysate Test or LAL test, to assess the pyrogen (endotoxin) content in water.

Table 5.1: Pharmacopeia specifications for water for injection (Wikipedia.org)

Properties	European United States	United States Pharmacopeia
Conductivity [B] (https://en.wikipedia.org/wiki/ Ultrapure_water#cite_note-14)	<1.3 μS/cm at 25 °C	<1.3 μS/cm at 25 °C
Total Organic Carbon (https://en.wikipedia.org/wiki/ Total_organic_carbon) (TOC)	<0.5 mg/L	<0.50 mg/L
Bacteria (guideline	<10 CFU (https://en.wikipedia.org/wiki/ Colony-forming_unit) /100 mL	<10 CFU (https://en.wikipedia.org/ wiki/ Colony-forming_unit) /100 mL
Endotoxin	<0.25 IU/mL	<0.25 EU/mL [C] (https://en.wikipedia.org/wiki/ Ultrapure_water#cite_note-15)
Nitrates	<0.2 ppm	N/A
Aluminum	<10 ppb	N/A

ISO Standards:

ISO also publishes and recognizes water standards. They have three grades of water with Grade I, as the purest. Their summary guidance can be found here, and provide international harmonization of water standards: https://www.iso.org/

files/live/sites/isoorg/files/archive/pdf/en/iso_and_water.pdf (https://www.iso.org/ files/live/sites/isoorg/files/archive/pdf/en/iso_and_water.pdf)

ASTM Standards:

The American Society for Testing and Materials (ASTM) publishes water standards as well (D1193-06). They have four grades of water, with Type I being the purest. Type I water is typically used for cell culture, and HPLC applications and is generated using a reverse osmosis system.

- ***Type I water*** is the highest class of purity and is used for most analytical procedures, tissue culture and instrumentation since it has very low levels of contaminants. It is routinely prepared by reverse osmosis (RO) in combination with distillation and deionization.
- ***Type II water*** is suitable for most routine lab work and is prepared by distillation or RO. Although it is not as reactive as type I, it should also be stored in non-reactive containers.
- ***Type III & IV water***, such as tap water, is useful for some applications such as rinsing glassware or preparing microscope slides.

ASTM further classifies water according to grade: A, B and C that are applied to these four types of water. These grades vary in the levels of bacterial and endotoxin contamination permitted. It's important to note, water used for making pharmaceuticals requires even stricter standards than Type I water.

PURIFICATION OF WATER

There are many different types of contaminants found in water and may include dissolved inorganic, dissolved organics, suspended particles, dissolved gasses, microorganisms, and most importantly for pharmaceutical companies, pyrogen, and endotoxins. There are a variety of methods used to remove these contaminants from water, and frequently many methods are used together to achieve the highest quality needed. It is important to note that 100% pure water does not exist. Although the techniques described below do an excellent job of removing contaminants, water is never truly free of all contaminants.

There are Five Major Ways Water is Purified:

Distillation is often used to purify water for the lab. Water is heated to boiling; the steam travels through a cooled condenser coil, where it condenses back to liquid form and is collected in a different container. Although distillation removes many contaminants, it is not effective at removing dissolved ionized gasses. One of the main disadvantages of distillation is that it is a relatively expensive and slow process.

In **ion exchange,** water passes through filters with bead-shaped resins, which remove ions. Cationic resins remove positive ions, and anionic resins remove negative ions. It is essential that lab water be deionized. Ions are highly reactive and can ruin experiments, interfere in the production process and contaminate the final product. In Central Texas, we have hard water, water in which many ions, such as, calcium, are dissolved. We can see this hard water in the low sudsing of our soaps and shampoos as well as the tough-to-remove ring around the tub. Water softeners are cationic resins with Na+ loosely attached to them. As our hard water is passed over the water softener, the hard water Ca^{++} ions are exchanged for the Na^{+} ions. Many homes use these water softeners. However, because of the presence of Na+, they are not sufficient for lab purification.

Deionization is accomplished using both a cationic exchange column with H+ (instead of Na+) ions as well as an anionic exchange column with OH-. The positive contaminants are removed by the first column and exchanged with the H+, and the negative contaminants exchanged for the OH-. The resulting ions combine to form more water molecules, and the water is purified. It is purer than tap or softened water but still contains many contaminants such as dissolved organics, bacteria, and pyrogens.

Carbon Adsorption is very effective in removing dissolved organic compounds from water. The water is passed over activated charcoal (carbon) made traditionally by burning wood. Most activated carbon is made from styrene beads today since this produces a purer carbon. The organic contaminants stick to the activated carbon and are removed. Organic removal is usually the preliminary step before deionization.

There are Many Types of Filtration Methods Used in Treating Water:

- The first of these are **depth filters** made of sand or matted fibers. They are often used at the beginning of the filtration system to remove large debris.
- **Microfiltration membrane filters** are filters with a pore size of 0.20 um, which filter out bacteria but not smaller dissolved molecules.
- Small dissolved molecules, including most organics, are filtered out of the water with an **ultrafiltration membrane** whose pore sizes are smaller than the microfilters. Molecules smaller than most proteins can be separated!
- **Reverse osmosis (RO)** filters are more restrictive than ultrafiltration membranes and with a range of 0.0001 um in diameter and smaller, removes molecules less than 300 Daltons. RO is successful at removing viruses, bacteria, and pyrogens. They also reject ions and polar molecules such as sugars. In RO, water passes through a thin membrane, which retains material based both on their size and on ionic charge. Water is usually

under pressure to speed up this slow process. Reverse osmosis is often used to make Type I water.

- **Nanofiltration** is a similar process to RO and can also remove viruses and pyrogens; however, its range is a little smaller at 0.001 um.

There are several other methods used to purify water. Two worth mentioning are **ultraviolet (UV) oxidation** and **ozone sterilization.** In UV oxidation, water is passed for about 30 minutes over a UV lamp with a wavelength of 185 nm. The organic compounds are oxidized to simple compounds such as carbon dioxide. A wavelength of 245 nm will kill bacteria and is sometimes used to sterilize water. In ozone sterilization, ozone kills bacteria by rupturing their membranes. Ozone is highly reactive and therefore must be removed in subsequent purification steps.

WATER MONITORING

Labs must monitor their water purification systems to ensure that the systems are working properly. Labs usually keep logs of the date, the monitoring system(s) used and the results. The **maintenance** of the water purification systems includes cleaning (usually daily for distillation systems), sanitation and regeneration of ion exchange systems, sanitation and checking for holes of filtration systems, and cleaning and recharging of activated carbon.

Five quality parameters are typically used to monitor the water system.

- **Resistance** is a measurement of the electrical current through the water. Since ions carry the current, water without ion contaminants should have a high resistance. There is usually an attached meter to read the water resistance of water purification systems. The acceptable value for Type I water is 17.0 megohm-cm. Resistance is only useful for monitoring ions in water.
- Bacterialcounts are used to track levels of microorganisms in water. There are typically three different ways a Heterotrophic Plate Count is performed: the pour plate technique, the spread plate technique, and the membrane filtration technique. In all three methods, water samples using various techniques are incubated 25-35oC for 48-72 hours on agar plate media. The numbers of colonies growing are counted to give the number of Colony Forming Units (CFU) per mL of water. Different standards provide the maximum CFU permitted.
- As mentioned in the introduction, **pyrogens** are tested with a Limulus Amebocyte Lysate (LAL) test. An extract from the blood of a horseshoe crab is mixed with different dilutions of water. Pyrogens will cause the blood extract to clot. The results are given in endotoxin units per milliliter (EU/ml).

- **Organic carbon contaminants** are monitored with instruments such as the mass spectrometer, which analyze the carbon compounds.
- The **pH** of ultrapure water exposed to the carbon dioxide in the air is about 5.7. In a covered container, water has a pH of about 6.0. The pH is easilymonitored with a pH meter.

CHAPTER-6

BIOTECHNICIAN INSTRUMENTS: FUNDAMENTAL LABORATORY EQUIPMENT

INTRODUCTION: AN OVERVIEW

In the intricate realm of biotechnology, the foundation of scientific exploration is laid by the utilization of fundamental laboratory equipment. These indispensable instruments serve as the backbone of biotechnician endeavours, enabling precise experimentation, analysis, and observation. From pipettes and centrifuges to incubators and microscopes, the array of fundamental tools in a biotechnician's laboratory not only facilitates essential procedures but also catalyzes the advancement of knowledge and innovation in the biological sciences. This exploration delves into the essential role played by these instruments, unravelling the significance of their application in shaping the frontiers of biotechnological research and discovery.

BASIC LAB EQUIPMENT

Basic lab equipment biotechnicians may use to prepare biological and chemical reagents.

Laboratory Equipment

Calibration and Operation of a pH Meter

Most solutions prepared in the biological laboratory must have a carefully controlled pH. The pH measurement is a log of the hydrogen ion (H+) concentration of a solution. By definition, the pH value of neutral solutions is any solution with a pH < 7 is considered acidic, and any solution with a pH > 7 is considered basic. There are several effective means a biotechnician monitor's pH in solutions: pH indicator, pH paper, and a pH meter.

A pH indicator is a chemical added to a solution that changes colour when it experiences different pH conditions. A chemical pH indicator is frequently used in cell culture media to provide biotechnicians quick assessment of cell growth. For example, using phenol red, when cells have grown rapidly and spent the media, the pH becomes acidic, changing the media from a bright red to deep yellow colour.

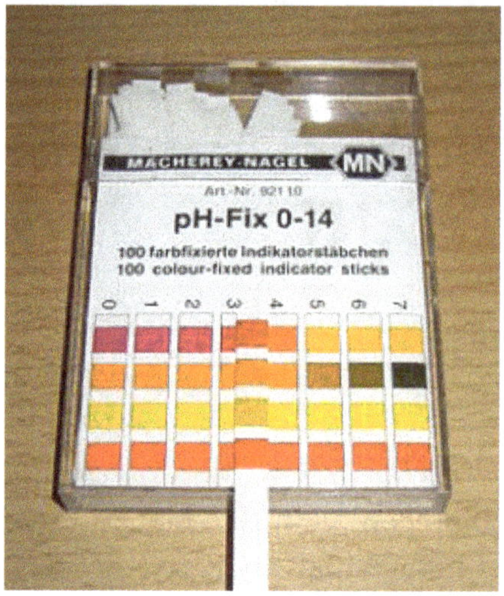

Figure 6.1 Plastic box of pH indicator paper. A single pH paper strip with four colored squares sits on top of the box. By Michael Krahe - CC BY-SA 3.0, https://en.wikipedia.org/wiki/PH_indicator#/media/File:PH_indicator_paper.jpg (https://en.wikipedia.org/wiki/PH_indicator#/media/File:PH_indicator_paper.jpg) .

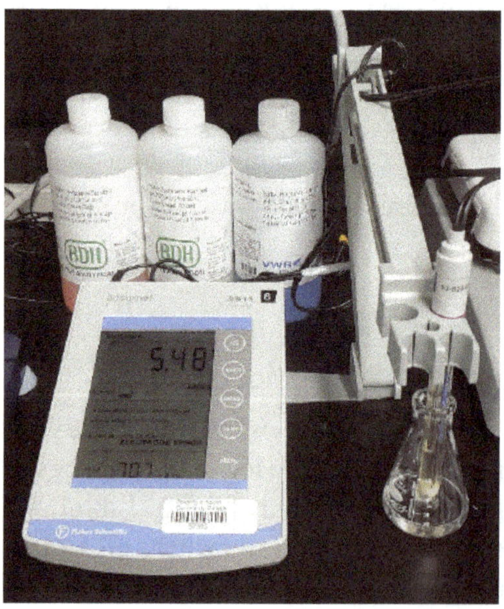

Figure 6.2 Digital pH meter. Photo credit: Jack O›Grady, Austin Community College

When looking for an estimate of pH, indicator paper can also be used. The indicator paper has an array of chemicals that exhibit a specific color when exposed to a particular pH (see figure). However, if a precise pH measurement is required, a pH meter (https://en.wikipedia.org/wiki/PH_meter) can be calibrated and used.

A ***pH meter*** is a scientific instrument that measures the hydrogen-ion activity in aqueous solutions, indicating its acidity or alkalinity expressed as pH. The pH meter measures the difference in electrical potential between a pH electrode and a reference electrode, and so the pH meter is sometimes referred to as a "potentiometric pH meter." The difference in electrical potential relates to the acidity or pH of the solution.

Very precise measurements require that the pH meter is calibrated frequently. Depending on the validation protocols, this may mean before each use, or once per day when turned on. Calibration is a critical process for pH meters, as the glass electrode does not provide reproducible electrostatic potentials over longer periods.

Consistent with principles of ***good laboratory practice***, calibration is performed with at least two standard buffer solutions that span the range of pH values to be measured. For general purposes, buffers at pH 4.00 and pH 10.00 are suitable. The pH meter has one calibration control to set the meter reading equal to the value of the first standard buffer and a second control to adjust the meter reading to the value of the second buffer. A third control allows the temperature to be set. More precise measurements sometimes require calibration at three different pH values. Some pH meters provide built-in temperature-coefficient correction, with temperature thermocouples in the electrode probes. The calibration process correlates the voltage produced by the probe (approximately 0.06 volts per pH unit) with the pH scale.

Good laboratory practice dictates that, after each measurement, the probes are rinsed with distilled water or deionized water to remove any traces of the solution being measured, blotted with a scientific wipe to absorb any remaining water, which could dilute the sample and thus alter the reading, and then immersed in a storage solution suitable for the probe type.

Watch the following video to learn how to calibrate a typical pH meter. (https://youtu.be/UfdKhja6u2I)

Electronic Balances

Instruments for weighing materials are called ***balances***, and most laboratories have more than one type of balance, depending on the amount of material being measured and the degree of accuracy required. ***Electronic balances*** usually have a digital readout and weighing dishes can be tarred to read zero mass before using.

Top loading electronic balances used for the preparation of solutions have a sensitivity of +/- 0.01 g, but *analytical balances* (see figure) can be sensitive to +/- 0.01mg or less. Electronic balances require routine maintenance and recalibration.

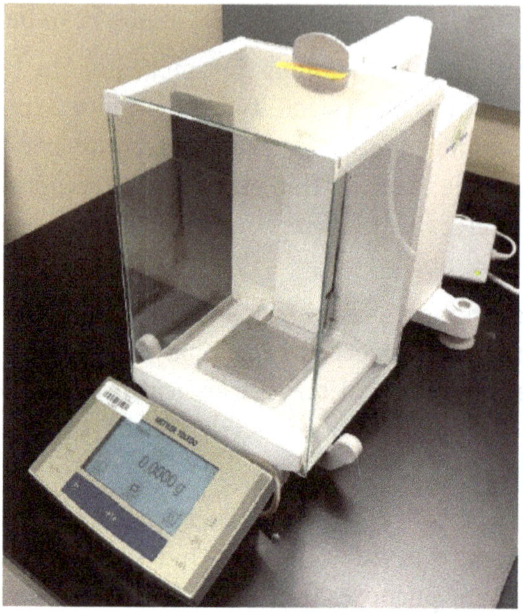

Figure 6.3 Mettler Analytical balance Photo credit: Jeremy Garza, Austin Community College.

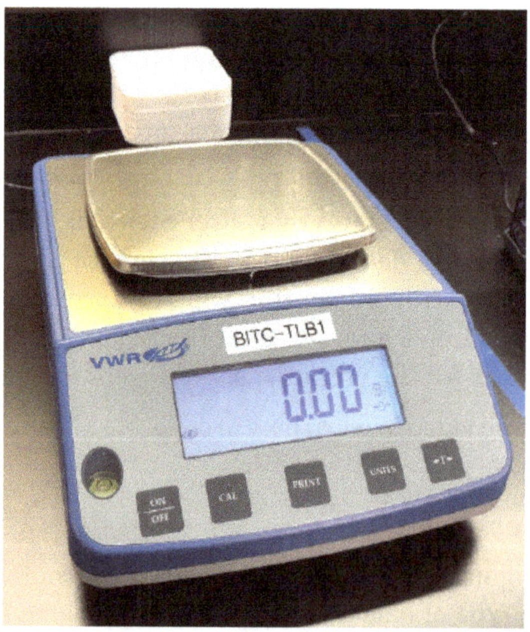

Figure 6.4 **VWR top loading electronic balance. Photo credit: Jeremy Garza, Austin Community College.**

Centrifugation

Many pieces of equipment are used to centrifuge biological samples for analysis. Centrifugation means to apply a centrifugal force to separate molecules by size (or shape, or density). Biotechnicians select the appropriate equipment for a task.

A ***preparative centrifuge*** has a balanced rotor that holds vessels and spins at high speed, up to 20,000 rpm. High-speed centrifugation will cause insoluble particles such as cells, and sometimes subcellular components, to form a pellet at the bottom of the vessel. Rotors are available that hold vessels as small as a few milliliters to as large as a liter. These centrifuges are often refrigerated so that heat-sensitive material, such as cells and proteins, are not damaged due to the high heat generated during centrifugation.

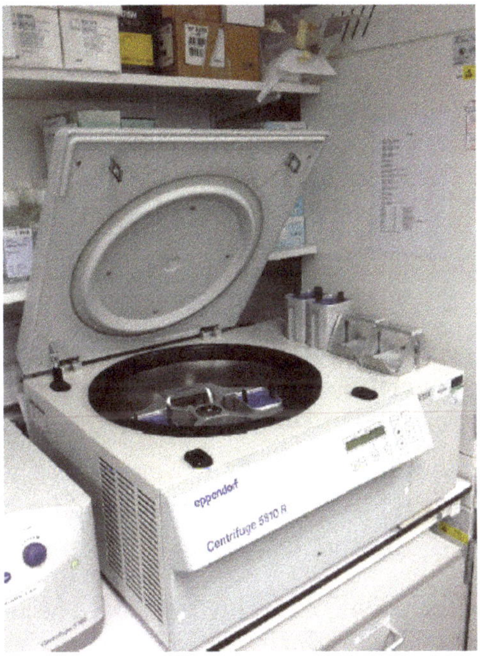

Figure 6.5 Tabletop refrigerated preparative centrifuge. Photo credits: https://en.wikipedia.org/wiki/ Laboratory_centrifuge (https://en.wikipedia.org/wiki/Laboratory_centrifuge)

A tabletop or ***clinical centrifuge*** is not refrigerated and spins at a much slower speed than a preparative centrifuge. Rotors for clinical centrifuges hold tubes with a capacity of 15 mL or less and a 'swing-bucket' rotor. These are frequently found in medical laboratory testing labs.

A ***microcentrifuge*** holds 2.0 mL (or smaller) microcentrifuge tubes with liquid and can centrifuge at high speeds (up to 14,000 rpm) to separate liquids and particulates in solutions. There are a variety of rotor sizes available, and there are refrigerated centrifuges available as well.

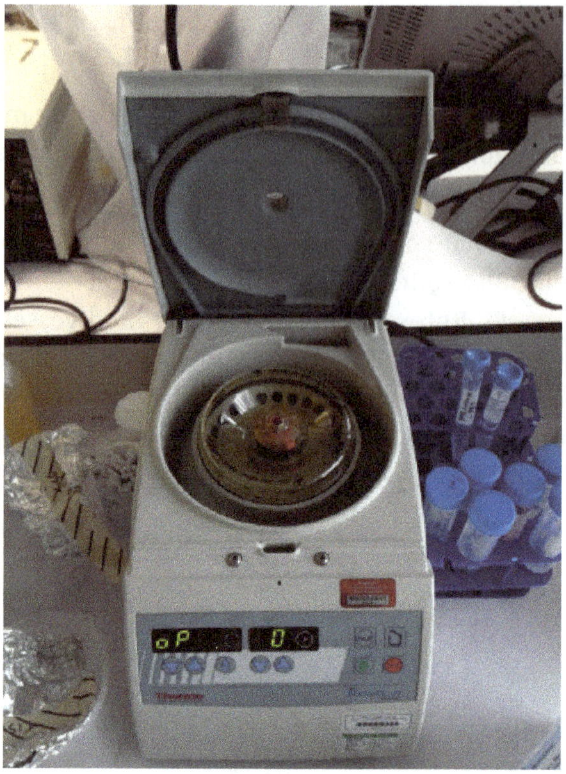

Figure 6.6 Refrigerated microcentrifuge. Photo credits: https://en.wikipedia.org/wiki/Laboratory_centrifuge(https://en.wikipedia.org/wiki/Laboratory_centrifuge)

A ***picofuge*** is a fixed low-speed microcentrifuge, which spins much slower than a microcentrifuge and functions mostly to move liquid from the sides and top of a microcentrifuge tube to the bottom.

BIOTECHNICIANS EQUIPMENT TO MEASURE VOLUME

Biotechnicians must select appropriate size measuring tool depending on the task. In this module is a summary of common laboratory tools for measuring volume including their relative accuracy.

Measurement of Volume

There are many different types of glassware used in a biotechnology lab. The glassware selected will depend on the purpose of its usage as different glassware have various levels of accuracy of measurement.

Erlenmeyer flasks are primarily for the preparation or storage of solutions, not an accurate volume adjustment. Although there are volume markings on these flasks, they are not calibrated and should not be relied upon for exact volume measurements.

Beakers are used for preparing solutions, such as dissolving powdered reagents in water. Beakers are useful when a pH adjustment requires access to the solution by a pH probe. The volumetric markings on beakers are not reliable - and is not an accurate measuring tool.

Graduated cylinders are calibrated with enough accuracy for most volume measurements when preparing solutions. For example, the calibration of most 100 mL graduated cylinder can be relied upon to measure to within +/- 0.6 mL accurately. Graduated cylinders are most frequently used to bring solutions to a final volume. When pouring from graduated cylinders, some error and volume is lost as liquid remains in the cylinder.

Volumetric flasks are used to measure one volume with the highest degree of accuracy and are used to make standard solutions for analytical assays. For example, the calibration of a 100 mL volumetric flask can have an accuracy of +/- 0.1 mL. Note, there are no graduated markings in a volumetric flask - you can only bring solutions to one final volume in that flask.

Figure 6.7 Various types of glassware; beakers, Erlenmeyer flask, graduated cylinder, and volumetric flask. Image credit: https://upload.wikimedia.org/wikipedia/commons/f/fc/Lab_glassware.jpg (https://upload.wikimedia.org/wikipedia/commons/f/fc/Lab_glassware.jpg)

Pipets

- ***Pipets*** are glass or plastic devices that are routinely used to measure and transfer liquids by drawing the liquid into the tube with a mechanical device.

- ***Pasteur pipets*** are small glass tubes used with a bulb to transfer volumes as little as a single drop and as large as a few milliliters. They are not graduated and are not used to measure volumes.
- ***Beral pipets*** (transfer pipets) are plastic pipettes with a bulb at one end used for transfer of liquids. Sometimes they have calibration marks, which have a low level of accuracy. They are often disposable, sterile and individually wrapped.
- ***Serological pipets*** are graduated tubes used to measure anywhere from 1 to 50 mL. When the liquid has drained from this pipet, the final drop in the tip is transferred with a puff of air. These are known as TC or *to contain* pipets. Serological pipets are most frequently used in bioscience labs with pipet controller (or pipet- aid), mechanical devices that aspirate liquid into and out of the pipet.
- ***Mohr pipets***, or "to deliver," pipets are like blowout pipets, but do not require a puff of air to deliver the desired volume accurately. They can be identified by the label "TD" on the top. These are sometimes called TD serological pipets.
- ***Volumetric pipets*** are not graduated but are calibrated to deliver a single, highly accurate, volume.
- ***Micropipettes*** are mechanical devices with disposable plastic tips, which deliver with a high degree of accuracy adjustable microliter volumes of liquid. There are several micropipette devices available of varying sizes, such as a 0.5-10ul, 2-20ul, 20-200ul, and 100-1000ul.
- ***Repeater Pipettes*** are mechanical devices that can be set to deliver, repeatedly, a precise microliter volume. The liquid is aspirated and dispensed from a disposable tip.
- ***Multichannel micropipettes*** can deliver the same volume from as many as 12 tips simultaneously.

Serological Pipets

Serological, or "blowout," pipets are graduated tubes used to measure anywhere from 0.1 to 50 mL. They are typically made of single-use plastic and have the top end plugged with cotton to prevent contamination and overflow. When filling a ***serological pipet***, bring the container with the liquid and the pipet to eye level. Hold the tapered end beneath the surface of the liquid. Draw the liquid up the pipet by suction until the level is just above the desired, then bring the level down to the meniscus of the volume you want. When reading the volume, ALWAYS view the pipet at eye level with the pipet held vertically, perpendicular to the ground.

Pipets are operated with a hand-held electronic pump (pipet-aid) or bulb, of which there are many varieties.

Micropipettes

The micropipette is one of the biotechnician's most frequently used tools. There are different brands of micropipette but all of them work in the same way. All micropipettes are essentially long tubes with a handle and an adjustable piston inside. A disposable tip is placed on the bottom of the micropipette. This tip is the only piece that is inserted into the liquid. In or near the handle is a screw/knob/button that adjusts the volume of the micropipette by moving the piston up and down. On top, there is a plunger button for filling and for dispensing the liquid. There is often a second button on top for ejecting the tip. Micropipettes come in a variety of sizes; for example, 1-20 uL, 20-200 uL, and 100-1000 uL. When choosing which micropipette to use, the rule of thumb is to select the smallest size micropipette that can deliver the desired volume.

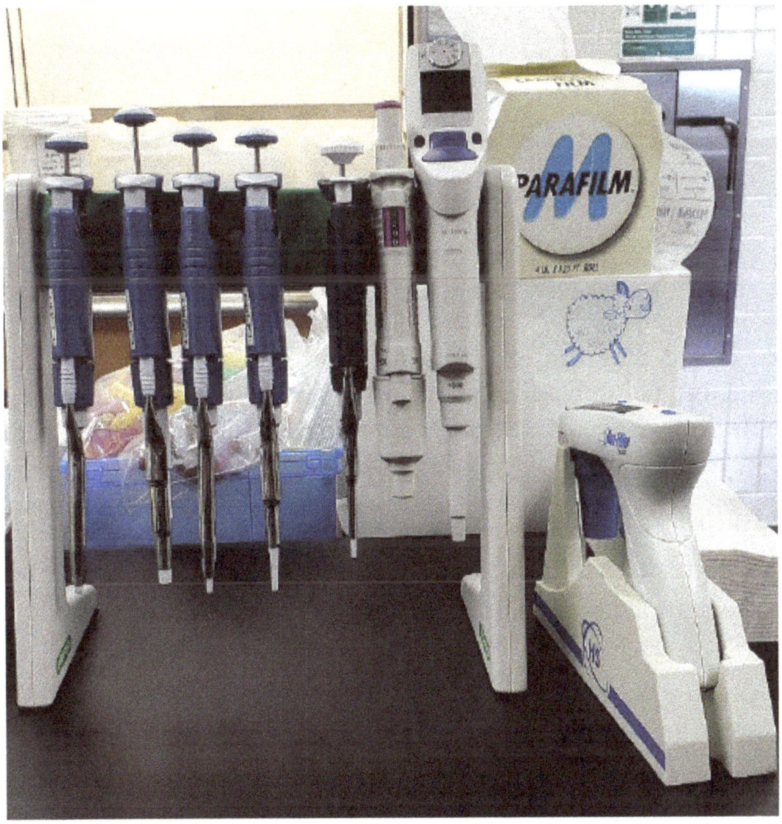

Figure 6.8 Various micropipetting devices. From left to right: Set of 6 micropipettes measuring capacities from 2 microliters to 5 mL; 1 mL ergonomic digital micropipette; electronic pipet-aid (to draw liquid into serological pipet). Image Credit: Jack O›Grady, Austin Community College

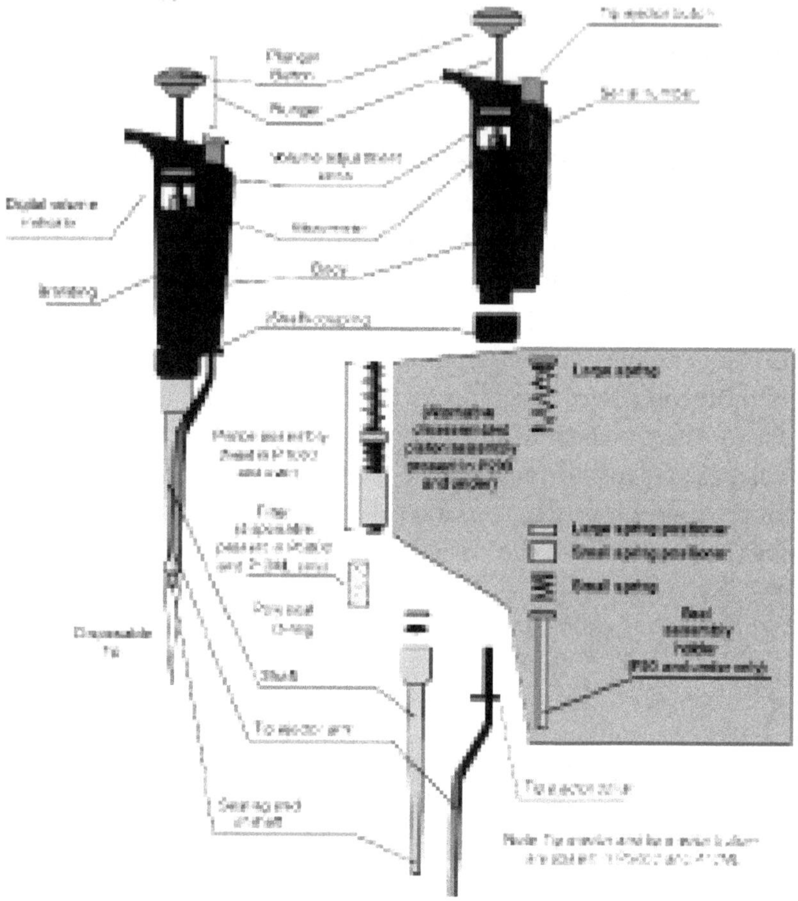

Figure 6.9 Labeled diagram of a Gilson brand micropipette. Note the plunger, tip ejector button, volume adjustment knob, digital volume indicator, shaft, and disposable tip. Image credit: https://commons.wikimedia.org/wiki/ File:Gilson_ schematic.svg

Micropipetting Technique

Of all the factors contributing to the performance of a micropipette, the skill of the operator is the most critical! Here are some tips to help improve your micropipette technique.

- Pre-wetting the tip by aspirating and expelling liquids three times increases the humidity within the tip and reduces sample loss due to evaporation.
- Allow liquids to ambient temperature before dispensing. Humidity and pressure are temperature-dependent and therefore will affect the volume dispensed from the tip.
- Examine the tip before and after dispensing. Remove droplets from the side of the tip and ensure there are no air bubbles in the tip before dispensing.

- Use standard mode. Depress the plunger to the first stop, immerse into liquid, aspirate by releasing the plunger slowly. Remove tip from the liquid and depress the plunger to the second stop to dispense the entire contents.
- Pause consistently after aspiration. Pause for one second before removing from the liquid. If aspirating a viscous liquid, pause for at least 3 seconds or until you can see no further liquid moving up the tip.
- Pull the pipette straight out. This is particularly important for small volumes. If you hold your tip at an angle, it can alter the volume aspirated.
- Minimize handling of the pipette and the tip with fingers. Touching the tip and tube with a gloved hand will warm them up resulting in inaccurate aspiration. Touching the tip with a bare hand can lead to contamination.
- Immerse the tip to the proper depth. Immerse a p1000 5-6mm, and smaller micropipettes 2-3mm only. Too little immersion will lead to aspiration of air, too much can result in liquids clinging to the side of the tip, or touching the bottom causing incorrect aspiration volumes.
- Use the correct pipette tip. Use the manufacturer's recommended tip for the micropipette. Remember to use barrier tips when using biohazards or to avoid cross-contamination of your experiment is necessary (such as working with human blood samples).
- Use consistent plunger pressure and speed. Depress and release the plunger smoothly, slowly, and consistently.

Equipment Validation

Process validation is the method by which companies demonstrate that their activities, procedures, and processes consistently produce a quality result. Processes almost always involve equipment, and therefore, ***equipment validation*** is a critical process where a specified set of procedures are used to check if the equipment meets requirements. Different pieces of lab equipment are designed to operate within varying degrees of accuracy.

CALIBRATION, VERIFICATION & VALIDATION

GMP and ISO certified laboratories have written policies for equipment calibration, verification and validation for all equipment used in a cGMP lab.

- ***Calibration*** is a process that compares a known (the "standard" device) against an unknown (the target device in question). During the calibration process, the offset between these two devices is quantified, and the target device is adjusted back into tolerance (if possible). A calibration report usually contains both "as found" and "as left" data.

- ***Verification*** is simply the process of "verifying" that a device is within tolerance (within an acceptable range). Verification usually results in "as found" data. If the instrument is not within tolerance, it is sent for recalibration.
- ***Validation*** is a detailed process of confirming that the device is installed correctly, that it is operating effectively, and that it is performing without error. Validation in the pharmaceutical industry emerged from problems in the 1960s and 1970s and went together with the QA/QC philosophy that quality is built into the product not tested into the product. The FDA states that quality, safety, and effectiveness are designed and built into the product.

For example, micropipettes are validated by the manufacturer before shipping. However, they become less accurate the more they are used by biotechnicians. Therefore, the performance of a micropipette must be verified periodically and calibrated regularly.

Equipment Validation

For a biomanufacturing process to proceed correctly, the equipment must be validated or qualified to ensure that it will function reliably under all the conditions that may occur during production. Equipment qualification may be performed separately from process validation, but it is also a requirement for process validation. Equipment Validation is a detailed process of confirming that the instrument is installed correctly, that it is operating effectively, and that it is performing without error.

Validation of equipment is a vital link in the quality chain. Equipment should be adequately inspected, cleaned, and maintained as well as, tested, calibrated, and standardized at an appropriate frequency. Laboratories usually establish schedules for such activities based on manufacturer's recommendations, and frequency of use. Equipment is one of the major areas of focus for FDA (or any other regulatory body) audits. All equipment must meet manufacturer or preset standards for operation and performance.

Every biotechnology company approaches equipment validation with a different set of evaluation tools. What remains constant is that validation of equipment occurs *before* the equipment is used to support an associated process validation. The level of detail associated with each step of equipment validation depends on the usage of equipment. It is important, though that each step of the validation, *Installation Qualification (IQ)*, *Operational Qualification (OQ)*, and *Process or Performance Qualification (PQ)* must pass before proceeding to the next. If a piece of equipment is not passing the requirements written in the validation, remove the equipment from use and documentation of these deviations is included with the records.

The Basic Principles of Equipment Validation Include:

- The equipment is installed by vendor requirements, intended use, and regulations. Requirements for calibration, maintenance, and cleaning have been developed as SOPs.
- Operating requirements are established, and tests are conducted to ensure equipment is functioning correctly, under ideal, and challenged conditions.
- Operator training requirements are established, completed, and recorded.

Equipment Validation is Divided into Three Processes:

Installation Qualification (IQ). First, the equipment item is checked to be sure that it meets its design and purchase specifications and is properly installed; this is called installation qualification. Installation qualification includes, for example, checking that instruction manuals, schematic diagrams, and spare parts lists are present; checking that all parts of the device are installed; checking that the materials used in construction were those specified; and making sure that fittings, attachments, cables, plumbing, and wiring are properly connected. IQ is documented proof that the building, wiring, installation and calibration of equipment, utilities, SOPs, spare parts, and specifications meet the design intention.

Operational Qualification (OQ). After installation, the equipment can be tested to verify that it performs within acceptable limits. For example, an autoclave might be tested to see that it reaches the proper temperature, plus or minus certain limits, in a set period; that it reaches the proper pressure, plus or minus certain limits, etc. The penetration of steam to all parts of the chamber, the pressure achieved at various settings, and so forth, would all be tested in the context of the operational qualification of an autoclave. OQ is documented proof that the equipment performs as specified.

Performance Qualification (PQ). Once all measuring instruments are calibrated, and all equipment is validated, process validation (or qualification) can be performed. The validation of the process involves assessing the process under all the conditions that can be expected to occur during production. Testing includes checking the process endpoint(s) under these circumstances and establishing that the process consistently meets its specifications. PQ is documented proof the equipment or systems operate as intended under challenge conditions.

PQ also involves challenging the system with unusual circumstances. FDA speaks of the "worst case" situation(s) that might be encountered during production. For example, a sterilization process might be challenged by placing large numbers of the heat-resistant bacterium in the corner of the autoclave known to be least accessible to steam. The effectiveness of bacterial killing under these "worst case" conditions must meet the specifications for the process.

Validation of Equipment: Considerations

It is crucial when validating laboratory equipment to do so with calibrated and certified equipment and standards. Examples of certified equipment include a certified timer, certified weights, a certified thermometer and a certified tachometer which accurately measures rotor speed.

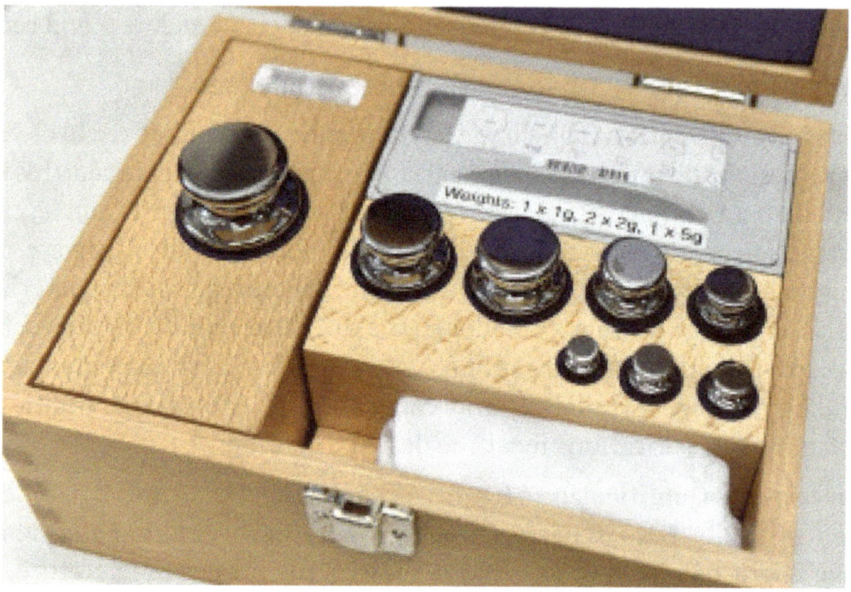

Figure 6.10 Electronic balance calibration weight set. Image credit: Dean Calma / IAEA https://www.flickr.com/ photos/iaea_imagebank/40785322521 (https://www.flickr.com/photos/iaea_imagebank/40785322521)

Validation of Temperature: Temperature is a key element of many biomanufacturing processes, from solution preparation to growing microorganisms. Therefore, it is important to measure and record the temperature accurately. Always keep in mind that "room temperature" (ambient temperature) is a figure of speech and not an actual temperature. Ideal room temperature is different for everyone!

Also note, conventional laboratory glass and mercury thermometers can be highly inaccurate. Most have an error of +/- 2°C, but some can have an error of up to +/- 4°C! Digital thermometers (LED outputs) can be just as poor in accuracy. Some labs calibrate their conventional glass thermometers with a certified thermometer, but the best way to measure temperature accurately is to use a certified validated thermometer and re-certify validated equipment before the expiration date.

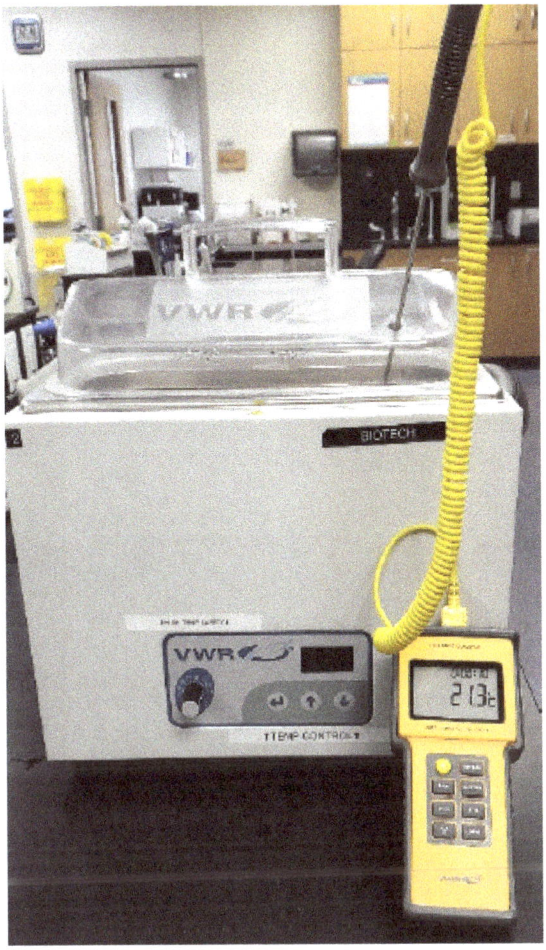

Figure 6.11 Calibrated digital thermometer can be used to accurately measure temperatures. In this figure, a (yellow) calibrated digital thermometer is being used to measure the temperature of a water bath. Image credit: Jack O'Grady, Austin Community College.

Deviations & Corrective Actions

After completing validation activities, the collected data is analyzed as described in the validation protocol and summarized in a report. *Successful validation demonstrates that a process is efficient and reliable.* With careful validation design, planning and implementation problems are easily avoided. However, even in the most carefully designed facilities, unplanned occurrences do happen. These unexpected occurrences are called ***deviations***, and every company must be prepared to deal with them. Typically, the validation plan will have a form for documenting the deviation. The supervisor and the quality department will review the deviation to determine the plan of action to correct the deviation. Along with the deviation, the ***corrective action*** is also carefully documented and implemented.

CHAPTER-7

STRUCTURE AND FUNCTION OF CELLS

INTRODUCTION: AN OVERVIEW

Close your eyes and picture a brick wall. What is the basic building block of that wall? It is a single brick, of course. Like a brick wall, your body is composed of basic building blocks, and the building blocks of your body are cells.

Your body has many kinds of cells, each specialized for a specific purpose. Just as a home is made from a variety of building materials, the human body is constructed from many cell types. For example, epithelial cells protect the surface of the body and cover the organs and body cavities within. Bone cells help to support and protect the body. Cells of the immune system fight invading bacteria. Additionally, red blood cells carry oxygen throughout the body. Each of these cell types plays a vital role during the growth, development, and day-to- day maintenance of the body. In spite of their enormous variety, however, all cells share certain fundamental characteristics.

COMPARING PROKARYOTIC AND EUKARYOTIC CELLS

Cells fall into one of two broad categories: prokaryotic and eukaryotic. The predominantly single-celled organisms of the domains Bacteria and Archaea are classified as prokaryotes (*pro-* = before; *-karyon-* = nucleus). Animal cells, plant cells, fungi, and protists are eukaryotes (*eu-* = true).

Components of Prokaryotic Cells

All cells share four common components: 1) a plasma membrane, an outer covering that separates the cell's interior from its surrounding environment; 2) cytoplasm, consisting of a jelly-like region within the cell in which other cellular components are found; 3) DNA, the genetic material of the cell; and 4) ribosomes, particles that synthesize proteins. However, prokaryotes differ from eukaryotic cells in several ways.

A prokaryotic cell is a simple, single-celled (unicellular) organism that lacks a nucleus, or any other membrane- bound organelle. We will shortly come to see that this is significantly different in eukaryotes. Prokaryotic DNA is found in the central part of the cell: a darkened region called the nucleoid (Figure 7.1).

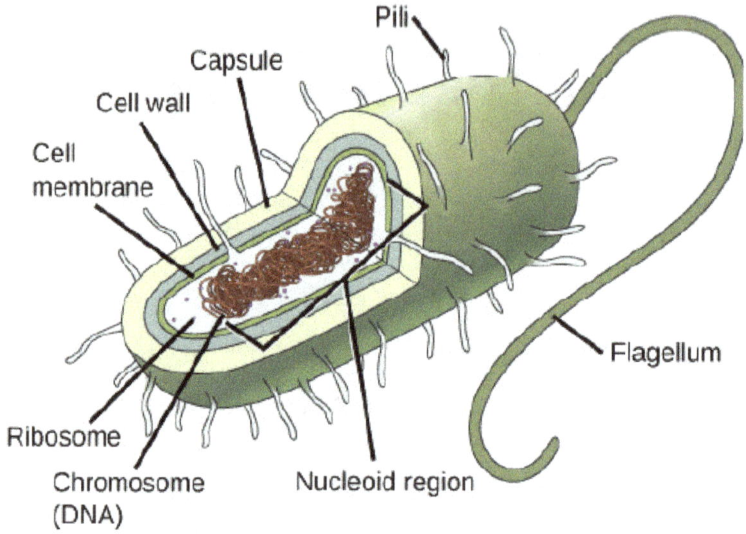

Figure 7.1 The generalized structure of a prokaryotic cell.

Unlike Archaea and eukaryotes, bacteria have a cell wall made of peptidoglycan, comprised of sugars and amino acids, and many have a polysaccharide capsule (Figure 7.1). The cell wall acts as an extra layer of protection, helps the cell maintain its shape, and prevents dehydration. The capsule enables the cell to attach to surfaces in its environment. Some prokaryotes have flagella, pili, or fimbriae. Flagella are used for locomotion, while most pili are used to exchange genetic material during a type of reproduction called conjugation.

EUKARYOTIC CELLS

In nature, the relationship between form and function is apparent at all levels, including the level of the cell, and this will become clear as we explore eukaryotic cells. The principle "form follows function" is found in many contexts. For example, birds and fish have streamlined bodies that allow them to move quickly through the medium in which they live, be it air or water. It means that, in general, one can deduce the function of a structure by looking at its form, because the two are matched.

A **eukaryotic cell** is a cell that has a membrane-bound nucleus and other membrane-bound compartments or sacs, called **organelles**, which have specialized functions. The word eukaryotic means "true kernel" or "true nucleus," alluding to the presence of the membrane-bound nucleus in these cells. The word "organelle" means "little organ," and, as already mentioned, organelles have specialized cellular functions, just as the organs of your body have specialized functions.

Cell Size

At 0.1–5.0 µm in diameter, prokaryotic cells are significantly smaller than eukaryotic cells, which have diameters ranging from 10–100 µm (Figure 7.2). The small size of prokaryotes allows ions and organic molecules that enter them to quickly spread to other parts of the cell. Similarly, any wastes produced within a prokaryotic cell can quickly move out. However, larger eukaryotic cells have evolved different structural adaptations to enhance cellular transport. Indeed, the large size of these cells would not be possible without these adaptations. In general, cell size is limited because volume increases much more quickly than does cell surface area. As a cell becomes larger, it becomes more and more difficult for the cell to acquire sufficient materials to support the processes inside the cell, because the relative size of the surface area across which materials must be transported declines.

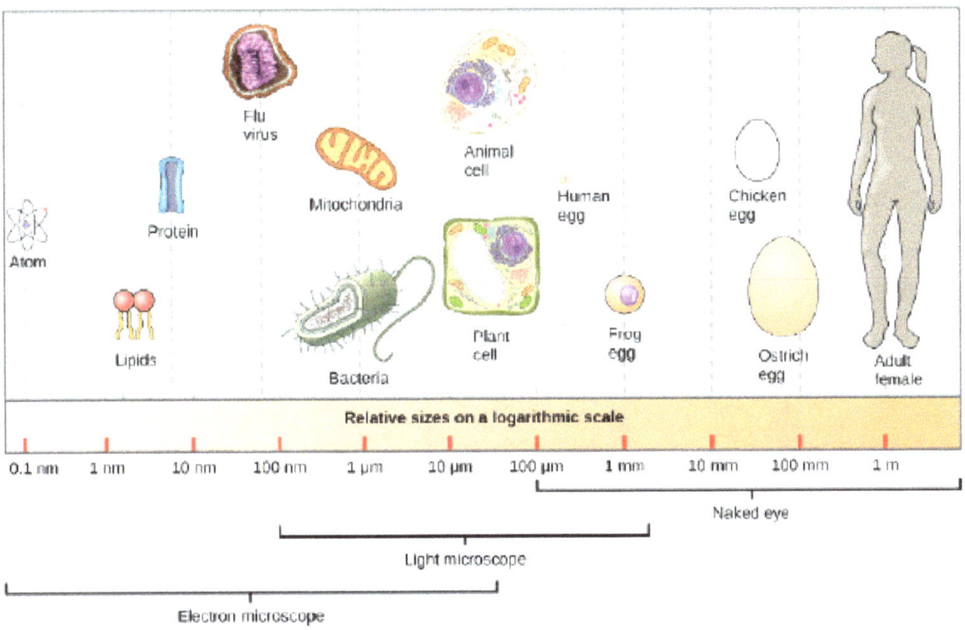

Figure 7.2 The relative sizes of different kinds of cells and cellular components. An adult human is shown for comparison.

Eukaryotic Cells

At this point, it should be clear that eukaryotic cells have a more complex structure than do prokaryotic cells. Organelles allow for various functions to occur in the cell at the same time. Before discussing the functions of organelles within a eukaryotic cell, let us first examine two important components of the cell: the plasma membrane and the cytoplasm.

Visual Connection

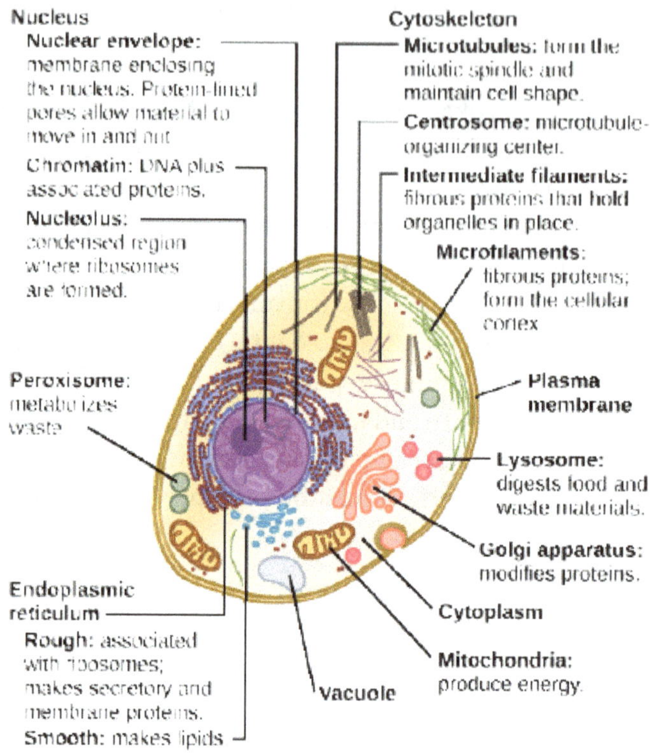

(a)

(a)

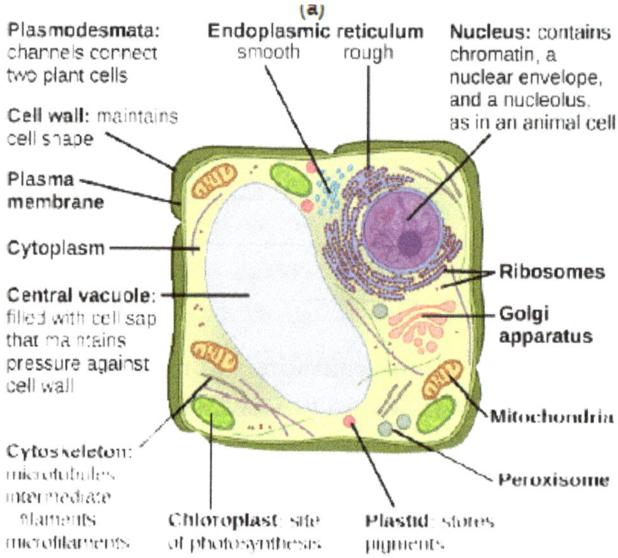

(b)

(b)

Figure 7.3 (a) a typical animal cell and (b) a typical plant cell.

What structures does a plant cell have that an animal cell does not have? What structures does an animal cell have that a plant cell does not have?

The Plasma Membrane

Like prokaryotes, eukaryotic cells have a plasma membrane made up of a phospholipid bilayer with embedded proteins that separates the internal contents of the cell from its surrounding environment. A phospholipid is a lipid molecule composed of two fatty acid chains, a glycerol backbone, and a phosphate group. The plasma membrane regulates the passage of some substances, such as organic molecules, ions, and water, preventing the passage of some to maintain internal conditions, while actively bringing in or removing others. Other compounds move passively across the membrane.

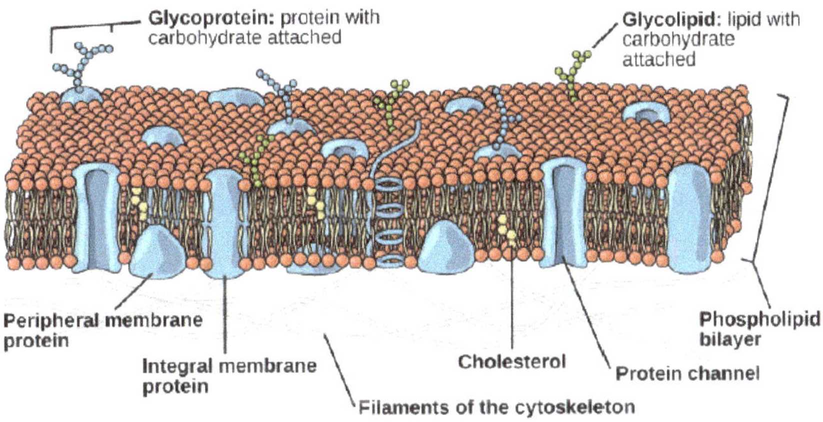

Figure 7.4 The plasma membrane is a phospholipid bilayer with embedded proteins. There are other components, such as cholesterol and carbohydrates, which can be found in the membrane in addition to phospholipids and protein.

The plasma membranes of cells that specialize in absorption are folded into fingerlike projections called microvilli (singular = microvillus). This folding increases the surface area of the plasma membrane. Such cells are typically found lining the small intestine, the organ that absorbs nutrients from digested food. This is an excellent example of form matching the function of a structure.

People with celiac disease have an immune response to gluten, which is a protein found in wheat, barley, and rye. The immune response damages microvilli, and thus, afflicted individuals cannot absorb nutrients. This leads to malnutrition, cramping, and diarrhoea. Patients suffering from celiac disease must follow a gluten-free diet.

The Cytoplasm

The **cytoplasm** comprises the contents of a cell between the plasma membrane and the nuclear envelope (a structure to be discussed shortly). It is made up of organelles suspended in the gel-like **cytosol**, the cytoskeleton, and various chemicals . Even though the cytoplasm consists of 70 to 80 percent water, it has a semi-solid consistency, which comes from the proteins within it. However, proteins are not the only organic molecules found in the cytoplasm. Glucose and other simple sugars, polysaccharides, amino acids, nucleic acids, fatty acids, and derivatives of glycerol are found there too. Ions of sodium, potassium, calcium, and many other elements are also dissolved in the cytoplasm. Many metabolic reactions, including protein synthesis, take place in the cytoplasm.

The Cytoskeleton

If you were to remove all the organelles from a cell, would the plasma membrane and the cytoplasm be the only components left? No. Within the cytoplasm, there would still be ions and organic molecules, plus a network of protein fibers that helps to maintain the shape of the cell, secures certain organelles in specific positions, allows cytoplasm and vesicles to move within the cell, and enables unicellular organisms to move independently. Collectively, this network of protein fibers is known as the **cytoskeleton.** There are three types of fibers within the cytoskeleton: microfilaments, also known as actin filaments, intermediate filaments, and microtubules (Figure 7.5).

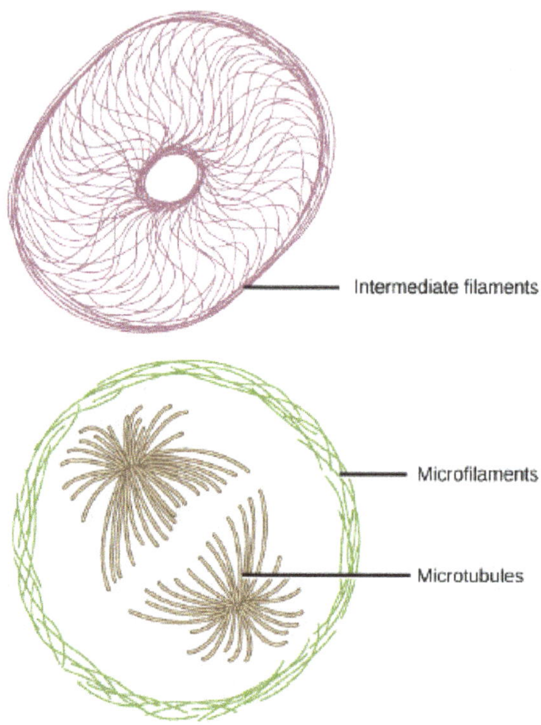

Figure 7.5 Microfilaments, intermediate filaments, and microtubules compose a cell's cytoskeleton.

Microfilaments are the thinnest of the cytoskeletal fibers and function in moving cellular components, for example, during cell division. They also maintain the structure of microvilli, the extensive folding of the plasma membrane found in cells dedicated to absorption. These components are also common in muscle cells and are responsible for muscle cell contraction. Intermediate filaments are of intermediate diameter and have structural functions, such as maintaining the shape of the cell and anchoring organelles. Keratin, the compound that strengthens hair and nails, forms one type of intermediate filament. Microtubules are the thickest of the cytoskeletal fibers. These are hollow tubes that can dissolve and reform quickly. Microtubules guide organelle movement and are the structures that pull chromosomes to their poles during cell division. They are also the structural components of flagella and cilia. In cilia and flagella, the microtubules are organized as a circle of nine double microtubules on the outside and two microtubules in the center.

The centrosome is a region near the nucleus of animal cells that functions as a microtubule-organizing center. It contains a pair of centrioles, two structures that lie perpendicular to each other. Each centriole is a cylinder of nine triplets of microtubules.

The centrosome replicates itself before a cell divides, and the centrioles play a role in pulling the duplicated chromosomes to opposite ends of the dividing cell. However, the exact function of the centrioles in cell division is not clear, since cells that have the centrioles removed can still divide, and plant cells, which lack centrioles, are capable of cell division.

Flagella and Cilia

Flagella (singular = flagellum) are long, hair-like structures that extend from the plasma membrane and are used to move an entire cell, (for example, sperm, *Euglena*). When present, the cell has just one flagellum or a few flagella. When **cilia** (singular = cilium) are present, however, they are many in number and extend along the entire surface of the plasma membrane. They are short, hair-like structures that are used to move entire cells (such as paramecium) or move substances along the outer surface of the cell (for example, the cilia of cells lining the fallopian tubes that move the ovum toward the uterus, or cilia lining the cells of the respiratory tract that move particulate matter toward the throat that mucus has trapped).

The Endomembrane System

The **endomembrane system** (endo = within) is a group of membranes and organelles (Figure 7.9) in eukaryotic cells that work together to modify, package, and transport lipids and proteins. It includes the nuclear envelope, lysosomes, and vesicles, the endoplasmic reticulum and Golgi apparatus, which we will cover shortly. Although not technically within the cell, the plasma membrane is included in the endomembrane system because, as you will see, it interacts with the other endomembranous organelles.

The Nucleus

Typically, the nucleus is the most prominent organelle in a cell (Figure 7.3). The nucleus (plural = nuclei) houses the cell's DNA in the form of chromatin and directs the synthesis of ribosomes and proteins. Let us look at it in more detail (Figure 7.6).

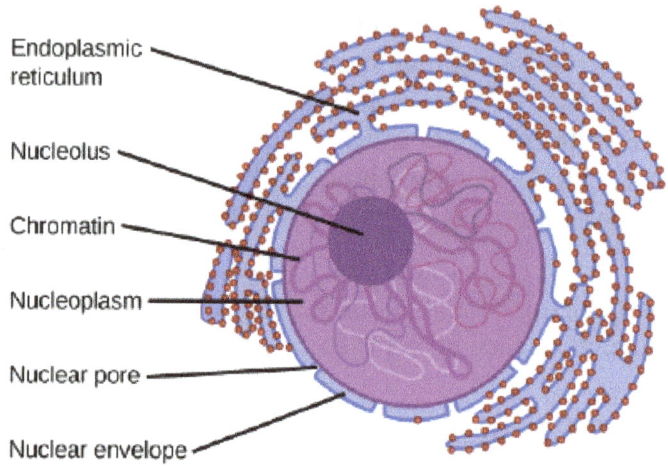

Figure 7.6 The outermost boundary of the nucleus is the nuclear envelope. Notice that the nuclear envelope consists of two phospholipid bilayers (membranes)—an outer membrane and an inner membrane—in contrast to the plasma membrane (Figure 7.4), which consists of only one phospholipid bilayer. (credit: modification of work by NIGMS, NIH)

The **nuclear envelope** is a double-membrane structure that constitutes the outermost portion of the nucleus (Figure 7.6). Both the inner and outer membranes of the nuclear envelope are phospholipid bilayers.

The nuclear envelope is punctuated with pores that control the passage of ions, molecules, and RNA between the nucleoplasm and the cytoplasm.

To understand **chromatin**, it is helpful to first consider chromosomes. A **chromosome** is a structure within the nucleus that is made up of DNA, the hereditary material, and proteins. This combination of DNA and proteins is called chromatin. In eukaryotes, chromosomes are linear structures. Every species has a specific number of chromosomes in the nucleus of its body cells. For example, in humans, the chromosome number is 46, whereas in fruit flies, the chromosome number is eight.

Chromosomes are only visible and distinguishable from one another when the cell is getting ready to divide. When the cell is in the growth and maintenance phases of its life cycle, the chromosomes resemble an unwound, jumbled bunch of threads.

We already know that the nucleus directs the synthesis of ribosomes, but how does it do this? Some chromosomes have sections of DNA that encode ribosomal RNA. A darkly staining area within the nucleus, called the **nucleolus** (plural = nucleoli), aggregates the ribosomal RNA with associated proteins to assemble the ribosomal subunits that are then transported through the nuclear pores into the cytoplasm.

The Endoplasmic Reticulum

The **endoplasmic reticulum** (ER) (Figure 7.9) is a series of interconnected membranous tubules that collectively modify proteins and synthesize lipids. However, these two functions are performed in separate areas of the endoplasmic reticulum: the rough endoplasmic reticulum and the smooth endoplasmic reticulum, respectively.

The hollow portion of the ER tubules is called the lumen or cisternal space. The membrane of the ER, which is a phospholipid bilayer embedded with proteins, is continuous with the nuclear envelope.

The **rough endoplasmic reticulum (RER)** is so named because the ribosomes attached to its cytoplasmic surface give it a studded appearance when viewed through an electron microscope.

The ribosomes synthesize proteins while attached to the ER, resulting in transfer of their newly synthesized proteins into the lumen of the RER where they undergo modifications such as folding or addition of sugars. The RER also makes phospholipids for cell membranes.

If the phospholipids or modified proteins are not destined to stay in the RER, they will be packaged within vesicles and transported from the RER by budding from the membrane (Figure 7.9). Since the RER is engaged in modifying proteins that will be secreted from the cell, it is abundant in cells that secrete proteins, such as the liver.

The **smooth endoplasmic reticulum (SER)** is continuous with the RER but has few or no ribosomes on its cytoplasmic surface (see Figure 7.4). The SER's functions include synthesis of carbohydrates, lipids (including phospholipids), and steroid hormones; detoxification of medications and poisons; alcohol metabolism; and storage of calcium ions.

The Golgi Apparatus

We have already mentioned that vesicles can bud from the ER, but where do the vesicles go? Before reaching their final destination, the lipids or proteins within the transport vesicles need to be sorted, packaged, and tagged so that they wind up in the right place. The sorting, tagging, packaging, and distribution of lipids and proteins take place in the **Golgi apparatus** (also called the Golgi body), a series of flattened membranous sacs (Figure 7.7).

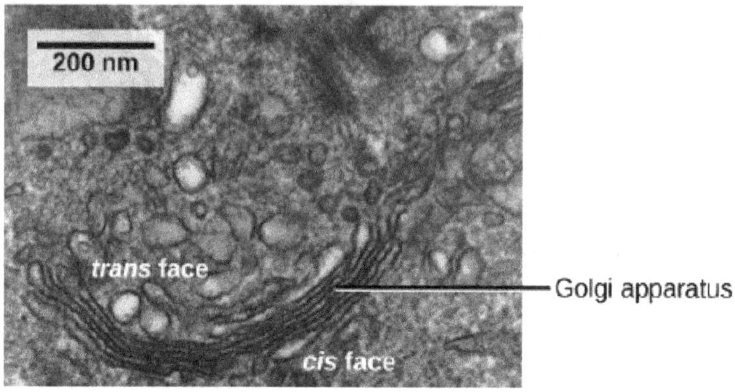

Figure 7.7 The Golgi apparatus in this transmission electron micrograph of a white blood cell is visible as a stack of semicircular flattened rings in the lower portion of this image. Several vesicles can be seen near the Golgi apparatus. (credit: modification of work by Louisa Howard; scale-bar data from Matt Russell)

The Golgi apparatus has a receiving face near the endoplasmic reticulum and a releasing face on the side away from the ER, toward the cell membrane. The transport vesicles that form from the ER travel to the receiving face, fuse with it, and empty their contents into the lumen of the Golgi apparatus. As the proteins and lipids travel through the Golgi, they undergo further modifications. The most frequent modification is the addition of short chains of sugar molecules. The newly modified proteins and lipids are then tagged with small molecular groups to enable them to be routed to their proper destinations.

Finally, the modified and tagged proteins are packaged into vesicles that bud from the opposite face of the Golgi. While some of these vesicles, transport vesicles, deposit their contents into other parts of the cell where they will be used, others, secretory vesicles, fuse with the plasma membrane and release their contents outside the cell.

The amount of Golgi in different cell types again illustrates that form follows function within cells. Cells that engage in a great deal of secretory activity (such as cells of the salivary glands that secrete digestive enzymes or cells of the immune system that secrete antibodies) have an abundant number of Golgi.

In plant cells, the Golgi has an additional role of synthesizing polysaccharides, some of which are incorporated into the cell wall and some of which are used in other parts of the cell.

Lysosomes

In animal cells, the **lysosomes** are the cell's "garbage disposal." Digestive enzymes within the lysosomes aid the breakdown of proteins, polysaccharides, lipids, nucleic acids, and even worn-out organelles. In single-celled eukaryotes, lysosomes are important for digestion of the food they ingest and the recycling of organelles. These enzymes are active at a much lower pH (more acidic) than those located in the cytoplasm. Many reactions that take place in the cytoplasm could not occur at a low pH, thus the advantage of compartmentalizing the eukaryotic cell into organelles is apparent.

Lysosomes also use their hydrolytic enzymes to destroy disease-causing organisms that might enter the cell. A good example of this occurs in a group of white blood cells called macrophages, which are part of your body's immune system. In a process known as phagocytosis, a section of the plasma membrane of the macrophage invaginates (folds in) and engulfs a pathogen. The invaginated section, with the pathogen inside, then pinches itself off from the plasma membrane and becomes a vesicle. The vesicle fuses with a lysosome. The lysosome's hydrolytic enzymes then destroy the pathogen (Figure 7.8).

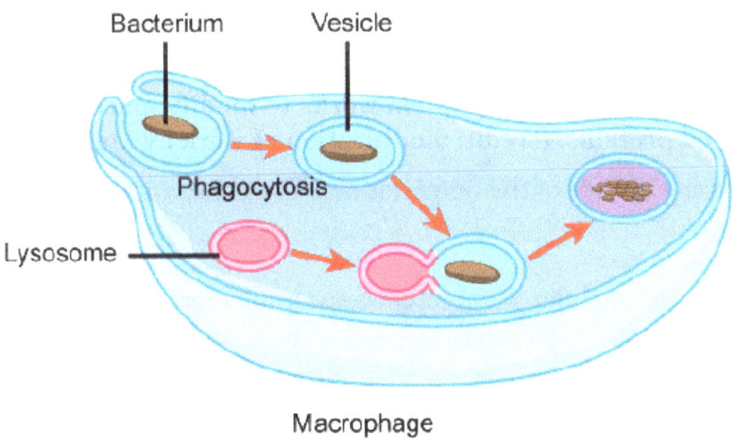

Figure 7.8 A macrophage has phagocytized a potentially pathogenic bacterium into a vesicle, which then fuses with a lysosome within the cell so that the pathogen can be destroyed. Other organelles are present in the cell, but for simplicity, are not shown.

Vesicles and Vacuoles

Vesicles and **vacuoles** are membrane-bound sacs that function in storage and transport. Vacuoles are somewhat larger than vesicles, and the membrane of a vacuole does not fuse with the membranes of other cellular components. Vesicles can fuse with other membranes within the cell system. Additionally, enzymes within plant vacuoles can break down macromolecules.

Visual Connection

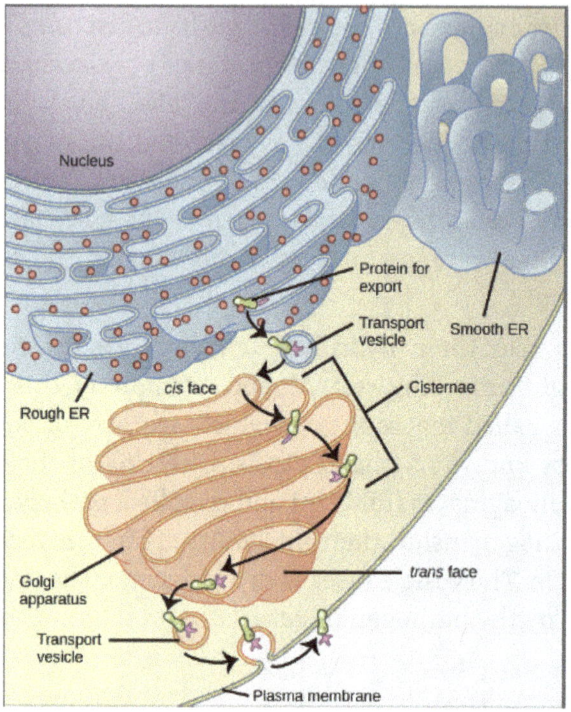

Figure 7.9 The endomembrane system works to modify, package, and transport lipids and proteins. (credit: modification of work by Magnus Manske)

Why does the *cis* face of the Golgi not face the plasma membrane?

Ribosomes

Ribosomes are the cellular structures responsible for protein synthesis. When viewed through an electron microscope, free ribosomes appear as either clusters or single tiny dots floating freely in the cytoplasm. Ribosomes may be attached to either the cytoplasmic side of the plasma membrane or the cytoplasmic side of the endoplasmic reticulum (Figure 7.3). Electron microscopy has shown that ribosomes consist of large and small subunits. Ribosomes are enzyme complexes that are responsible for protein synthesis.

Because protein synthesis is essential for all cells, ribosomes are found in practically every cell, although they are smaller in prokaryotic cells. They are particularly abundant in immature red blood cells for the synthesis of hemoglobin, which functions in the transport of oxygen throughout the body.

Mitochondria

Mitochondria (singular = mitochondrion) are often called the "powerhouses" or "energy factories" of a cell because they are responsible for making adenosine triphosphate (ATP), the cell's main energy-carrying molecule. The formation of ATP from the breakdown of glucose is known as cellular respiration. Mitochondria are oval-shaped, double-membrane organelles (Figure 7.10) that have their own ribosomes and DNA. Each membrane is a phospholipid bilayer embedded with proteins. The inner layer has folds called cristae, which increase the surface area of the inner membrane. The area surrounded by the folds is called the mitochondrial matrix. The cristae and the matrix have different roles in cellular respiration.

In keeping with our theme of form following function, it is important to point out that muscle cells have a very high concentration of mitochondria because muscle cells need a lot of energy to contract.

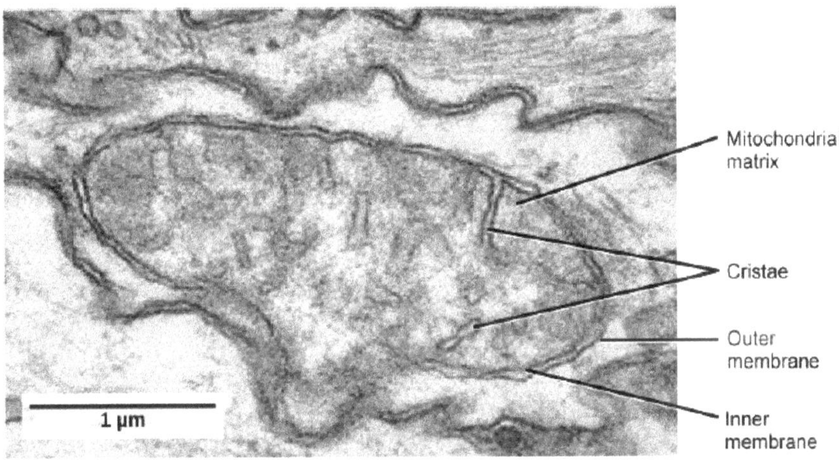

Figure 7.10 This transmission electron micrograph shows a mitochondrion as viewed with an electron microscope. Notice the inner and outer membranes, the cristae, and the mitochondrial matrix. (credit: modification of work by Matthew Britton; scale-bar data from Matt Russell)

Peroxisomes

Peroxisomes are small, round organelles enclosed by single membranes. They carry out oxidation reactions that break down fatty acids and amino acids. They also detoxify many poisons that may enter the body. Alcohol is detoxified by peroxisomes in liver cells. A byproduct of these oxidation reactions is hydrogen peroxide, H_2O_2, which is contained within the peroxisomes to prevent the chemical from causing damage to cellular components outside of the organelle. Hydrogen peroxide is safely broken down by peroxisomal enzymes into water and oxygen.

Animal Cells Versus Plant Cells

Despite their fundamental similarities, there are some striking differences between animal and plant cells (see Table 7.1). Animal cells have centrioles, centrosomes (discussed under the cytoskeleton), and lysosomes, whereas plant cells do not.

Plant cells have a cell wall, chloroplasts, plasmodesmata, and plastids used for storage, and a large central vacuole, whereas animal cells do not.

The Cell Wall

In Figure 7.3b, the diagram of a plant cell, you see a structure external to the plasma membrane called the cell wall. The cell wall is a rigid covering that protects the cell, provides structural support, and gives shape to the cell. Fungal and protist cells also have cell walls.

While the chief component of prokaryotic cell walls is peptidoglycan, the major organic molecule in the plant cell wall is cellulose, a polysaccharide made up of long, straight chains of glucose units. When nutritional information refers to dietary fiber, it is referring to the cellulose content of food.

Chloroplasts

Like mitochondria, chloroplasts also have their own DNA and ribosomes. **Chloroplasts** function in photosynthesis and can be found in eukaryotic cells such as plants and algae. In photosynthesis, carbon dioxide, water, and light energy are used to make glucose and oxygen. This is the major difference between plants and animals: Plants (autotrophs) are able to make their own food, like glucose, whereas animals (heterotrophs) must rely on other organisms for their organic compounds or food source.

Like mitochondria, chloroplasts have outer and inner membranes, but within the space enclosed by a chloroplast's inner membrane is a set of interconnected and stacked, fluid-filled membrane sacs called thylakoids (Figure 7.11). Each stack of thylakoids is called a granum (plural = grana). The fluid enclosed by the inner membrane and surrounding the grana is called the stroma.

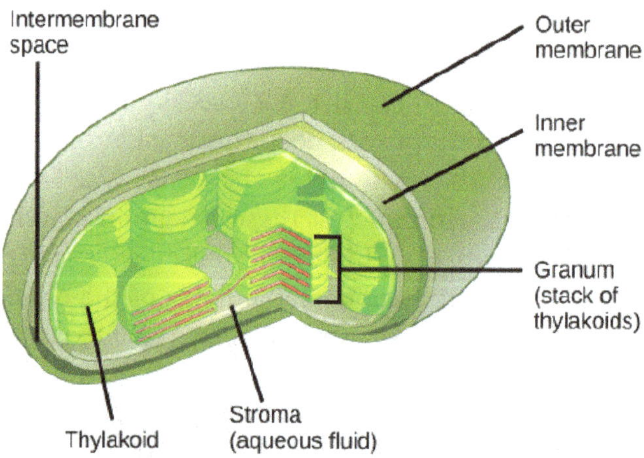

Figure 7.11 This simplified diagram of a chloroplast shows the outer membrane, inner membrane, thylakoids, grana, and stroma.

The chloroplasts contain a green pigment called chlorophyll, which captures the energy of sunlight for photosynthesis. Like plant cells, photosynthetic protists also have chloroplasts. Some bacteria also perform photosynthesis, but they do not have chloroplasts. Their photosynthetic pigments are located in the thylakoid membrane within the cell itself.

Endosymbiosis

We have mentioned that both mitochondria and chloroplasts contain DNA and ribosomes. Have you wondered why? Strong evidence points to endosymbiosis as the explanation.

Symbiosis is a relationship in which organisms from two separate species live in close association and typically exhibit specific adaptations to each other. Endosymbiosis (*endo-*= within) is a relationship in which one organism lives inside the other. Endosymbiotic relationships abound in nature. Microbes that produce vitamin K live inside the human gut. This relationship is beneficial for us because we are unable to synthesize vitamin K. It is also beneficial for the microbes because they are protected from other organisms and are provided a stable habitat and abundant food by living within the large intestine.

Scientists have long noticed that bacteria, mitochondria, and chloroplasts are similar in size and have other similar features. In the 1950s and 1960s, scientists discovered that mitochondria and chloroplasts have their own DNA and ribosomes, just as bacteria do. In 1967, Lynn Margulis used microbial evidence in her proposal of endosymbiotic theory, which indicated that these organelles originated from separate organisms. Although Margulis's work was met with resistance, this basic component of this once- revolutionary hypothesis is now widely accepted. Scientists believe that host cells and bacteria formed a mutually beneficial endosymbiotic relationship when the host cells ingested aerobic bacteria and cyanobacteria but did not destroy them. Through evolution, these ingested bacteria became more specialized in their functions, with the aerobic bacteria becoming mitochondria and the photosynthetic bacteria becoming chloroplasts.

The Central Vacuole

Previously, we mentioned vacuoles as essential components of plant cells. If you look at Figure 7.3, you will see that plant cells each have a large, central vacuole that occupies most of the cell. The **central vacuole** plays a key role in regulating the cell's concentration of water in changing environmental conditions. In plant cells, the liquid inside the central vacuole provides turgor pressure, which is the outward pressure caused by the fluid inside the cell. Have you ever noticed that if you forget to water a plant for a few days, it wilts? That is because as the water concentration in the soil becomes lower than the water concentration in the plant, water moves out of the central vacuoles and cytoplasm and into the soil. As the central vacuole shrinks, it leaves the cell wall unsupported. This loss of support to the cell walls of a plant results in the wilted appearance. Additionally, this fluid has a very bitter taste, which discourages consumption by insects and animals. The central vacuole also functions to store proteins in developing seed cells.

Extracellular Matrix of Animal Cells

Most animal cells release materials into the extracellular space. The primary components of these materials are glycoproteins and the protein collagen. Collectively, these materials are called the **extracellular matrix** (Figure 7.12). Not only does the extracellular matrix hold the cells together to form a tissue, but it also allows the cells within the tissue to communicate with each other.

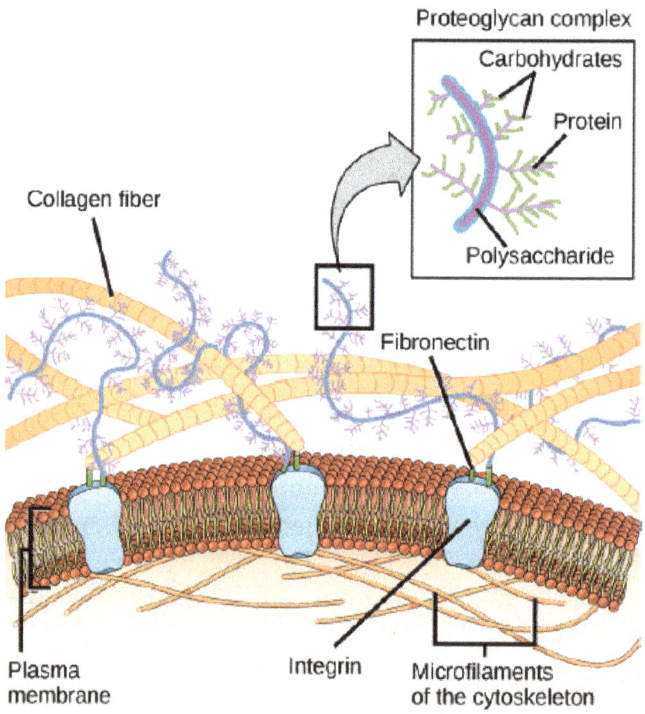

Figure 7.12 The extracellular matrix consists of a network of substances secreted by cells.

Blood clotting provides an example of the role of the extracellular matrix in cell communication. When the cells lining a blood vessel are damaged, they display a protein receptor called tissue factor. When tissue factor binds with another factor in the extracellular matrix, it causes platelets to adhere to the wall of the damaged blood vessel, stimulates adjacent smooth muscle cells in the blood vessel to contract (thus constricting the blood vessel), and initiates a series of steps that stimulate the platelets to produce clotting factors.

Intercellular Junctions

Cells can also communicate with each other by direct contact, referred to as intercellular junctions. There are some differences in the ways that plant and animal cells do this. **Plasmodesmata** (singular = plasmodesma) are junctions between plant cells, whereas animal cell contacts include tight and gap junctions, and desmosomes.

In general, long stretches of the plasma membranes of neighbouring plant cells cannot touch one another because they are separated by the cell walls surrounding each cell. Plasmodesmata are numerous channels that pass between the cell walls of adjacent plant cells, connecting their cytoplasm and enabling signal molecules and nutrients to be transported from cell to cell (Figure 7.13a).

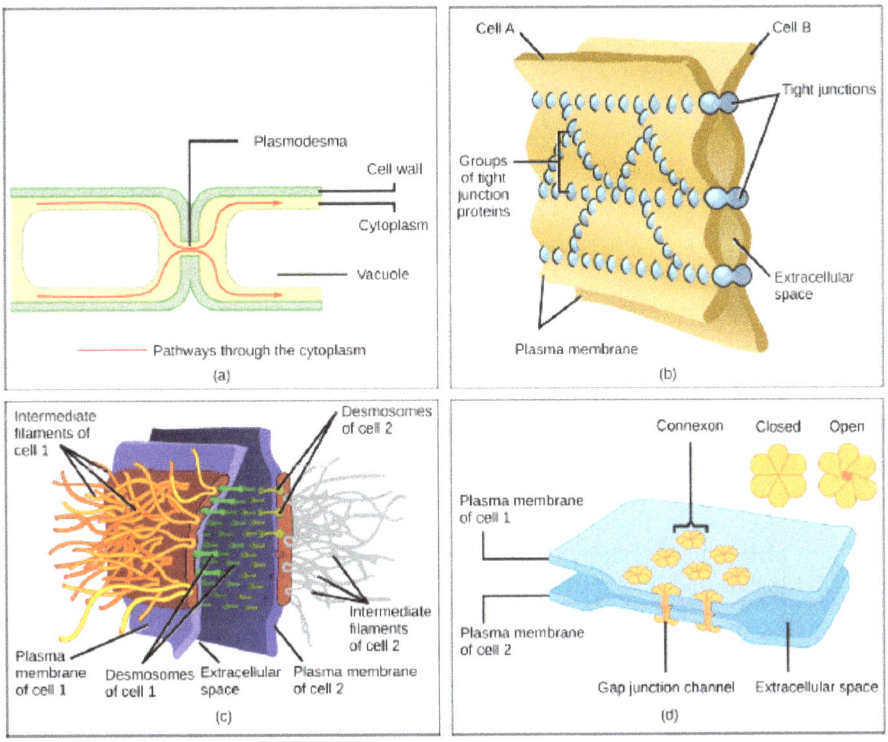

Figure 7.13 There are four kinds of connections between cells. (a) A plasmodesma is a channel between the cell walls of two adjacent plant cells. (b) Tight junctions join adjacent animal cells. (c) Desmosomes join two animal cells together. (d) Gap junctions act as channels between animal cells. (credit b, c, d: modification of work by Mariana Ruiz Villareal)

A tight junction is a watertight seal between two adjacent animal cells (Figure 7.13b). Proteins hold the cells tightly against each other. This tight adhesion prevents materials from leaking between the cells. Tight junctions are typically found in the epithelial tissue that lines internal organs and cavities, and composes most of the skin. For example, the tight junctions of the epithelial cells lining the urinary bladder prevent urine from leaking into the extracellular space.

Also found only in animal cells are **desmosomes**, which act like spot welds between adjacent epithelial cells (Figure 7.13c). They keep cells together in a sheet-like formation in organs and tissues that stretch, like the skin, heart, and muscles.

Gap junctions in animal cells are like plasmodesmata in plant cells in that they are channels between adjacent cells that allow for the transport of ions, nutrients, and other substances that enable cells to communicate (Figure 7.13d). Structurally, however, gap junctions and plasmodesmata differ.

Components of Prokaryotic and Eukaryotic Cells and Their Functions

Table 7.1

Cell Component	Function	Present in Prokaryotes?	Present in Animal Cells?	Present in Plant Cells?
Plasma membrane	Separates cell from external environment; controls passage of organic molecules, ions, water, oxygen, and wastes into and out of the cell	Yes	Yes	Yes
Cytoplasm	Provides structure to cell; site of many metabolic reactions; medium in which organelles are found	Yes	Yes	Yes
Nucleoid	Location of DNA	Yes	No	No
Nucleus	Cell organelle that houses DNA and directs synthesis of ribosomes and proteins	No	Yes	Yes
Ribosomes	Protein synthesis	Yes	Yes	Yes
Mitochondria	ATP production/cellular respiration	No	Yes	Yes
Peroxisomes	Oxidizes and breaks down fatty acids and amino acids, and detoxifies poisons	No	Yes	Yes
Vesicles and vacuoles	Storage and transport; digestive function in plant cells	No	Yes	Yes
Centrosome	Unspecified role in cell division in animal cells; organizing center of microtubules in animal cells	No	Yes	No

Cell Components	Function	Present in Prokaryotes?	Present in Animal Cells?	Present in Plant Cells?
Lysosomes	Digestion of macromolecules; recycling of worn-out organelles	No	Yes	No
Cell wall	Protection, structural support and maintenance of cell shape	Yes, primarily peptidoglycan in bacteria but not Archaea	No	Yes, primarily cellulose
Chloroplasts	Photosynthesis	No	No	Yes
Endoplasmic reticulum	Modifies proteins and synthesizes lipids	No	Yes	Yes
Golgi apparatus	Modifies, sorts, tags, packages, and distributes lipids and proteins	No	Yes	Yes
Cytoskeleton	Maintains cell's shape, secures organelles in specific positions, allows cytoplasm and vesicles to move within the cell, and enables unicellular organisms to move independently	Yes	Yes	Yes
Flagella	Cellular locomotion	Some	Some	No, except for some plant sperm.
Cilia	Cellular locomotion, movement of particles along extracellular surface of plasma membrane, and filtration	No	Some	No

This table provides the components of prokaryotic and eukaryotic cells and their respective functions.

THE CELL MEMBRANE

A cell's plasma membrane defines the boundary of the cell and determines the nature of its contact with the environment. Cells exclude some substances, take in others, and excrete still others, all in controlled quantities. Plasma membranes enclose the borders of cells, but rather than being a static bag, they are dynamic and constantly in flux. The plasma membrane must be sufficiently flexible to allow certain cells, such as red blood cells and white blood cells, to change shape as they pass through narrow capillaries. These are the more obvious functions of a plasma membrane. In addition, the surface of the plasma membrane carries markers that allow cells to recognize one another, which is vital as tissues and organs form during early development, and which later plays a role in the “self” versus “non-self” distinction of the immune response.

The plasma membrane also carries receptors, which are attachment sites for specific substances that interact with the cell. Each receptor is structured to bind with a specific substance. For example, surface receptors of the membrane create changes in the interior, such as changes in enzymes of metabolic pathways. These metabolic pathways might be vital for providing the cell with energy, making specific substances for the cell, or breaking down cellular waste or toxins for disposal. Receptors on the plasma membrane’s exterior surface interact with hormones or neurotransmitters, and allow their messages to be transmitted into the cell. Some recognition sites are used by viruses as attachment points. Although they are highly specific, pathogens like viruses may evolve to exploit receptors to gain entry to a cell by mimicking the specific substance that the receptor is meant to bind. This specificity helps to explain why human immunodeficiency virus (HIV) or any of the five types of hepatitis viruses invade only specific cells.

Fluid Mosaic Model

In 1972, S. J. Singer and Garth L. Nicolson proposed a new model of the plasma membrane that, compared to earlier understanding, better explained both microscopic observations and the function of the plasma membrane. This was called the **fluid mosaic model**. The model has evolved somewhat over time, but still best accounts for the structure and functions of the plasma membrane as we now understand them. The fluid mosaic model describes the structure of the plasma membrane as a mosaic of components—including phospholipids, cholesterol, proteins, and carbohydrates—in which the components are able to flow and change position, while maintaining the basic integrity of the membrane. Both phospholipid molecules and embedded proteins are able to diffuse rapidly and laterally in the membrane. The fluidity of the plasma membrane is necessary for the activities of certain enzymes and transport molecules within the membrane. Plasma membranes range from 5–10 nm thick. As a comparison, human red blood cells, visible via light microscopy, are approximately 8 μm thick, or approximately 1,000 times thicker than a plasma membrane. (Figure 7.14)

The plasma membrane is made up primarily of a bilayer of phospholipids with embedded proteins, carbohydrates, glycolipids, and glycoproteins, and, in animal cells, cholesterol. The amount of cholesterol in animal plasma membranes regulates the fluidity of the membrane and changes based on the temperature of the cell’s environment. In other words, cholesterol acts as antifreeze in the cell membrane and is more abundant in animals that live in cold climates.

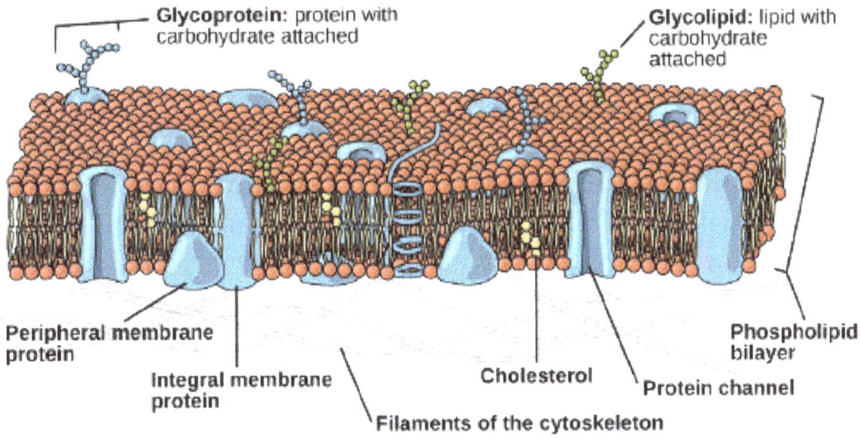

Figure 7.14 The fluid mosaic model of the plasma membrane structure describes the plasma membrane as a fluid combination of phospholipids, cholesterol, proteins, and carbohydrates.

The main fabric of the membrane is composed of two layers of phospholipid molecules, and the polar ends of these molecules (which look like a collection of balls in an artist's rendition of the model) (Figure 7.14) are in contact with aqueous fluid both inside and outside the cell. Thus, both surfaces of the plasma membrane are hydrophilic. In contrast, the interior of the membrane, between its two surfaces, is a hydrophobic or nonpolar region because of the fatty acid tails. This region has no attraction for water or other polar molecules.

Proteins make up the second major chemical component of plasma membranes. Integral proteins are embedded in the plasma membrane and may span all or part of the membrane. Integral proteins may serve as channels or pumps to move materials into or out of the cell. Peripheral proteins are found on the exterior or interior surfaces of membranes, attached either to integral proteins or to phospholipid molecules. Both integral and peripheral proteins may serve as enzymes, as structural attachments for the fibers of the cytoskeleton, or as part of the cell's recognition sites.

Carbohydrates are the third major component of plasma membranes. They are always found on the exterior surface of cells and are bound either to proteins (forming glycoproteins) or to lipids (forming glycolipids). These carbohydrate chains may consist of 2–60 monosaccharide units and may be either straight or branched. Along with peripheral proteins, carbohydrates form specialized sites on the cell surface that allow cells to recognize each other.

How Viruses Infect Specific Organs

Specific glycoprotein molecules exposed on the surface of the cell membranes of host cells are exploited by many viruses to infect specific organs. For example, HIV is able to penetrate the plasma membranes of specific kinds of white blood cells called T-helper cells and monocytes, as well as some cells of the central nervous system. The hepatitis virus attacks only liver cells.

These viruses are able to invade these cells, because the cells have binding sites on their surfaces that the viruses have exploited with equally specific glycoproteins in their coats. (Figure 7.15). The cell is tricked by the mimicry of the virus coat molecules, and the virus is able to enter the cell. Other recognition sites on the virus's surface interact with the human immune system, prompting the body to produce antibodies. Antibodies are made in response to the antigens (or proteins associated with invasive pathogens). These same sites serve as places for antibodies to attach, and either destroy or inhibit the activity of the virus. Unfortunately, these sites on HIV are encoded by genes that change quickly, making the production of an effective vaccine against the virus very difficult. The virus population within an infected individual quickly evolves through mutation into different populations, or variants, distinguished by differences in these recognition sites. This rapid change of viral surface markers decreases the effectiveness of the person's immune system in attacking the virus, because the antibodies will not recognize the new variations of the surface patterns.

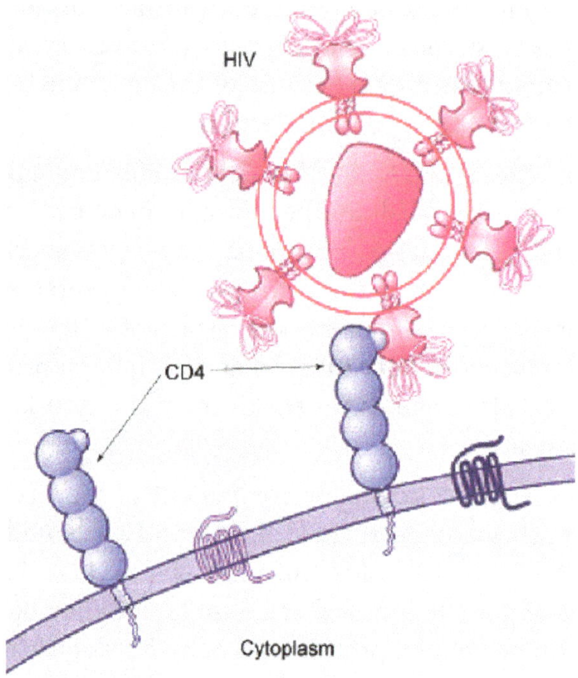

Figure 7.15 HIV docks at and binds to the CD4 receptor, a glycoprotein on the surface of T cells, before entering, or infecting, the cell. (credit: modification of work by US National Institutes of Health/National Institute of Allergy and Infectious Diseases)

PASSIVE TRANSPORT

Plasma membranes must allow certain substances to enter and leave a cell, while preventing harmful material from entering and essential material from leaving. In other words, plasma membranes are **selectively permeable** (semipermeable)—they allow some substances through but not others. If they were to lose this selectivity, the cell would no longer be able to sustain itself, and it would be destroyed. Some cells require larger amounts of specific substances than do other cells; they must have a way of obtaining these materials from the extracellular fluids. This may happen passively, as certain materials move back and forth, or the cell may have special mechanisms that ensure transport. Most cells expend most of their energy, in the form of adenosine triphosphate (ATP), to create and maintain an uneven distribution of ions on the opposite sides of their membranes. The structure of the plasma membrane contributes to these functions, but it also presents some problems.

The most direct forms of membrane transport are passive. **Passive transport** is a naturally occurring phenomenon and does not require the cell to expend energy to accomplish the movement. In passive transport, substances move from an area of higher concentration to an area of lower concentration in a process called diffusion. A physical space in which there is a different concentration of a single substance is said to have a **concentration gradient.**

Selective Permeability

Plasma membranes are asymmetric, meaning that despite the mirror image formed by the phospholipids, the interior of the membrane is not identical to the exterior of the membrane. Integral proteins that act as channels or pumps work in one direction. Carbohydrates, attached to lipids or proteins, are also found on the exterior surface of the plasma membrane. These carbohydrate complexes help the cell bind substances that the cell needs in the extracellular fluid. This adds considerably to the selective nature of plasma membranes.

Recall that plasma membranes have hydrophilic and hydrophobic regions. This characteristic helps the movement of certain materials through the membrane and hinders the movement of others. Lipid-soluble material can easily slip through the hydrophobic lipid core of the membrane. Substances such as the fat-soluble vitamins A, D, E, and K readily pass through the plasma membranes in the digestive tract and other tissues. Fat-soluble drugs also gain easy entry into cells and are readily transported into the body's tissues and organs. Molecules of oxygen and carbon dioxide have no charge and pass through by simple diffusion.

Polar substances present problems for the membrane. While some polar molecules connect easily with the outside of a cell, they cannot readily pass through the lipid core of the plasma membrane. Additionally, whereas small ions could easily slip through the spaces in the mosaic of the membrane, their charge prevents them from doing so. Ions such as sodium, potassium, calcium, and chloride must have a special means of penetrating plasma membranes. Simple sugars and amino acids also need help with transport across plasma membranes.

Diffusion

Diffusion is a passive process of transport. A single substance tends to move from an area of high concentration to an area of low concentration until the concentration is equal across the space. You are familiar with diffusion of substances through the air. For example, think about someone opening a bottle of perfume in a room filled with people. The perfume is at its highest concentration in the bottle and is at its lowest at the edges of the room. The perfume vapor will diffuse, or spread away, from the bottle, and gradually, more and more people will smell the perfume as it spreads. Materials move within the cell's cytosol by diffusion, and certain materials move through the plasma membrane by diffusion (Figure 7.16). Diffusion expends no energy. Rather the different concentrations of materials in different areas are a form of potential energy, and diffusion is the dissipation of that potential energy as materials move down their concentration gradients, from high to low.

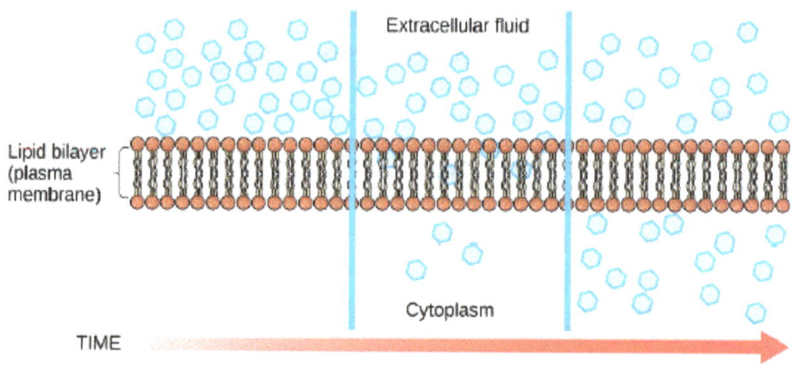

Figure 7.16 Diffusion through a permeable membrane follows the concentration gradient of a substance, moving the substance from an area of high concentration to one of low concentration. (credit: modification of work by Mariana Ruiz Villarreal)

Each separate substance in a medium, such as the extracellular fluid, has its own concentration gradient, independent of the concentration gradients of other materials. Additionally, each substance will diffuse according to that gradient.

Several Factors Affect the Rate of Diffusion.

- Extent of the concentration gradient: The greater the difference in concentration, the more rapid the diffusion. The closer the distribution of the material gets to equilibrium, the slower the rate of diffusion becomes.

- Mass of the molecules diffusing: More massive molecules move more slowly, because it is more difficult for them to move between the molecules of the substance they are moving through; therefore, they diffuse more slowly.
- Temperature: Higher temperatures increase the energy and therefore the movement of the molecules, increasing the rate of diffusion.
- Solvent density: As the density of the solvent increases, the rate of diffusion decreases. The molecules slow down because they have a more difficult time getting through the denser medium.

Facilitated Transport

In **facilitated transport**, also called facilitated diffusion, material moves across the plasma membrane with the assistance of trans-membrane proteins down a concentration gradient (from high to low concentration) without the expenditure of cellular energy. However, the substances that undergo facilitated transport would otherwise not diffuse easily or quickly across the plasma membrane. The solution to moving polar substances and other substances across the plasma membrane rests in the proteins that span its surface. The material being transported is first attached to protein or glycoprotein receptors on the exterior surface of the plasma membrane. This allows the material that is needed by the cell to be removed from the extracellular fluid. The substances are then passed to specific integral proteins that facilitate their passage, because they form channels or pores that allow certain substances to pass through the membrane. The integral proteins involved in facilitated transport are collectively referred to as transport proteins, and they function as either channels for the material or carriers.

Osmosis

Osmosis is the movement of free water molecules through a semipermeable membrane according to the water's concentration gradient across the membrane, which is inversely proportional to the solutes' concentration. Whereas diffusion transports material across membranes and within cells, osmosis transports *only water* across a membrane and the membrane limits the diffusion of solutes in the water. Osmosis is a special case of diffusion. Water, like other substances, moves from an area of high concentration of free water molecules to one of low free water molecule concentration. Imagine a beaker with a semipermeable membrane, separating the two sides or halves (Figure 7.17). On both sides of the membrane, the water level is the same, but there are different concentrations on each side of a dissolved substance, or **solute**, that cannot cross the membrane. If the volume of the water is the same, but the concentrations of solute are different, then there are also different concentrations of water, the solvent, on either side of the membrane.

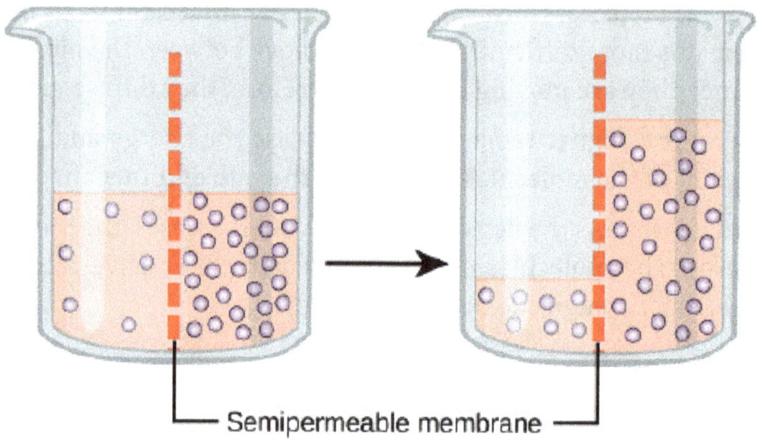

Figure 7.17 In osmosis, water always moves from an area of higher concentration (of water) to one of lower concentration (of water). In this system, the solute cannot pass through the selectively permeable membrane.

A principle of diffusion is that the molecules move around and will spread evenly throughout the medium if they can. However, only the material capable of getting through the membrane will diffuse through it. In this example, the solute cannot diffuse through the membrane, but the water can. Water has a concentration gradient in this system. Therefore, water will diffuse down its concentration gradient, crossing the membrane to the side where it is less concentrated. This diffusion of water through the membrane—osmosis—will continue until the concentration gradient of water goes to zero. Osmosis proceeds constantly in living systems.

Tonicity

Tonicity describes the amount of solute in a solution. The measure of the tonicity of a solution, or the total amount of solutes dissolved in a specific amount of solution, is called its **osmolarity**. Three terms—hypotonic, isotonic, and hypertonic—are used to relate the osmolarity of a cell to the osmolarity of the extracellular fluid that contains the cells. In a **hypotonic** solution, such as tap water, the extracellular fluid has a lower concentration of solutes than the fluid inside the cell, and water enters the cell. (In living systems, the point of reference is always the cytoplasm, so the prefix *hypo-* means that the extracellular fluid has a lower concentration of solutes, or a lower osmolarity, than the cell cytoplasm.) It also means that the extracellular fluid has a higher concentration of water than does the cell. In this situation, water will follow its concentration gradient and enter the cell. This may cause an animal cell to burst, or lyse.

In a **hypertonic** solution (the prefix *hyper-* refers to the extracellular fluid having a higher concentration of solutes than the cell's cytoplasm), the fluid contains less water than the cell does, such as seawater. Because the cell has a lower concentration of solutes, the water will leave the cell. In effect, the solute is drawing the water out of the cell. This may cause an animal cell to shrivel, or crenate.

In an **isotonic** solution, the extracellular fluid has the same osmolarity as the cell. If the concentration of solutes of the cell matches that of the extracellular fluid, there will be no net movement of water into or out of the cell. Blood cells in hypertonic, isotonic, and hypotonic solutions take on characteristic appearances (Figure 7.18).

Visual Connection

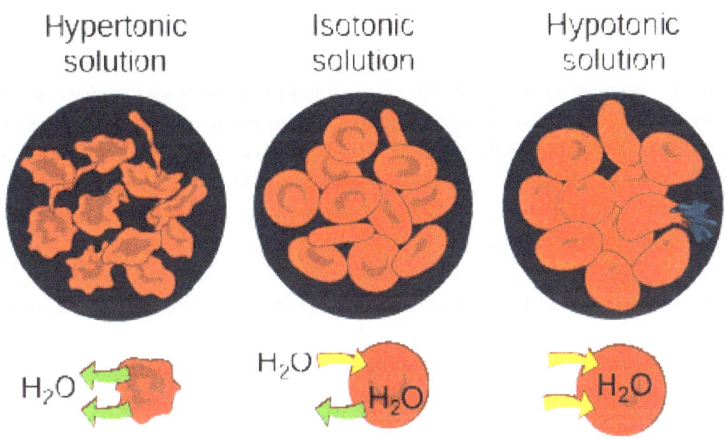

Figure 7.18 Osmotic pressure changes the shape of red blood cells in hypertonic, isotonic, and hypotonic solutions. (credit: modification of work by Mariana Ruiz Villarreal)

A doctor injects a patient with what the doctor thinks is isotonic saline solution. The patient dies, and autopsy reveals that many red blood cells have been destroyed. Do you think the solution the doctor injected was really isotonic?

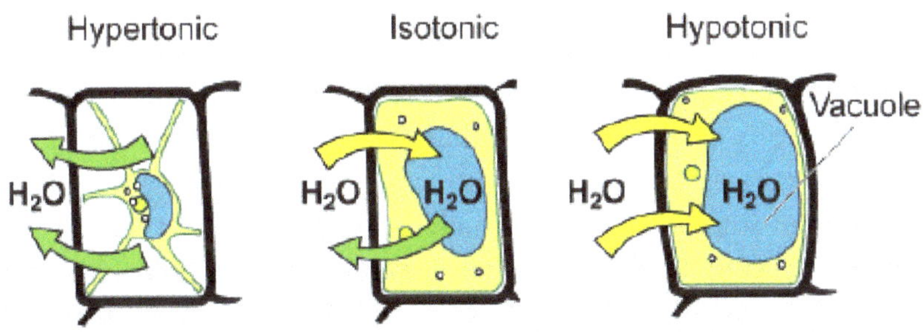

Figure 7.19 The turgor pressure within a plant cell depends on the tonicity of the solution that it is bathed in. (credit: modification of work by Mariana Ruiz Villarreal)

Some organisms, such as plants, fungi, bacteria, and some protists, have cell walls that surround the plasma membrane and prevent cell lysis. The plasma membrane can only expand to the limit of the cell wall, so the cell will not lyse. In fact, the cytoplasm in plants is always slightly hypertonic compared to the cellular environment, and water will always enter a cell if water is available. This influx of water produces turgor pressure, which stiffens the cell walls of the plant (Figure 7.19). In non-woody plants, turgor pressure supports the plant. If the plant cells become hypertonic, as occurs in drought or if a plant is not watered adequately, water will leave the cell. Plants lose turgor pressure in this condition and wilt.

ACTIVE TRANSPORT

Active transport mechanisms require the use of the cell's energy, usually in the form of adenosine triphosphate (ATP). If a substance must move into the cell against its concentration gradient, that is, if the concentration of the substance inside the cell must be greater than its concentration in the extracellular fluid, the cell must use energy to move the substance. Some active transport mechanisms move small-molecular weight material, such as ions, through the membrane.

In addition to moving small ions and molecules through the membrane, cells also need to remove and take in larger molecules and particles. Some cells are even capable of engulfing entire unicellular microorganisms. You might have correctly hypothesized that the uptake and release of large particles by the cell requires energy. A large particle, however, cannot pass through the membrane, even with energy supplied by the cell.

Electrochemical Gradient

We have discussed simple concentration gradients—differential concentrations of a substance across a space or a membrane—but in living systems, gradients are more complex. Because cells contain proteins, most of which are negatively charged, and because ions move into and out of cells, there is an electrical gradient, a difference of charge, across the plasma membrane. The interior of living cells is electrically negative with respect to the extracellular fluid in which they are bathed; at the same time, cells have higher concentrations of potassium (K^+) and lower concentrations of sodium (Na^+) than does the extracellular fluid. Thus, in a living cell, the concentration gradient and electrical gradient of Na^+ promotes diffusion of the ion into the cell, and the electrical gradient of Na^+ (a positive ion) tends to drive it inward to the negatively charged interior. The situation is more complex, however, for other elements such as potassium. The electrical gradient of K^+ promotes diffusion of the ion *into* the cell, but the concentration gradient of K^+ promotes diffusion *out* of the cell (Figure 7.20). The combined gradient that affects an ion is called its **electrochemical gradient**, and it is especially important to muscle and nerve cells.

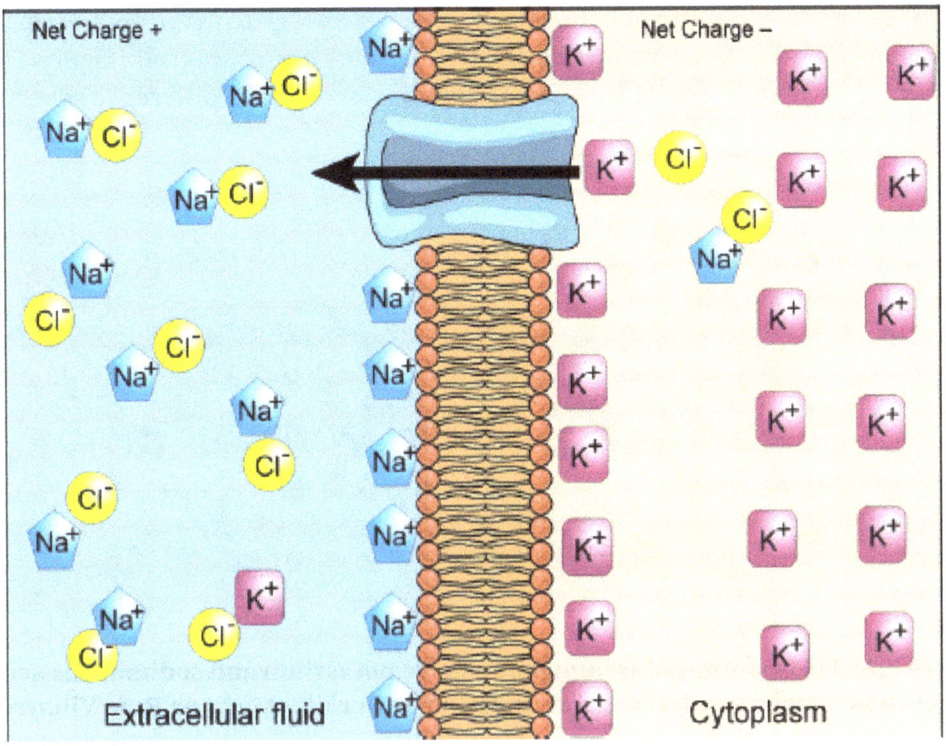

Figure 7.20 Electrochemical gradients arise from the combined effects of concentration gradients and electrical gradients. (credit: modification of work by "Synaptitude"/Wikimedia Commons)

Moving Against a Gradient

To move substances against a concentration or an electrochemical gradient, the cell must use energy. This energy is harvested from ATP that is generated through cellular metabolism. Active transport mechanisms, collectively called pumps or carrier proteins, work against electrochemical gradients. With the exception of ions, small substances constantly pass through plasma membranes. Active transport maintains concentrations of ions and other substances needed by living cells in the face of these passive changes. Much of a cell's supply of metabolic energy may be spent maintaining these processes. Because active transport mechanisms depend on cellular metabolism for energy, they are sensitive to many metabolic poisons that interfere with the supply of ATP.

Two mechanisms exist for the transport of small-molecular weight material and macromolecules. Primary active transport moves ions across a membrane and creates a difference in charge across that membrane. The primary active transport system uses ATP to move a substance, such as an ion, into the cell, and often at the same time, a second substance is moved out of the cell. The sodium-potassium pump, an important pump in animal cells, expends energy to move potassium ions into the cell and a different number of sodium ions out of the cell (Figure 7.21). The action of this pump results in a concentration and charge difference across the membrane.

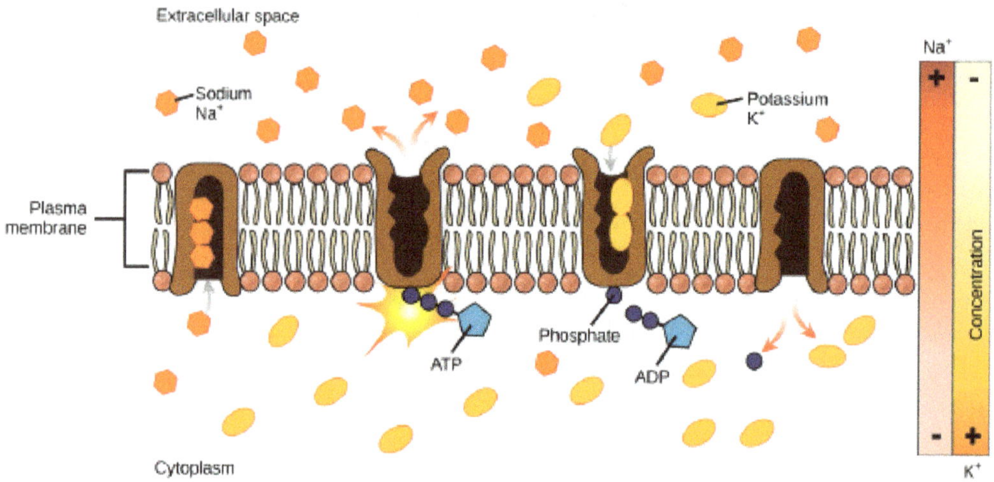

Figure 7.21 The sodium-potassium pump move potassium and sodium ions across the plasma membrane. (credit: modification of work by Mariana Ruiz Villarreal)

Secondary active transport describes the movement of material using the energy of the electrochemical gradient established by primary active transport. Using the energy of the electrochemical gradient created by the primary active transport system, other substances such as amino acids and glucose can be brought into the cell through membrane channels. ATP itself is formed through secondary active transport using a hydrogen ion gradient in the mitochondrion.

Endocytosis

Endocytosis is a type of active transport that moves particles, such as large molecules, parts of cells, and even whole cells, into a cell. There are different variations of endocytosis, but all share a common characteristic: The plasma membrane of the cell invaginates, forming a pocket around the target particle. The pocket pinches off, resulting in the particle being contained in a newly created vacuole that is formed from the plasma membrane.

Phagocytosis is the process by which large particles, such as cells, are taken in by a cell. For example, when microorganisms invade the human body, a type of white blood cell called a neutrophil removes the invader through this process, surrounding and engulfing the microorganism, which is then destroyed by the neutrophil (Figure 7.22).

A variation of endocytosis is called **pinocytosis**. This literally means "cell drinking" and was named at a time when the assumption was that the cell was purposefully taking in extracellular fluid. In reality, this process takes in solutes that the cell needs from the extracellular fluid (Figure 7.22).

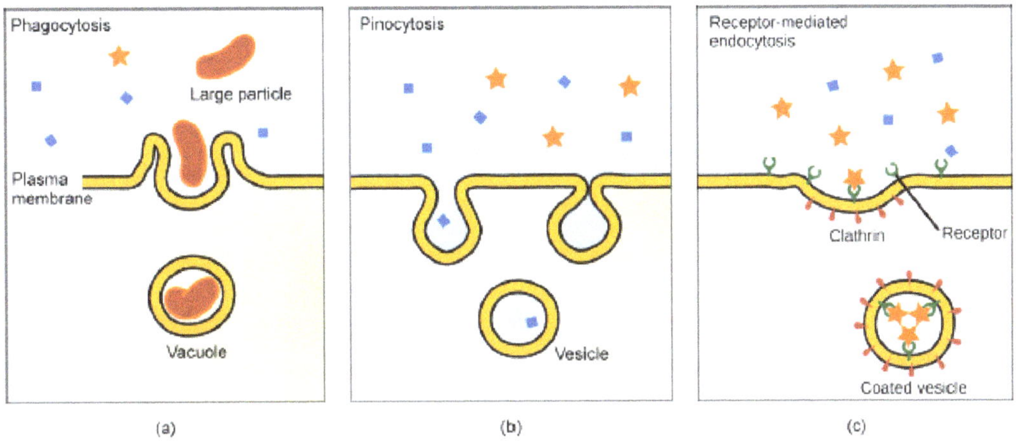

Figure 7.22 Three variations of endocytosis are shown. (a) In one form of endocytosis, phagocytosis, the cell membrane surrounds the particle and pinches off to form an intracellular vacuole. (b) In another type of endocytosis, pinocytosis, the cell membrane surrounds a small volume of fluid and pinches off, forming a vesicle. (c) In receptor- mediated endocytosis, uptake of substances by the cell is targeted to a single type of substance that binds at the receptor on the external cell membrane. (credit: modification of work by Mariana Ruiz Villarreal)

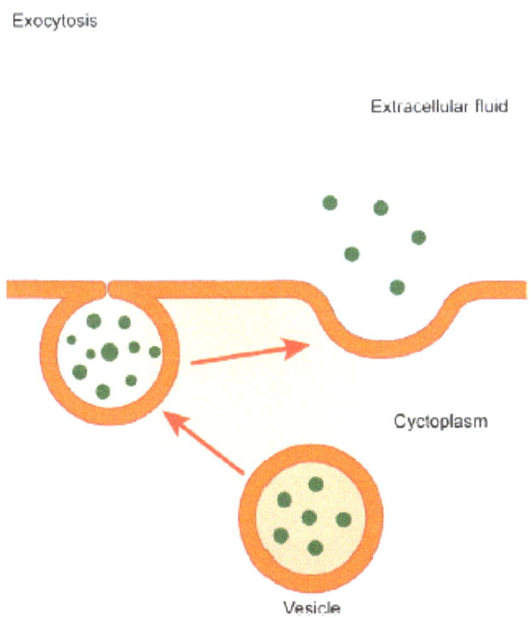

Figure 7.23 In exocytosis, a vesicle migrates to the plasma membrane, binds, and releases its contents to the outside of the cell. (credit: modification of work by Mariana Ruiz Villarreal)

A targeted variation of endocytosis employs binding proteins in the plasma membrane that are specific for certain substances (Figure 7.22). The particles bind to the proteins and the plasma membrane invaginates, bringing the substance and the proteins into the cell. If passage across the membrane of the target of **receptor-mediated endocytosis** is ineffective, it will not be removed from the tissue fluids or blood. Instead, it will stay in those fluids and increase in concentration. Some human diseases are caused by a failure of receptor-mediated endocytosis. For example, the form of cholesterol termed low-density lipoprotein or LDL (also referred to as "bad" cholesterol) is removed from the blood by receptor-mediated endocytosis. In the human genetic disease familial hypercholesterolemia, the LDL receptors are defective or missing entirely. People with this condition have life-threatening levels of cholesterol in their blood, because their cells cannot clear the chemical from their blood.

Exocytosis

In contrast to these methods of moving material into a cell is the process of exocytosis. **Exocytosis** is the opposite of the processes discussed above in that its purpose is to expel material from the cell into the extracellular fluid. A particle enveloped in membrane fuses with the interior of the plasma membrane. This fusion opens the membranous envelope to the exterior of the cell, and the particle is expelled into the extracellular space (Figure 7.23).

CHAPTER-8

INTRODUCTION TO MICROBES

INTRODUCTION: AN OVERVIEW

Microbes, often invisible to the naked eye yet omnipresent in our environment, constitute a diverse and fascinating world of microscopic organisms. From bacteria and viruses to fungi and protozoa, these tiny life forms play pivotal roles in shaping ecosystems, influencing human health, and driving key biogeochemical processes. The study of microbes, known as microbiology, unveils the intricacies of their structure, behaviour, and functions. As we embark on an exploration of this microbial realm, we discover the profound impact that these seemingly minute entities exert on our planet and our lives. This introduction serves as a portal into the captivating universe of microbes, emphasizing their significance in both the macroscopic and microscopic landscapes of biology.

From boiling thermal hot springs to deep beneath the Antarctic ice, microorganisms can be found almost everywhere on earth in great quantities. Microorganisms (or microbes, as they are also called) are small organisms. Most are so small that they cannot be seen without a microscope.

Most microorganisms are harmless to humans and, in fact, many are helpful. They play fundamental roles in ecosystems everywhere on earth, forming the backbone of many food webs. People use them to make biofuels, medicines, and even foods. Without microbes, there would be no bread, cheese, or beer. Our bodies are filled with microbes, and our skin alone is home to trillions of them. Some of them we can't live without; others cause diseases that can make us sick or even kill us.

Although much more is known today about microbial life than ever before, the vast majority of this invisible world remains unexplored. Microbiologists continue to identify new ways that microbes benefit and threaten humans.

FERMENTED FOODS AND BEVERAGES

People across the world have enjoyed fermented foods and beverages like beer, wine, bread, yogurt, cheese, and pickled vegetables for all of recorded history. Discoveries from several archaeological sites suggest that even prehistoric people took advantage of fermentation to preserve and enhance the taste of food. Archaeologists studying pottery jars from a Neolithic village in China found that people were making a fermented beverage from rice, honey, and fruit as early as 7000 BC.

Production of these foods and beverages requires microbial fermentation, a process that uses bacteria, mold, or yeast to convert sugars (carbohydrates) to alcohol, gases, and organic acids (Figure 8.1). While it is likely that people first learned about fermentation by accident—perhaps by drinking old milk that had curdled or old grape juice that had fermented—they later learned to harness the power of fermentation to make products like bread, cheese, and wine.

Figure 8.1 A microscopic view of *Saccharomyces cerevisiae*, the yeast responsible for making bread rise (left). Yeast is a microorganism. Its cells metabolize the carbohydrates in flour (middle) and produce carbon dioxide, which causes the bread to rise (right). (credit middle: modification of work by Janus Sandsgaard; credit right: modification of work by "MDreibelbis"/Flickr)

The Iceman Treateth

Prehistoric humans had a very limited understanding of the causes of disease, and various cultures developed different beliefs and explanations. While many believed that illness was punishment for angering the gods or was simply the result of fate, archaeological evidence suggests that prehistoric people attempted to treat illnesses and infections. One example of this is Ötzi the Iceman, a 5300-year-old mummy found frozen in the ice of the Ötzal Alps on the Austrian-Italian border in 1991. Because Ötzi was so well preserved by the ice, researchers discovered that he was infected with the eggs of the parasite Trichuris trichiura, which may have caused him to have abdominal pain and anemia. Researchers also found evidence of Borrelia burgdorferi, a bacterium that causes Lyme disease. Some researchers think Ötzi may have been trying to treat his infections with the woody fruit of the Fomitopsis betulinus fungus, which was discovered tied to his belongings. This fungus has both laxative and antibiotic properties. Ötzi was also covered in tattoos that were made by cutting incisions into his skin, filling them with herbs, and then burning the herbs· There is speculation that this may have been another attempt to treat his health ailments.

Early Notions of Disease, Contagion, and Containment

Several ancient civilizations appear to have had some understanding that disease could be transmitted by things they could not see. This is especially evident in historical attempts to contain the spread of disease. For example, the Bible refers to the practice of quarantining people with leprosy and other diseases, suggesting that people understood that diseases could be communicable. Ironically, while leprosy is communicable, it is also a disease that progresses slowly. This means that people were likely quarantined after they had already spread the disease to others.

The ancient Greeks attributed disease to bad air, *mal'aria*, which they called "miasmatic odors." They developed hygiene practices that built on this idea. The Romans also believed in the miasma hypothesis and created a complex sanitation infrastructure to deal with sewage. In Rome, they built aqueducts, which brought fresh water into the city, and a giant sewer, the *Cloaca* Maxima, which carried waste away and into the river Tiber (Figure 8.2). Some researchers believe that this infrastructure helped protect the Romans from epidemics of waterborne illnesses.

Even before the invention of the microscope, some doctors, philosophers, and scientists made great strides in understanding the invisible forces—what we now know as microbes—that can cause infection, disease, and death.

The Greek physician Hippocrates (460–370 BC) is considered the "father of Western medicine" (Figure 8.3). Unlike many of his ancestors and contemporaries, he dismissed the idea that disease was caused by supernatural forces. Instead, he posited that diseases had natural causes from within patients or their environments. Hippocrates and his heirs are believed to have written the *Hippocratic Corpus*, a collection of texts that make up some of the oldest surviving medical books. Hippocrates is also often credited as the author of the Hippocratic Oath, taken by new physicians to pledge their dedication to diagnosing and treating patients without causing harm.

While Hippocrates is considered the father of Western medicine, the Greek philosopher and historian Thucydides (460–395 BC) is considered the father of scientific history because he advocated for evidence- based analysis of cause-and-effect reasoning (Figure 8.3). Among his most important contributions are his observations regarding the Athenian plague that killed one-third of the population of Athens between 430 and 410 BC. Having survived the epidemic himself, Thucydides made the important observation that survivors did not get re-infected with the disease, even when taking care of actively sick people. This observation shows an early understanding of the concept of immunity.

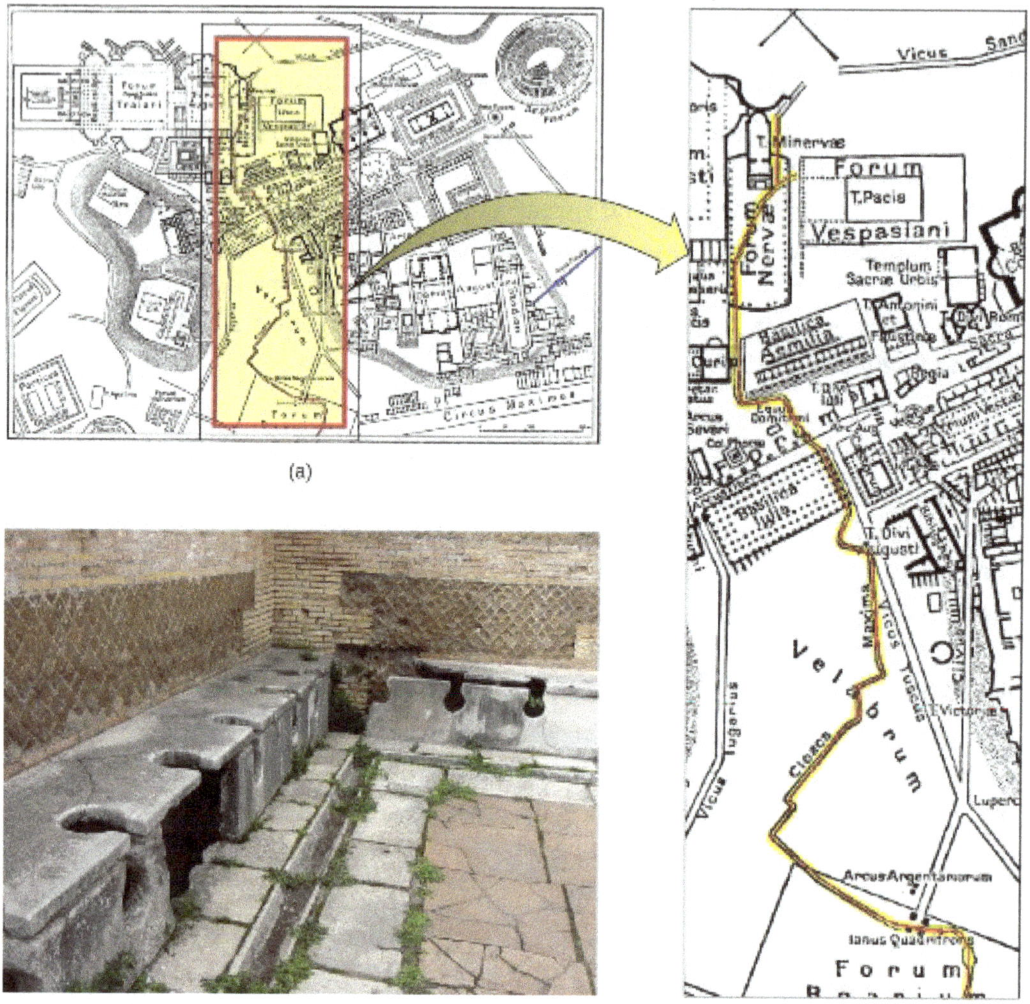

(a)

(b)

Figure 8.2 (a) The *Cloaca Maxima*, or "Greatest Sewer" (shown in red), ran through ancient Rome. It was an engineering marvel that carried waste away from the city and into the river Tiber. (b) These ancient latrines emptied into the *Cloaca Maxima*.

Marcus Terentius Varro (116–27 BC) was a prolific Roman writer who was one of the first people to propose the concept that things we cannot see (what we now call microorganisms) can cause disease (Figure 8.3). In *Res Rusticae* (*On Farming*), published in 36 BC, he said that "precautions must also be taken in neighbourhood swamps. . . because certain minute creatures [*animalia minuta*] grow there which cannot be seen by the eye, which float in the air and enter the body through the mouth and nose and there cause serious diseases."[9]

(a) (b) (c)

Figure 8.3 (a) Hippocrates, the "father of Western medicine," believed that diseases had natural, not supernatural, causes. (b) The historian Thucydides observed that survivors of the Athenian plague were subsequently immune to the infection. (c) Marcus Terentius Varro proposed that disease could be caused by "certain minute creatures . . . which cannot be seen by the eye." (credit c: modification of work by Alessandro Antonelli)

The Birth of Microbiology

While the ancients may have suspected the existence of invisible "minute creatures," it wasn't until the invention of the microscope that their existence was definitively confirmed. While it is unclear who exactly invented the microscope, a Dutch cloth merchant named Antonie van Leeuwenhoek (1632–1723) was the first to develop a lens powerful enough to view microbes. In 1675, using a simple but powerful microscope, Leeuwenhoek was able to observe single-celled organisms, which he described as "animalcules" or "wee little beasties," swimming in a drop of rain water. From his drawings of these little organisms, we now know he was looking at bacteria and protists.

Nearly 200 years after van Leeuwenhoek got his first glimpse of microbes, the "Golden Age of Microbiology" spawned a host of new discoveries between 1857 and 1914. Two famous microbiologists, Louis Pasteur and Robert Koch, were especially active in advancing our understanding of the unseen world of microbes (Figure 8.4). Pasteur, a French chemist, showed that individual microbial strains had unique properties and demonstrated that fermentation is caused by microorganisms. He also invented pasteurization, a process used to kill microorganisms responsible for spoilage, and developed vaccines for the treatment of diseases, including rabies, in animals and humans. Koch, a German physician, was the first to demonstrate the connection between a single, isolated microbe and a known human disease. For example, he discovered the bacteria that cause anthrax (*Bacillus anthracis*), cholera (*Vibrio cholera*), and tuberculosis (*Mycobacterium tuberculosis*). We will discuss these famous microbiologists, and others, in later chapters.

(a) (b)

Figure 8.4 (a) Louis Pasteur (1822–1895) is credited with numerous innovations that advanced the fields of microbiology and immunology. (b) Robert Koch (1843–1910) identified the specific microbes that cause anthrax, cholera, and tuberculosis.

As microbiology has developed, it has allowed the broader discipline of biology to grow and flourish in previously unimagined ways. Much of what we know about human cells comes from our understanding of microbes, and many of the tools we use today to study cells and their genetics derive from work with microbes.

Microbiology Toolbox

Because individual microbes are generally too small to be seen with the naked eye, the science of microbiology is dependent on technology that can artificially enhance the capacity of our natural senses of perception. Early microbiologists like Pasteur and Koch had fewer tools at their disposal than are found in modern laboratories, making their discoveries and innovations that much more impressive. Later chapters of this text will explore many applications of technology in depth, but for now, here is a brief overview of some of the fundamental tools of the microbiology lab.

- **Microscopes** produce magnified images of microorganisms, human cells and tissues, and many other types of specimens too small to be observed with the naked eye.
- **Stains and dyes** are used to add colour to microbes so they can be better observed under a microscope. Some dyes can be used on living microbes, whereas others require that the specimens be fixed with chemicals or heat before staining. Some stains only work on certain types of microbes because of differences in their cellular chemical composition.
- **Growth media** are used to grow microorganisms in a lab setting. Some media are liquids; others are more solid or gel-like. A growth medium provides nutrients, including water, various salts, a source of carbon (like glucose), and a source of nitrogen and amino acids (like yeast extract) so microorganisms can grow and reproduce. Ingredients in a growth medium can be modified to grow unique types of microorganisms.

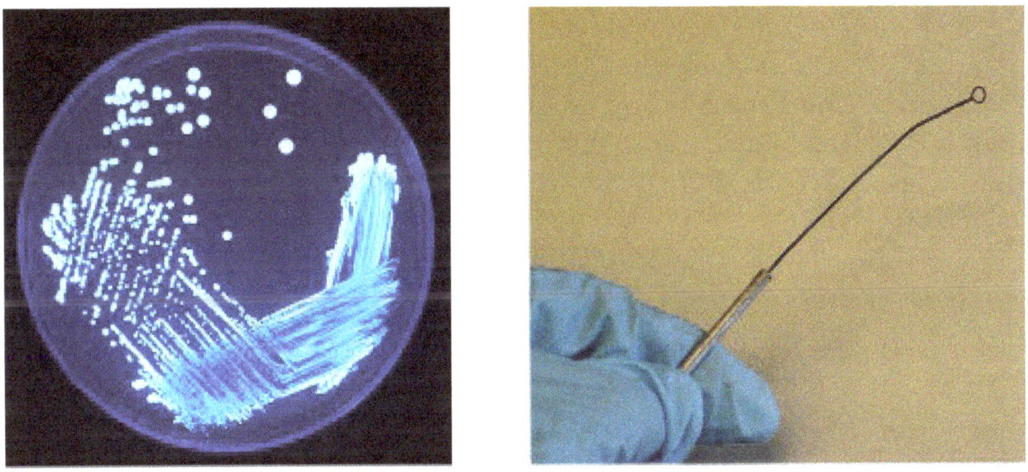

(a) (b)

Figure 8.5 (a) This Petri dish filled with agar has been streaked with *Legionella*, the bacterium responsible for causing Legionnaire's disease. (b) An inoculation loop like this one can be used to streak bacteria on agar in a Petri dish. (credit a: modification of work by Centers for Disease Control and Prevention; credit b: modification of work by Jeffrey M. Vinocur)

- **A Petri** dish is a flat-lidded dish that is typically 10–11 centimeters (cm) in diameter and 1–1.5 cm high. Petri dishes made out of either plastic or glass are used to hold growth media (Figure 8.5).
- **Test tubes** are cylindrical plastic or glass tubes with rounded bottoms and open tops. They can be used to grow microbes in broth, or semisolid or solid growth media.

- A **Bunsen burner** is a metal apparatus that creates a flame that can be used to sterilize pieces of equipment. A rubber tube carries gas (fuel) to the burner. In many labs, Bunsen burners are being phased out in favour of infrared **microincinerators**, which serve a similar purpose without the safety risks of an open flame.
- **An inoculation loop** is a handheld tool that ends in a small wire loop (Figure 8.5). The loop can be used to streak microorganisms on agar in a Petri dish or to transfer them from one test tube to another. Before each use, the inoculation loop must be sterilized so cultures do not become contaminated.

THE SCIENCE OF TAXONOMY

Taxonomy is the classification, description, identification, and naming of living organisms. Classification is the practice of organizing organisms into different groups based on their shared characteristics. The most famous early taxonomist was a Swedish botanist, zoologist, and physician named Carolus Linnaeus (1701–1778). In 1735, Linnaeus published *Systema Naturae*, an 11-page booklet in which he proposed the Linnaean taxonomy, a system of categorizing and naming organisms using a standard format so scientists could discuss organisms using consistent terminology. He continued to revise and add to the book, which grew into multiple volumes (Figure 8.6).

Figure 8.6 Swedish botanist, zoologist, and physician Carolus Linnaeus developed a new system for categorizing plants and animals. In this 1853 portrait by Hendrik Hollander, Linnaeus is holding a twinflower, named *Linnaea borealis* in his honor.

In his taxonomy, Linnaeus divided the natural world into three kingdoms: animal, plant, and mineral (the mineral kingdom was later abandoned). Within the animal and plant kingdoms, he grouped organisms using a hierarchy of increasingly specific levels and sublevels based on their similarities. The names of the levels in Linnaeus's original taxonomy were kingdom, class, order, family, genus (plural: genera), and species. Species was, and continues to be, the most specific and basic taxonomic unit.

Evolving Trees of Life (Phylogenies)

With advances in technology, other scientists gradually made refinements to the Linnaean system and eventually created new systems for classifying organisms. In the 1800s, there was a growing interest in developing taxonomies that took into account the evolutionary relationships, or **phylogenies**, of all different species of organisms on earth. One way to depict these relationships is via a diagram called a phylogenetic tree (or tree of life). In these diagrams, groups of organisms are arranged by how closely related they are thought to be. In early phylogenetic trees, the relatedness of organisms was inferred by their visible similarities, such as the presence or absence of hair or the number of limbs. Now, the analysis is more complicated. Today, phylogenic analyses include genetic, biochemical, and embryological comparisons, as will be discussed later in this chapter.

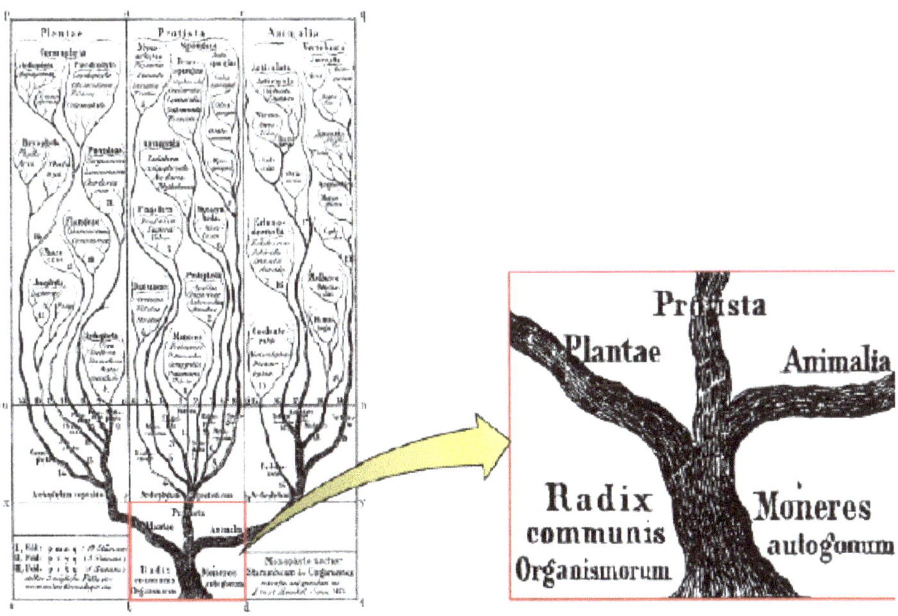

Figure 8.7 Ernst Haeckel's rendering of the tree of life, from his 1866 book *General Morphology of Organisms*, contained three kingdoms: Plantae, Protista, and Animalia. He later added a fourth kingdom, Monera, for unicellular organisms lacking a nucleus.

Linnaeus's tree of life contained just two main branches for all living things: the animal and plant kingdoms. In 1866, Ernst Haeckel, a German biologist, philosopher, and physician, proposed another kingdom, Protista, for unicellular organisms (Figure 8.7). He later proposed a fourth kingdom, Monera, for unicellular organisms whose cells lack nuclei, like bacteria.

Nearly 100 years later, in 1969, American ecologist Robert Whittaker (1920–1980) proposed adding another kingdom—Fungi—in his tree of life. Whittaker's tree also contained a level of categorization above the kingdom level—the empire or superkingdom level—to distinguish between organisms that have membrane-bound nuclei in their cells (**eukaryotes**) and those that do not (**prokaryotes**). Empire Prokaryota contained just the Kingdom Monera. The Empire Eukaryota contained the other four kingdoms: Fungi, Protista, Plantae, and Animalia. Whittaker's five-kingdom tree was considered the standard phylogeny for many years.

Figure 8.8 shows how the tree of life has changed over time. Note that viruses are not found in any of these trees. That is because they are not made up of cells and thus it is difficult to determine where they would fit into a tree of life.

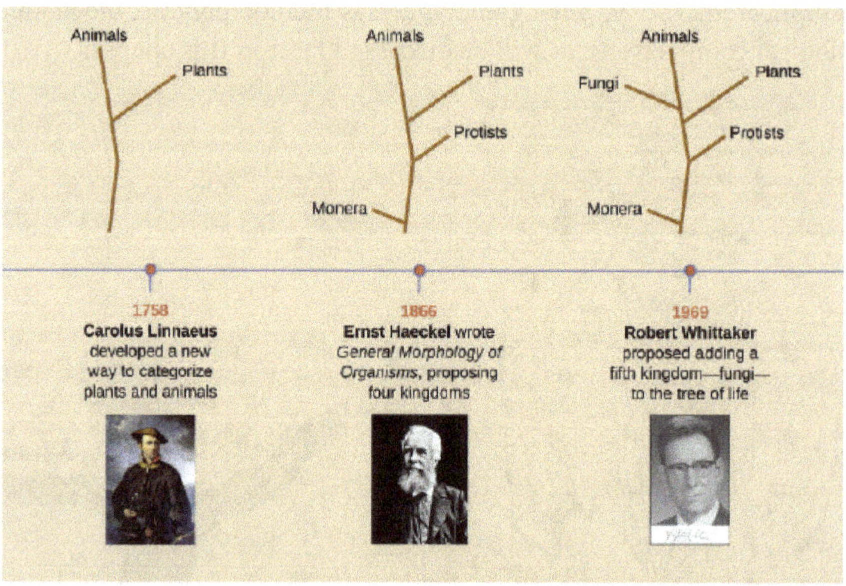

Figure 8.8 This timeline shows how the shape of the tree of life has changed over the centuries. Even today, the taxonomy of living organisms is continually being re-evaluated and refined with advances in technology.

The Role of Genetics in Modern Taxonomy

Haeckel's and Whittaker's trees presented hypotheses about the phylogeny of different organisms based on readily observable characteristics. But the advent of molecular genetics in the late 20th century revealed other ways to organize phylogenetic trees. Genetic methods allow for a standardized way to compare all living organisms without relying on observable characteristics that can often be subjective. Modern taxonomy relies heavily on comparing the nucleic acids (deoxyribonucleic acid [DNA] or ribonucleic acid [RNA]) or proteins from different organisms. The more similar the nucleic acids and proteins are between two organisms, the more closely related they are considered to be.

In the 1970s, American microbiologist Carl Woese discovered what appeared to be a "living record" of the evolution of organisms. He and his collaborator George Fox created a genetics-based tree of life based on similarities and differences they observed in the gene sequences coding for small subunit ribosomal RNA (rRNA) of different organisms. In the process, they discovered that a certain type of bacteria, called archaebacteria (now known simply as archaea), were significantly different from other bacteria and eukaryotes in terms of their small subunit rRNA gene sequences. To accommodate this difference, they created a tree with three Domains above the level of Kingdom: Archaea, Bacteria, and Eukarya (Figure 8.9). Analysis of small subunit rRNA gene sequences suggests archaea, bacteria, and eukaryotes all evolved from a common ancestral cell type. The tree is skewed to show a closer evolutionary relationship between Archaea and Eukarya than they have to Bacteria.

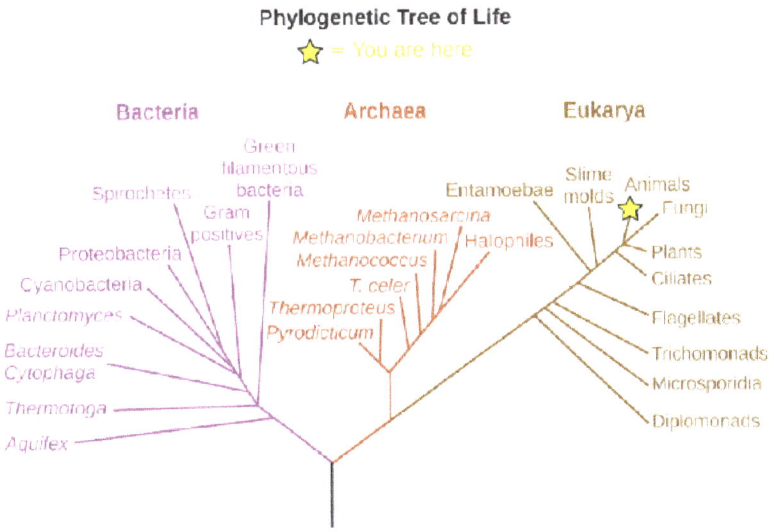

Figure 8.9 Woese and Fox's phylogenetic tree contains three domains: Bacteria, Archaea, and Eukarya. Domains Archaea and Bacteria contain all prokaryotic organisms, and Eukarya contains all eukaryotic organisms. (credit: modification of work by Eric Gaba)

Scientists continue to use analysis of RNA, DNA, and proteins to determine how organisms are related. One interesting, and complicating, discovery is that of horizontal gene transfer—when a gene of one species is absorbed into another organism's genome. Horizontal gene transfer is especially common in microorganisms and can make it difficult to determine how organisms are evolutionarily related. Consequently, some scientists now think in terms of "webs of life" rather than "trees of life."

- In modern taxonomy, how do scientists determine how closely two organisms are related?
- Explain why the branches on the "tree of life" all originate from a single "trunk."

Naming Microbes

In developing his taxonomy, Linnaeus used a system of **binomial nomenclature**, a two-word naming system for identifying organisms by genus and specific epithet. For example, modern humans are in the genus *Homo* and have the specific epithet name *sapiens*, so their scientific name in binomial nomenclature is *Homo sapiens*. In binomial nomenclature, the genus part of the name is always capitalized; it is followed by the specific epithet name, which is not capitalized. Both names are italicized. When referring to the species of humans, the binomial nomenclature would be *Homo sapiens*.

Taxonomic names in the 18th through 20th centuries were typically derived from Latin, since that was the common language used by scientists when taxonomic systems were first created. Today, newly discovered organisms can be given names derived from Latin, Greek, or English. Sometimes these names reflect some distinctive trait of the organism; in other cases, microorganisms are named after the scientists who discovered them. The archaeon *Haloquadratum walsbyi* is an example of both of these naming schemes. The genus, *Haloquadratum*, describes the microorganism's saltwater habitat (*halo* is derived from the Greek word for "salt") as well as the arrangement of its square cells, which are arranged in square clusters of four cells (*quadratum* is Latin for "foursquare"). The species, *walsbyi*, is named after Anthony Edward Walsby, the microbiologist who discovered *Haloquadratum walsbyi* in 1980. While it might seem easier to give an organism a common descriptive name—like a red-headed woodpecker—we can imagine how that could become problematic. What happens when another species of woodpecker with red head colouring is discovered? The systematic nomenclature scientists use eliminates this potential problem by assigning each organism a single, unique two-word name that is recognized by scientists all over the world.

In this text, we will typically abbreviate an organism's genus and species after its first mention. The abbreviated form is simply the first initial of the genus, followed by a period and the full name of the species. For example, the bacterium

Escherichia coli is shortened to *E. coli* in its abbreviated form. You will encounter this same convention in other scientific texts as well.

Bergey's Manuals

Whether in a tree or a web, microbes can be difficult to identify and classify. Without easily observable macroscopic features like feathers, feet, or fur, scientists must capture, grow, and devise ways to study their biochemical properties to differentiate and classify microbes. Despite these hurdles, a group of microbiologists created and updated a set of manuals for identifying and classifying microorganisms. First published in 1923 and since updated many times, *Bergey's Manual of Determinative Bacteriology* and *Bergey's Manual of Systematic Bacteriology* are the standard references for identifying and classifying different prokaryotes. (Appendix D (https://legacy.cnx.org/content/m58949/latest/) of this textbook is partly based on Bergey's manuals; it shows how the organisms that appear in this textbook are classified.) Because so many bacteria look identical, methods based on nonvisual characteristics must be used to identify them. For example, biochemical tests can be used to identify chemicals unique to certain species. Likewise, serological tests can be used to identify specific antibodies that will react against the proteins found in certain species. Ultimately, DNA and rRNA sequencing can be used both for identifying a particular bacterial species and for classifying newly discovered species.

Same Name, Different Strain

Within one species of microorganism, there can be several subtypes called strains. While different strains may be nearly identical genetically, they can have very different attributes. The bacterium *Escherichia coli* is infamous for causing food poisoning and traveler's diarrhea. However, there are actually many different strains of *E. coli*, and they vary in their ability to cause disease.

One pathogenic (disease-causing) *E. coli* strain that you may have heard of is *E. coli* O157:H7. In humans, infection from *E. coli* O157:H7 can cause abdominal cramps and diarrhea. Infection usually originates from contaminated water or food, particularly raw vegetables and undercooked meat. In the 1990s, there were several large outbreaks of *E. coli* O157:H7 thought to have originated in undercooked hamburgers.

While *E. coli* O157:H7 and some other strains have given *E. coli* a bad name, most *E. coli* strains do not cause disease. In fact, some can be helpful. Different strains of *E. coli* found naturally in our gut help us digest our food, provide us with some needed chemicals, and fight against pathogenic microbes.

Learn more about phylogenetic trees by exploring the Wellcome Trust's interactive Tree of Life. The website (https://www.openstax.org/l/22wellcome)

contains information, photos, and animations about many different organisms. Select two organisms to see how they are evolutionarily related.

TYPES OF MICROORGANISMS

Most microbes are unicellular and small enough that they require artificial magnification to be seen. However, there are some unicellular microbes that are visible to the naked eye, and some multicellular organisms that are microscopic. An object must measure about 100 micrometers (μm) to be visible without a microscope, but most microorganisms are many times smaller than that. For some perspective, consider that a typical animal cell measures roughly 10 μm across but is still microscopic. Bacterial cells are typically about 1 μm, and viruses can be 10 times smaller than bacteria (Figure 8.10). See Table 8.1 for units of length used in microbiology.

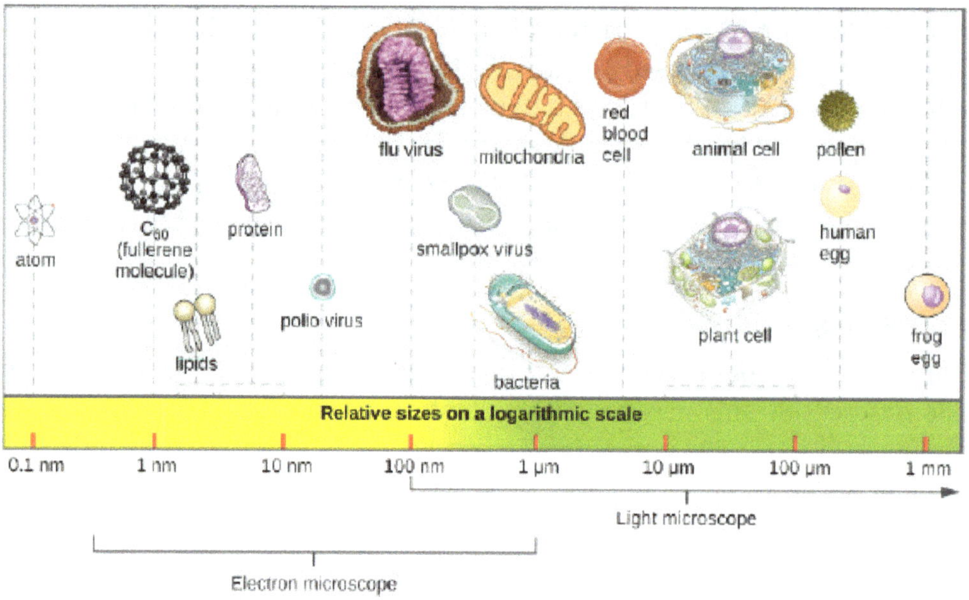

Figure 8.10 The relative sizes of various microscopic and nonmicroscopic objects. Note that a typical virus measures about 100 nm, 10 times smaller than a typical bacterium (~1 μm), which is at least 10 times smaller than a typical plant or animal cell (~10–100 μm). An object must measure about 100 μm to be visible without a microscope.

Microorganisms differ from each other not only in size, but also in structure, habitat, metabolism, and many other characteristics. While we typically think of microorganisms as being unicellular, there are also many multicellular organisms that are too small to be seen without a microscope. Some microbes, such as viruses, are even **acellular** (not composed of cells).

Table 8.1 Units of Length Commonly Used in Microbiology

Metric Unit	Meaning of Prefix	Metric Equivalent
meter (m)	—	1 m = 10^0 m
decimeter (dm)	1/10	1 dm = 0.1 m = 10^{-1} m
centimeter (cm)	1/100	1 cm = 0.01 m = 10^{-2} m
millimeter (mm)	1/1000	1 mm = 0.001 m = 10^{-3} m
micrometer (μm)	1/1,000,000	1 μm = 0.000001 m = 10^{-6} m
nanometer (nm)	1/1,000,000,000	1 nm = 0.000000001 m = 10^{-9} m

Microorganisms are found in each of the three domains of life: Archaea, Bacteria, and Eukarya. Microbes within the domains Bacteria and Archaea are all prokaryotes (their cells lack a nucleus), whereas microbes in the domain Eukarya are eukaryotes (their cells have a nucleus). Some microorganisms, such as viruses, do not fall within any of the three domains of life. In this section, we will briefly introduce each of the broad groups of microbes. Later chapters will go into greater depth about the diverse species within each group.

Prokaryotic Microorganisms

Bacteria are found in nearly every habitat on earth, including within and on humans. Most bacteria are harmless or helpful, but some are **pathogens**, causing disease in humans and other animals. Bacteria are prokaryotic because their genetic material (DNA) is not housed within a true nucleus. Most bacteria have cell walls that contain peptidoglycan.

Bacteria are often described in terms of their general shape. Common shapes include spherical (coccus), rod- shaped (bacillus), or curved (spirillum, spirochete, or vibrio). Figure 8.11 shows examples of these shapes.

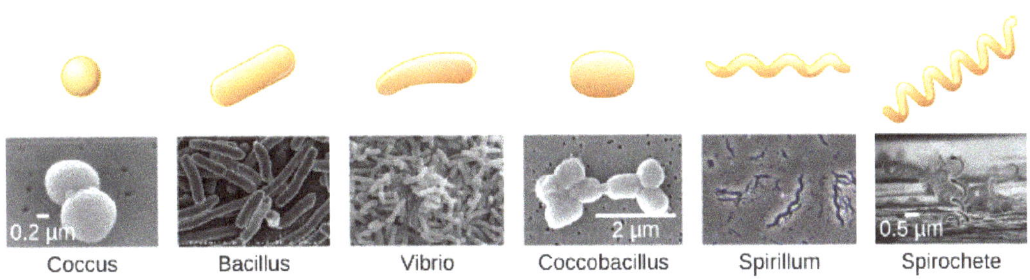

Figure 8.11 Common bacterial shapes. Note how coccobacillus is a combination of spherical (coccus) and rod- shaped (bacillus). (credit "Coccus": modification of work by Janice Haney Carr, Centers for Disease Control and Prevention; credit "Coccobacillus": modification of work by Janice Carr, Centers for Disease Control and Prevention; credit "Spirochete": Centers for Disease Control and Prevention)

They have a wide range of metabolic capabilities and can grow in a variety of environments, using different combinations of nutrients. Some bacteria are photosynthetic, such as oxygenic cyanobacteria and anoxygenic green sulfur and green nonsulfur bacteria; these bacteria use energy derived from sunlight, and fix carbon dioxide for growth. Other types of bacteria are non-photosynthetic, obtaining their energy from organic or inorganic compounds in their environment.

Archaea are also unicellular prokaryotic organisms. Archaea and bacteria have different evolutionary histories, as well as significant differences in genetics, metabolic pathways, and the composition of their cell walls and membranes. Unlike most bacteria, archaeal cell walls do not contain peptidoglycan, but their cell walls are often composed of a similar substance called pseudopeptidoglycan. Like bacteria, archaea are found in nearly every habitat on earth, even extreme environments that are very cold, very hot, very basic, or very acidic (Figure 8.12). Some archaea live in the human body, but none have been shown to be human pathogens.

Figure 8.12 Some archaea live in extreme environments, such as the Morning Glory pool, a hot spring in Yellowstone National Park. The colour differences in the pool result from the different communities of microbes that are able to thrive at various water temperatures.

Eukaryotic Microorganisms

The domain Eukarya contains all eukaryotes, including uni- or multicellular eukaryotes such as protists, fungi, plants, and animals. The major defining characteristic of eukaryotes is that their cells contain a nucleus.

Protists

Protists are an informal grouping of eukaryotes that are not plants, animals, or fungi. Some algae are protists and others are bacteria; all protozoa are examples of protists.

Algae (singular: alga) are mostly made up of protists that can be either unicellular or multicellular and vary widely in size, appearance, and habitat (Figure 8.13). Algal protists are surrounded by cell walls made of cellulose, a type of carbohydrate. Algae are photosynthetic organisms that extract energy from the sun and release oxygen and carbohydrates into their environment. Cyanobacteria, a type of bacteria, is also considered an algae, but these organisms are bacterial prokaryotes and therefore have a peptidoglycan-based cell wall, unlike the cellulose-based cell wall of the algal protists. Because other organisms can use the waste products of all algae for energy, algae are important parts of many ecosystems. Many consumer products contain ingredients derived from algae, such as carrageenan or alginic acid, which are found in some brands of ice cream, salad dressing, beverages, lipstick, and toothpaste. A derivative of algae also plays a prominent role in the microbiology laboratory. Agar, a gel derived from algae, can be mixed with various nutrients and used to grow microorganisms in a Petri dish. Algae are also being developed as a possible source for biofuels.

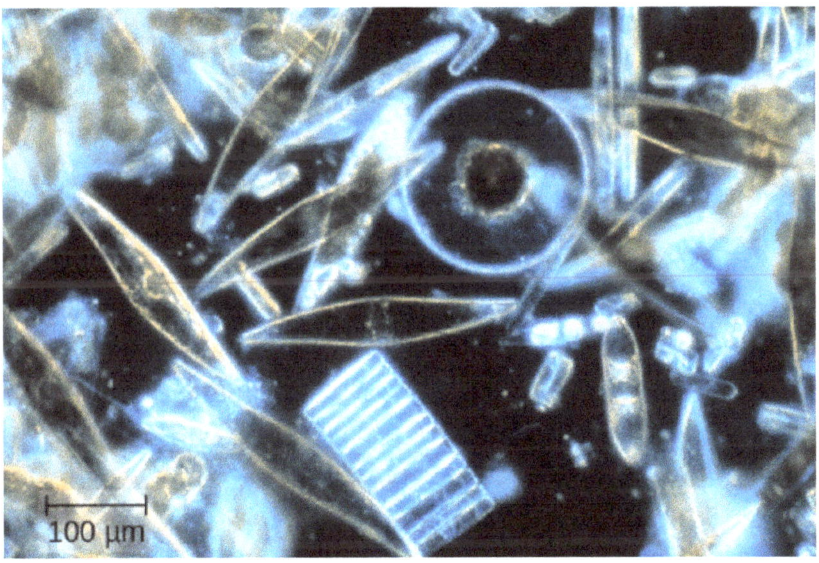

Figure 8.13 Assorted diatoms, a kind of algae, live in annual sea ice in McMurdo Sound, Antarctica. Diatoms range in size from 2 µm to 200 µm and are visualized here using light microscopy. (credit: modification of work by National Oceanic and Atmospheric Administration)

Protozoa (singular: protozoan) are protists that make up the backbone of many food webs by providing nutrients for other organisms. Protozoa are very diverse. Some protozoa move with help from hair-like structures called cilia or whip-like structures called flagella. Others extend part of their cell membrane and cytoplasm to propel themselves forward. These cytoplasmic extensions are called pseudopods ("false feet"). Some protozoa are photosynthetic; others feed on organic material. Some are free-living, whereas others are parasitic, only able to survive by extracting nutrients from a host organism. Most protozoa are harmless, but some are pathogens that can cause disease in animals or humans (Figure 8.14).

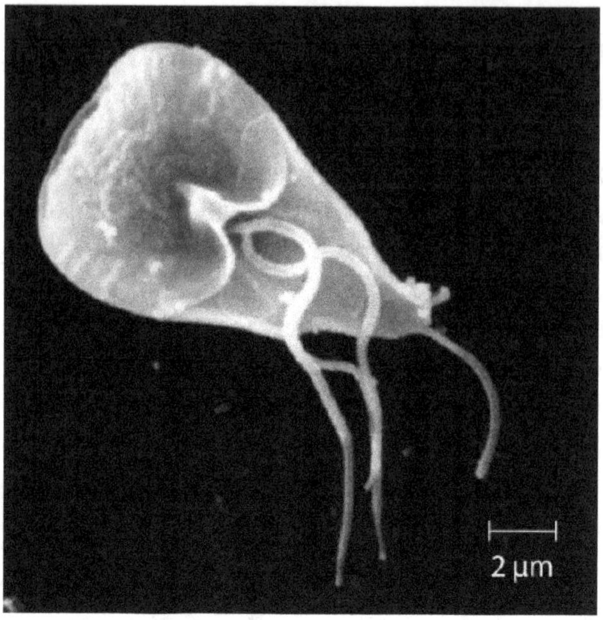

Figure 8.14 ***Giardia lamblia*****, an intestinal protozoan parasite that infects humans and other mammals, causing severe diarrhea. (credit: modification of work by Centers for Disease Control and Prevention)**

Fungi

Fungi (singular: fungus) are also eukaryotes. Some multicellular fungi, such as mushrooms, resemble plants, but they are actually quite different. Fungi are not photosynthetic, and their cell walls are usually made out of chitin rather than cellulose.

Unicellular fungi—yeasts—are included within the study of microbiology. There are more than 1000 known species. Yeasts are found in many different environments, from the deep sea to the human navel. Some yeasts have beneficial uses, such as causing bread to rise and beverages to ferment; but yeasts can also cause food to spoil. Some even cause diseases, such as vaginal yeast infections and oral thrush (Figure 8.15).

Other fungi of interest to microbiologists are multicellular organisms called molds. Molds are made up of long filaments that form visible colonies (Figure 8.16). Molds are found in many different environments, from soil to rotting food to dank bathroom corners. Molds play a critical role in the decomposition of dead plants and animals. Some molds can cause allergies, and others produce disease-causing metabolites called mycotoxins. Molds have been used to make pharmaceuticals, including penicillin, which is one of the most commonly prescribed antibiotics, and cyclosporine, used to prevent organ rejection following a transplant.

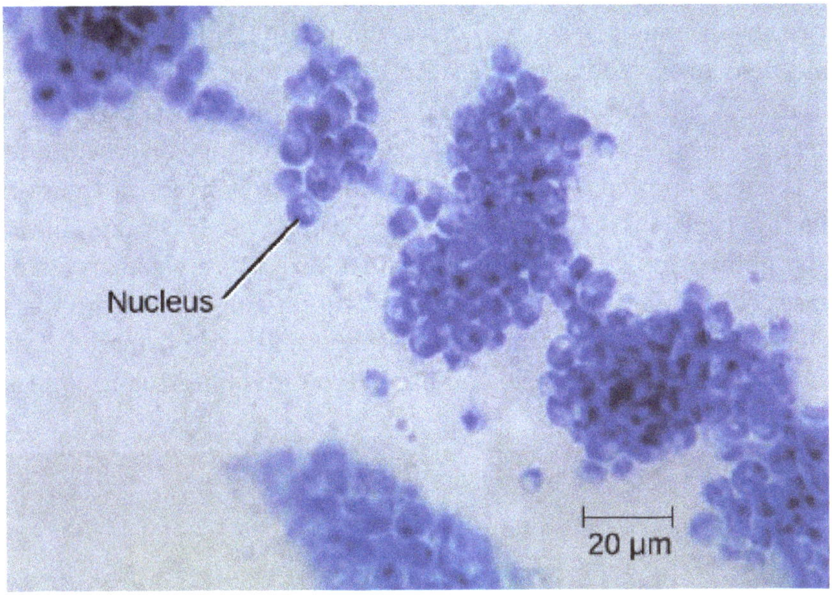

Figure 8.15 ***Candida albicans*** **is a unicellular fungus, or yeast. It is the causative agent of vaginal yeast infections as well as oral thrush, a yeast infection of the mouth that commonly afflicts infants.** ***C. albicans*** **has a morphology similar to that of coccus bacteria; however, yeast is a eukaryotic organism (note the nuclei) and is much larger. (credit: modification of work by Centers for Disease Control and Prevention)**

Figure 8.16 Large colonies of microscopic fungi can often be observed with the naked eye, as seen on the urface of these moldy oranges.

Helminths

Multicellular parasitic worms called **helminths** are not technically microorganisms, as most are large enough to see without a microscope. However, these worms fall within the field of microbiology because diseases caused by helminths involve microscopic eggs and larvae. One example of a helminth is the guinea worm, or *Dracunculus medinensis*, which causes dizziness, vomiting, diarrhea, and painful ulcers on the legs and feet when the worm works its way out of the skin (Figure 8.17). Infection typically occurs after a person drinks water containing water fleas infected by guinea-worm larvae. In the mid-1980s, there were an estimated 3.5 million cases of guinea-worm disease, but the disease has been largely eradicated. In 2014, there were only 126 cases reported, thanks to the coordinated efforts of the World Health Organization (WHO) and other groups committed to improvements in drinking water sanitation.

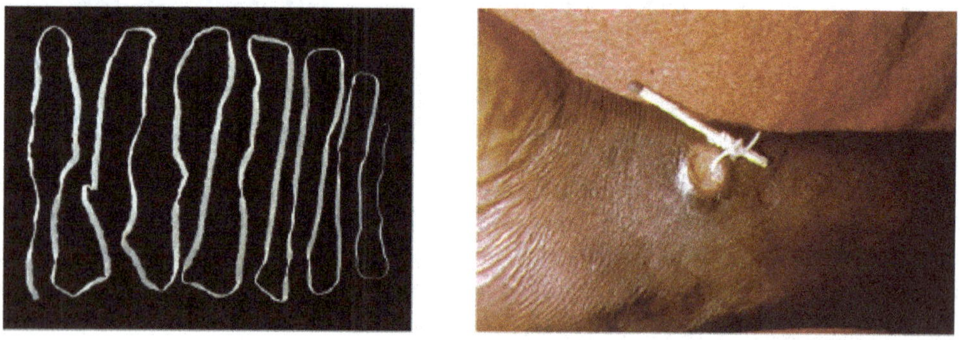

(a) (b)

Figure 8.17 (a) The beef tapeworm, *Taenia saginata*, infects both cattle and humans. *T. saginata* eggs are microscopic (around 50 μm), but adult worms like the one shown here can reach 4–10 m, taking up residence in the digestive system. (b) An adult guinea worm, *Dracunculus medinensis*, is removed through a lesion in the patient's skin by winding it around a matchstick. (credit a, b: modification of work by Centers for Disease Control and Prevention)

Viruses

Viruses are **acellular** microorganisms, which means they are not composed of cells. Essentially, a virus consists of proteins and genetic material—either DNA or RNA, but never both—that are inert outside of a host organism. However, by incorporating themselves into a host cell, viruses are able to co-opt the host's cellular mechanisms to multiply and infect other hosts.

Viruses can infect all types of cells, from human cells to the cells of other microorganisms. In humans, viruses are responsible for numerous diseases, from the common cold to deadly Ebola (Figure 8.18). However, many viruses do not cause disease.

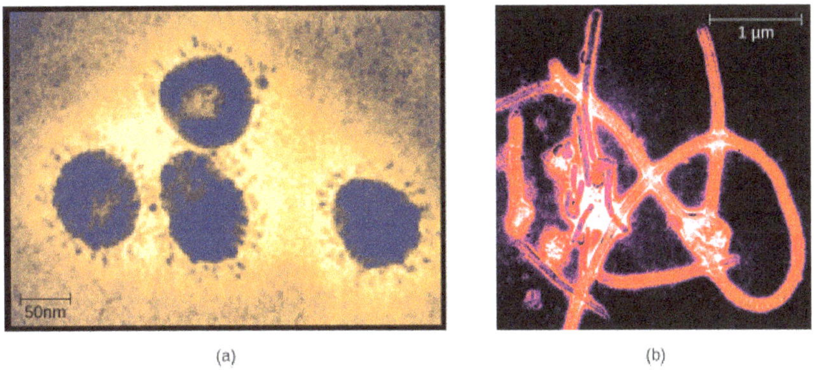

Figure 8.18 (a) Members of the Corona virus family can cause respiratory infections like the common cold, severe acute respiratory syndrome (SARS), and Middle East respiratory syndrome (MERS). Here they are viewed under a transmission electron microscope (TEM). (b) Ebolavirus, a member of the Filovirus family, as visualized using a TEM. (credit a: modification of work by Centers for Disease Control and Prevention; credit b: modification of work by Thomas W. Geisbert)

Microbiology as a Field of Study

Microbiology is a broad term that encompasses the study of all different types of microorganisms. But in practice, microbiologists tend to specialize in one of several subfields. For example, **bacteriology** is the study of bacteria; **mycology** is the study of fungi; **protozoology** is the study of protozoa; **parasitology** is the study of helminths and other parasites; and **virology** is the study of viruses (Figure 8.19).

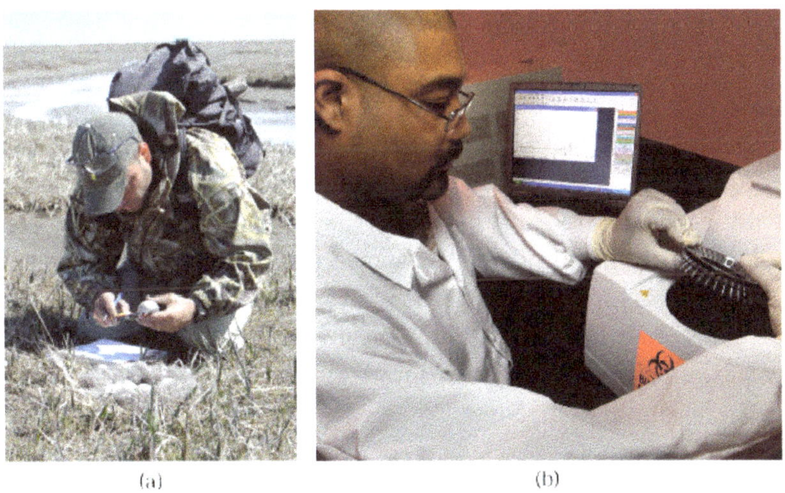

Figure 8.19 (a) A virologist samples eggs from this nest to be tested for the influenza A virus, which causes avian flu in birds. (b) A biologist performs a procedure to identify an organism that causes ulcerations in humans (credit a: U.S. Fish and Wildlife Service; credit b: James Gathany / CDC; Public Domain)

Immunology, the study of the immune system, is often included in the study of microbiology because host–pathogen interactions are central to our understanding of infectious disease processes. Microbiologists can also specialize in certain areas of microbiology, such as clinical microbiology, environmental microbiology, applied microbiology, or food microbiology.

In this textbook, we are primarily concerned with clinical applications of microbiology, but since the various subfields of microbiology are highly interrelated, we will often discuss applications that are not strictly clinical.

Bioethics in Microbiology

In the 1940s, the U.S. government was looking for a solution to a medical problem: the prevalence of sexually transmitted diseases (STDs) among soldiers. Several now-infamous government-funded studies used human subjects to research common STDs and treatments. In one such study, American researchers intentionally exposed more than 1300 human subjects in Guatemala to syphilis, gonorrhea, and chancroid to determine the ability of penicillin and other antibiotics to combat these diseases. Subjects of the study included Guatemalan soldiers, prisoners, prostitutes, and psychiatric patients—none of whom were informed that they were taking part in the study. Researchers exposed subjects to STDs by various methods, from facilitating intercourse with infected prostitutes to inoculating subjects with the bacteria known to cause the diseases. This latter method involved making a small wound on the subject's genitals or elsewhere on the body, and then putting bacteria directly into the wound. In 2011, a U.S. government commission tasked with investigating the experiment revealed that only some of the subjects were treated with penicillin, and 83 subjects died by 1953, likely as a result of the study.

Unfortunately, this is one of many horrific examples of microbiology experiments that have violated basic ethical standards. Even if this study had led to a life-saving medical breakthrough (it did not), few would argue that its methods were ethically sound or morally justifiable. But not every case is so clear cut. Professionals working in clinical settings are frequently confronted with ethical dilemmas, such as working with patients who decline a vaccine or life-saving blood transfusion. These are just two examples of life- and-death decisions that may intersect with the religious and philosophical beliefs of both the patient and the health-care professional.

No matter how noble the goal, microbiology studies and clinical practice must be guided by a certain set of ethical principles. Studies must be done with integrity. Patients and research subjects provide informed consent (not only agreeing to be treated or studied but demonstrating an understanding of the purpose of the study and any risks involved). Patients' rights must be respected. Procedures must be approved by an institutional review board. When working with patients, accurate record-keeping, honest communication, and confidentiality are paramount. Animals used for research must be treated humanely, and all protocols must be approved by an institutional animal care and use committee. These are just a few of the ethical principles explored in the *Eye on Ethics* boxes throughout this book.

Resolution

Cora's CSF samples show no signs of inflammation or infection, as would be expected with a viral infection. However, there is a high concentration of a particular protein, 14-3-3 protein, in her CSF. An electroencephalogram (EEG) of her brain function is also abnormal. The EEG resembles that of a patient with a neurodegenerative disease like Alzheimer's or Huntington's, but Cora's rapid cognitive decline is not consistent with either of these. Instead, her doctor concludes that Cora has Creutzfeldt-Jakob disease (CJD), a type of transmissible spongiform encephalopathy (TSE).

CJD is an extremely rare disease, with only about 300 cases in the United States each year. It is not caused by a bacterium, fungus, or virus, but rather by prions—which do not fit neatly into any particular category of microbe. Like viruses, prions are not found on the tree of life because they are acellular. Prions are extremely small, about one-tenth the size of a typical virus. They contain no genetic material and are composed solely of a type of abnormal protein.

CJD can have several different causes. It can be acquired through exposure to the brain or nervous-system tissue of an infected person or animal. Consuming meat from an infected animal is one way such exposure can occur. There have also been rare cases of exposure to CJD through contact with contaminated surgical equipment and from cornea and growth-hormone donors who unknowingly had CJD. In rare cases, the disease results from a specific genetic mutation that can sometimes be hereditary. However, in approximately 85% of patients with CJD, the cause of the disease is spontaneous (or sporadic) and has no identifiable cause.Based on her symptoms and their rapid progression, Cora is diagnosed with sporadic CJD.

Unfortunately for Cora, CJD is a fatal disease for which there is no approved treatment. Approximately 90% of patients die within 1 year of diagnosis.Her doctors focus on limiting her pain and cognitive symptoms as her disease progresses. Eight months later, Cora dies. Her CJD diagnosis is confirmed with a brain autopsy.

BENEFICIAL PROKARYOTES

Fortunately, only a few species of prokaryotes are pathogenic! Prokaryotes also interact with humans and other organisms in a number of ways that are beneficial. For example, prokaryotes are major participants in the carbon and nitrogen cycles. They produce or process nutrients in the digestive tracts of humans and other animals. Prokaryotes are used in the production of some human foods, and also have been recruited for the degradation of hazardous materials. In fact, our life would not be possible without prokaryotes!

Cooperation Between Bacteria and Eukaryotes: Nitrogen Fixation

Nitrogen is a very important element to living things, because it is part of nucleotides and amino acids that are the building blocks of nucleic acids and proteins, respectively. Nitrogen is usually the most limiting element in terrestrial ecosystems, with atmospheric nitrogen, N_2, providing the largest pool of available nitrogen. However, eukaryotes cannot use atmospheric, gaseous nitrogen to synthesize macromolecules. Fortunately, nitrogen can be "fixed," meaning it is converted into a more accessible form—ammonia (NH_3)—either biologically or abiotically.

Abiotic nitrogen fixation occurs as a result of physical processes such as lightning or by industrial processes. **Biological nitrogen fixation** (BNF) is exclusively carried out by prokaryotes: soil bacteria, cyanobacteria, and *Frankia* spp. (filamentous bacteria interacting with actinorhizal plants such as alder, bayberry, and sweet fern). After photosynthesis, BNF is the most important biological process on Earth. The overall nitrogen fixation equation below represents a series of *redox reactions* (Pi stands for inorganic phosphate).

$$N_2 + 16ATP + 8e^- + 8H^+ \rightarrow 2NH_3 + 16ADP + 16Pi + H_2$$

The total fixed nitrogen through BNF is about 100 to 180 million metric tons per year, which contributes about 65 percent of the nitrogen used in agriculture.

Cyanobacteria are the most important nitrogen fixers in aquatic environments. In soil, members of the genera *Clostridium* and *Azotobacter* are examples of free-living, nitrogen-fixing bacteria. Other bacteria live symbiotically with legume plants, providing the most important source of fixed nitrogen. Symbionts may fix more nitrogen in soils than free-living organisms by a factor of 10. Soil bacteria, collectively called rhizobia, are able to symbiotically interact with legumes to form **nodules**, specialized structures where nitrogen fixation occurs (Figure 8.20). *Nitrogenase*, the enzyme that fixes nitrogen, is inactivated by oxygen, so the nodule provides an oxygen-free area for nitrogen fixation to take place. The oxygen is sequestered by a form of plant hemoglobin called *leghemoglobin*, which protects the *nitrogenase*, but releases enough oxygen to support respiratory activity.

Symbiotic nitrogen fixation provides a natural and inexpensive plant fertilizer: It reduces atmospheric nitrogen to ammonia, which is easily usable by plants. The use of legumes is an excellent alternative to chemical fertilization and is of special interest to *sustainable agriculture*, which seeks to minimize the use of chemicals and conserve natural resources. Through symbiotic nitrogen fixation, the plant benefits from using an endless source of nitrogen: the atmosphere.

The bacteria benefit from using photosynthates (carbohydrates produced during photosynthesis) from the plant and having a protected niche. In addition, the soil benefits from being naturally fertilized. Therefore, the use of rhizobia as biofertilizers is a sustainable practice.

Why are legumes so important? Some, like soybeans, are key sources of agricultural protein. Some of the most important legumes consumed by humans are soybeans, peanuts, peas, chickpeas, and beans. Other legumes, such as alfalfa, are used to feed cattle.

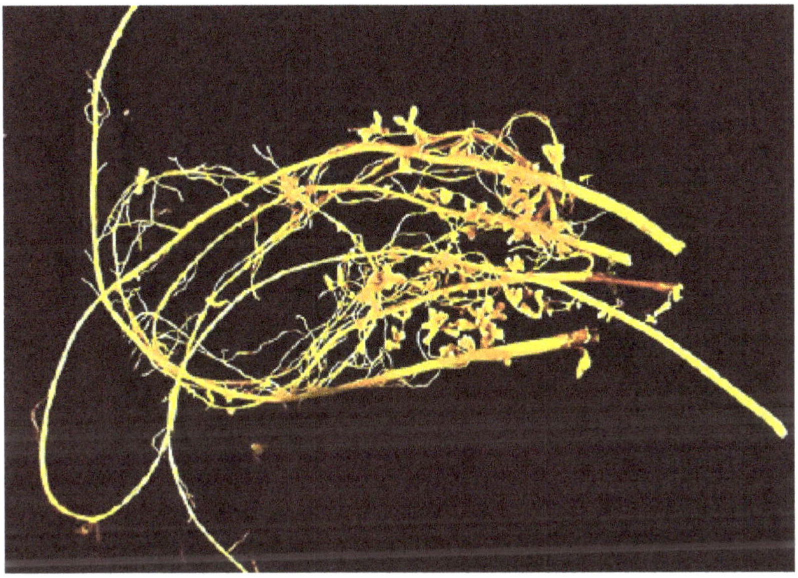

Figure 8.20 Nitrogen-fixation nodules on soybean roots. Soybean (*Glycine max*) is a legume that interacts symbiotically with the soil bacterium *Bradyrhizobium japonicum* to form specialized structures on the roots called *nodules* where nitrogen fixation occurs. (credit: USDA)

Microbes on the Human Body

The commensal bacteria that inhabit our skin and gastrointestinal tract do a host of good things for us. They protect us from pathogens, help us digest our food, and produce some of our vitamins and other nutrients. These activities have been known for a long time. More recently, scientists have gathered evidence that these bacteria may also help regulate our moods, influence our activity levels, and even help control weight by affecting our food choices and absorption patterns. The Human Microbiome Project has begun the process of cataloging our normal bacteria (and archaea) so we can better understand these functions.

A particularly fascinating example of our normal flora relates to our digestive systems. People who take high doses of antibiotics tend to lose many of their normal gut bacteria, allowing a naturally antibiotic- resistant species called *Clostridium difficile* to overgrow and cause severe gastric problems, especially chronic diarrhea (Figure 8.21). Obviously, trying to treat this problem with antibiotics only makes it worse. However, it has been successfully treated by giving the patients fecal transplants from healthy donors to reestablish the normal intestinal microbial community. Clinical trials are underway to ensure the safety and effectiveness of this technique.

Figure 8.21 *Clostridium difficile*. This scanning electron micrograph shows *Clostridium difficile*, a Gram-positive, rod-shaped bacterium that causes severe diarrhea. Infection commonly occurs after the normal gut fauna are eradicated by antibiotics, and in the hospital can be deadly to seriously ill patients. (credit: modification of work by CDC, HHS; scale-bar data from Matt Russell)

Scientists are also discovering that the absence of certain key microbes from our intestinal tract may set us up for a variety of problems. This seems to be particularly true regarding the appropriate functioning of the immune system. There are intriguing findings that suggest that the absence of these microbes is an important contributor to the development of allergies and some autoimmune disorders. Research is currently underway to test whether adding certain microbes to our internal ecosystem may help in the treatment of these problems, as well as in treating some forms of autism.

Early Biotechnology: Cheese, Bread, Wine, Beer, and Yogurt

According to the United Nations Convention on Biological Diversity, **biotechnology** is "any technological application that uses biological systems, living organisms, or derivatives thereof, to make or modify products or processes for specific use." The concept of "specific use" involves some sort of commercial application. Genetic engineering, artificial selection, antibiotic production, and cell culture are current topics of study in biotechnology and will be described in later chapters. However, humans were using prokaryotes before the term biotechnology was even coined. Some of the products of this early biotechnology are as familiar as cheese, bread, wine, beer, and yogurt, which employ both bacteria and other microbes, such as yeast, a fungus (Figure 8.22).

Cheese production began around 4,000 to 7,000 years ago when humans began to breed animals and process their milk. Fermentation in this case preserves nutrients: Milk will spoil relatively quickly, but when processed as cheese, it is more stable. As for beer, the oldest records of brewing are about 6,000 years old and were an integral part of the Sumerian culture. Evidence indicates that the Sumerians discovered fermentation by chance. Wine has been produced for about 4,500 years, and evidence suggests that cultured milk products, like yogurt, have existed for at least 4,000 years.

Figure 8.22 Some foods produced by microorganisms. Some of the products derived from the use of prokaryotes in early biotechnology include (a) cheese, (b) wine, (c) beer and bread, and (d) yogurt. (credit bread: modification of work by F. Rodrigo/Wikimedia Commons; credit wine: modification of work by Jon Sullivan; credit beer and bread: modification of work by Kris Miller; credit yogurt: modification of work by Jon Sullivan)

Using Prokaryotes to Clean up Our Planet: Bioremediation

Microbial **bioremediation** is the use of prokaryotes (or microbial metabolism) to remove pollutants. Bioremediation has been used to remove agricultural chemicals (e.g., pesticides, fertilizers) that leach from soil into groundwater and the subsurface. Certain toxic metals and oxides, such as selenium and arsenic compounds, can also be removed from water by bioremediation. The reduction of SeO^{-2} to SeO^{-2} and to Se^0 (metallic selenium) is a method used to remove selenium ions from water. Mercury (Hg) is an example of a toxic metal that can be removed from an environment by bioremediation. As an active ingredient of some pesticides, mercury is used in industry and is also a by-product of certain processes, such as battery production. Methyl mercury is usually present in very low concentrations in natural environments, but it is highly toxic because it accumulates in living tissues. Several species of bacteria can carry out the biotransformation of toxic mercury into nontoxic forms. These bacteria, such as *Pseudomonas aeruginosa*, can convert Hg^{+2} into Hg^{0}, which is nontoxic to humans.

One of the most useful and interesting examples of the use of prokaryotes for bioremediation purposes is the cleanup of oil spills. The significance of prokaryotes to petroleum bioremediation has been demonstrated in several oil spills in recent years, such as the Exxon Valdez spill in Alaska (1989) (Figure 8.23), the Prestige oil spill in Spain (2002), the spill into the Mediterranean from a Lebanon power plant (2006), and more recently, the BP oil spill in the Gulf of Mexico (2010). In the case of oil spills in the ocean, ongoing natural bioremediation tends to occur, since there are oil-consuming bacteria in the ocean prior to the spill. In addition to these naturally occurring oil-degrading bacteria, humans select and engineer bacteria that possess the same capability with increased efficacy and spectrum of hydrocarbon compounds that can be processed. Bioremediation is enhanced by the addition of inorganic nutrients that help bacteria to grow.

Some hydrocarbon-degrading bacteria feed on hydrocarbons in the oil droplet, breaking down the hydrocarbons into smaller subunits. Some species, such as *Alcanivorax borkumensis*, produce surfactants that *solubilize* the oil (making it soluble in water), whereas other bacteria degrade the oil into carbon dioxide. Under ideal conditions, it has been reported that up to 80 percent of the nonvolatile components in oil can be degraded within one year of the spill. Other oil fractions containing aromatic and highly branched hydrocarbon chains are more difficult to remove and remain in the environment for longer periods of time.

(a) (b)

Figure 8.23 Prokaryotes and bioremediation. (a) Cleaning up oil after the Exxon Valdez spill in Alaska, workers hosed oil from beaches and then used a floating boom to corral the oil, which was finally skimmed from the water surface. Some species of bacteria are able to solubilize and degrade the oil. (b) One of the most catastrophic consequences of oil spills is the damage to fauna. (credit a: modification of work by NOAA; credit b: modification of work by GOLUBENKOV, NGO: Saving Taman)

CHAPTER-9

MICROBIAL TECHNIQUES

INTRODUCTION: AN OVERVIEW

In the dynamic field of microbiology, the ability to manipulate and study microbes effectively is a cornerstone of scientific discovery and technological advancement. Microbial techniques encompass a diverse array of methodologies designed to explore the intricacies of microscopic life, from bacteria and archaea to fungi and viruses. These techniques not only unravel the mysteries of microbial structure, function, and behaviour but also contribute substantially to fields such as medicine, agriculture, and environmental science. As we delve into the realm of microbial techniques, we embark on a journey that unveils the tools and methodologies enabling scientists to unlock the secrets of the microbial world, ultimately paving the way for ground breaking innovations and a deeper understanding of the role microbes play in shaping our world.

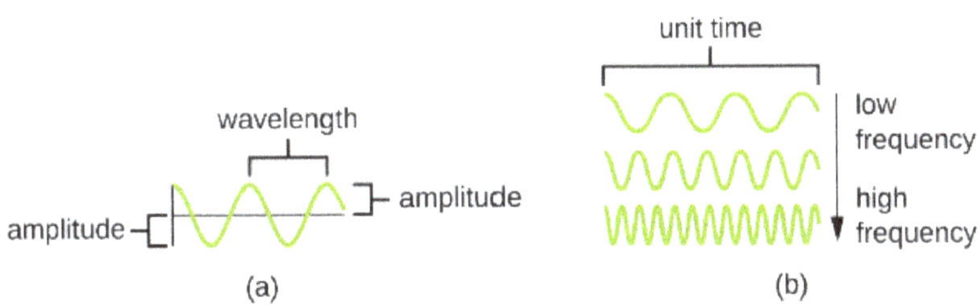

Figure 9.1 (a) The amplitude is the height of a wave, whereas the wavelength is the distance between one peak and the next. (b) These waves have different frequencies, or rates of vibration. The wave at the top has the lowest frequency, since it has the fewest peaks per unit time. The wave at the bottom has the highest frequency.

THE PROPERTIES OF LIGHT

Visible light consists of electromagnetic waves that behave like other waves. Hence, many of the properties of light that are relevant to microscopy can be understood in terms of light's behaviour as a wave. An important property of light waves is the

wavelength, or the distance between one peak of a wave and the next peak. The height of each peak (or depth of each trough) is called the **amplitude.** In contrast, the **frequency** of the wave is the rate of vibration of the wave, or the number of wavelengths within a specified time period (Figure 9.1).

Interactions of Light

Light waves interact with materials by being reflected, absorbed, or transmitted. **Reflection** occurs when a wave bounces off of a material. For example, a red piece of cloth may reflect red light to our eyes while absorbing other colours of light. **Absorbance** occurs when a material captures the energy of a light wave. In the case of glow-in-the-dark plastics, the energy from light can be absorbed and then later re-emitted as another form of phosphorescence. Transmission occurs when a wave travels through a material, like light through glass (the process of transmission is called **transmittance**). When a material allows a large proportion of light to be transmitted, it may do so because it is thinner, or more transparent (having more **transparency** and less **opacity**). Figure 9.2 illustrates the difference between transparency and opacity.

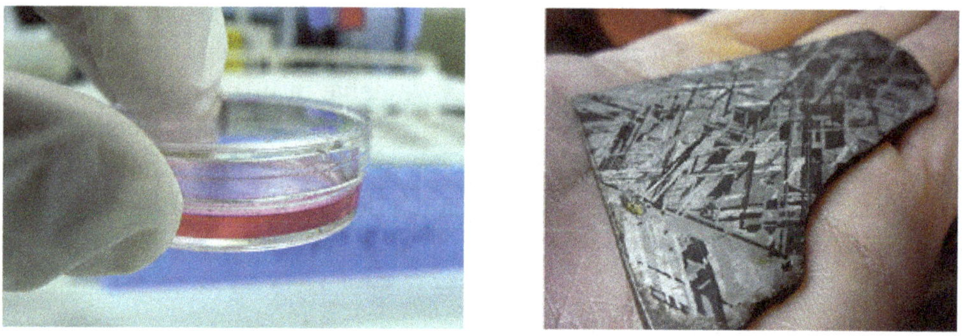

(a) (b)

Figure 9.2 (a) A Petri dish is made of transparent plastic or glass, which allows transmission of a high proportion of light. This transparency allows us to see through the sides of the dish to view the contents. (b) This slice of an iron meteorite is opaque (i.e., it has opacity). Light is not transmitted through the material, making it impossible to see the part of the hand covered by the object. (credit a: modification of work by Umberto Salvagnin; credit b: modification of work by "Waifer X"/Flickr)

Light waves can also interact with each other by **interference**, creating complex patterns of motion. Dropping two pebbles into a puddle causes the waves on the puddle's surface to interact, creating complex interference patterns. Light waves can interact in the same way.

In addition to interfering with each other, light waves can also interact with small objects or openings by bending or scattering. This is called **diffraction.**

Diffraction is larger when the object is smaller relative to the wavelength of the light (the distance between two consecutive peaks of a light wave). Often, when waves diffract in different directions around an obstacle or opening, they will interfere with each other.

Lenses and Refraction

In the context of microscopy, **refraction** is perhaps the most important behaviour exhibited by light waves. Refraction occurs when light waves change direction as they enter a new medium (Figure 9.3). Different transparent materials transmit light at different speeds; thus, light can change speed when passing from one material to another. This change in speed usually also causes a change in direction (refraction), with the degree of change dependent on the angle of the incoming light.

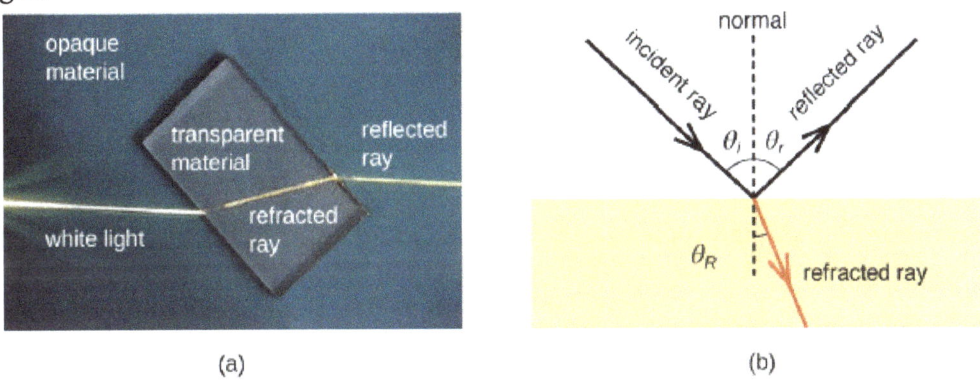

Figure 9.3 (a) Refraction occurs when light passes from one medium, such as air, to another, such as glass, changing the direction of the light rays. (b) As shown in this diagram, light rays passing from one medium to another may be either refracted or reflected.

The extent to which a material slows transmission speed relative to empty space is called the **refractive index** of that material. Large differences between the refractive indices of two materials will result in a large amount of refraction when light passes from one material to the other. For example, light moves much more slowly through water than through air, so light entering water from air can change direction greatly. We say that the water has a higher refractive index than air (Figure 9.4).

When light crosses a boundary into a material with a higher refractive index, its direction turns to be closer to perpendicular to the boundary (i.e., more toward a normal to that boundary; see Figure 9.4). This is the principle behind lenses. We can think of a lens as an object with a curved boundary (or a collection of prisms) that collects all of the light that strikes it and refracts it so that it all meets at a single point called the **image point (focus)**. A convex lens can be used to magnify because it can focus at closer range than the human eye, producing a larger image. Concave lenses and mirrors can also be used in microscopes to redirect the light path.

Figure 9.4 This straight pole appears to bend at an angle as it enters the water. This optical illusion is due to the large difference between the refractive indices of air and water.

Figure 9.5 shows the **focal point** (the image point when light entering the lens is parallel) and the **focal length** (the distance to the focal point) for convex and concave lenses.

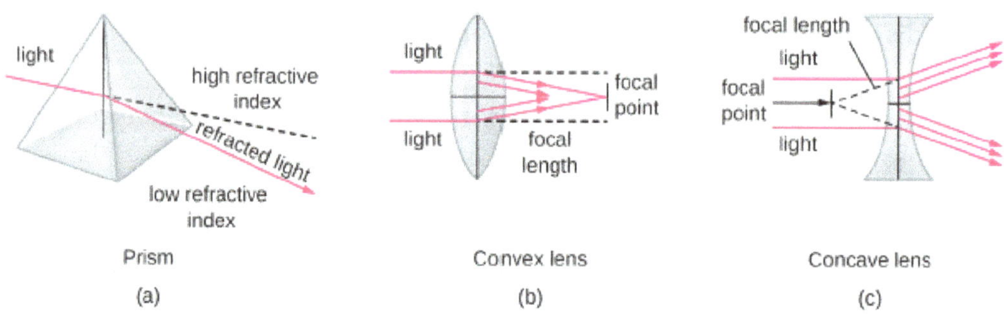

Figure 9.5 (a) A lens is like a collection of prisms, such as the one shown here. (b) When light passes through a convex lens, it is refracted toward a focal point on the other side of the lens. The focal length is the distance to the focal point. (c) Light passing through a concave lens is refracted away from a focal point in front of the lens.

The human eye contains a lens that enables us to see images. This lens focuses the light reflecting off of objects in front of the eye onto the surface of the retina, which is like a screen in the back of the eye. Artificial lenses placed in front of the eye (contact lenses, glasses, or microscopic lenses) focus light before it is focused (again) by the lens of the eye, manipulating the image that ends up on the retina (e.g., by making it appear larger).

Images are commonly manipulated by controlling the distances between the object, the lens, and the screen, as well as the curvature of the lens. For example, for a given amount of curvature, when an object is closer to the lens, the focal points are farther from the lens. As a result, it is often necessary to manipulate these distances to create a focused image on a screen. Similarly, more curvature creates image points closer to the lens and a larger image when the image is in focus. This property is often described in terms of the focal distance, or distance to the focal point.

Electromagnetic Spectrum and Colour

Visible light is just one form of electromagnetic radiation (EMR), a type of energy that is all around us. Other forms of EMR include microwaves, X-rays, and radio waves, among others. The different types of EMR fall on the electromagnetic spectrum, which is defined in terms of wavelength and frequency. The spectrum of visible light occupies a relatively small range of frequencies between infrared and ultraviolet light (Figure 9.6).

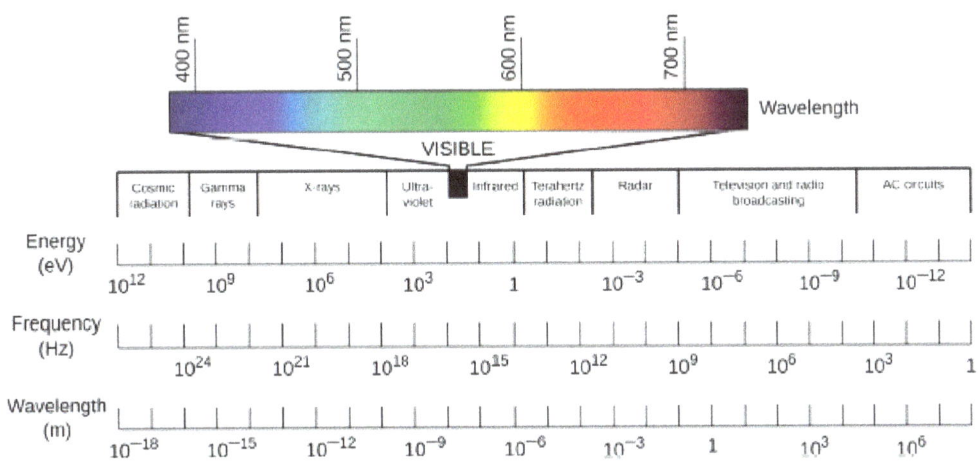

Figure 9.6 The electromagnetic spectrum ranges from high-frequency gamma rays to low-frequency radio waves. Visible light is the relatively small range of electromagnetic frequencies that can be sensed by the human eye. On the electromagnetic spectrum, visible light falls between ultraviolet and infrared light. (credit: modification of work by Johannes Ahlmann)

Whereas wavelength represents the distance between adjacent peaks of a light wave, frequency, in a simplified definition, represents the rate of oscillation. Waves with higher frequencies have shorter wavelengths and, therefore, have more oscillations per unit time than lower-frequency waves. Higher-frequency waves also contain more energy than lower-frequency waves. This energy is delivered as elementary particles called photons. Higher-frequency waves deliver more energetic photons than lower-frequency waves.

Photons with different energies interact differently with the retina. In the spectrum of visible light, each colour corresponds to a particular frequency and wavelength (Figure 9.6).The lowest frequency of visible light appears as the colour red, whereas the highest appears as the colour violet. When the retina receives visible light of many different frequencies, we perceive this as white light. However, white light can be separated into its component colours using refraction. If we pass white light through a prism, different colours will be refracted in different directions, creating a rainbow-like spectrum on a screen behind the prism. This separation of colours is called **dispersion**, and it occurs because, for a given material, the refractive index is different for different frequencies of light.

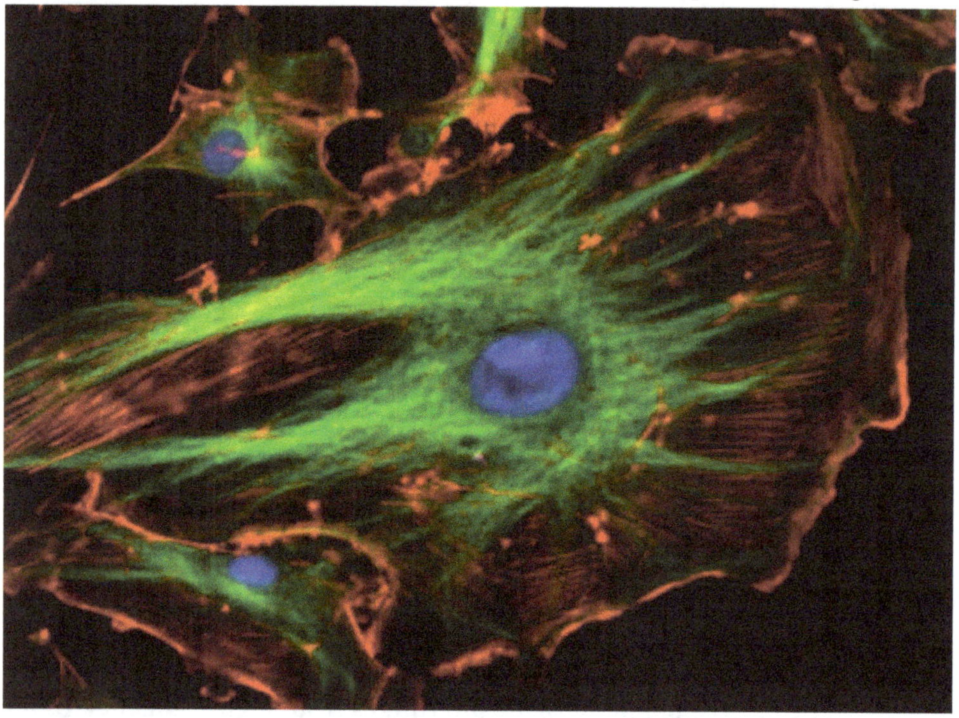

Figure 9.7 The fluorescent dyes absorbed by these bovine pulmonary artery endothelial cells emit brilliant colours when excited by ultraviolet light under a fluorescence microscope. Various cell structures absorb different dyes. The nuclei are stained blue with 4',6-diamidino-2-phenylindole (DAPI); microtubles are marked green by an antibody bound to FITC; and actin filaments are labeled red with phalloidin bound to tetramethylrhodamine (TRITC). (credit: National Institutes of Health)

Certain materials can refract non-visible forms of EMR and, in effect, transform them into visible light. Certain **fluorescent** dyes, for instance, absorb ultraviolet or blue light and then use the energy to emit photons of a different colour, giving off light rather than simply vibrating. This occurs because the energy absorption causes electrons to jump to higher energy states, after which they then almost immediately fall back down to their ground states, emitting specific amounts of energy as photons. Not all of the energy is emitted in a given photon, so the emitted photons will be of lower energy and, thus, of lower frequency than the absorbed ones. Thus, a dye such as Texas red may be excited by blue light, but emit red light; or a dye such as fluorescein isothiocyanate (FITC) may absorb (invisible) high-energy ultraviolet light and emit green light (Figure 9.7). In some materials, the photons may be emitted following a delay after absorption; in this case, the process is called **phosphorescence**. Glow-in-the-dark plastic works by using phosphorescent material.

Magnification, Resolution, and Contrast

Microscopes magnify images and use the properties of light to create useful images of small objects.

Magnification is defined as the ability of a lens to enlarge the image of an object when compared to the real object. For example, a magnification of 10× means that the image appears 10 times the size of the object as viewed with the naked eye.

Greater magnification typically improves our ability to see details of small objects, but magnification alone is not sufficient to make the most useful images. It is often useful to enhance the **resolution** of objects: the ability to tell that two separate points or objects are separate. A low-resolution image appears fuzzy, whereas a high- resolution image appears sharp. Two factors affect resolution. The first is wavelength. Shorter wavelengths are able to resolve smaller objects; thus, an electron microscope has a much higher resolution than a light microscope, since it uses an electron beam with a very short wavelength, as opposed to the long-wavelength visible light used by a light microscope. The second factor that affects resolution is **numerical aperture**, which is a measure of a lens's ability to gather light. The higher the numerical aperture, the better the resolution.

Even when a microscope has high resolution, it can be difficult to distinguish small structures in many specimens because microorganisms are relatively transparent. It is often necessary to increase **contrast** to detect different structures in a specimen. Various types of microscopes use different features of light or electrons to increase contrast—visible differences between the parts of a specimen (see Instruments of Microscopy). Additionally, dyes that bind to some structures but not others can be used to improve the contrast between images of relatively transparent objects (see Staining Microscopic Specimens).

INSTRUMENTS OF MICROSCOPY

Light Microscopy

Many types of microscopes fall under the category of light microscopes, which use light to visualize images. Examples of light microscopes include brightfield microscopes, darkfield microscopes, phase-contrast microscopes, differential interference contrast microscopes, fluorescence microscopes, confocal scanning laser microscopes, and two-photon microscopes. These various types of light microscopes can be used to complement each other in diagnostics and research.

Brightfield Microscopes

The **brightfield microscope**, perhaps the most commonly used type of microscope, is a compound microscope with two or more lenses that produce a dark image on a bright background. Some brightfield microscopes are **monocular** (having a single eyepiece), though most newer brightfield microscopes are **binocular** (having two eyepieces), like the one shown in Figure 9.8; in either case, each eyepiece contains a lens called an ocular lens. The **ocular lenses** typically magnify images 10 times (10×). At the other end of the body tube are a set of objective lenses on a rotating nosepiece. The magnification of these objective lenses typically ranges from 4 × to 100 ×, with the magnification for each lens designated on the metal casing of the lens. The ocular and objective lenses work together to create a magnified image. The **total magnification** is the product of the ocular magnification times the objective magnification:

$$\text{ocular magnification} \times \text{objective magnification}$$

For example, if a 40× objective lens is selected and the ocular lens is 10×, the total magnification would be

$$(40\times)(10\times) = 400\times$$

The item being viewed is called a specimen. The specimen is placed on a glass slide, which is then clipped into place on the **stage** (a platform) of the microscope. Once the slide is secured, the specimen on the slide is positioned over the light using the **x-y mechanical stage knobs.** These knobs move the slide on the surface of the stage, but do not raise or lower the stage. Once the specimen is centered over the light, the stage position can be raised or lowered to focus the image. The **coarse focusing knob** is used for large-scale movements with 4× and 10× objective lenses; the **fine focusing knob** is used for small-scale movements, especially with 40× or 100× objective lenses.

When images are magnified, they become dimmer because there is less light per unit area of image. Highly magnified images produced by microscopes, therefore, require intense lighting. In a brightfield microscope, this light is provided by an

illuminator, which is typically a high-intensity bulb below the stage. Light from the illuminator passes up through **condenser lens** (located below the stage), which focuses all of the light rays on the specimen to maximize illumination. The position of the condenser can be optimized using the attached condenser focus knob; once the optimal distance is established, the condenser should not be moved to adjust the brightness. If less-than-maximal light levels are needed, the amount of light striking the specimen can be easily adjusted by opening or closing a **diaphragm** between the condenser and the specimen. In some cases, brightness can also be adjusted using the **rheostat**, a dimmer switch that controls the intensity of the illuminator.

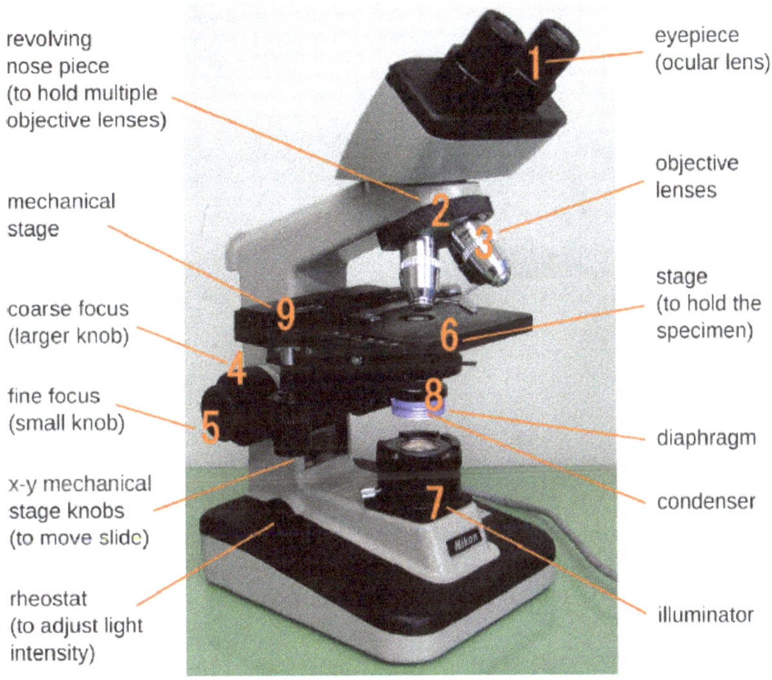

Figure 9.8 Components of a typical brightfield microscope.

A brightfield microscope creates an image by directing light from the illuminator at the specimen; this light is differentially transmitted, absorbed, reflected, or refracted by different structures. Different colors can behave differently as they interact with **chromophores** (pigments that absorb and reflect particular wavelengths of light) in parts of the specimen. Often, chromophores are artificially added to the specimen using stains, which serve to increase contrast and resolution. In general, structures in the specimen will appear darker, to various extents, than the bright background, creating maximally sharp images at magnifications up to about 1000×. Further magnification would create a larger image, but without increased resolution. This allows us to see objects as small as bacteria, which are visible at about 400× or so, but not smaller objects such as viruses.

At very high magnifications, resolution may be compromised when light passes through the small amount of air between the specimen and the lens. This is due to the large difference between the refractive indices of air and glass; the air scatters the light rays before they can be focused by the lens. To solve this problem, a drop of oil can be used to fill the space between the specimen and an **oil immersion lens**, a special lens designed to be used with immersion oils. Since the oil has a refractive index very similar to that of glass, it increases the maximum angle at which light leaving the specimen can strike the lens. This increases the light collected and, thus, the resolution of the image (Figure 9.9). A variety of oils can be used for different types of light.

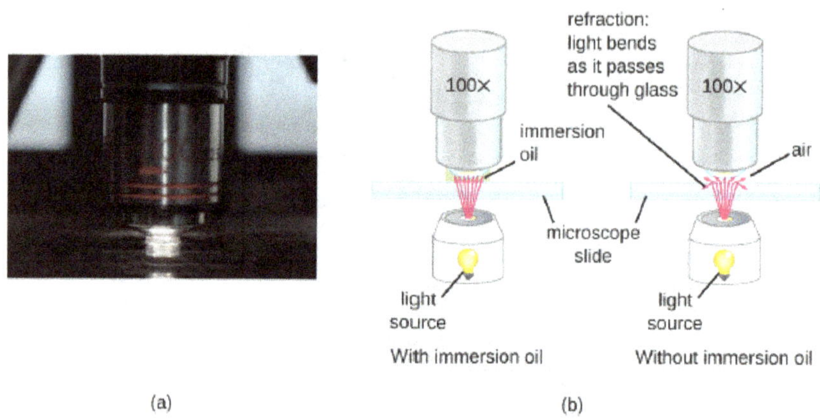

Figure 9.9 (a) Oil immersion lenses like this one are used to improve resolution. (b) Because immersion oil and glass have very similar refractive indices, there is a minimal amount of refraction before the light reaches the lens. Without immersion oil, light scatters as it passes through the air above the slide, degrading the resolution of the image.

Microscope Maintenance: Best Practices

Even a very powerful microscope cannot deliver high-resolution images if it is not properly cleaned and maintained. Since lenses are carefully designed and manufactured to refract light with a high degree of precision, even a slightly dirty or scratched lens will refract light in unintended ways, degrading the image of the specimen. In addition, microscopes are rather delicate instruments, and great care must be taken to avoid damaging parts and surfaces. Among other things, proper care of a microscope includes the following:

- Cleaning the lenses with lens paper
- Not allowing lenses to contact the slide (e.g., by rapidly changing the focus)
- Protecting the bulb (if there is one) from breakage
- Not pushing an objective into a slide

- Not using the coarse focusing knob when using the 40× or greater objective lenses
- Only using immersion oil with a specialized oil objective, usually the 100× objective
- Cleaning oil from immersion lenses after using the microscope
- Cleaning any oil accidentally transferred from other lenses
- Covering the microscope or placing it in a cabinet when not in use

Visit the online resource linked below for simulations and demonstrations involving the use of microscopes. Keep in mind that execution of specific techniques and procedures can vary depending on the specific instrument you are using. Thus, it is important to learn and practice with an actual microscope in a laboratory setting under expert supervision.

Darkfield Microscopy

A **darkfield microscope** is a brightfield microscope that has a small but significant modification to the condenser. A small, opaque disk (about 1 cm in diameter) is placed between the illuminator and the condenser lens. This opaque light stop, as the disk is called, blocks most of the light from the illuminator as it passes through the condenser on its way to the objective lens, producing a hollow cone of light that is focused on the specimen. The only light that reaches the objective is light that has been refracted or reflected by structures in the specimen. The resulting image typically shows bright objects on a dark background (Figure 9.10).

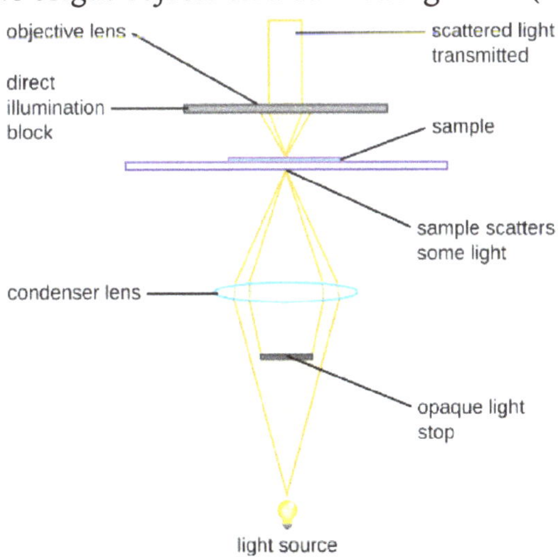

Figure 9.10 An opaque light stop inserted into a brightfield microscope is used to produce a darkfield image. The light stop blocks light traveling directly from the illuminator to the objective lens, allowing only light reflected or refracted off the specimen to reach the eye.

Darkfield microscopy can often create high-contrast, high-resolution images of specimens without the use of stains, which is particularly useful for viewing live specimens that might be killed or otherwise compromised by the stains. For example, thin spirochetes like Treponema pallidum, the causative agent of syphilis, can be best viewed using a darkfield microscope (Figure 9.11).

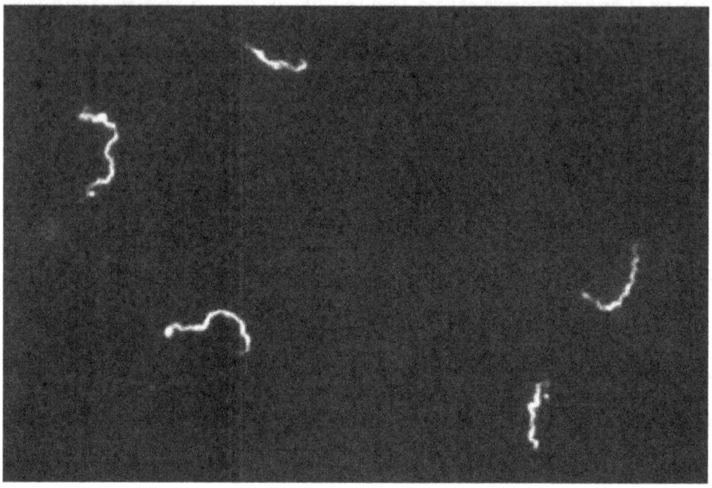

Figure 9.11 Use of a darkfield microscope allows us to view living, unstained samples of the spirochete *Treponema pallidum*. Similar to a photographic negative, the spirochetes appear bright against a dark background. (credit: Centers for Disease Control and Prevention)

Phase-Contrast Microscopes

Phase-contrast microscopes use refraction and interference caused by structures in a specimen to create high-contrast, high-resolution images without staining. It is the oldest and simplest type of microscope that creates an image by altering the wavelengths of light rays passing through the specimen. To create altered wavelength paths, an annular stop is used in the condenser. The annular stop produces a hollow cone of light that is focused on the specimen before reaching the objective lens. The objective contains a phase plate containing a phase ring. As a result, light travelling directly from the illuminator passes through the phase ring while light refracted or reflected by the specimen passes through the plate. This causes waves travelling through the ring to be about one-half of a wavelength out of phase with those passing through the plate. Because waves have peaks and troughs, they can add together (if in phase together) or cancel each other out (if out of phase). When the wavelengths are out of phase, wave troughs will cancel out wave peaks, which is called destructive interference. Structures that refract light then appear dark against a bright background of only unrefracted light. More generally, structures that differ in features such as refractive index will differ in levels of darkness (Figure 9.12).

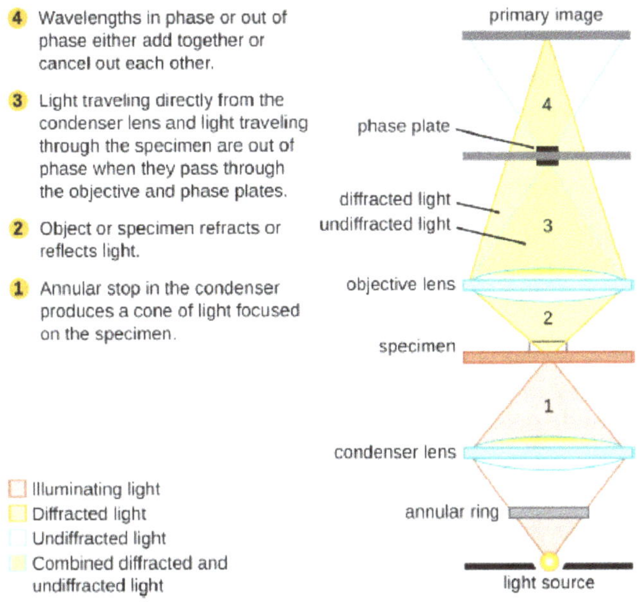

Figure 9.12 This diagram of a phase-contrast microscope illustrates phase differences between light passing through the object and background. These differences are produced by passing the rays through different parts of a phase plate. The light rays are superimposed in the image plane, producing contrast due to their interference.

Because it increases contrast without requiring stains, phase-contrast microscopy is often used to observe live specimens. Certain structures, such as organelles in eukaryotic cells and endospores in prokaryotic cells, are especially well visualized with phase-contrast microscopy (Figure 9.13).

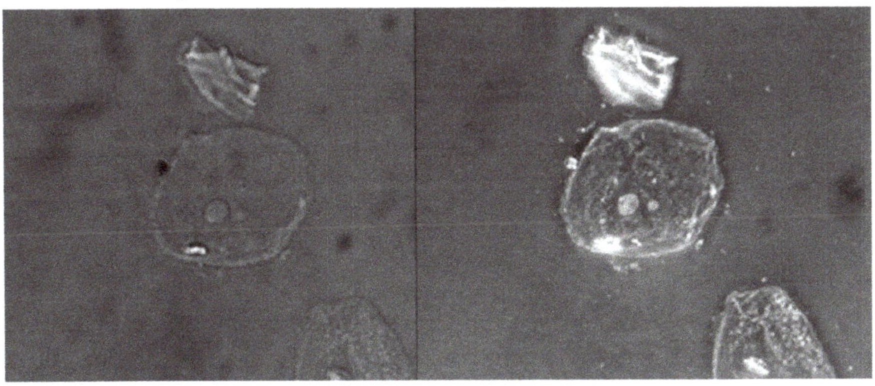

Figure 9.13 This figure compares a brightfield image (left) with a phase-contrast image (right) of the same unstained simple squamous epithelial cells. The cells are in the center and bottom right of each photograph (the irregular item above the cells is acellular debris). Notice that the unstained cells in the brightfield image are almost invisible against the background, whereas the cells in the phase-contrast image appear to glow against the background, revealing far more detail.

Differential Interference Contrast Microscopes

Differential interference contrast (DIC) microscopes (also known as Nomarski optics) are similar to phase- contrast microscopes in that they use interference patterns to enhance contrast between different features of a specimen. In a DIC microscope, two beams of light are created in which the direction of wave movement (polarization) differs. Once the beams pass through either the specimen or specimen-free space, they are recombined and effects of the specimens cause differences in the interference patterns generated by the combining of the beams. This results in high-contrast images of living organisms with a three-dimensional appearance. These microscopes are especially useful in distinguishing structures within live, unstained specimens. (Figure 9.14)

Figure 9.14 A DIC image of *Fonsecaea pedrosoi* grown on modified Leonian's agar. This fungus causes chromoblastomycosis, a chronic skin infection common in tropical and subtropical climates.

What are some advantages of phase-contrast and DIC microscopy?

Fluorescence Microscopes

A **fluorescence microscope** uses fluorescent chromophores called **fluorochromes,** which are capable of absorbing energy from a light source and then emitting this energy as visible light. Fluorochromes include naturally fluorescent substances (such as chlorophylls) as well as fluorescent stains that are added to the specimen to create contrast. Dyes such as Texas red and FITC are examples of fluorochromes. Other examples include the nucleic acid dyes 4',6'-diamidino-2-phenylindole (DAPI) and acridine orange.

The microscope transmits an excitation light, generally a form of EMR with a short wavelength, such as ultraviolet or blue light, toward the specimen; the chromophores absorb the excitation light and emit visible light with longer wavelengths. The excitation light is then filtered out (in part because ultraviolet light is harmful to the eyes) so that only visible light passes through the ocular lens. This produces an image of the specimen in bright colours against a dark background.

Fluorescence microscopes are especially useful in clinical microbiology. They can be used to identify pathogens, to find particular species within an environment, or to find the locations of particular molecules and structures within a cell. Approaches have also been developed to distinguish living from dead cells using fluorescence microscopy based upon whether they take up particular fluorochromes. Sometimes, multiple fluorochromes are used on the same specimen to show different structures or features.

One of the most important applications of fluorescence microscopy is a technique called **immunofluorescence**, which is used to identify certain disease-causing microbes by observing whether antibodies bind to them. (Antibodies are protein molecules produced by the immune system that attach to specific pathogens to kill or inhibit them.) There are two approaches to this technique: direct immunofluorescence assay (DFA) and indirect immunofluorescence assay (IFA). In DFA, specific antibodies (e.g., those that the target the rabies virus) are stained with a fluorochrome. If the specimen contains the targeted pathogen, one can observe the antibodies binding to the pathogen under the fluorescent microscope. This is called a primary antibody stain because the stained antibodies attach directly to the pathogen.

In IFA, secondary antibodies are stained with a fluorochrome rather than primary antibodies. Secondary antibodies do not attach directly to the pathogen, but they do bind to primary antibodies. When the unstained primary antibodies bind to the pathogen, the fluorescent secondary antibodies can be observed binding to the primary antibodies. Thus, the secondary antibodies are attached indirectly to the pathogen. Since multiple secondary antibodies can often attach to a primary antibody, IFA increases the number of fluorescent antibodies attached to the specimen, making it easier visualize features in the specimen (Figure 9.15).

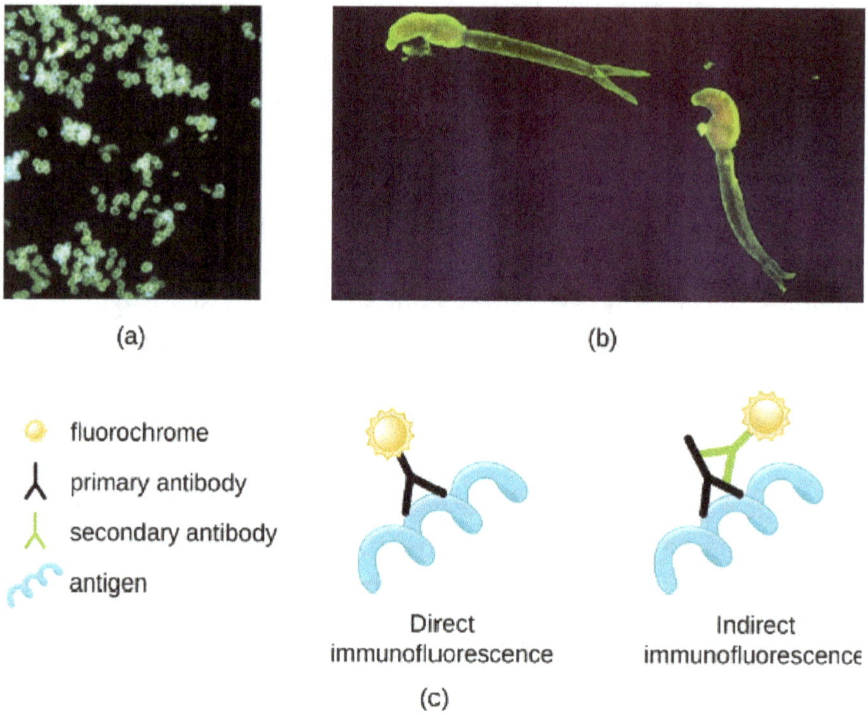

Figure 9.15 (a) A direct immunofluorescent stain is used to visualize *Neisseria gonorrhoeae*, the bacterium that causes gonorrhea. (b) An indirect immunofluorescent stain is used to visualize larvae of *Schistosoma mansoni*, a parasitic worm that causes schistosomiasis, an intestinal disease common in the tropics. (c) In direct immunofluorescence, the stain is absorbed by a primary antibody, which binds to the antigen. In indirect immunofluorescence, the stain is absorbed by a secondary antibody, which binds to a primary antibody, which, in turn, binds to the antigen. (credit a: modification of work by Centers for Disease Control and Prevention; credit b: modification of work by Centers for Disease Control and Prevention)

Confocal Microscopes

Whereas other forms of light microscopy create an image that is maximally focused at a single distance from the observer (the depth, or z-plane), a **confocal microscope** uses a laser to scan multiple z-planes successively. This produces numerous two-dimensional, high-resolution images at various depths, which can be constructed into a three-dimensional image by a computer. As with fluorescence microscopes, fluorescent stains are generally used to increase contrast and resolution. Image clarity is further enhanced by a narrow aperture that eliminates any light that is not from the z-plane. Confocal microscopes are thus very useful for examining thick specimens such as biofilms, which can be examined alive and unfixed (Figure 9.16).

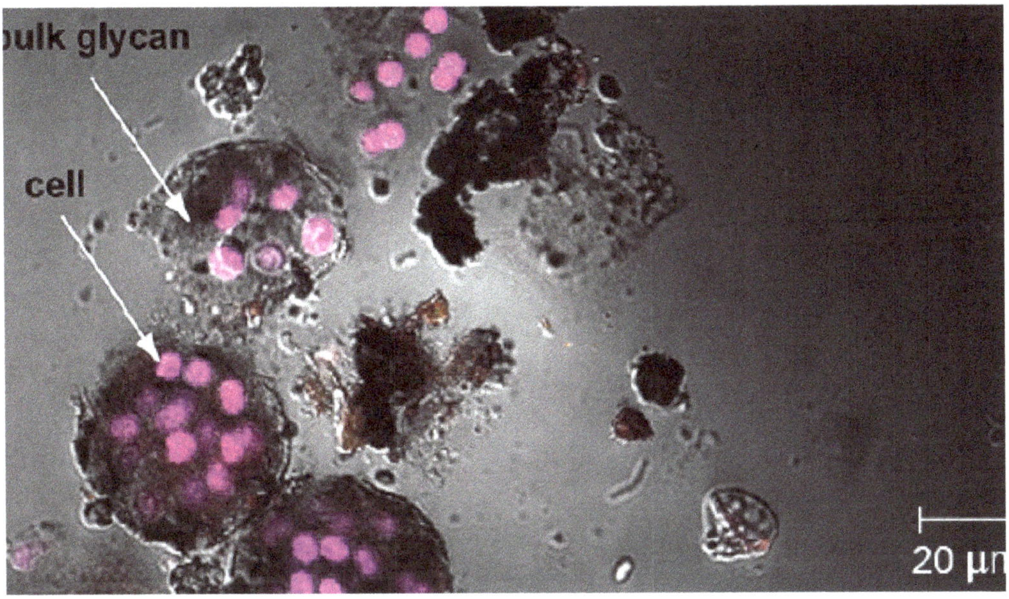

Figure 9.16 Confocal microscopy can be used to visualize structures such as this roof-dwelling cyanobacterium biofilm. (credit: modification of work by American Society for Microbiology)

Two-Photon Microscopes

While the original fluorescent and confocal microscopes allowed better visualization of unique features in specimens, there were still problems that prevented optimum visualization. The effective sensitivity of fluorescence microscopy when viewing thick specimens was generally limited by out-of-focus flare, which resulted in poor resolution. This limitation was greatly reduced in the confocal microscope through the use of a confocal pinhole to reject out-of-focus background fluorescence with thin (<1 μm), unblurred optical sections. However, even the confocal microscopes lacked the resolution needed for viewing thick tissue samples. These problems were resolved with the development of the **two-photon microscope**, which uses a scanning technique, fluorochromes, and long-wavelength light (such as infrared) to visualize specimens. The low energy associated with the long-wavelength light means that two photons must strike a location at the same time to excite the fluorochrome. The low energy of the excitation light is less damaging to cells, and the long wavelength of the excitation light more easily penetrates deep into thick specimens. This makes the two-photon microscope useful for examining living cells within intact tissues—brain slices, embryos, whole organs, and even entire animals.

Currently, use of two-photon microscopes is limited to advanced clinical and research laboratories because of the high costs of the instruments. A single two-photon microscope typically costs between $300,000 and $500,000, and the lasers used to excite the dyes used on specimens are also very expensive. However, as technology improves, two-photon microscopes may become more readily available in clinical settings.

Electron Microscopy

The maximum theoretical resolution of images created by light microscopes is ultimately limited by the wavelengths of visible light. Most light microscopes can only magnify 1000×, and a few can magnify up to 1500×, but this does not begin to approach the magnifying power of an **electron microscope (EM)**, which uses short-wavelength electron beams rather than light to increase magnification and resolution.

Electrons, like electromagnetic radiation, can behave as waves, but with wavelengths of 0.005 nm, they can produce much better resolution than visible light. An EM can produce a sharp image that is magnified up to 100,000 ×. Thus, EMs can resolve subcellular structures as well as some molecular structures (e.g., single strands of DNA); however, electron microscopy cannot be used on living material because of the methods needed to prepare the specimens.

There are two basic types of EM: the transmission **electron microscope (TEM) and the scanning electron microscope (SEM)** (Figure 9.17). The TEM is somewhat analogous to the brightfield light microscope in terms of the way it functions. However, it uses an electron beam from above the specimen that is focused using a magnetic lens (rather than a glass lens) and projected through the specimen onto a detector. Electrons pass through the specimen, and then the detector captures the image (Figure 9.18).

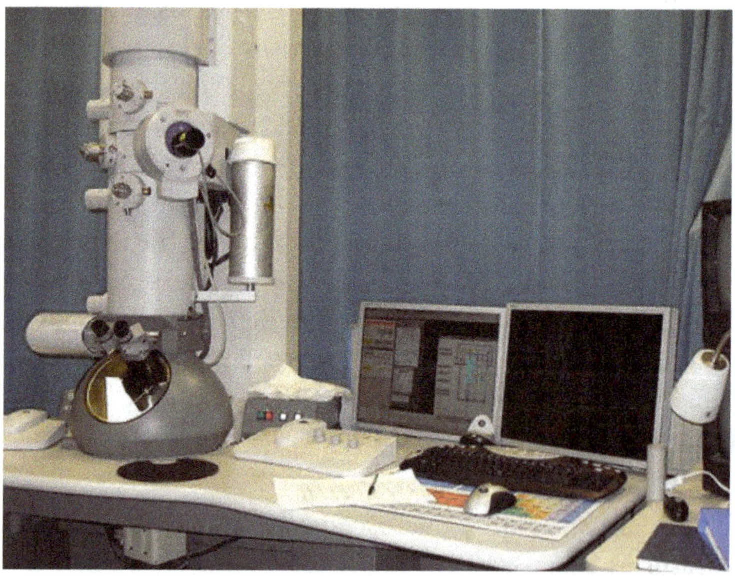

Figure 9.17 A transmission electron microscope (TEM).

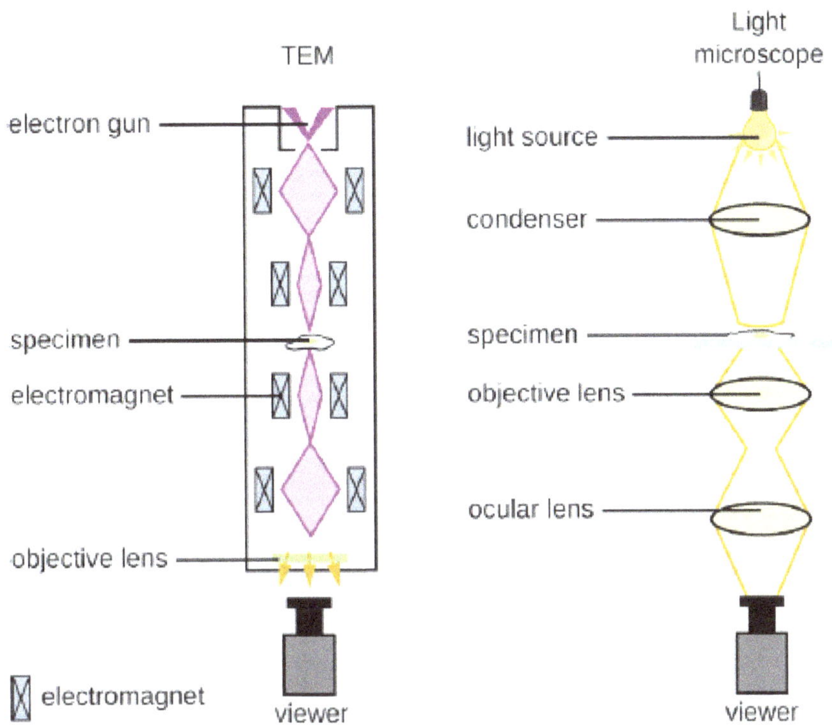

Figure 9.18 Electron microscopes use magnets to focus electron beams similarly to the way that light microscopes use lenses to focus light.

For electrons to pass through the specimen in a TEM, the specimen must be extremely thin (20–100 nm thick). The image is produced because of varying opacity in various parts of the specimen. This opacity can be enhanced by staining the specimen with materials such as heavy metals, which are electron dense. TEM requires that the beam and specimen be in a vacuum and that the specimen be very thin and dehydrated. The specific steps needed to prepare a specimen for observation under an EM are discussed in detail in the next section.

SEMs form images of surfaces of specimens, usually from electrons that are knocked off of specimens by a beam of electrons. This can create highly detailed images with a three-dimensional appearance that are displayed on a monitor (Figure 9.19). Typically, specimens are dried and prepared with fixatives that reduce artifacts, such as shriveling, that can be produced by drying, before being sputter-coated with a thin layer of metal such as gold. Whereas transmission electron microscopy requires very thin sections and allows one to see internal structures such as organelles and the interior of membranes, scanning electron microscopy can be used to view the surfaces of larger objects (such as a pollen grain) as well as the surfaces of very small samples (Figure 9.20). Some EMs can magnify an image up to 2,000,000×.

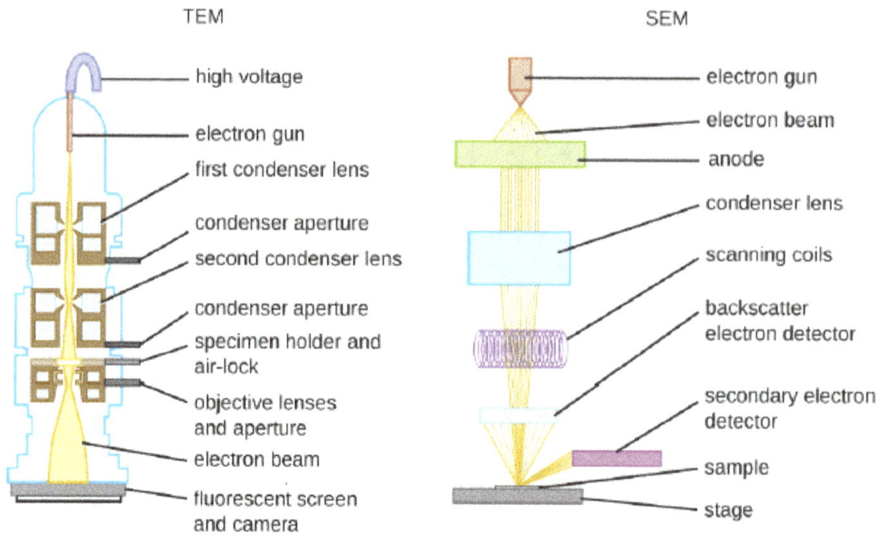

Figure 9.19 These schematic illustrations compare the components of transmission electron microscopes and scanning electron microscopes.

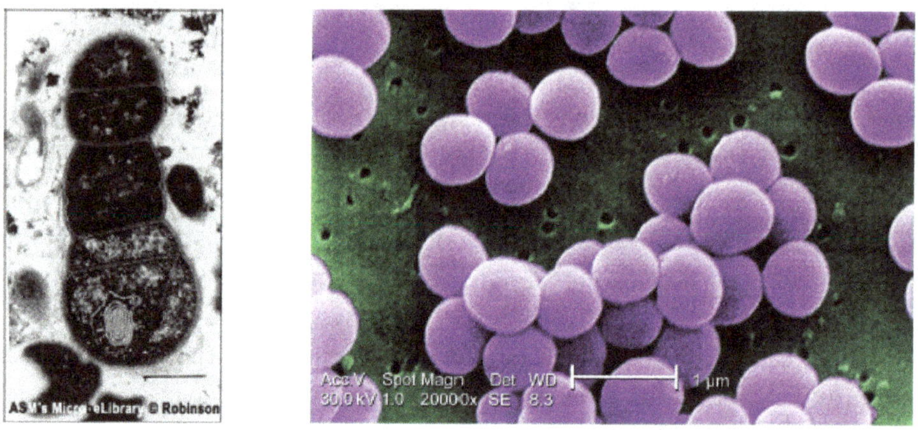

(a) (b)

Figure 9.20 (a) This TEM image of cells in a biofilm shows well-defined internal structures of the cells because of varying levels of opacity in the specimen. (b) This colour-enhanced SEM image of the bacterium *Staphylococcus aureus* illustrates the ability of scanning electron microscopy to render three-dimensional images of the surface structure of cells. (credit a: modification of work by American Society for Microbiology; credit b: modification of work by Centers for Disease Control and Prevention)

Using Microscopy to Study Biofilms

A biofilm is a complex community of one or more microorganism species, typically forming as a slimy coating attached to a surface because of the production of an extra polymeric substance (EPS) that attaches to a surface or at the interface between surfaces (e.g., between air and water). In nature, biofilms are abundant and frequently occupy complex niches within ecosystems (Figure 9.21). In medicine, biofilms can coat medical devices and exist within the body. Because they possess unique characteristics, such as increased resistance against the immune system and to antimicrobial drugs, biofilms are of particular interest to microbiologists and clinicians alike.

Because biofilms are thick, they cannot be observed very well using light microscopy; slicing a biofilm to create a thinner specimen might kill or disturb the microbial community. Confocal microscopy provides clearer images of biofilms because it can focus on one z-plane at a time and produce a three-dimensional image of a thick specimen. Fluorescent dyes can be helpful in identifying cells within the matrix. Additionally, techniques such as immunofluorescence and fluorescence in situ hybridization (FISH), in which fluorescent probes are used to bind to DNA, can be used.

Electron microscopy can be used to observe biofilms, but only after dehydrating the specimen, which produces undesirable artifacts and distorts the specimen. In addition to these approaches, it is possible to follow water currents through the shapes (such as cones and mushrooms) of biofilms, using video of the movement of fluorescently coated beads (Figure 9.22).

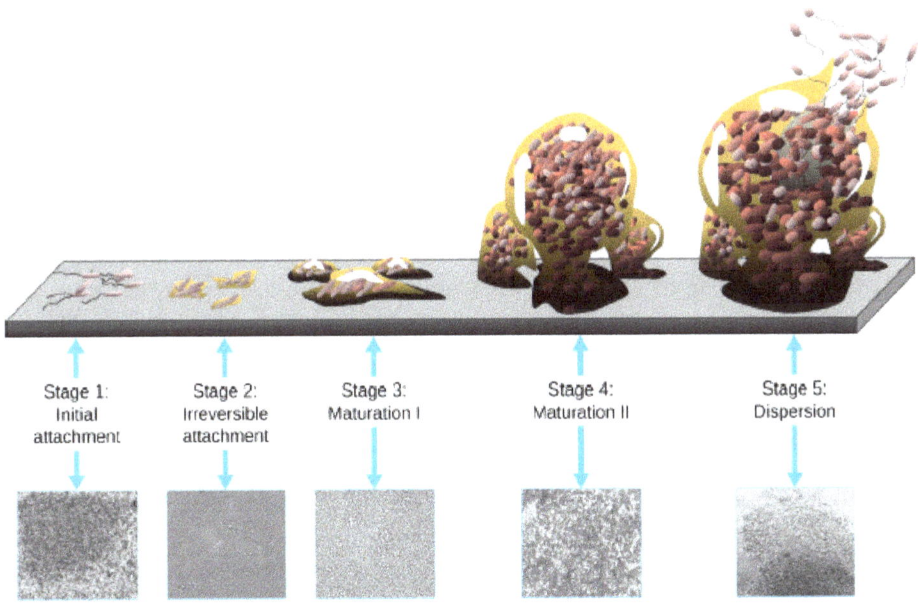

Figure 9.21 A biofilm forms when planktonic (free-floating) bacteria of one or more species adhere to a surface, produce slime, and form a colony. (credit: Public Library of Science)

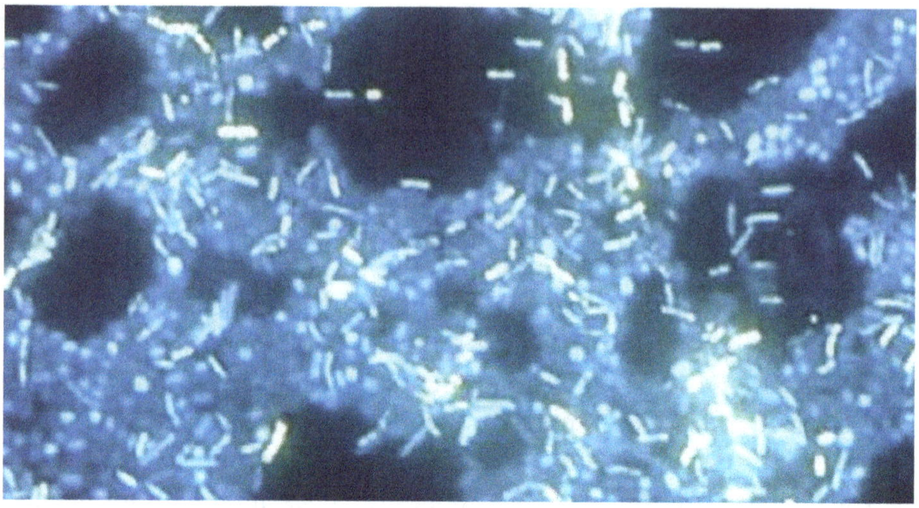

Figure 9.22 In this image, multiple species of bacteria grow in a biofilm on stainless steel (stained with DAPI for epifluorescence miscroscopy). (credit: Ricardo Murga, Rodney Donlan)

Scanning Probe Microscopy

A **scanning probe microscope** does not use light or electrons, but rather very sharp probes that are passed over the surface of the specimen and interact with it directly. This produces information that can be assembled into images with magnifications up to 100,000,000×. Such large magnifications can be used to observe individual atoms on surfaces. To date, these techniques have been used primarily for research rather than for diagnostics.

There are two types of scanning probe microscope: the **scanning tunneling microscope (STM)** and the **atomic force microscope (AFM)**. An STM uses a probe that is passed just above the specimen as a constant voltage bias creates the potential for an electric current between the probe and the specimen. This current occurs via quantum tunneling of electrons between the probe and the specimen, and the intensity of the current is dependent upon the distance between the probe and the specimen. The probe is moved horizontally above the surface and the intensity of the current is measured. Scanning tunneling microscopy can effectively map the structure of surfaces at a resolution at which individual atoms can be detected.

Similar to an STM, AFMs have a thin probe that is passed just above the specimen. However, rather than measuring variations in the current at a constant height above the specimen, an AFM establishes a constant current and measures variations in the height of the probe tip as it passes over the specimen. As the probe tip is passed over the specimen, forces between the atoms (van der Waals forces, capillary forces, chemical bonding, electrostatic forces, and others) cause it to move up and down. Deflection of the probe tip is determined and measured using Hooke's law of elasticity, and this information is used to construct images of the surface of the specimen with resolution at the atomic level (Figure 9.23).

Figure 9.24, Figure 9.25, and Figure 9.26 summarize the microscopy techniques for light microscopes, electron microscopes, and scanning probe microscopes, respectively.

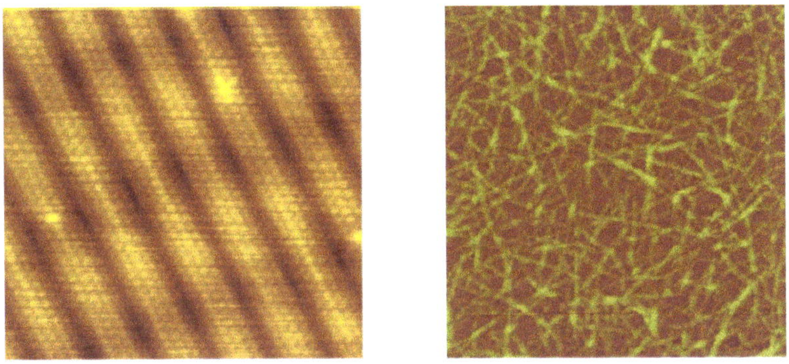

(a) (b)

Figure 9.23 STMs and AFMs allow us to view images at the atomic level. (a) This STM image of a pure gold surface shows individual atoms of gold arranged in columns. (b) This AFM image shows long, strand-like molecules of nanocellulose, a laboratory-created substance derived from plant fibers. (credit a: modification of work by "Erwinrossen"/Wikimedia Commons)

LIGHT MICROSCOPES Magnification: up to about 1000× Use visible or ultraviolet light to produce an image		
Microscope Type	**Key Uses**	**Sample Images**
Brightfield	Commonly used in a wide variety of laboratory applications as the standard microscope; produces an image on a bright background. **Example:** *Bacillus* sp. showing endospores.	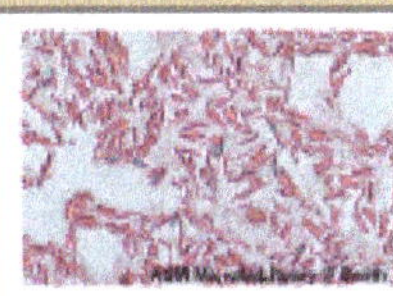
Darkfield	Increases contrast without staining by producing a bright image on a darker background; especially useful for viewing live specimens. **Example:** *Borrelia burgdorferi*	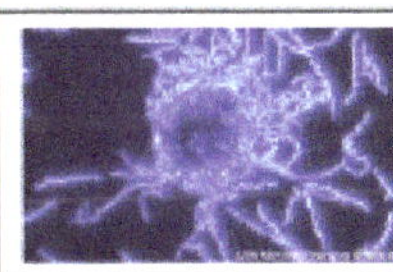
Phase contrast	Uses refraction and interference caused by structures in the specimen to create high-contrast, high-resolution images without staining, making it useful for viewing live specimens, and structures such as endospores and organelles. **Example:** *Pseudomonas* sp.	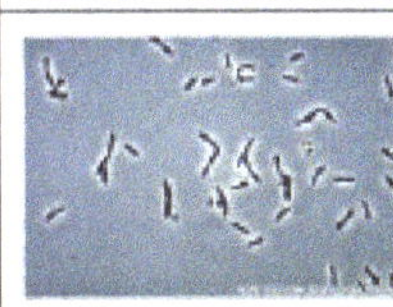
Differential interference contrast (DIC)	Uses interference patterns to enhance contrast between different features of a specimen to produce high-contrast images of living organisms with a three-dimensional appearance, making it especially useful in distinguishing structures within live, unstained specimens; images viewed reveal detailed structures within cells. **Example:** *Escherichia coli* O157:H7	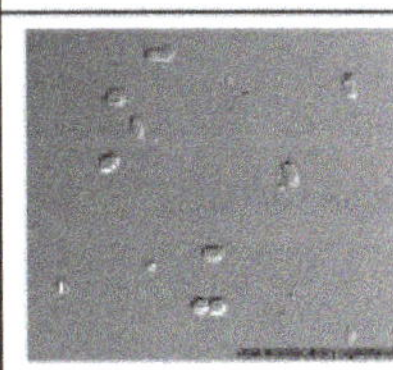

Fluorescence	Uses fluorescent stains to produce an image; can be used to identify pathogens, to find particular species, to distinguish living from dead cells, or to find locations of particular molecules within a cell; also used for immunofluorescence. **Example:** *P. putida* stained with fluorescent dyes to visualize the capsule.	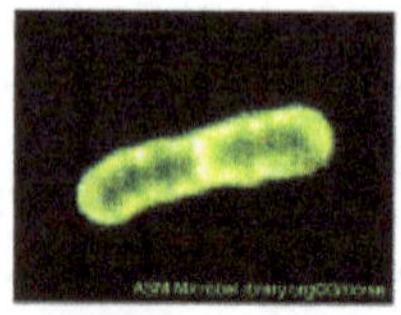
Confocal	Uses a laser to scan mutiple z-planes successively, producing numerous two-dimensional, high-resolution images at various depths that can be constructed into a three-dimensional image by a computer, making this useful for examining thick specimens such as biofilms. **Example:** *Escherichia coli* stained with acridine orange dye to show the nucleoid regions of the cells.	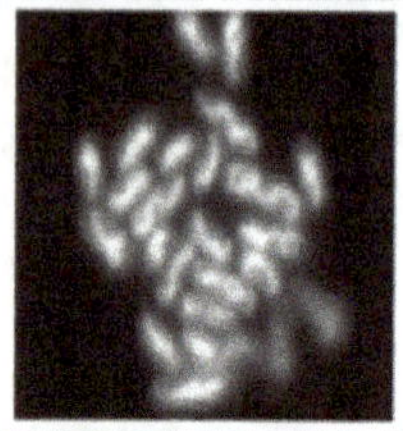
Two-photon	Uses a scanning technique, fluorochromes, and long-wavelength light (such as infrared) to penetrate deep into thick specimens such as biofilms. **Example:** Mouse intestine cells stained with fluorescent dye.	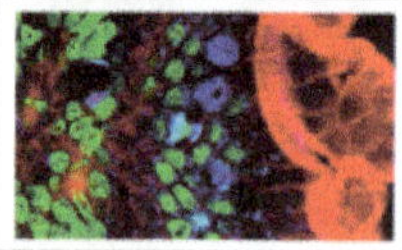

Figure 9.24 (credit "Brightfield": modification of work by American Society for Microbiology; credit "Darkfield": modification of work by American Society for Microbiology; credit "Phase contrast": modification of work by American Society for Microbiology; credit "DIC": modification of work by American Society for Microbiology; credit "Fluorescence": modification of work by American Society for Microbiology; credit "Confocal": modification of work by American Society for Microbiology; credit "Two-photon": modification of work by Alberto Diaspro, Paolo Bianchini, Giuseppe Vicidomini, Mario Faretta, Paola Ramoino, Cesare Usai)

ELECTRON MICROSCOPES Magnification: 20–100,000× or more Use electron beams focused with magnets to produce an image.		
Microscope Type	**Key Uses**	**Sample Images**
Transmission (TEM)	Uses electron beams that pass through a specimen to visualize small images; useful to observe small, thin specimens such as tissue sections and subcellular structures. **Example:** *Ebola* virus	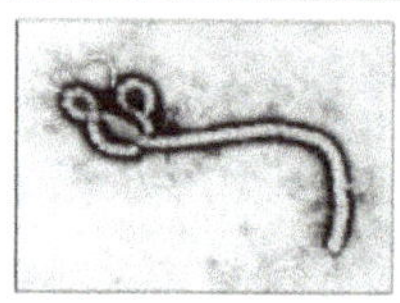
Scanning (SEM)	Uses electron beams to visualize surfaces; useful to observe the three-dimensional surface details of specimens. **Example:** *Campylobacter jejuni*	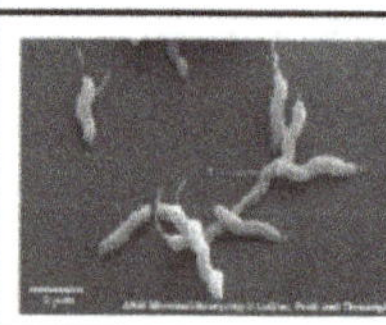

Figure 9.25 (credit "TEM": modification of work by American Society for Microbiology; credit "SEM": modification of work by American Society for Microbiology

SCANNING PROBE MICROSCOPES Magnification: 100–100,000,000× or more Use very short probes that are passed over the surface of the specimen and interact with it directly.		
Microscope Type	**Key Uses**	**Sample Images**
Scanning tunneling (STM)	Uses a probe passed horizontally at a constant distance just above the specimen while the intensity of the current is measured; can map the structure of surfaces at the atomic level; works best on conducting materials but can also be used to examine organic materials such as DNA, if fixed on a surface. **Example:** Image of surface reconstruction on a clean gold [Au(100)] surface, as visualized using scanning tunneling microscopy.	
Atomic force (AFM)	Can be used in several ways, including using a laser focused on a cantilever to measure the bending of the tip or a probe passed above the specimen while the height needed to maintain a constant current is measured; useful to observe specimens at the atomic level and can be more easily used with nonconducting samples. **Example:** AFM height image of carboxymethylated nanocellulose adsorbed on a silica surface.	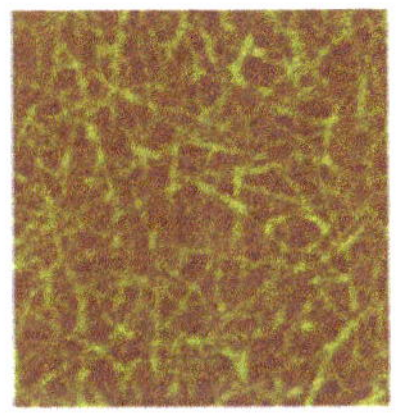

Figure 9.26

STAINING MICROSCOPIC SPECIMENS

In their natural state, most of the cells and microorganisms that we observe under the microscope lack colour and contrast. This makes it difficult, if not impossible, to detect important cellular structures and their distinguishing characteristics without artificially treating specimens. We have already alluded to certain techniques involving stains and fluorescent dyes, and in this section we will discuss specific techniques for sample preparation in greater detail. Indeed, numerous methods have been developed to identify specific microbes, cellular structures, DNA sequences, or indicators of infection in tissue samples, under the microscope. Here, we will focus on the most clinically relevant techniques.

Preparing Specimens for Light Microscopy

In clinical settings, light microscopes are the most commonly used microscopes. There are two basic types of preparation used to view specimens with a light microscope: wet mounts and fixed specimens.

The simplest type of preparation is the **wet mount**, in which the specimen is placed on the slide in a drop of liquid. Some specimens, such as a drop of urine, are already in a liquid form and can be deposited on the slide using a dropper. Solid specimens, such as a skin scraping, can be placed on the slide before adding a drop of liquid to prepare the wet mount. Sometimes the liquid used is simply water, but often stains are added to enhance contrast. Once the liquid has been added to the slide, a coverslip is placed on top and the specimen is ready for examination under the microscope.

The second method of preparing specimens for light microscopy is **fixation.** The "fixing" of a sample refers to the process of attaching cells to a slide. Fixation is often achieved either by heating (heat fixing) or chemically treating the specimen. In addition to attaching the specimen to the slide, fixation also kills microorganisms in the specimen, stopping their movement and metabolism while preserving the integrity of their cellular components for observation.

To heat-fix a sample, a thin layer of the specimen is spread on the slide (called a **smear**), and the slide is then briefly heated over a heat source (Figure 9.27). Chemical fixatives are often preferable to heat for tissue specimens. Chemical agents such as acetic acid, ethanol, methanol, formaldehyde (formalin), and glutaraldehyde can denature proteins, stop biochemical reactions, and stabilize cell structures in tissue samples (Figure 9.27).

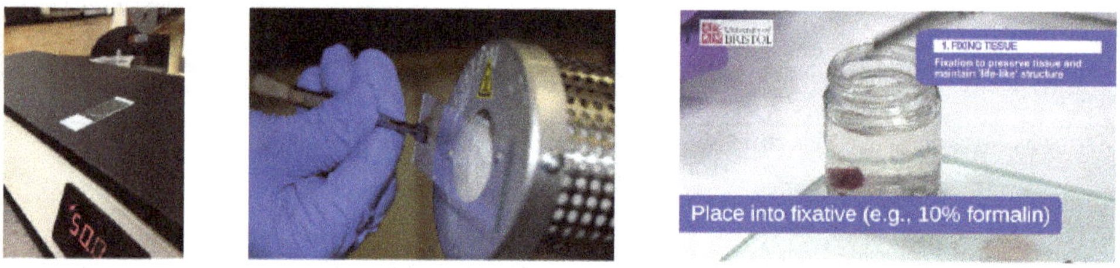

(a) (b) (c)

Figure 9.27 (a) A specimen can be heat-fixed by using a slide warmer like this one. (b) Another method for heat- fixing a specimen is to hold a slide with a smear over a microincinerator. (c) This tissue sample is being fixed in a solution of formalin (also known as formaldehyde). Chemical fixation kills microorganisms in the specimen, stopping degradation of the tissues and preserving their structure so that they can be examined later under the microscope. (credit a: modification of work by Nina Parker; credit b: modification of work by Nina Parker; credit c: modification of work by "University of Bristol"/YouTube)

In addition to fixation, **staining** is almost always applied to colour certain features of a specimen before examining it under a light microscope. Stains, or dyes, contain salts made up of a positive ion and a negative ion. Depending on the type of dye, the positive or the negative ion may be the chromophore (the colored ion); the other, uncolored ion is called the counterion. If the chromophore is the positively charged ion, the stain is classified as a **basic dye**; if the negative ion is the chromophore, the stain is considered an **acidic dye**.

Dyes are selected for staining based on the chemical properties of the dye and the specimen being observed, which determine how the dye will interact with the specimen. In most cases, it is preferable to use a **positive stain**, a dye that will be absorbed by the cells or organisms being observed, adding colour to objects of interest to make them stand out against the background. However, there are scenarios in which it is advantageous to use a **negative stain**, which is absorbed by the background but not by the cells or organisms in the specimen. Negative staining produces an outline or silhouette of the organisms against a colourful background (Figure 9.28).

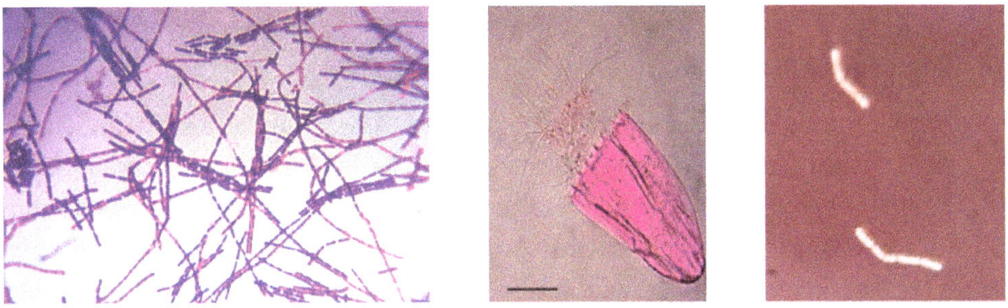

(a) (b) (c)

Figure 9.28 (a) These *Bacillus anthracis* cells have absorbed crystal violet, a basic positive stain. (b) This specimen of *Spinoloricus*, a microscopic marine organism, has been stained with rose bengal, a positive acidic stain. (c) These *megaterium* appear to be white because they have not absorbed the negative red stain applied to the slide. (credit a: modification of work by Centers for Disease Control and Prevention; credit b: modification of work by Roberto Danovaro, Antonio Pusceddu, Cristina Gambi, Iben Heiner, Reinhardt Mobjerg Kristensen; credit c: modification of work by Anh-Hue Tu)

Because cells typically have negatively charged cell walls, the positive chromophores in basic dyes tend to stick to the cell walls, making them positive stains. Thus, commonly used basic dyes such as basic fuchsin, crystal violet, malachite green, methylene blue, and safranin typically serve as positive stains. On the other hand, the negatively charged chromophores in acidic dyes are repelled by negatively charged cell walls, making them negative stains. Commonly used acidic dyes include acid fuchsin, eosin, and rose bengal.

Some staining techniques involve the application of only one dye to the sample; others require more than one dye. In **simple staining**, a single dye is used to emphasize particular structures in the specimen. A simple stain will generally make all of the organisms in a sample appear to be the same colour, even if the sample contains more than one type of organism. In contrast, **differential staining** distinguishes organisms based on their interactions with multiple stains. In other words, two organisms in a differentially stained sample may appear to be different colours. Differential staining techniques commonly used in clinical settings include Gram staining, acid-fast staining, endospore staining, flagella staining, and capsule staining.

Gram Staining

The **Gram stain procedure** is a differential staining procedure that involves multiple steps. It was developed by Danish microbiologist Hans Christian Gram in

1884 as an effective method to distinguish between bacteria with different types of cell walls, and even today it remains one of the most frequently used staining techniques. The steps of the Gram stain procedure are listed below:

- First, crystal violet, a **primary stain**, is applied to a heat-fixed smear, giving all of the cells a purple colour.
- Next, Gram's iodine, a **mordant**, is added. A mordant is a substance used to set or stabilize stains or dyes; in this case, Gram's iodine acts like a trapping agent that complexes with the crystal violet, making the crystal violet–iodine complex clump and stay contained in thick layers of peptidoglycan in the cell walls.
- Next, a **decolorizing agent** is added, usually ethanol or an acetone/ethanol solution. Cells that have thick peptidoglycan layers in their cell walls are much less affected by the decolorizing agent; they generally retain the crystal violet dye and remain purple. However, the decolorizing agent more easily washes the dye out of cells with thinner peptidoglycan layers, making them again colorless.
- Finally, a secondary **counterstain**, usually safranin, is added. This stains the decolorized cells pink and is less noticeable in the cells that still contain the crystal violet dye.

Gram stain process			
Gram staining steps	Cell effects	Gram-positive	Gram-negative
Step 1 **Crystal violet** *primary stain added to specimen smear.*	Stains cells purple or blue.		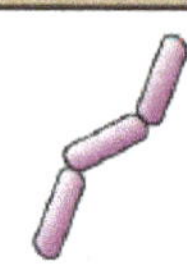
Step 2 **Iodine** *mordant makes dye less soluble so it adheres to cell walls.*	Cells remain purple or blue.	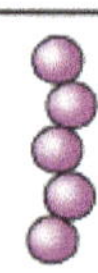	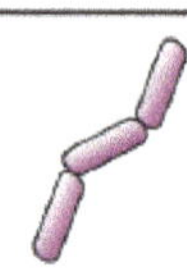
Step 3 **Alcohol** *decolorizer washes away stain from gram-negative cell walls.*	Gram-positive cells remain purple or blue. Gram-negative cells are colorless.	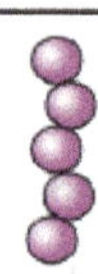	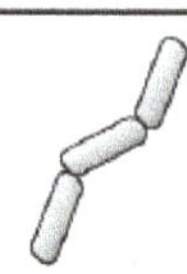
Step 4 **Safranin** *counterstain allows dye adherance to gram-negative cells.*	Gram-positive cells remain purple or blue. Gram-negative cells appear pink or red.	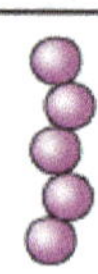	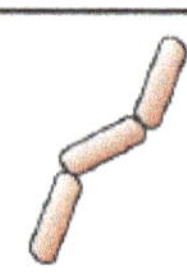

Figure 9.30 Gram-staining is a differential staining technique that uses a primary stain and a secondary counterstain to distinguish between gram-positive and gram-negative bacteria.

The purple, crystal-violet stained cells are referred to as gram-positive cells, while the red, safranin-dyed cells are gram-negative (Figure 9.31). However, there are several important considerations in interpreting the results of a Gram stain. First, older bacterial cells may have damage to their cell walls that causes them to appear gram-negative even if the species is gram-positive. Thus, it is best to use fresh bacterial cultures for Gram staining. Second, errors such as leaving on decolorizer too long can affect the results. In some cases, most cells will appear gram-positive while a few appear gram-negative (as in Figure 9.31). This suggests damage to the individual cells or that decolorizer was left on for too long; the cells should still be classified as gram-positive if they are all the same species rather than a mixed culture.

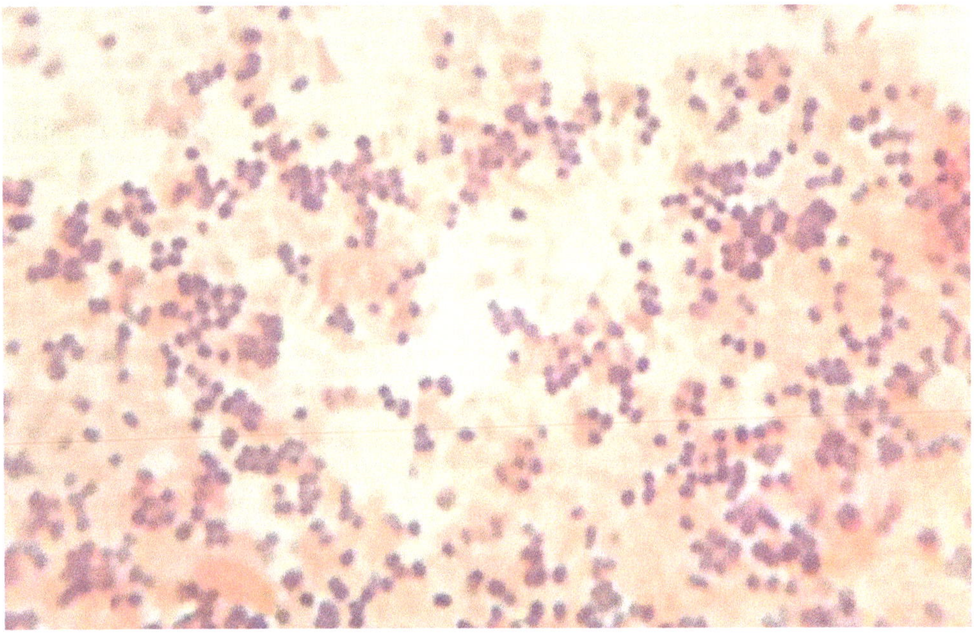

Figure 9.31 In this specimen, the gram-positive bacterium *Staphylococcus aureus* retains crystal violet dye even after the decolorizing agent is added. Gram-negative *Escherichia coli*, the most common Gram stain quality-control bacterium, is decolorized, and is only visible after the addition of the pink counterstain safranin. (credit: modification of work by Nina Parker)

Besides their differing interactions with dyes and decolorizing agents, the chemical differences between gram- positive and gram-negative cells have other implications with clinical relevance. For example, Gram staining can help clinicians classify bacterial pathogens in a sample into categories associated with specific properties. Gram- negative bacteria tend to be more resistant to certain antibiotics than gram-positive bacteria. We will discuss this and other applications of Gram staining in more detail in later chapters.

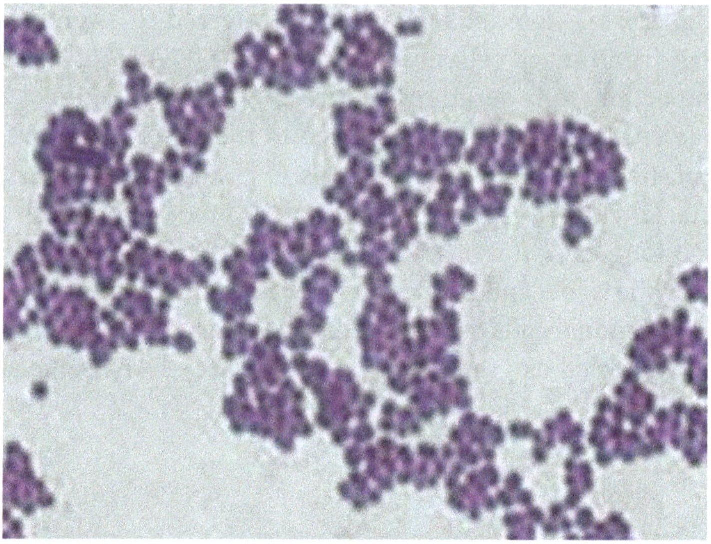

Figure 9.32 (credit: modification of work by American Society for Microbiology)

Acid-Fast Stains

Acid-fast staining is another commonly used, differential staining technique that can be an important diagnostic tool. An **acid-fast stain** is able to differentiate two types of gram-positive cells: those that have waxy mycolic acids in their cell walls, and those that do not. Two different methods for acid-fast staining are the **Ziehl-Neelsen technique** and the **Kinyoun technique**. Both use carbolfuchsin as the primary stain. The waxy, acid-fast cells retain the carbolfuchsin even after a decolorizing agent (an acid-alcohol solution) is applied. A secondary counterstain, methylene blue, is then applied, which renders non–acid-fast cells blue.

The fundamental difference between the two carbolfuchsin-based methods is whether heat is used during the primary staining process. The Ziehl-Neelsen method uses heat to infuse the carbolfuchsin into the acid-fast cells, whereas the Kinyoun method does not use heat. Both techniques are important diagnostic tools because a number of specific diseases are caused by acid-fast bacteria (AFB). If AFB are present in a tissue sample, their red or pink color can be seen clearly against the blue background of the surrounding tissue cells (Figure 9.33).

USING MICROSCOPY TO DIAGNOSE TUBERCULOSIS

Mycobacterium tuberculosis, the bacterium that causes tuberculosis, can be detected in specimens based on the presence of acid-fast bacilli. Often, a smear is prepared from a sample of the patient's sputum and then stained using the Ziehl-Neelsen technique (Figure 9.33). If acid-fast bacteria are confirmed, they are generally cultured to make a positive identification. Variations of this approach can be used as a first step in determining whether M. tuberculosis or other acid-fast bacteria are present, though samples from elsewhere in the body (such as urine) may contain other Mycobacterium species.

An alternative approach for determining the presence of M. tuberculosis is immunofluorescence. In this technique, fluorochrome-labeled antibodies bind to M. tuberculosis, if present. Antibody-specific fluorescent dyes can be used to view the mycobacteria with a fluorescence microscope.

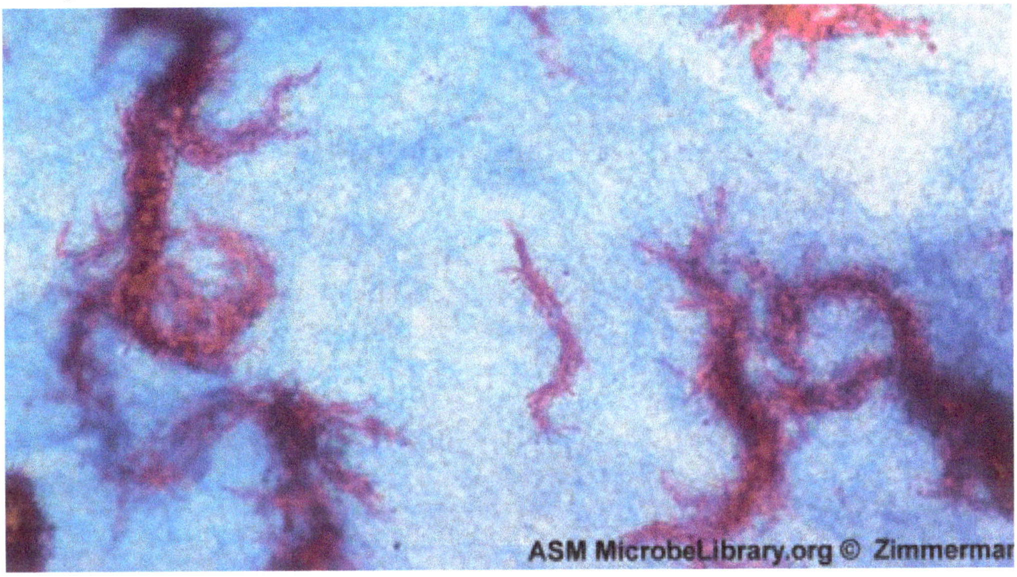

Figure 9.33 Ziehl-Neelsen staining has rendered these *Mycobacterium tuberculosis* cells red and the surrounding growth indicator medium blue. (credit: modification of work by American Socicty for Microbiology)

Capsule Staining

Certain bacteria and yeasts have a protective outer structure called a capsule. Since the presence of a capsule is directly related to a microbe's virulence (its ability to cause disease), the ability to determine whether cells in a sample have capsules is an important diagnostic tool. Capsules do not absorb most basic dyes; therefore, a negative staining technique (staining around the cells) is typically used for **capsule staining.** The dye stains the background but does not penetrate the capsules, which appear like halos around the borders of the cell. The specimen does not need to be heat-fixed prior to negative staining.

One common negative staining technique for identifying encapsulated yeast and bacteria is to add a few drops of India ink or nigrosin to a specimen. Other capsular stains can also be used to negatively stain encapsulated cells (Figure 9.34). Alternatively, positive and negative staining techniques can be combined to visualize capsules: The positive stain colours the body of the cell, and the negative stain colours the background but not the capsule, leaving halo around each cell.

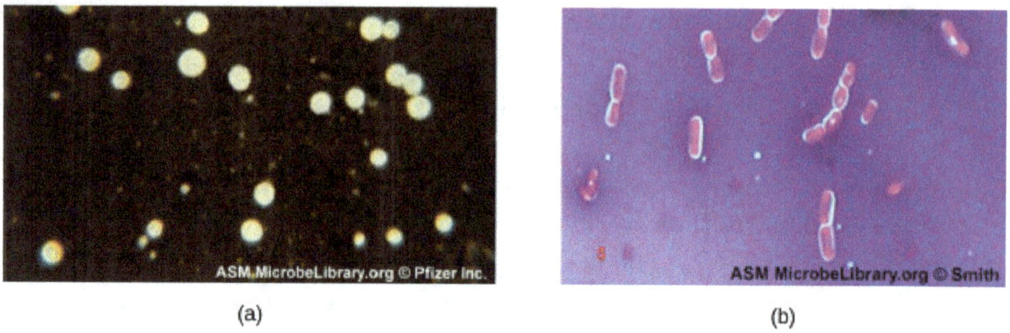

Figure 9.34 (a) India-ink was used to stain the background around these cells of the yeast *Cryptococcus neoformans*. The halos surrounding the cells are the polysaccharide capsules. (b) Crystal violet and copper sulfate dyes cannot penetrate the encapsulated *Bacillus* cells in this negatively stained sample. Encapsulated cells appear to have a light-blue halo. (credit a: modification of work by American Society for Microbiology; credit b: modification of work by American Society for Microbiology)

Endospore Staining

Endospores are structures produced within certain bacterial cells that allow them to survive harsh conditions.

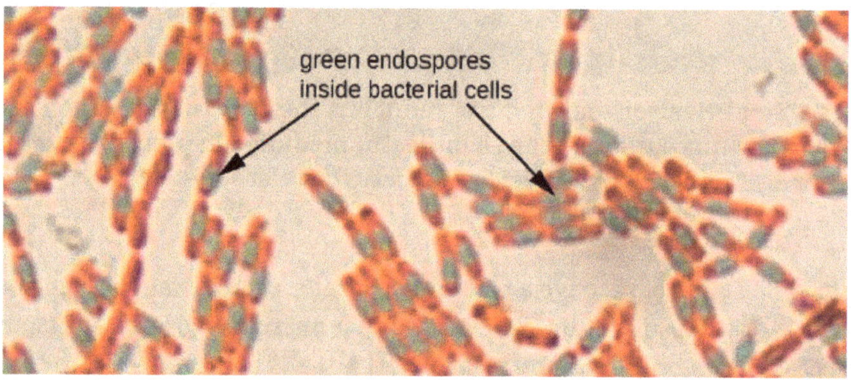

Figure 9.35 A stained preparation of *Bacillus subtilis* showing endospores as green and the vegetative cells as pink. (credit: modification of work by American Society for Microbiology)

Gram staining alone cannot be used to visualize endospores, which appear clear when Gram-stained cells are viewed. **Endospore staining** uses two stains to differentiate endospores from the rest of the cell. The Schaeffer- Fulton method (the most commonly used endospore-staining technique) uses heat to push the primary stain (malachite green) into the endospore. Washing with water decolorizes the cell, but the endospore retains the green stain. The cell is then counterstained pink with safranin. The resulting image reveals the shape and location of endospores, if they are present. The green endospores will appear either within the pink vegetative cells or as separate from the pink cells altogether. If no endospores are present, then only the pink vegetative cells will be visible (Figure 9.35).

Endospore-staining techniques are important for identifying *Bacillus* and *Clostridium*, two genera of endospore- producing bacteria that contain clinically significant species. Among others, *B. anthracis* (which causes anthrax) has been of particular interest because of concern that its spores could be used as a bioterrorism agent. *C. difficile* is a particularly important species responsible for the typically hospital-acquired infection known as "C. diff."

Flagella Staining

Flagella (singular: flagellum) are tail-like cellular structures used for locomotion by some bacteria, archaea, and eukaryotes. Because they are so thin, flagella typically cannot be seen under a light microscope without a specialized **flagella staining** technique. Flagella staining thickens the flagella by first applying mordant (generally tannic acid, but sometimes potassium alum), which coats the flagella; then the specimen is stained with pararosaniline (most commonly) or basic fuchsin (Figure 9.36).

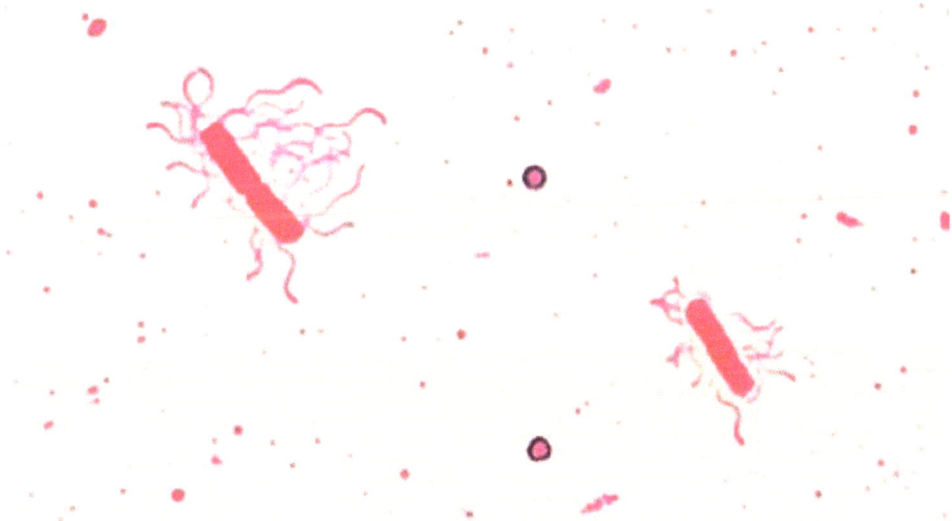

Figure 9.36 A flagella stain of *Bacillus cereus*, a common cause of foodborne illness, reveals that the cells have numerous flagella, used for locomotion. (credit: modification of work by Centers for Disease Control and Prevention)

Though flagella staining is uncommon in clinical settings, the technique is commonly used by microbiologists, since the location and number of flagella can be useful in classifying and identifying bacteria in a sample. When using this technique, it is important to handle the specimen with great care; flagella are delicate structures that can easily be damaged or pulled off, compromising attempts to accurately locate and count the number of flagella.

SIMPLE STAINS				
Stain Type	**Specific Dyes**	**Purpose**	**Outcome**	**Sample Images**
Basic stains	Methylene blue, crystal violet, malachite green, basic fuchsin, carbolfuchsin, safranin	Stain negatively charged molecules and structures, such as nucleic acids and proteins	Positive stain	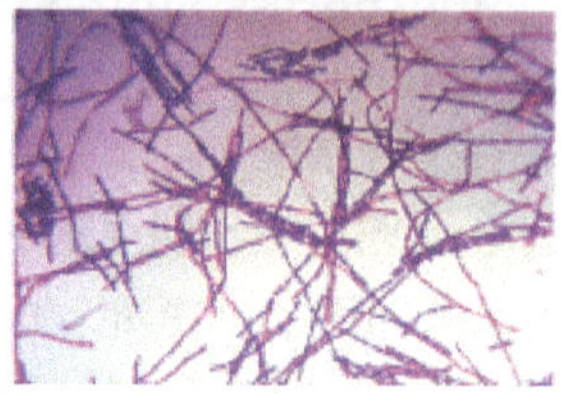
Acidic stains	Eosin, acid fuchsin, rose bengal, Congo red	Stain positively charged molecules and structures, such as proteins	Can be either a positive or negative stain, depending on the cell's chemistry.	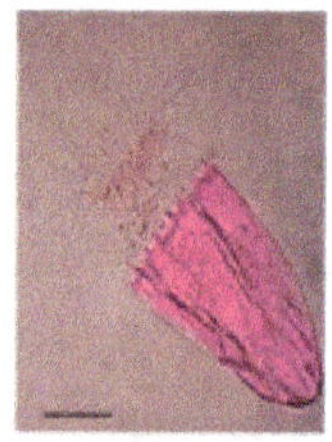
Negative stains	India ink, nigrosin	Stains background, not specimen	Dark background with light specimen	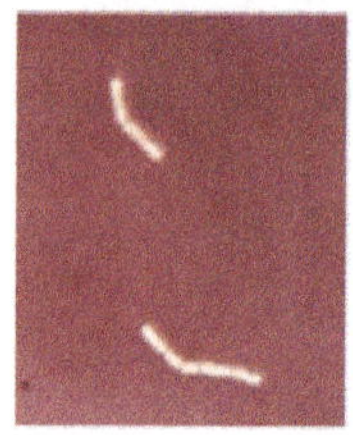

Figure 9.37 (credit "basic stains": modification of work by Centers for Disease Control and Prevention; credit "Acidic stains": modification of work by Roberto Danovaro, Antonio Dell'Anno, Antonio Pusceddu, Cristina Gambi, Iben Heiner, Reinhardt Mobjerg Kristensen; credit "Negative stains": modification of work by Anh-Hue Tu)

DIFFERENTIAL STAINS				
Stain Type	**Specific Dyes**	**Purpose**	**Outcome**	**Sample Images**
Gram stain	Uses crystal violet, Gram's iodine, ethanol (decolorizer), and safranin	Used to distinguish cells by cell-wall type (gram-positive, gram-negative)	Gram-positive cells stain purple/violet. Gram-negative cells stain pink.	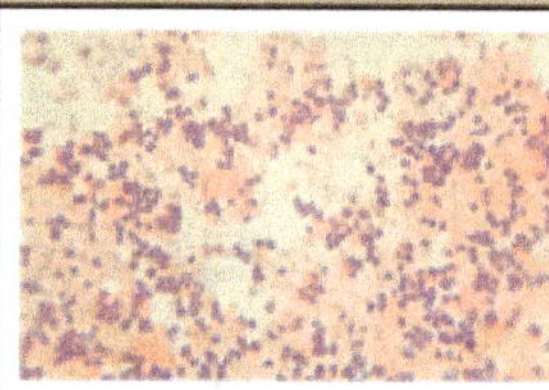
Acid-fast stain	After staining with basic fuchsin, acid-fast bacteria resist decolorization by acid-alcohol. Non acid-fast bacteria are counterstained with methylene blue.	Used to distinguish acid-fast bacteria such as *M. tuberculosis*, from non–acid-fast cells	Acid-fast bacteria are red; non–acid-fast cells are blue.	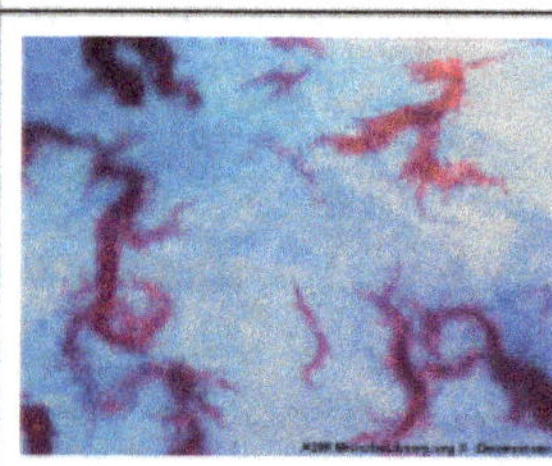

Endospore stain	Uses heat to stain endospores with malachite green (Schaeffer-Fulton procedure), then cell is washed and counterstained with safranin.	Used to distinguish organisms with endospores from those without; used to study the endospore.	Endospores appear bluish-green; other structures appear pink to red.	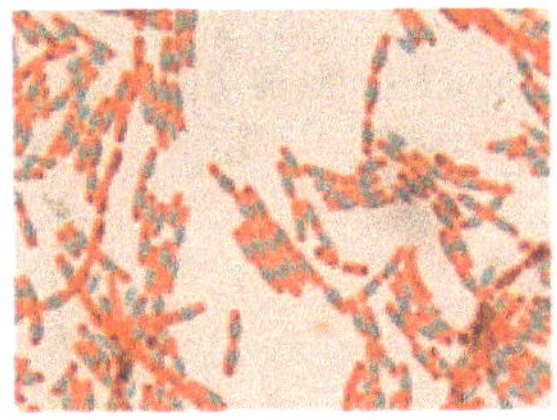
Flagella stain	Flagella are coated with a tannic acid or potassium alum mordant, then stained using either pararosaline or basic fuchsin.	Used to view and study flagella in bacteria that have them.	Flagella are visible if present.	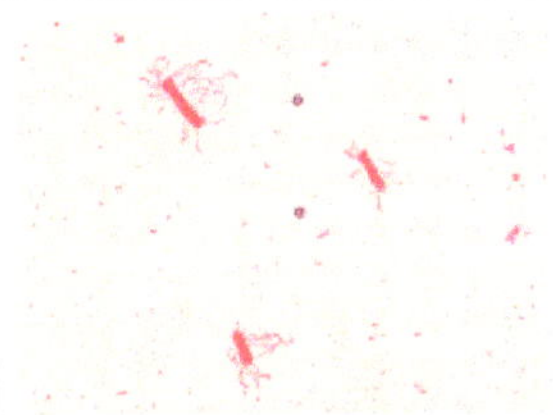
Capsule stain	Negative staining with India ink or nigrosin is used to stain the background, leaving a clear area of the cell and the capsule. Counterstaining can be used to stain the cell while leaving the capsule clear.	Used to distinguish cells with capsules from those without.	Capsules appear clear or as halos if present.	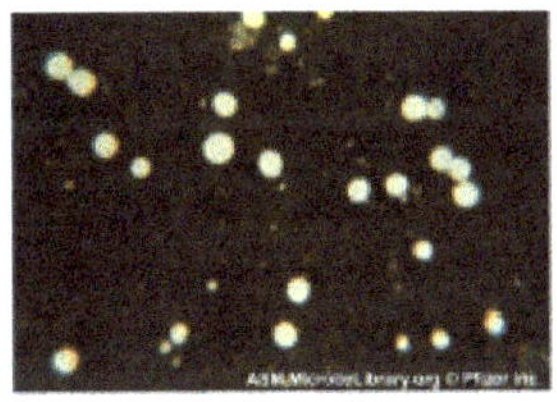

Figure 9.38 (credit "Gram stain": modification of work by Nina Parker; credit "Acid-fast stain": modification of work by American Society for Microbiology; credit "Endospore stain": modification of work by American Society for Microbiology; credit "Capsule stain" : modification of work by American Society for Microbiology; credit "Flagella stain": modification of work by Centers for Disease Control and Prevention)

Preparing Specimens for Electron Microscopy

Samples to be analyzed using a TEM must have very thin sections. But cells are too soft to cut thinly, even with diamond knives. To cut cells without damage, the cells must be embedded in plastic resin and then dehydrated through a series of soaks in ethanol solutions (50%, 60%, 70%, and so on). The ethanol replaces the water in the cells, and the resin dissolves in ethanol and enters the cell, where it solidifies. Next, **thin sections** are cut using a specialized device called an **ultramicrotome** (Figure 9.39). Finally, samples are fixed to fine copper wire or carbon-fiber grids and stained—not with colored dyes, but with substances like uranyl acetate or osmium tetroxide, which contain electron-dense heavy metal atoms.

When samples are prepared for viewing using an SEM, they must also be dehydrated using an ethanol series. However, they must be even drier than is necessary for a TEM. Critical point drying with inert liquid carbon dioxide under pressure is used to displace the water from the specimen. After drying, the specimens are sputter- coated with metal by knocking atoms off of a palladium target, with energetic particles. Sputter-coating prevents specimens from becoming charged by the SEM's electron beam.

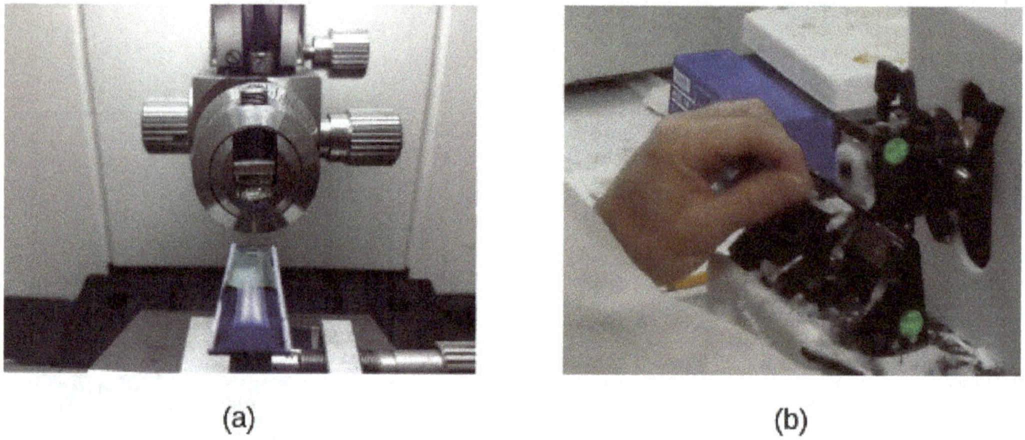

Figure 9.39 (a) An ultramicrotome used to prepare specimens for a TEM. (b) A technician uses an ultramicrotome to slice a specimen into thin sections. (credit a: modification of work by "Frost Museum"/Flickr; credit b: modification of work by U.S. Fish and Wildlife Service Northeast Region)

Using Microscopy to Diagnose Syphilis

The causative agent of syphilis is *Treponema pallidum*, a flexible, spiral cell (spirochete) that can be very thin (<0.15 μm) and match the refractive index of the medium, making it difficult to view using brightfield microscopy. Additionally, this species has not been successfully cultured in the laboratory on an artificial medium; therefore, diagnosis depends upon successful identification using microscopic techniques and serology (analysis of body fluids, often looking for antibodies to a pathogen). Since fixation and staining would kill the cells, darkfield microscopy is typically used for observing live specimens and viewing their movements. However, other approaches can also be used. For example, the cells can be thickened with silver particles (in tissue sections) and observed using a light microscope. It is also possible to use fluorescence or electron microscopy to view *Treponema* (Figure 9.40).

In clinical settings, indirect immunofluorescence is often used to identify *Treponema.* A primary, unstained antibody attaches directly to the pathogen surface, and secondary antibodies "tagged" with a fluorescent stain attach to the primary antibody. Multiple secondary antibodies can attach to each primary antibody, amplifying the amount of stain attached to each *Treponema* cell, making them easier to spot (Figure 9.41).

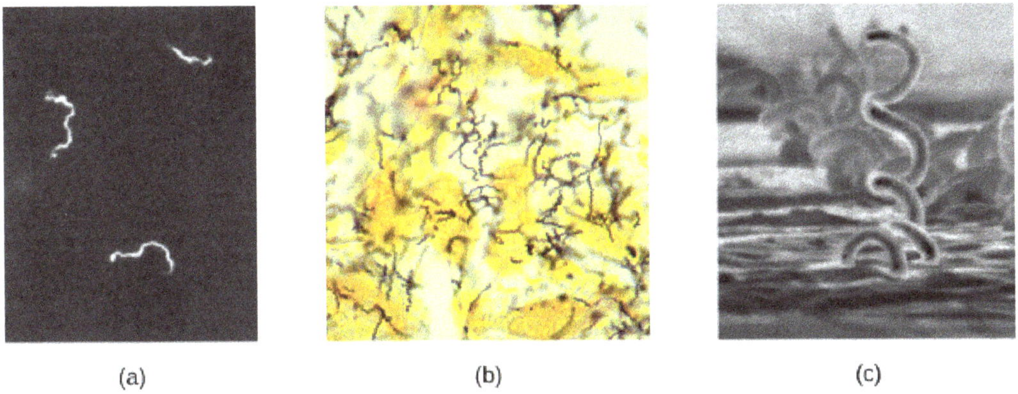

Figure 9.40 (a) Living, unstained *Treponema pallidum* spirochetes can be viewed under a darkfield microscope. In this brightfield image, a modified Steiner silver stain is used to visualized *T. pallidum* spirochetes. Though the stain kills the cells, it increases the contrast to make them more visible. (c) While not used for standard diagnostic testing, *T. pallidum* can also be examined using scanning electron microscopy. (credit a: modification of work by Centers for Disease Control and Prevention; credit b: modification of work by Centers for Disease Control and Prevention; credit c: modification of work by Centers for Disease Control and Prevention)

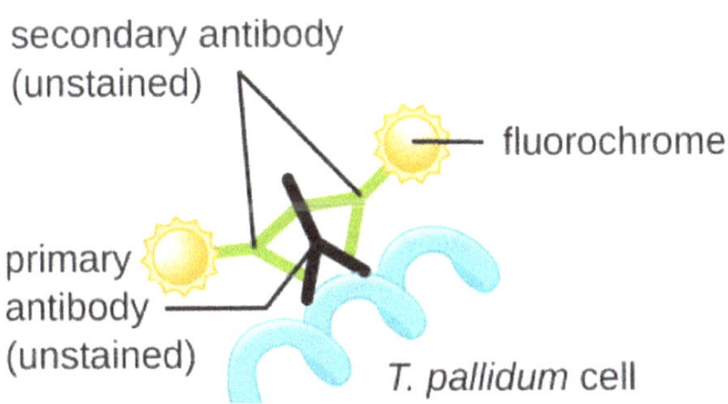

Figure 9.41 Indirect immunofluorescence can be used to identify *T. pallidum*, the causative agent of syphilis, in a specimen.

Preparation and Staining for Other Microscopes

Samples for fluorescence and confocal microscopy are prepared similarly to samples for light microscopy, except that the dyes are fluorochromes. Stains are often diluted in liquid before applying to the slide. Some dyes attach to an antibody to stain specific proteins on specific types of cells (immunofluorescence); others may attach to DNA molecules in a process called fluorescence in situ hybridization (FISH), causing cells to be stained based on whether they have a specific DNA sequence.

Sample preparation for two-photon microscopy is similar to fluorescence microscopy, except for the use of infrared dyes. Specimens for STM need to be on a very clean and atomically smooth surface. They are often mica coated with Au(111). Toluene vapor is a common fixative.

Resolution

From the results of the Gram stain, the technician now knows that Cindy's infection is caused by spherical, gram-positive bacteria that form grape-like clusters, which is typical of staphylococcal bacteria. After some additional testing, the technician determines that these bacteria are the medically important species known as *Staphylococcus aureus*, a common culprit in wound infections. Because some strains of *S. aureus* are resistant to many antibiotics, skin infections may spread to other areas of the body and become serious, sometimes even resulting in amputations or death if the correct antibiotics are not used.

After testing several antibiotics, the lab is able to identify one that is effective against this particular strain of *S. aureus*. Cindy's doctor quickly prescribes the medication and emphasizes the importance of taking the entire course of antibiotics, even if the infection appears to clear up before the last scheduled dose. This reduces the risk that any especially resistant bacteria could survive, causing a second infection or spreading to another person.

Microscopy and Antibiotic Resistance

As the use of antibiotics has proliferated in medicine, as well as agriculture, microbes have evolved to become more resistant. Strains of bacteria such as methicillin-resistant *S. aureus* (MRSA), which has developed a high level of resistance to many antibiotics, are an increasingly worrying problem, so much so that research is underway to develop new and more diversified antibiotics.

Fluorescence microscopy can be useful in testing the effectiveness of new antibiotics against resistant strains like MRSA. In a test of one new antibiotic derived from a marine bacterium, MC21-A (bromophene), researchers used the fluorescent dye SYTOX Green to stain samples of MRSA. SYTOX Green is often used to distinguish dead cells from living cells, with fluorescence microscopy. Live cells will not absorb the dye, but cells killed by an antibiotic will absorb the dye, since the antibiotic has damaged the bacterial cell membrane. In this particular case, MRSA bacteria that had been exposed to MC21-A did, indeed, appear green under the fluorescence microscope, leading researchers to conclude that it is an effective antibiotic against MRSA.

Of course, some argue that developing new antibiotics will only lead to even more antibiotic-resistant microbes, so-called superbugs that could spawn epidemics before new treatments can be developed. For this reason, many health professionals are beginning to exercise more discretion in prescribing antibiotics. Whereas antibiotics were once routinely prescribed for common illnesses without a definite diagnosis, doctors and hospitals are much more likely to conduct additional testing to determine whether an antibiotic is necessary and appropriate before prescribing.

A sick patient might reasonably object to this stingy approach to prescribing antibiotics. To the patient who simply wants to feel better as quickly as possible, the potential benefits of taking an antibiotic may seem to outweigh any immediate health risks that might occur if the antibiotic is ineffective. But at what point do the risks of widespread antibiotic use supersede the desire to use them in individual cases?

CHAPTER-10

PROCESS OF MICROBIAL GROWTH

INTRODUCTION: AN OVERVIEW

We are all familiar with the slimy layer on a pond surface or that makes rocks slippery. These are examples of biofilms—microorganisms embedded in thin layers of matrix material. Biofilms were long considered random assemblages of cells and had little attention from researchers. Recently, progress in visualization and biochemical methods has revealed that biofilms are an organized ecosystem within which many cells, usually of different species of bacteria, fungi, and algae, interact through cell signaling and coordinated responses. The biofilm provides a protected environment in harsh conditions and aids colonization by microorganisms. Biofilms also have clinical importance. They form on medical devices, resist routine cleaning and sterilization, and cause health-acquired infections. Within the body, biofilms form on the teeth as plaque, in the lungs of patients with cystic fibrosis, and on the cardiac tissue of patients with endocarditis. The slime layer helps protect the cells from host immune defenses and antibiotic treatments.

Studying biofilms requires new approaches. Because of the cells' adhesion properties, many of the methods for culturing and counting cells that are explored in this chapter are not easily applied to biofilms. This is the beginning of a new era of challenges and rewarding insight into the ways that microorganisms grow and thrive in nature.

HOW MICROBES GROW

The bacterial cell cycle involves the formation of new cells through the replication of DNA and partitioning of cellular components into two daughter cells. In prokaryotes, reproduction is always asexual, although extensive genetic recombination in the form of horizontal gene transfer takes place, as will be explored in a different chapter. Most bacteria have a single circular chromosome; however, some exceptions exist. For example, *Borrelia burgdorferi*, the causative agent of Lyme disease, has a linear chromosome.

Binary Fission

The most common mechanism of cell replication in bacteria is a process called **binary fission**, which is depicted in Figure 10.1. Before dividing, the cell grows and increases its number of cellular components. Next, the replication of DNA starts at a location on the circular chromosome called the origin of replication, where the chromosome is attached to the inner cell membrane. Replication continues in opposite directions along the chromosome until the terminus is reached.

The center of the enlarged cell constricts until two daughter cells are formed, each offspring receiving a complete copy of the parental genome and a division of the cytoplasm (cytokinesis). This process of cytokinesis and cell division is directed by a protein called FtsZ. FtsZ assembles into a Z ring on the cytoplasmic membrane (Figure 10.2). The Z ring is anchored by FtsZ-binding proteins and defines the division plane between the two daughter cells. Additional proteins required for cell division are added to the Z ring to form a structure called the divisome. The divisome activates to produce a peptidoglycan cell wall and build a **septum** that divides the two daughter cells. The daughter cells are separated by the division septum, where all of the cells' outer layers (the cell wall and outer membranes, if present) must be remodeled to complete division. For example, we know that specific enzymes break bonds between the monomers in peptidoglycans and allow addition of new subunits along the division septum.

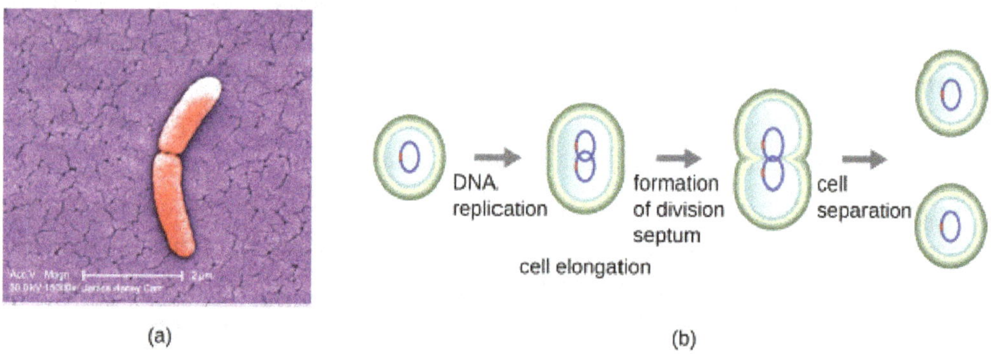

Figure 10.1 (a) The electron micrograph depicts two cells of *Salmonella typhimurium* after a binary fission event. (b) Binary fission in bacteria starts with the replication of DNA as the cell elongates. A division septum forms in the center of the cell. Two daughter cells of similar size form and separate, each receiving a copy of the original chromosome. (credit a: modification of work by Centers for Disease Control and Prevention)

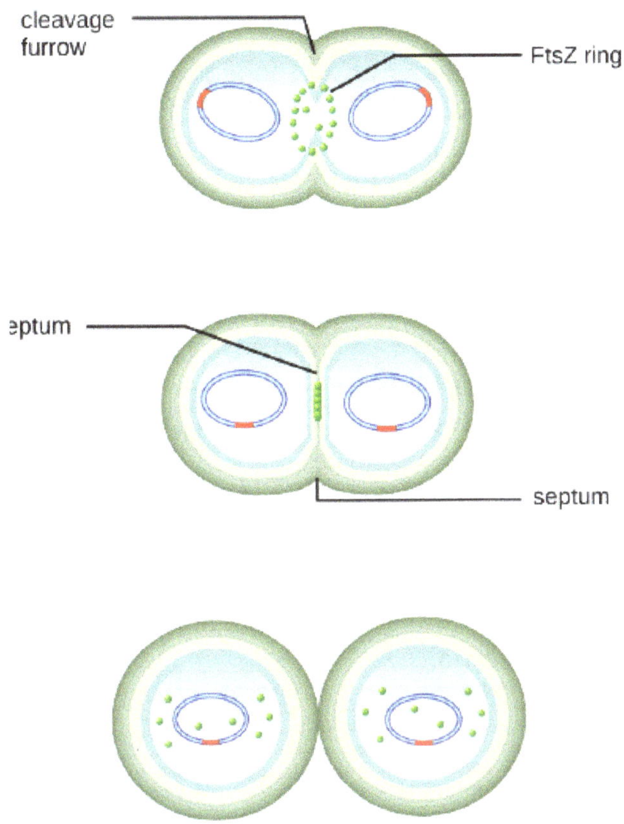

Figure 10.2 FtsZ proteins assemble to form a Z ring that is anchored to the plasma membrane. The Z ring pinches the cell envelope to separate the cytoplasm of the new cells.

Generation Time

In eukaryotic organisms, the generation time is the time between the same points of the life cycle in two successive generations. For example, the typical generation time for the human population is 25 years. This definition is not practical for bacteria, which may reproduce rapidly or remain dormant for thousands of years. In prokaryotes (Bacteria and Archaea), the **generation time** is also called the **doubling time** and is defined as the time it takes for the population to double through one round of binary fission. Bacterial doubling times vary enormously. Whereas *Escherichia coli* can double in as little as 20 minutes under optimal growth conditions in the laboratory, bacteria of the same species may need several days to double in especially harsh environments. Most pathogens grow rapidly, like *E. coli*, but there are exceptions. For example, *Mycobacterium tuberculosis*, the causative agent of tuberculosis, has a generation time of between 15 and 20 hours. On the other hand, *M. leprae*, which causes Hansen's disease (leprosy), grows much more slowly, with a doubling time of 14 days.

Calculating Number of Cells

It is possible to predict the number of cells in a population when they divide by binary fission at a constant rate. As an example, consider what happens if a single cell divides every 30 minutes for 24 hours. The diagram in Figure 10.3 shows the increase in cell numbers for the first three generations.

The number of cells increases exponentially and can be expressed as 2^n, where n is the number of generations. If cells divide every 30 minutes, after 24 hours, 48 divisions would have taken place. If we apply the formula 2^n, where n is equal to 48, the single cell would give rise to 2^{48} or 281,474,976,710,656 cells at 48 generations (24 hours). When dealing with such huge numbers, it is more practical to use scientific notation. Therefore, we express the number of cells as 2.8×10^{14} cells.

In our example, we used one cell as the initial number of cells. For any number of starting cells, the formula is adapted as follows:

$$N_n = N_0\, 2^n$$

N_n is the number of cells at any generation n, N_0 is the initial number of cells, and n is the number of generations.

Number of generations (n)	Number of cells	Each division adds two new cells
0	1	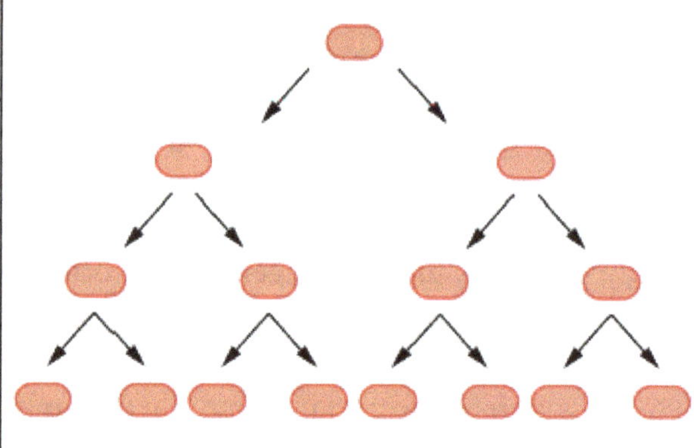
1	2	
2	4	
3	8	

Figure 10.3 The parental cell divides and gives rise to two daughter cells. Each of the daughter cells, in turn, divides, giving a total of four cells in the second generation and eight cells in the third generation. Each division doubles the number of cells.

The Growth Curve

Microorganisms grown in closed culture (also known as a batch culture), in which no nutrients are added and most waste is not removed, follow a reproducible growth pattern referred to as the **growth curve**. An example of a batch culture in nature is a pond in which a small number of cells grow in a closed environment. The **culture density** is defined as the number of cells per unit volume. In a closed environment, the culture density is also a measure of the number of cells in the population. Infections of the body do not always follow the growth curve, but correlations can exist depending upon the site and type of infection. When the number of live cells is plotted against time, distinct phases can be observed in the curve (Figure 10.4).

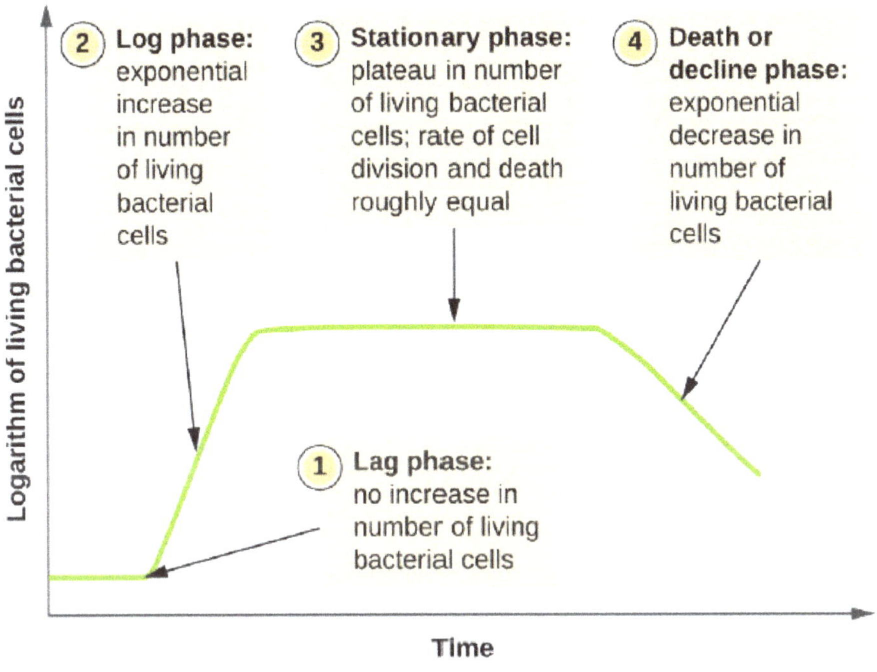

Figure 10.4 The growth curve of a bacterial culture is represented by the logarithm of the number of live cells plotted as a function of time. The graph can be divided into four phases according to the slope, each of which matches events in the cell. The four phases are lag, log, stationary, and death.

The Lag Phase

The beginning of the growth curve represents a small number of cells, referred to as an **inoculum**, that are added to a fresh **culture medium**, a nutritional broth that supports growth. The initial phase of the growth curve is called the **lag phase**, during which cells are gearing up for the next phase of growth. The number of cells does not change during the lag phase; however, cells grow larger and are metabolically active, synthesizing proteins needed to grow within the medium. If any cells were damaged or shocked during the transfer to the new medium, repair takes place during the lag phase. The duration of the lag phase is determined by many factors, including the species and genetic make-up of the cells, the composition of the medium, and the size of the original inoculum.

The Log Phase

In the **logarithmic (log) growth phase**, sometimes called exponential growth phase, the cells are actively dividing by binary fission and their number increases exponentially. For any given bacterial species, the generation time under specific growth conditions (nutrients, temperature, pH, and so forth) is genetically determined, and this generation time is called the **intrinsic growth rate.** During the log phase, the relationship between time and number of cells is not linear but exponential; however, the growth curve is often plotted on a semilogarithmic graph, as shown in Figure 10.5, which gives the appearance of a linear relationship.

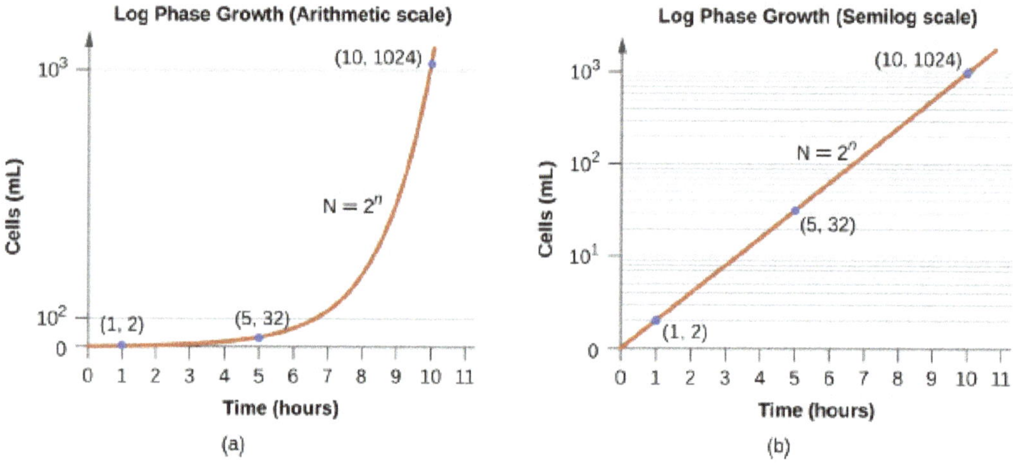

Figure 10.5 Both graphs illustrate population growth during the log phase for a bacterial sample with an initial population of one cell and a doubling time of 1 hour. (a) When plotted on an arithmetic scale, the growth rate resembles a curve. (b) When plotted on a semilogarithmic scale (meaning the values on the *y*-axis are logarithmic), the growth rate appears linear.

Cells in the log phase show constant growth rate and uniform metabolic activity. For this reason, cells in the log phase are preferentially used for industrial applications and research work. The log phase is also the stage where bacteria are the most susceptible to the action of disinfectants and common antibiotics that affect protein, DNA, and cell-wall synthesis.

Stationary Phase

As the number of cells increases through the log phase, several factors contribute to a slowing of the growth rate. Waste products accumulate and nutrients are gradually used up. In addition, gradual depletion of oxygen begins to limit aerobic cell growth. This combination of unfavourable conditions slows and finally stalls population growth. The total number of live cells reaches a plateau referred to as the **stationary phase** (Figure 10.4). In this phase, the number of new cells created by cell division is now equivalent to the number of cells dying; thus, the total population of living cells is relatively stagnant. The culture density in a stationary culture is constant. The culture's carrying capacity, or maximum culture density, depends on the types of microorganisms in the culture and the specific conditions of the culture; however, carrying capacity is constant for a given organism grown under the same conditions.

During the stationary phase, cells switch to a survival mode of metabolism. As growth slows, so too does the synthesis of peptidoglycans, proteins, and nucleic-acids; thus, stationary cultures are less susceptible to antibiotics that disrupt these processes. In bacteria capable of producing endospores, many cells undergo sporulation during the stationary phase. Secondary metabolites, including antibiotics, are synthesized in the stationary phase. In certain pathogenic bacteria, the stationary phase is also associated with the expression of virulence factors, products that contribute to a microbe's ability to survive, reproduce, and cause disease in a host organism. For example, quorum sensing in *Staphylococcus aureus* initiates the production of enzymes that can break down human tissue and cellular debris, clearing the way for bacteria to spread to new tissue where nutrients are more plentiful.

The Death Phase

As a culture medium accumulates toxic waste and nutrients are exhausted, cells die in greater and greater numbers. Soon, the number of dying cells exceeds the number of dividing cells, leading to an exponential decrease in the number of cells (Figure 10.4). This is the aptly named **death phase**, sometimes called the decline phase. Many cells lyse and release nutrients into the medium, allowing surviving cells to maintain viability and form endospores. A few cells, the so-called **persisters**, are characterized by a slow metabolic rate. Persister cells are medically important because they are associated with certain chronic infections, such as tuberculosis, that do not respond to antibiotic treatment.

Sustaining Microbial Growth

The growth pattern shown in Figure 10.4 takes place in a closed environment; nutrients are not added and waste and dead cells are not removed. In many cases, though, it is advantageous to maintain cells in the logarithmic phase of growth. One example is in industries that harvest microbial products. A chemostat (Figure 10.6) is used to maintain a continuous culture in which nutrients are supplied at a steady rate. A controlled amount of air is mixed in for aerobic processes. Bacterial suspension is removed at the same rate as nutrients flow in to maintain an optimal growth environment.

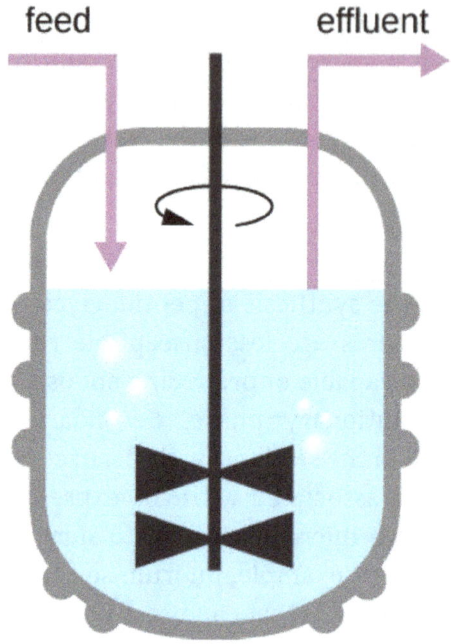

Figure 10.6 A chemostat is a culture vessel fitted with an opening to add nutrients (feed) and an outlet to remove contents (effluent), effectively diluting toxic wastes and dead cells. The addition and removal of fluids is adjusted to maintain the culture in the logarithmic phase of growth. If aerobic bacteria are grown, suitable oxygen levels are maintained.

Measurement of Bacterial Growth

Estimating the number of bacterial cells in a sample, known as a bacterial count, is a common task performed by microbiologists. The number of bacteria in a clinical sample serves as an indication of the extent of an infection. Quality control of drinking water, food, medication, and even cosmetics relies on estimates of bacterial counts to detect contamination and prevent the spread of disease. Two major approaches are used to measure cell number. The direct methods involve counting cells, whereas the indirect methods depend on the measurement of cell presence or activity without actually counting individual cells. Both direct and indirect methods have advantages and disadvantages for specific applications.

Direct Cell Count

Direct cell count refers to counting the cells in a liquid culture or colonies on a plate. It is a direct way of estimating how many organisms are present in a sample. Let's look first at a simple and fast method that requires only a specialized slide and a compound microscope.

The simplest way to count bacteria is called the **direct microscopic cell count,** which involves transferring a known volume of a culture to a calibrated slide and counting the cells under a light microscope. The calibrated slide is called a **Petroff-Hausser chamber** (**Figure 10.7**) and is similar to a hemocytometer used to count red blood cells. The central area of the counting chamber is etched into squares of various sizes. A sample of the culture suspension is added to the chamber under a coverslip that is placed at a specific height from the surface of the grid. It is possible to estimate the concentration of cells in the original sample by counting individual cells in a number of squares and determining the volume of the sample observed. The area of the squares and the height at which the coverslip is positioned are specified for the chamber. The concentration must be corrected for dilution if the sample was diluted before enumeration.

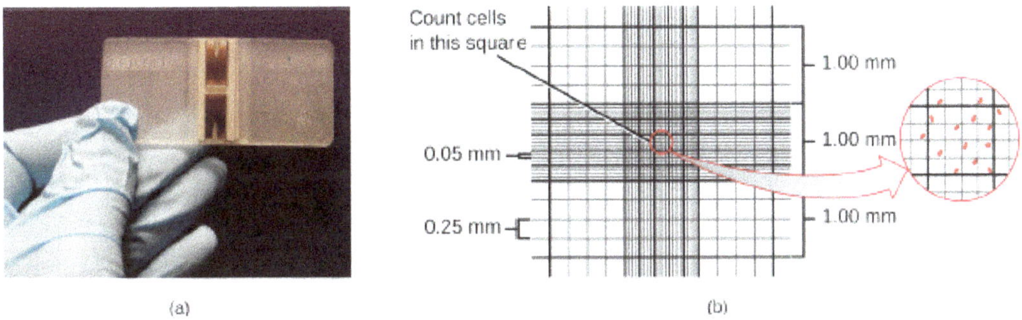

Figure 10.7 (a) A Petroff-Hausser chamber is a special slide designed for counting the bacterial cells in a measured volume of a sample. A grid is etched on the slide to facilitate precision in counting. (b) This diagram illustrates the grid of a Petroff-Hausser chamber, which is made up of squares of known areas. The enlarged view shows the square within which bacteria (red cells) are counted. If the coverslip is 0.2 mm above the grid and the square has an area of mm^2, then the volume is 0.008 mm^3, or 0.000008 mL. Since there are 10 cells inside the square, the density of bacteria is 10 cells/0.000008 mL, which equates to 1,250,000 cells/mL. (credit a: modification of work by Jeffrey M. Vinocur)

Cells in several small squares must be counted and the average taken to obtain a reliable measurement. The advantages of the chamber are that the method is easy to use, relatively fast, and inexpensive. On the downside, the counting chamber does not work well with dilute cultures because there may not be enough cells to count.

Using a counting chamber does not necessarily yield an accurate count of the number of live cells because it is not always possible to distinguish between live cells, dead cells, and debris of the same size under the microscope. However, newly developed fluorescence staining techniques make it possible to distinguish viable and dead bacteria. These viability stains (or live stains) bind to nucleic acids, but the primary and secondary stains differ in their ability to cross the cytoplasmic membrane. The primary stain, which fluoresces green, can penetrate intact cytoplasmic membranes, staining both live and dead cells. The secondary stain, which fluoresces red, can stain a cell only if the cytoplasmic membrane is considerably damaged. Thus, live cells fluoresce green because they only absorb the green stain, whereas dead cells appear red because the red stain displaces the green stain on their nucleic acids (Figure 10.8).

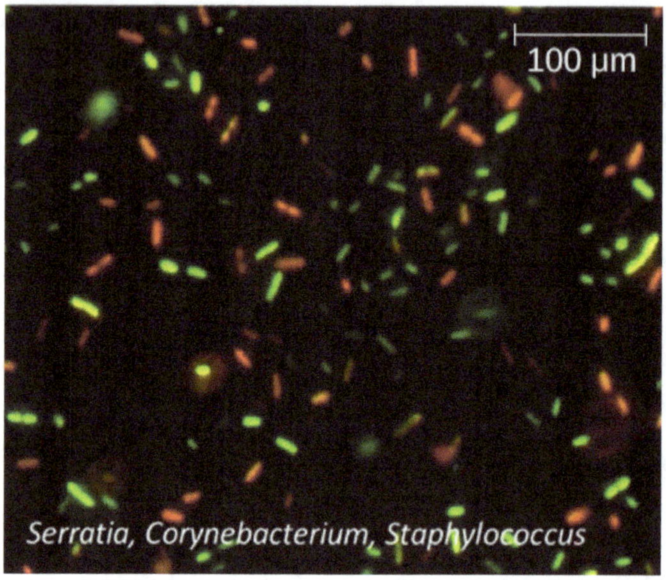

Figure 10.8 Fluorescence staining can be used to differentiate between viable and dead bacterial cells in a sample for purposes of counting. Viable cells are stained green, whereas dead cells are stained red. (credit: modification of work by Emerson J, Adams R, Bentancourt Román C, Brooks B, Coil D, Dahlhousen K, Ganz H, et al.)

Another technique uses an electronic cell counting device (Coulter counter) to detect and count the changes in electrical resistance in a saline solution. A glass tube with a small opening is immersed in an electrolyte solution. A first electrode is suspended in the glass tube. A second electrode is located outside of the tube. As cells are drawn through the small aperture in the glass tube, they briefly change the resistance measured between the two electrodes and the change is recorded by an electronic sensor (Figure 10.9); each resistance change represents a cell. The method is rapid and accurate within a range of concentrations; however, if the culture is too concentrated, more than one cell may pass through the aperture at any given time and skew the results. This method also does not differentiate between live and dead cells.

Direct counts provide an estimate of the total number of cells in a sample. However, in many situations, it is important to know the number of live, or **viable**, cells. Counts of live cells are needed when assessing the extent of an infection, the effectiveness of antimicrobial compounds and medication, or contamination of food and water.

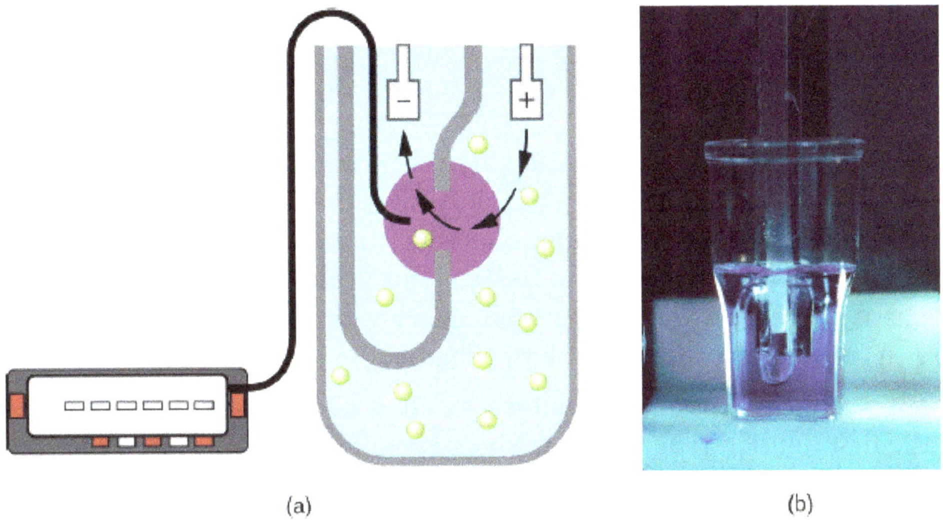

Figure 10.9 A Coulter counter is an electronic device that counts cells. It measures the change in resistance in an electrolyte solution that takes place when a cell passes through a small opening in the inside container wall. A detector automatically counts the number of cells passing through the opening. (credit b: modification of work by National Institutes of Health)

Plate Count

The **viable plate count**, or simply plate count, is a count of viable or live cells. It is based on the principle that viable cells replicate and give rise to visible colonies when incubated under suitable conditions for the specimen. The results are usually expressed as **colony-forming units** per milliliter (CFU/mL) rather than cells per milliliter because more than one cell may have landed on the same spot to give rise to a single colony. Furthermore, samples of bacteria that grow in clusters or chains are difficult to disperse and a single colony may represent several cells. Some cells are described as viable but nonculturable and will not form colonies on solid media. For all these reasons, the viable plate count is considered a low estimate of the actual number of live cells. These limitations do not detract from the usefulness of the method, which provides estimates of live bacterial numbers.

Microbiologists typically count plates with 30–300 colonies. Samples with too few colonies (<30) do not give statistically reliable numbers, and overcrowded plates (>300 colonies) make it difficult to accurately count individual colonies. Also, counts in this range minimize occurrences of more than one bacterial cell forming a single colony. Thus, the calculated CFU is closer to the true number of live bacteria in the population.

There are two common approaches to inoculating plates for viable counts: the pour plate and the spread plate methods. Although the final inoculation procedure differs between these two methods, they both start with a serial dilution of the culture.

Serial Dilution

The serial dilution of a culture is an important first step before proceeding to either the pour plate or spread plate method. The goal of the serial dilution process is to obtain plates with CFUs in the range of 30–300, and the process usually involves several dilutions in multiples of 10 to simplify calculation. The number of serial dilutions is chosen according to a preliminary estimate of the culture density. Figure 10.10 illustrates the serial dilution method.

A fixed volume of the original culture, 1.0 mL, is added to and thoroughly mixed with the first dilution tube solution, which contains 9.0 mL of sterile broth. This step represents a dilution factor of 10, or 1:10, compared with the original culture. From this first dilution, the same volume, 1.0 mL, is withdrawn and mixed with a fresh tube of 9.0 mL of dilution solution. The dilution factor is now 1:100 compared with the original culture. This process continues until a series of dilutions is produced that will bracket the desired cell concentration for accurate counting. From each tube, a sample is plated on solid medium using either the pour plate method (Figure 10.11) or the spread plate method (Figure 10.12). The plates are incubated until colonies appear. Two to three plates are usually prepared from each dilution and the numbers of colonies counted on each plate are averaged. In all cases, thorough mixing of samples with the dilution medium (to ensure the cell distribution in the tube is random) is paramount to obtaining reliable results.

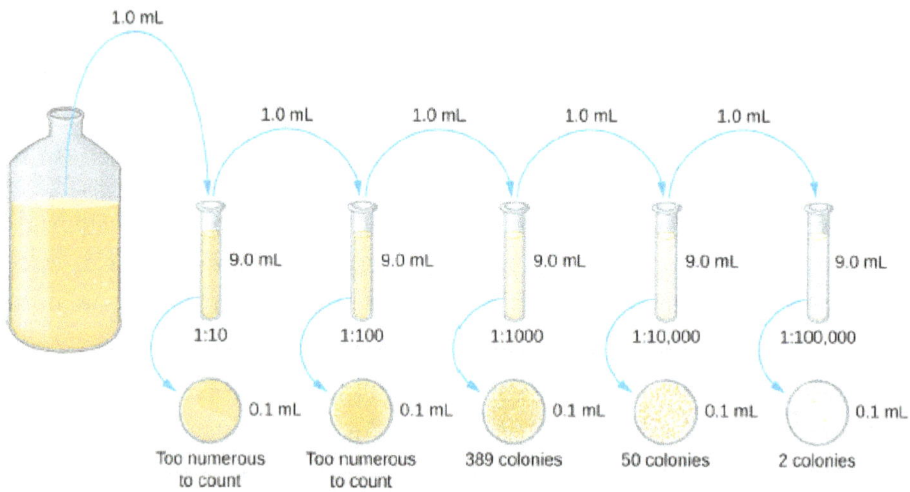

Figure 10.10 Serial dilution involves diluting a fixed volume of cells mixed with dilution solution using the previous dilution as an inoculum. The result is dilution of the original culture by an exponentially growing factor. (credit: modification of work by "Leberechtc"/Wikimedia Commons)

The dilution factor is used to calculate the number of cells in the original cell culture. In our example, an average of 50 colonies was counted on the plates obtained from the 1:10,000 dilution. Because only 0.1 mL of suspension was pipetted on the plate, the multiplier required to reconstitute the original concentration is 10 × 10,000. The number of CFU per mL is equal to 50 × 10 × 10,000 = 5,000,000. The number of bacteria in the culture is estimated as 5 million cells/mL. The colony count obtained from the 1:1000 dilution was 389, well below the expected 500 for a 10-fold difference in dilutions. This highlights the issue of inaccuracy when colony counts are greater than 300 and more than one bacterial cell grows into a single colony.

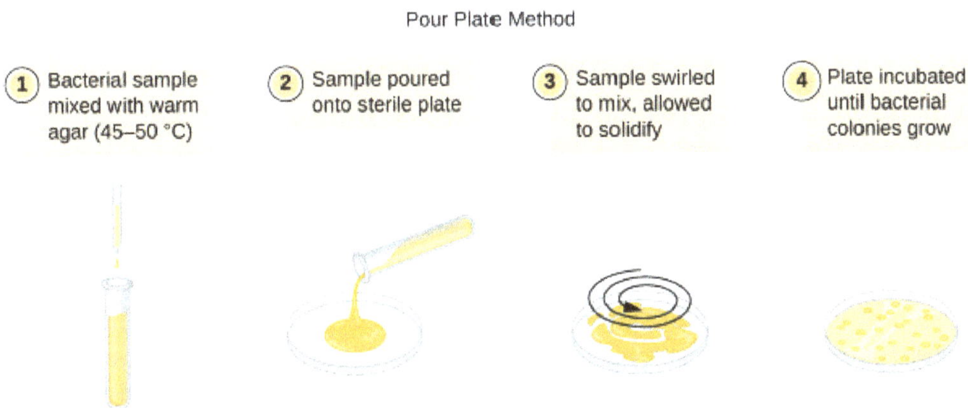

Figure 10.11 In the pour plate method of cell counting, the sample is mixed in liquid warm agar (45–50 °C) poured into a sterile Petri dish and further mixed by swirling. This process is repeated for each serial dilution prepared. The resulting colonies are counted and provide an estimate of the number of cells in the original volume sampled.

Figure 10.12 In the spread plate method of cell counting, the sample is poured onto solid agar and then spread using a sterile spreader. This process is repeated for each serial dilution prepared. The resulting colonies are counted and provide an estimate of the number of cells in the original volume samples.

A very dilute sample—drinking water, for example—may not contain enough organisms to use either of the plate count methods described. In such cases, the original sample must be concentrated rather than diluted before plating. This can be accomplished using a modification of the plate count technique called the **membrane filtration technique.** Known volumes are vacuum-filtered aseptically through a membrane with a pore size small enough to trap microorganisms. The membrane is transferred to a Petri plate containing an appropriate growth medium. Colonies are counted after incubation. Calculation of the cell density is made by dividing the cell count by the volume of filtered liquid.

The Most Probable Number

The number of microorganisms in dilute samples is usually too low to be detected by the plate count methods described thus far. For these specimens, microbiologists routinely use the **most probable number (MPN) method**, a statistical procedure for estimating of the number of viable microorganisms in a sample. Often used for water and food samples, the MPN method evaluates detectable growth by observing changes in turbidity or colour due to metabolic activity.

A typical application of MPN method is the estimation of the number of coliforms in a sample of pond water. Coliforms are gram-negative rod bacteria that ferment lactose. The presence of coliforms in water is considered a sign of contamination by fecal matter. For the method illustrated in Figure 10.13, a series of three dilutions of the water sample is tested by inoculating five lactose broth tubes with 10 mL of sample, five lactose broth tubes with 1 mL of sample, and five lactose broth tubes with 0.1 mL of sample. The lactose broth tubes contain a pH indicator that changes colour from red to yellow when the lactose is fermented. After inoculation and incubation, the tubes are examined for an indication of coliform growth by a colour change in media from red to yellow. The first set of tubes (10-mL sample) showed growth in all the tubes; the second set of tubes (1 mL) showed growth in two tubes out of five; in the third set of tubes, no growth is observed in any of the tubes (0.1-mL dilution). The numbers 5, 2, and 0 are compared with Figure B1 in Appendix B, which has been constructed using a probability model of the sampling procedure. From our reading of the table, we conclude that 49 is the most probable number of bacteria per 100 mL of pond water.

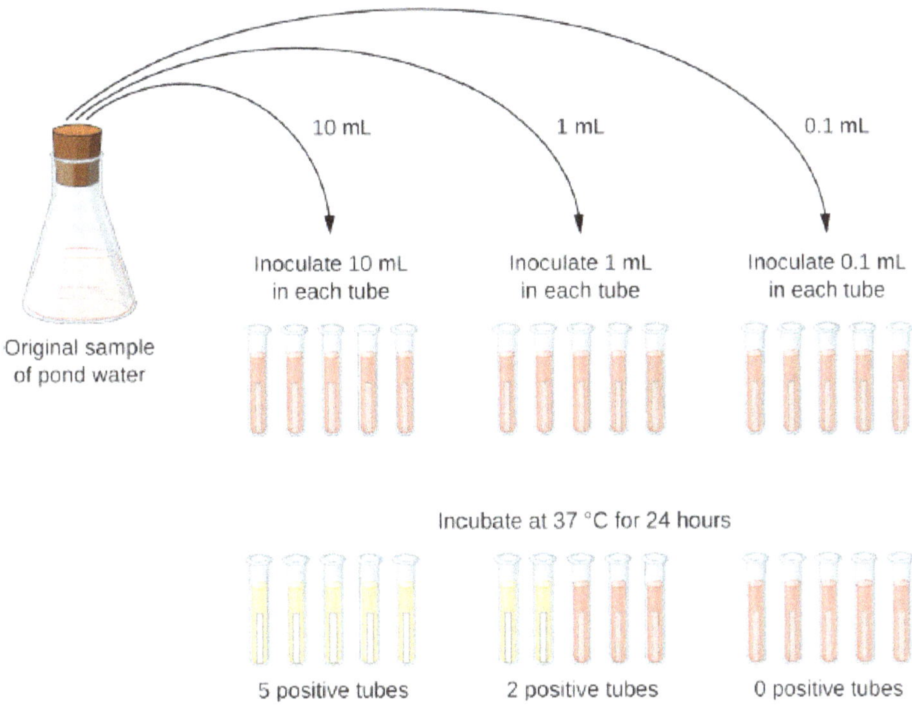

Figure 10.13 In the most probable number method, sets of five lactose broth tubes are inoculated with three different volumes of pond water: 10 mL, 1 mL, and 0.1mL. Bacterial growth is assessed through a change in the colour of the broth from red to yellow as lactose is fermented.

Indirect Cell Counts

Besides direct methods of counting cells, other methods, based on an indirect detection of cell density, are commonly used to estimate and compare cell densities in a culture. The foremost approach is to measure the **turbidity** (cloudiness) of a sample of bacteria in a liquid suspension. The laboratory instrument used to measure turbidity is called a spectrophotometer (Figure 10.14). In a spectrophotometer, a light beam is transmitted through a bacterial suspension, the light passing through the suspension is measured by a detector, and the amount of light passing through the sample and reaching the detector is converted to either percent transmission or a logarithmic value called absorbance (optical density). As the numbers of bacteria in a suspension increase, the turbidity also increases and causes less light to reach the detector. The decrease in light passing through the sample and reaching the detector is associated with a decrease in percent transmission and increase in absorbance measured by the spectrophotometer.

Measuring turbidity is a fast method to estimate cell density as long as there are enough cells in a sample to produce turbidity. It is possible to correlate turbidity readings to the actual number of cells by performing a viable plate count of samples taken from cultures having a range of absorbance values. Using these values, a calibration curve is generated by plotting turbidity as a function of cell density. Once the calibration curve has been produced, it can be used to estimate cell counts for all samples obtained or cultured under similar conditions and with densities within the range of values used to construct the curve.

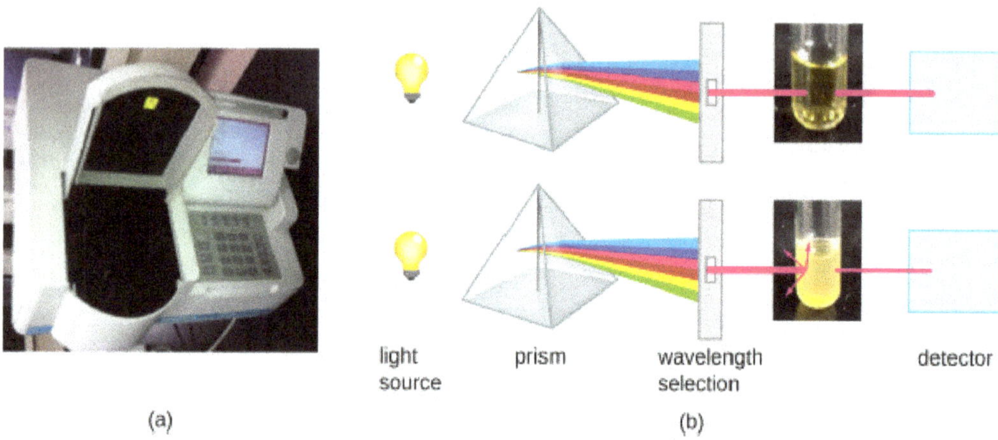

Figure 10.14 (a) A spectrophotometer is commonly used to measure the turbidity of a bacterial cell suspension as an indirect measure of cell density. (b) A spectrophotometer works by splitting white light from a source into a spectrum. The spectrophotometer allows choice of the wavelength of light to use for the measurement. The optical density (turbidity) of the sample will depend on the wavelength, so once one wavelength is chosen, it must be used consistently. The filtered light passes through the sample (or a control with only medium) and the light intensity is measured by a detector. The light passing into a suspension of bacteria is scattered by the cells in such a way that some fraction of it never reaches the detector. This scattering happens to a far lesser degree in the control tube with only the medium. (credit a: modification of work by Hwang HS, Kim MS; credit b "test tube photos": modification of work by Suzanne Wakim)

Measuring dry weight of a culture sample is another indirect method of evaluating culture density without directly measuring cell counts. The cell suspension used for weighing must be concentrated by filtration or centrifugation, washed, and then dried before the measurements are taken. The degree of drying must be standardized to account for residual water content. This method is especially useful for filamentous microorganisms, which are difficult to enumerate by direct or viable plate count.

As we have seen, methods to estimate viable cell numbers can be labour intensive and take time because cells must be grown. Recently, indirect ways of measuring live cells have been developed that are both fast and easy to implement. These methods measure cell activity by following the production of metabolic products or disappearance of reactants. Adenosine triphosphate (ATP) formation, biosynthesis of proteins and nucleic acids, and consumption of oxygen can all be monitored to estimate the number of cells.

Alternative Patterns of Cell Division

Binary fission is the most common pattern of cell division in prokaryotes, but it is not the only one. Other mechanisms usually involve asymmetrical division (as in budding) or production of spores in aerial filaments.

In some cyanobacteria, many nucleoids may accumulate in an enlarged round cell or along a filament, leading to the generation of many new cells at once. The new cells often split from the parent filament and float away in a process called **fragmentation** (Figure 10.15). Fragmentation is commonly observed in the Actinomycetes, a group of gram-positive, anaerobic bacteria commonly found in soil. Another curious example of cell division in prokaryotes, reminiscent of live birth in animals, is exhibited by the giant bacterium *Epulopiscium*. Several daughter cells grow fully in the parent cell, which eventually disintegrates, releasing the new cells to the environment. Other species may form a long narrow extension at one pole in a process called **budding**. The tip of the extension swells and forms a smaller cell, the bud that eventually detaches from the parent cell. Budding is most common in yeast (Figure 10.15), but it is also observed in prosthecate bacteria and some cyanobacteria.

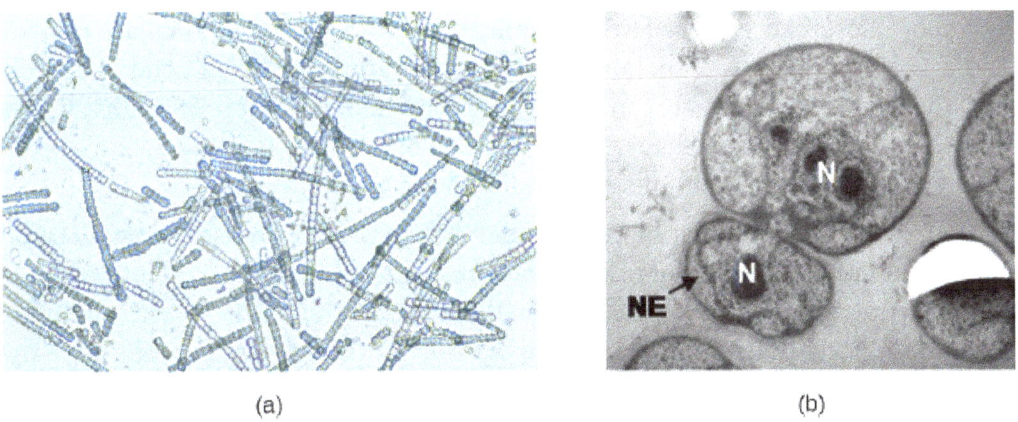

Figure 10.15 (a) Filamentous cyanobacteria, like those pictured here, replicate by fragmentation. (b) In this electron micrograph, cells of the bacterium *Gemmata obscuriglobus* are budding. The larger cell is the mother cell. Labels indicate the nucleoids (N) and the still-forming nuclear envelope (NE) of the daughter cell. (credit a: modification of work by CSIRO; credit b: modification of work by Kuo-Chang Lee, Rick I Webb and John A Fuerst)

The soil bacteria *Actinomyces* grow in long filaments divided by septa, similar to the mycelia seen in fungi, resulting in long cells with multiple nucleoids. Environmental signals, probably related to low nutrient availability, lead to the formation of aerial filaments. Within these aerial filaments, elongated cells divide simultaneously. The new cells, which contain a single nucleoid, develop into spores that give rise to new colonies.

Biofilms

In nature, microorganisms grow mainly in **biofilms**, complex and dynamic ecosystems that form on a variety of environmental surfaces, from industrial conduits and water treatment pipelines to rocks in river beds. Biofilms are not restricted to solid surface substrates, however. Almost any surface in a liquid environment containing some minimal nutrients will eventually develop a biofilm. Microbial mats that float on water, for example, are biofilms that contain large populations of photosynthetic microorganisms. Biofilms found in the human mouth may contain hundreds of bacterial species. Regardless of the environment where they occur, biofilms are not random collections of microorganisms; rather, they are highly structured communities that provide a selective advantage to their constituent microorganisms.

Biofilm Structure

Observations using confocal microscopy have shown that environmental conditions influence the overall structure of biofilms. Filamentous biofilms called streamers form in rapidly flowing water, such as freshwater streams, eddies, and specially designed laboratory flow cells that replicate growth conditions in fast-moving fluids. The streamers are anchored to the substrate by a "head" and the "tail" floats downstream in the current. In still or slow-moving water, biofilms mainly assume a mushroom-like shape. The structure of biofilms may also change with other environmental conditions such as nutrient availability.

Detailed observations of biofilms under confocal laser and scanning electron microscopes reveal clusters of microorganisms embedded in a matrix interspersed with open water channels. The extracellular matrix consists of **extracellular polymeric substances (EPS)** secreted by the organisms in the biofilm. The extracellular matrix represents a large fraction of the biofilm, accounting for 50%–90% of the total dry mass. The properties of the EPS vary according to the resident organisms and environmental conditions.

EPS is a hydrated gel composed primarily of polysaccharides and containing other macromolecules such as proteins, nucleic acids, and lipids. It plays a key role in maintaining the integrity and function of the biofilm. Channels in the EPS allow movement of nutrients, waste, and gases throughout the biofilm. This keeps the cells hydrated, preventing desiccation. EPS also shelters organisms in the biofilm from predation by other microbes or cells (e.g., protozoans, white blood cells in the human body).

Biofilm Formation

Free-floating microbial cells that live in an aquatic environment are called **planktonic** cells. The formation of a biofilm essentially involves the attachment of planktonic cells to a substrate, where they become **sessile** (attached to a surface). This occurs in stages, as depicted in Figure 10.16. The first stage involves the attachment of planktonic cells to a surface coated with a conditioning film of organic material. At this point, attachment to the substrate is reversible, but as cells express new phenotypes that facilitate the formation of EPS, they transition from a planktonic to a sessile lifestyle. The biofilm develops characteristic structures, including an extensive matrix and water channels. Appendages such as fimbriae, pili, and flagella interact with the EPS, and microscopy and genetic analysis suggest that such structures are required for the establishment of a mature biofilm. In the last stage of the biofilm life cycle, cells on the periphery of the biofilm revert to a planktonic lifestyle, sloughing off the mature biofilm to colonize new sites. This stage is referred to as dispersal.

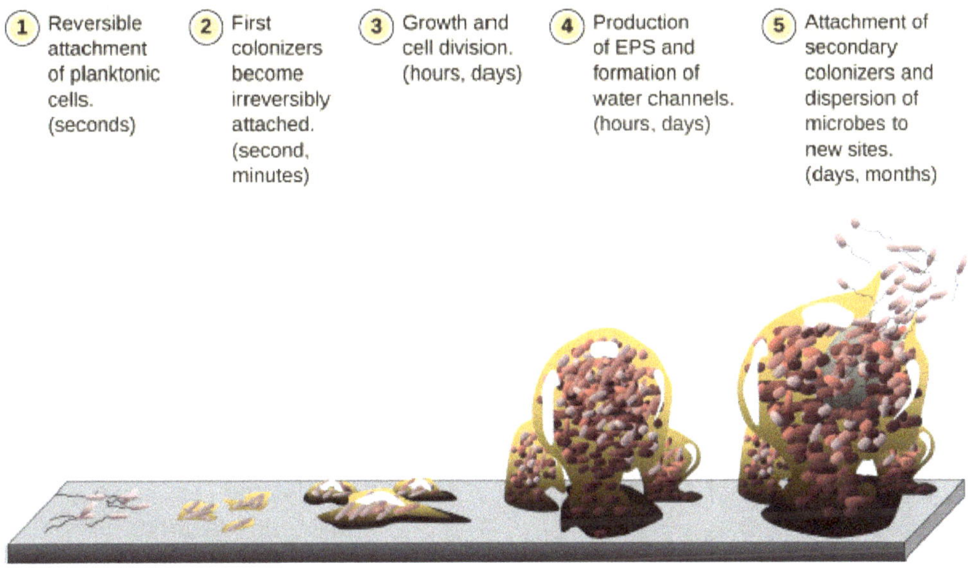

Figure 10.16 Stages in the formation and life cycle of a biofilm. (credit: modification of work by Public Library of Science and American Society for Microbiology)

Within a biofilm, different species of microorganisms establish metabolic collaborations in which the waste product of one organism becomes the nutrient for another. For example, aerobic microorganisms consume oxygen, creating anaerobic regions that promote the growth of anaerobes. This occurs in many polymicrobial infections that involve both aerobic and anaerobic pathogens.

The mechanism by which cells in a biofilm coordinate their activities in response to environmental stimuli is called **quorum sensing.** Quorum sensing—

which can occur between cells of different species within a biofilm—enables microorganisms to detect their cell density through the release and binding of small, diffusible molecules called **autoinducers**. When the cell population reaches a critical threshold (a quorum), these autoinducers initiate a cascade of reactions that activate genes associated with cellular functions that are beneficial only when the population reaches a critical density. For example, in some pathogens, synthesis of virulence factors only begins when enough cells are present to overwhelm the immune defenses of the host. Although mostly studied in bacterial populations, quorum sensing takes place between bacteria and eukaryotes and between eukaryotic cells such as the fungus *Candida albicans*, a common member of the human microbiota that can cause infections in immune-compromised individuals.

The signaling molecules in quorum sensing belong to two major classes. Gram-negative bacteria communicate mainly using N-acylated homoserine lactones, whereas gram-positive bacteria mostly use small peptides (Figure 10.17). In all cases, the first step in quorum sensing consists of the binding of the autoinducer to its specific receptor only when a threshold concentration of signaling molecules is reached. Once binding to the receptor takes place, a cascade of signaling events leads to changes in gene expression. The result is the activation of biological responses linked to quorum sensing, notably an increase in the production of signaling molecules themselves, hence the term autoinducer.

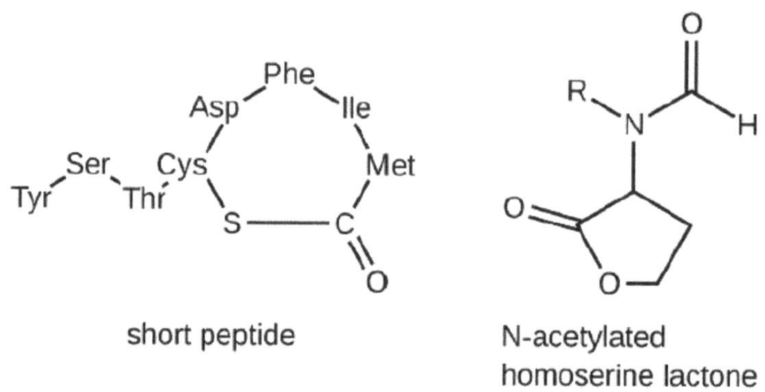

Figure 10.17 Short peptides in gram-positive bacteria and N-acetylated homoserine lactones in gram-negative bacteria act as autoinducers in quorum sensing and mediate the coordinated response of bacterial cells. The R side chain of the N-acetylated homoserine lactone is specific for the species of gram-negative bacteria. Some secreted homoserine lactones are recognized by more than one species.

Biofilms and Human Health

The human body harbors many types of biofilms, some beneficial and some harmful. For example, the layers of normal microbiota lining the intestinal and respiratory mucosa play a role in warding off infections by pathogens. However, other biofilms in the body can have a detrimental effect on health. For example, the plaque that forms on teeth is a biofilm that can contribute to dental and periodontal disease. Biofilms can also form in wounds, sometimes causing serious infections that can spread. The bacterium *Pseudomonas aeruginosa* often colonizes biofilms in the airways of patients with cystic fibrosis, causing chronic and sometimes fatal infections of the lungs. Biofilms can also form on medical devices used in or on the body, causing infections in patients with in-dwelling catheters, artificial joints, or contact lenses.

Pathogens embedded within biofilms exhibit a higher resistance to antibiotics than their free-floating counterparts. Several hypotheses have been proposed to explain why. Cells in the deep layers of a biofilm are metabolically inactive and may be less susceptible to the action of antibiotics that disrupt metabolic activities. The EPS may also slow the diffusion of antibiotics and antiseptics, preventing them from reaching cells in the deeper layers of the biofilm. Phenotypic changes may also contribute to the increased resistance exhibited by bacterial cells in biofilms. For example, the increased production of efflux pumps, membrane-embedded proteins that actively extrude antibiotics out of bacterial cells, have been shown to be an important mechanism of antibiotic resistance among biofilm-associated bacteria. Finally, biofilms provide an ideal environment for the exchange of extra chromosomal DNA, which often includes genes that confer antibiotic resistance.

OXYGEN REQUIREMENTS FOR MICROBIAL GROWTH

Ask most people "What are the major requirements for life?" and the answers are likely to include water and oxygen. Few would argue about thc need for water, but what about oxygen? Can there be life without oxygen?

The answer is that molecular oxygen (O_2) is not always needed. The earliest signs of life are dated to a period when conditions on earth were highly reducing and free oxygen gas was essentially nonexistent. Only after cyanobacteria started releasing oxygen as a byproduct of photosynthesis and the capacity of iron in the oceans for taking up oxygen was exhausted did oxygen levels increase in the atmosphere. This event, often referred to as the Great Oxygenation Event or the Oxygen Revolution, caused a massive extinction. Most organisms could not survive the powerful oxidative properties of **reactive oxygen species** (ROS), highly unstable ions and molecules derived from partial reduction of oxygen that can damage virtually any macromolecule or structure with which they come in contact. Singlet oxygen (O2•), superoxide ($O2^-$), peroxides (H_2O_2), hydroxyl radical (OH•), and hypochlorite ion (OCl^-), the active ingredient of household bleach, are all examples of ROS. The organisms that were able to detoxify reactive oxygen species harnessed the high electro negativity of oxygen to produce free energy for their metabolism and thrived in the new environment.

Oxygen Requirements of Microorganisms

Many ecosystems are still free of molecular oxygen. Some are found in extreme locations, such as deep in the ocean or in earth's crust; others are part of our everyday landscape, such as marshes, bogs, and sewers. Within the bodies of humans and other animals, regions with little or no oxygen provide an anaerobic environment for microorganisms. (Figure 10.18).

(a) (b)

Figure 10.18 Anaerobic environments are still common on earth. They include environments like (a) a bog where undisturbed dense sediments are virtually devoid of oxygen, and (b) the rumen (the first compartment of a cow's stomach), which provides an oxygen-free incubator for methanogens and other obligate anaerobic bacteria. (credit a: modification of work by National Park Service; credit b: modification of work by US Department of Agriculture)

We can easily observe different requirements for molecular oxygen by growing bacteria in **thioglycolate tube cultures**. A test-tube culture starts with autoclaved **thioglycolate medium** containing a low percentage of agar to allow motile bacteria to move throughout the medium. Thioglycolate has strong reducing properties and autoclaving flushes out most of the oxygen. The tubes are inoculated with the bacterial cultures to be tested and incubated at an appropriate temperature. Over time, oxygen slowly diffuses throughout the thioglycolate tube culture from the top. Bacterial density increases in the area where oxygen concentration is best suited for the growth of that particular organism.

The growth of bacteria with varying oxygen requirements in thioglycolate tubes is illustrated in Figure 10.19. In tube A, all the growth is seen at the top of the tube. The bacteria are **obligate (strict) aerobes** that cannot grow without an abundant supply of oxygen. Tube B looks like the opposite of tube A. Bacteria grow at the bottom of tube B. Those are **obligate anaerobes**, which are killed by oxygen. Tube C shows heavy growth at the top of the tube and growth throughout the tube, a typical result with **facultative anaerobes**. Facultative anaerobes are organisms that thrive in the presence of oxygen but also grow in its absence by relying on fermentation or anaerobic respiration, if there is a suitable electron acceptor other than oxygen and the organism is able to perform anaerobic respiration. The **aerotolerant anaerobes** in tube D are indifferent to the presence of oxygen. They do not use oxygen because they usually have a fermentative metabolism, but they are not harmed by the presence of oxygen as obligate anaerobes are. Tube E on the right shows a "Goldilocks" culture. The oxygen level has to be just right for growth, not too much and not too little. These **microaerophiles** are bacteria that require a minimum level of oxygen for growth, about 1%–10%, well below the 21% found in the atmosphere.

Examples of obligate aerobes are *Mycobacterium tuberculosis*, the causative agent of tuberculosis and *Micrococcus luteus*, a gram-positive bacterium that colonizes the skin. *Neisseria meningitidis*, the causative agent of severe bacterial meningitis, and *N. gonorrhoeae*, the causative agent of sexually transmitted gonorrhea, are also obligate aerobes.

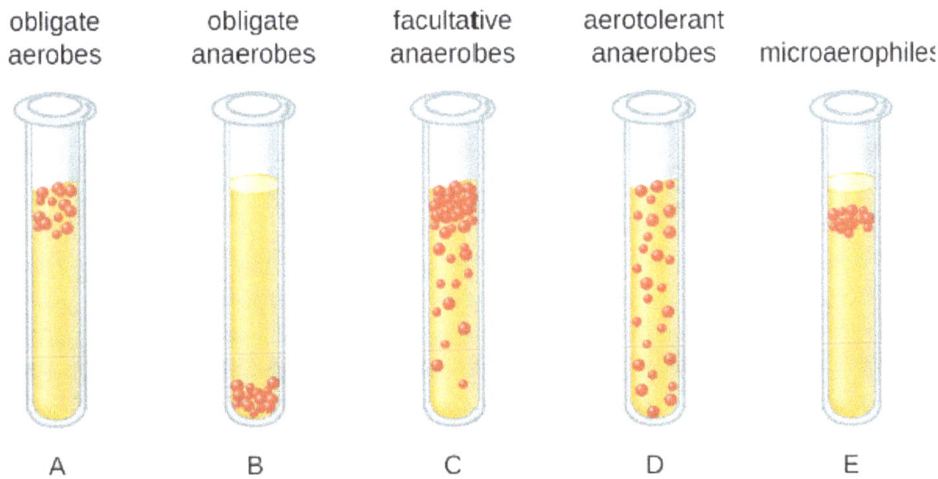

Figure 10.19 Diagram of bacterial cell distribution in thioglycolate tubes.

Many obligate anaerobes are found in the environment where anaerobic conditions exist, such as in deep sediments of soil, still waters, and at the bottom of the deep ocean where there is no photosynthetic life. Anaerobic conditions also exist naturally in the intestinal tract of animals. Obligate anaerobes, mainly *Bacteroidetes*, represent a large fraction of the microbes in the human gut. Transient anaerobic conditions exist when tissues are not supplied with blood circulation; they die and become an ideal breeding ground for obligate anaerobes. Another type of obligate anaerobe encountered in the human body is the gram-positive, rod-shaped *Clostridium* spp. Their ability to form endospores allows them to survive in the presence of oxygen. One of the major causes of health-acquired infections is *C. difficile*, known as C. diff. Prolonged use of antibiotics for other infections increases the probability of a patient developing a secondary *C. difficile* infection. Antibiotic treatment disrupts the balance of microorganisms in the intestine and allows the colonization of the gut by *C. difficile*, causing a significant inflammation of the colon.

Other clostridia responsible for serious infections include *C. tetani*, the agent of tetanus, and *C. perfringens*, which causes gas gangrene. In both cases, the infection starts in necrotic tissue (dead tissue that is not supplied with oxygen by blood circulation). This is the reason that deep puncture wounds are associated with tetanus. When tissue death is accompanied by lack of circulation, gangrene is always a danger.

The study of obligate anaerobes requires special equipment. Obligate anaerobic bacteria must be grown under conditions devoid of oxygen. The most common approach is culture in an **anaerobic jar** (Figure 10.20). Anaerobic jars include chemical packs that remove oxygen and release carbon dioxide (CO_2). An **anaerobic chamber** is an enclosed box from which all oxygen is removed. Gloves sealed to openings in the box allow handling of the cultures without exposing the culture to air (Figure 10.20).

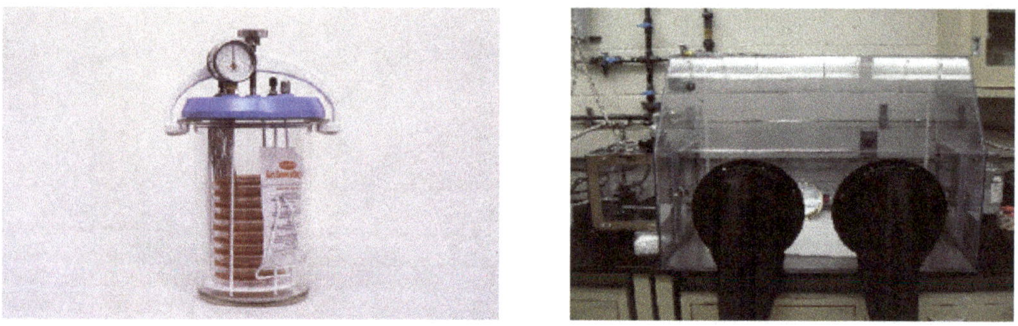

(a) (b)

Figure 10.20 (a) An anaerobic jar is pictured that is holding nine Petri plates supporting cultures. (b) Openings in the side of an anaerobic box are sealed by glove-like sleeves that allow for the handling of cultures inside the box. (credit a: modification of work by Centers for Disease Control and Prevention; credit b: modification of work by NIST)

Staphylococci and Enterobacteriaceae are examples of facultative anaerobes. Staphylococci are found on the skin and upper respiratory tract. Enterobacteriaceae are found primarily in the gut and upper respiratory tract but can sometimes spread to the urinary tract, where they are capable of causing infections. It is not unusual to see mixed bacterial infections in which the facultative anaerobes use up the oxygen, creating an environment for the obligate anaerobes to flourish.

Examples of aerotolerant anaerobes include lactobacilli and streptococci, both found in the oral microbiota. *Campylobacter jejuni*, which causes gastrointestinal infections, is an example of a microaerophile and is grown under low-oxygen conditions.

The **optimum oxygen concentration**, as the name implies, is the ideal concentration of oxygen for a particular microorganism. The lowest concentration of oxygen that allows growth is called the **minimum permissive oxygen concentration**. The highest tolerated concentration of oxygen is the **maximum permissive oxygen concentration**. The organism will not grow outside the range of oxygen levels found between the minimum and maximum permissive oxygen concentrations.

An Unwelcome Anaerobe

Charles is a retired bus driver who developed type 2 diabetes over 10 years ago. Since his retirement, his lifestyle has become very sedentary and he has put on a substantial amount of weight. Although he has felt tingling and numbness in his left foot for a while, he has not been worried because he thought his foot was simply "falling asleep." Recently, a scratch on his foot does not seem to be healing and is becoming increasingly ugly. Because the sore did not bother him much, Charles figured it could not be serious until his daughter noticed a purplish discoloration spreading on the skin and oozing (Figure 10.21). When he was finally seen by his physician, Charles was rushed to the operating room. His open sore, or ulcer, is the result of a diabetic foot.

The concern here is that gas gangrene may have taken hold in the dead tissue. The most likely agent of gas gangrene is *Clostridium perfringens*, an endospore-forming, gram-positive bacterium. It is an obligate anaerobe that grows in tissue devoid of oxygen. Since dead tissue is no longer supplied with oxygen by the circulatory system, the dead tissue provides pockets of ideal environment for the growth of *C. perfringens*.

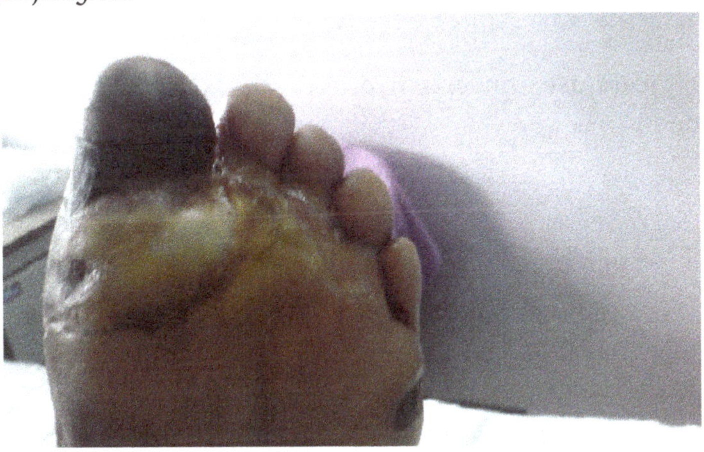

Figure 10.21 This clinical photo depicts ulcers on the foot of a diabetic patient. Dead tissue accumulating in ulcers can provide an ideal growth environment for the anaerobe *C. perfringens*, a causative agent of gas gangrene. (Credit: Phalinn Ooi / Wikimedia Commons (CC-BY))

A surgeon examines the ulcer and radiographs of Charles's foot and determines that the bone is not yet infected. The wound will have to be surgically debrided (debridement refers to the removal of dead and infected tissue) and a sample sent for microbiological lab analysis, but Charles will not have to have his foot amputated. Many diabetic patients are not so lucky. In 2008, nearly 70,000 diabetic patients in the United States lost a foot or limb to amputation, according to statistics from the Centers for Disease Control and Prevention.

- Which growth conditions would you recommend for the detection of *C. perfringens*?

Detoxification of Reactive Oxygen Species

Aerobic respiration constantly generates reactive oxygen species (ROS), byproducts that must be detoxified. Even organisms that do not use aerobic respiration need some way to break down some of the ROS that may form from atmospheric oxygen. Three main enzymes break down those toxic byproducts: superoxide dismutase, peroxidase, and catalase. Each one catalyzes a different reaction. Reactions of type seen in Reaction 1 are catalyzed by **peroxidases**.

$$(1)\ X - (2H^+) + H_2O_2 \rightarrow \text{oxidized-}X + 2H_2O$$

In these reactions, an electron donor (reduced compound; e.g., reduced nicotinamide adenine dinucleotide

[NADH]) oxidizes hydrogen peroxide, or other peroxides, to water. The enzymes play an important role by limiting the damage caused by peroxidation of membrane lipids. Reaction 2 is mediated by the enzyme **superoxide dismutase** (SOD) and breaks down the powerful superoxide anions generated by aerobic metabolism:

$$(2)\ 2O_2^- + 2H^+ \rightarrow H_2O_2 + O_2$$

The enzyme **catalase** converts hydrogen peroxide to water and oxygen as shown in Reaction 3.

$$(3)\ 2H_2O_2 \rightarrow 2H_2O + O_2$$

Obligate anaerobes usually lack all three enzymes. Aerotolerant anaerobes do have SOD but no catalase.

Reaction 3, shown occurring in Figure 10.22, is the basis of a useful and rapid test to distinguish streptococci, which are aerotolerant and do not possess catalase, from staphylococci, which are facultative anaerobes. A sample of culture rapidly mixed in a drop of 3% hydrogen peroxide will release bubbles if the culture is catalase positive.

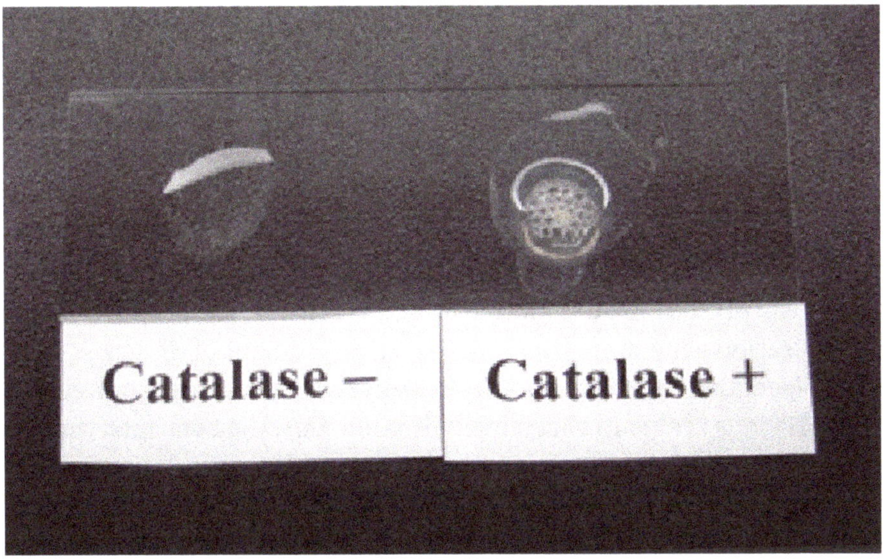

Figure 10.22 The catalase test detects the presence of the enzyme catalase by noting whether bubbles are released when hydrogen peroxide is added to a culture sample. Compare the positive result (right) with the negative result (left). (credit: Centers for Disease Control and Prevention)

Bacteria that grow best in a higher concentration of CO_2 and a lower concentration of oxygen than present in the atmosphere are called **capnophiles**. One common approach to grow capnophiles is to use a **candle jar**. A candle jar consists of a jar with a tight-fitting lid that can accommodate the cultures and a candle. After the cultures are added to the jar, the candle is lit and the lid closed. As the candle burns, it consumes most of the oxygen present and releases CO_2.

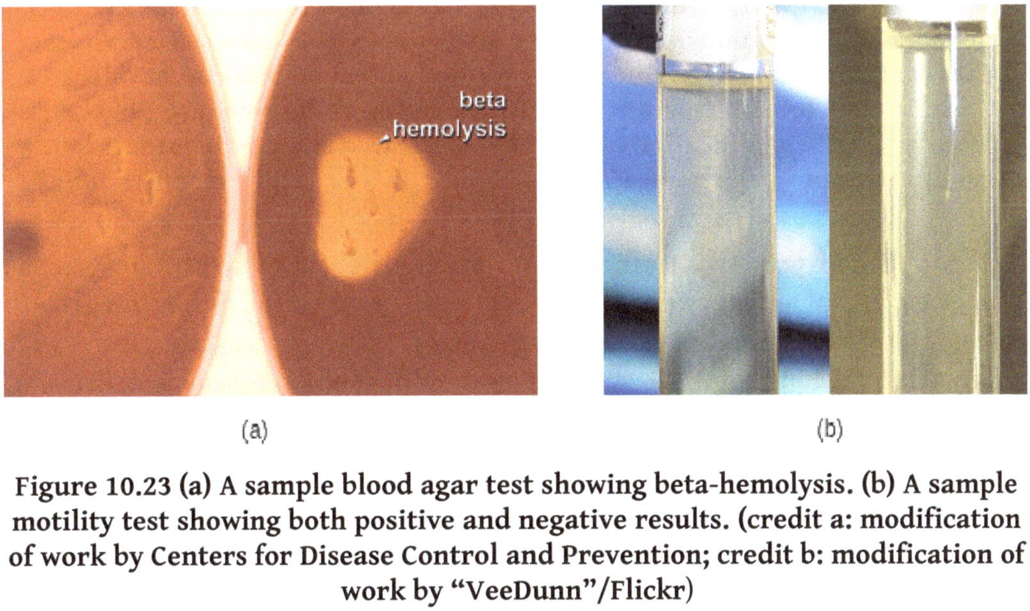

Figure 10.23 (a) A sample blood agar test showing beta-hemolysis. (b) A sample motility test showing both positive and negative results. (credit a: modification of work by Centers for Disease Control and Prevention; credit b: modification of work by "VeeDunn"/Flickr)

THE EFFECTS OF pH ON MICROBIAL GROWTH

Yogurt, pickles, sauerkraut, and lime-seasoned dishes all owe their tangy taste to a high acid content (Figure 10.24). Recall that acidity is a function of the concentration of hydrogen ions [H^+] and is measured as pH. Environments with pH values below 7.0 are considered acidic, whereas those with pH values above 7.0 are considered basic. Extreme pH affects the structure of all macromolecules. The hydrogen bonds holding together strands of DNA break up at high pH. Lipids are hydrolyzed by an extremely basic pH. The proton motive force responsible for production of ATP in cellular respiration depends on the concentration gradient of H+ across the plasma membrane (see Cellular Respiration (https://legacy.cnx.org/content/m58821/latest/)). If H+ ions are neutralized by hydroxide ions, the concentration gradient collapses and impairs energy production. But the component most sensitive to pH in the cell is its workhorse, the protein. Moderate changes in pH modify the ionization of amino-acid functional groups and disrupt hydrogen bonding, which, in turn, promotes changes in the folding of the molecule, promoting denaturation and destroying activity.

Figure 10.24 Lactic acid bacteria that ferment milk into yogurt or transform vegetables in pickles thrive at a pH close to 4.0. Sauerkraut and dishes such as pico de gallo owe their tangy flavor to their acidity. Acidic foods have been a mainstay of the human diet for centuries, partly because most microbes that cause food spoilage grow best at a near neutral pH and do not tolerate acidity well. (credit "yogurt": modification of work by "nina.jsc"/Flickr; credit "pickles": modification of work by Noah Sussman; credit "sauerkraut": modification of work by Jesse LaBuff; credit "pico de gallo": modification of work by "regan76"/Flickr)

The **optimum growth pH** is the most favorable pH for the growth of an organism. The lowest pH value that an organism can tolerate is called the **minimum growth pH** and the highest pH is the **maximum growth pH**. These values can cover a wide range, which is important for the preservation of food and to microorganisms' survival in the stomach. For example, the optimum growth pH of *Salmonella* spp. is 7.0–7.5, but the minimum growth pH is closer to 4.2.

Most bacteria are **neutrophiles**, meaning they grow optimally at a pH within one or two pH units of the neutral pH of 7 (see Figure 10.25). Most familiar bacteria, like *Escherichia coli,* staphylococci, and *Salmonella* spp. are neutrophiles and do not fare well in the acidic pH of the stomach. However, there are pathogenic strains of *E. coli, S. typhi,* and other species of intestinal pathogens that are much more resistant to stomach acid. In comparison, fungi thrive at slightly acidic pH values of 5.0–6.0.

Microorganisms that grow optimally at pH less than 5.55 are called **acidophiles**. For example, the sulfur- oxidizing *Sulfolobus* spp. isolated from sulfur mud fields and hot springs in Yellowstone National Park are extreme acidophiles. These archaea survive at pH values of 2.5–3.5. Species of the archaean genus *Ferroplasma* live in acid mine drainage at pH values of 0–2.9. *Lactobacillus* bacteria, which are an important part of the normal microbiota of the vagina, can tolerate acidic environments at pH values 3.5–6.8 and also contribute to the acidity of the vagina (pH of 4, except at the onset of menstruation) through their metabolic production of lactic acid. The vagina's acidity plays an important role in inhibiting other microbes that are less tolerant of acidity. Acidophilic microorganisms display a number of adaptations to survive in strong acidic environments. For example, proteins show increased negative surface charge that stabilizes them at low pH. Pumps actively eject H^+ ions out of the cells. The changes in the composition of membrane phospholipids probably reflect the need to maintain membrane fluidity at low pH.

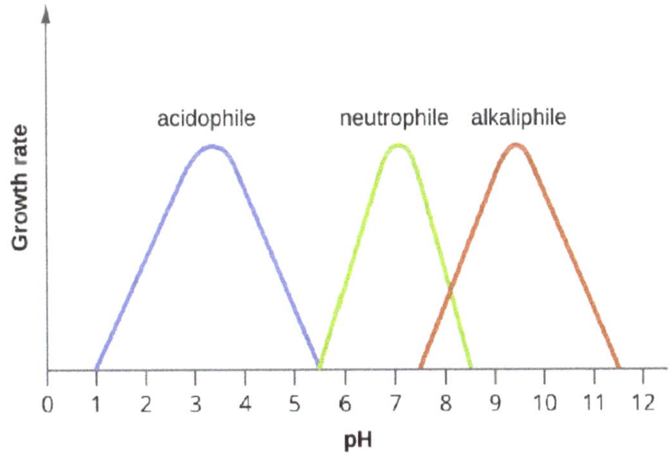

Figure 10.25 The curves show the approximate pH ranges for the growth of the different classes of pH-specific prokaryotes. Each curve has an optimal pH and extreme pH values at which growth is much reduced. Most bacteria are neutrophiles and grow best at near-neutral pH (center curve). Acidophiles have optimal growth at pH values near 3 and alkaliphiles have optimal growth at pH values above 9.

At the other end of the spectrum are **alkaliphiles**, microorganisms that grow best at pH between 8.0 and 10.5. *Vibrio cholerae*, the pathogenic agent of cholera, grows best at the slightly basic pH of 8.0; it can survive pH values of 11.0 but is inactivated by the acid of the stomach. When it comes to survival at high pH, the bright pink archaean *Natronobacterium*, found in the soda lakes of the African Rift Valley, may hold the record at a pH of 10.5 (Figure 10.26). Extreme alkaliphiles have adapted to their harsh environment through evolutionary modification of lipid and protein structure and compensatory mechanisms to maintain the proton motive force in an alkaline environment. For example, the alkaliphile *Bacillus firmus* derives the energy for transport reactions and motility from a Na^+ ion gradient rather than a proton motive force. Many enzymes from alkaliphiles have a higher isoelectric point, due to an increase in the number of basic amino acids, than homologous enzymes from neutrophiles.

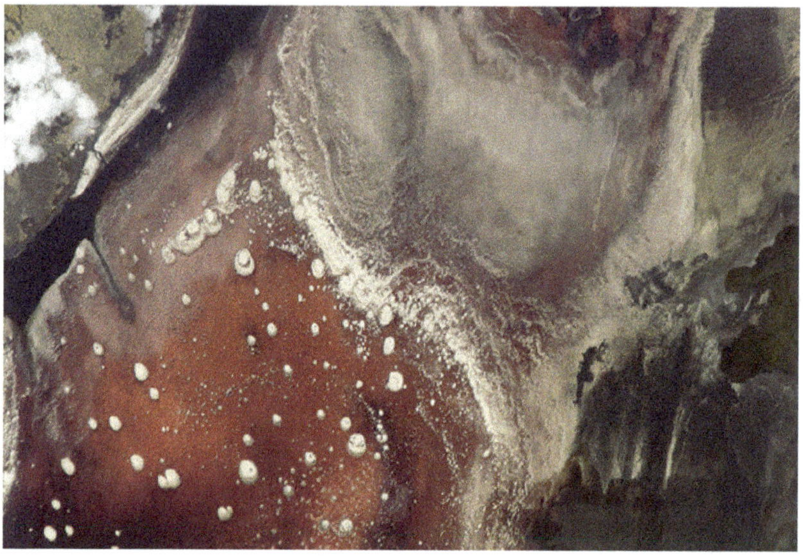

Figure 10.26 View from space of Lake Natron in Tanzania. The pink color is due to the pigmentation of the extreme alkaliphilic and halophilic microbes that colonize the lake. (credit: NASA)

Survival at the Low pH of the Stomach

Peptic ulcers (or stomach ulcers) are painful sores on the stomach lining. Until the 1980s, they were believed to be caused by spicy foods, stress, or a combination of both. Patients were typically advised to eat bland foods, take anti-acid medications, and avoid stress. These remedies were not particularly effective, and the condition often recurred. This all changed dramatically when the real cause of most peptic ulcers was discovered to be a slim, corkscrew-shaped bacterium, *Helicobacter pylori*. This organism was identified and isolated by Barry Marshall and Robin Warren, whose discovery earned them the Nobel Prize in Medicine in 2005.

The ability of *H. pylori* to survive the low pH of the stomach would seem to suggest that it is an extreme acidophile. As it turns out, this is not the case. In fact, *H. pylori* is a neutrophile. So, how does it survive in the stomach? Remarkably, *H. pylori* creates a microenvironment in which the pH is nearly neutral. It achieves this by producing large amounts of the enzyme urease, which breaks down urea to form NH_4^+ and CO_2. The ammonium ion raises the pH of the immediate environment.

This metabolic capability of *H. pylori* is the basis of an accurate, noninvasive test for infection. The patient is given a solution of urea containing radioactively labeled carbon atoms. If *H. pylori* is present in the stomach, it will rapidly break down the urea, producing radioactive CO_2 that can be detected in the patient's breath. Because peptic ulcers may lead to gastric cancer, patients who are determined to have *H. pylori* infections are treated with antibiotics.

EFFECT OF TEMPERATURE ON MICROBIAL GROWTH

When the exploration of Lake Whillans started in Antarctica, researchers did not expect to find much life. Constant subzero temperatures and lack of obvious sources of nutrients did not seem to be conditions that would support a thriving ecosystem. To their surprise, the samples retrieved from the lake showed abundant microbial life. In a different but equally harsh setting, bacteria grow at the bottom of the ocean in sea vents (Figure 10.27), where temperatures can reach 340 °C (700 °F).

Microbes can be roughly classified according to the range of temperature at which they can grow. The growth rates are the highest at the **optimum growth temperature** for the organism. The lowest temperature at which the organism can survive and replicate is its **minimum growth temperature**. The highest temperature at which growth can occur is its **maximum growth temperature**. The following ranges of permissive growth temperatures are approximate only and can vary according to other environmental factors.

Organisms categorized as **mesophiles** ("middle loving") are adapted to moderate temperatures, with optimal growth temperatures ranging from room temperature (about 20 °C) to about 45 °C. As would be expected from the core temperature of the human body, 37 °C (98.6 °F), normal human microbiota and pathogens (e.g., *E. coli*, *Salmonella* spp., and *Lactobacillus* spp.) are mesophiles.

Organisms called **psychrotrophs**, also known as psychrotolerant, prefer cooler environments, from a high temperature of 25 °C to refrigeration temperature about 4 °C. They are found in many natural environments in temperate climates. They are also responsible for the spoilage of refrigerated food.

OTHER ENVIRONMENTAL CONDITIONS THAT AFFECT GROWTH

Microorganisms interact with their environment along more dimensions than pH, temperature, and free oxygen levels, although these factors require significant adaptations. We also find microorganisms adapted to varying levels of salinity, barometric pressure, humidity, and light.

Osmotic and Barometric Pressure

Most natural environments tend to have lower solute concentrations than the cytoplasm of most microorganisms. Rigid cell walls protect the cells from bursting in a dilute environment. Not much protection is available against high osmotic pressure. In this case, water, following its concentration gradient, flows out of the cell. This results in plasmolysis (the shrinking of the protoplasm away from the intact cell wall) and cell death. This fact explains why brines and layering meat and fish in salt are time-honored methods of preserving food. Microorganisms called **halophiles** ("salt loving") actually require high salt concentrations for growth. These organisms are found in marine environments where salt concentrations hover at 3.5%. Extreme halophilic microorganisms, such as the red alga *Dunaliella salina* and the archaeal species *Halobacterium* in Figure 10.31, grow in hypersaline lakes such as the Great Salt Lake, which is 3.5–8 times saltier than the ocean, and the Dead Sea, which is 10 times saltier than the ocean.

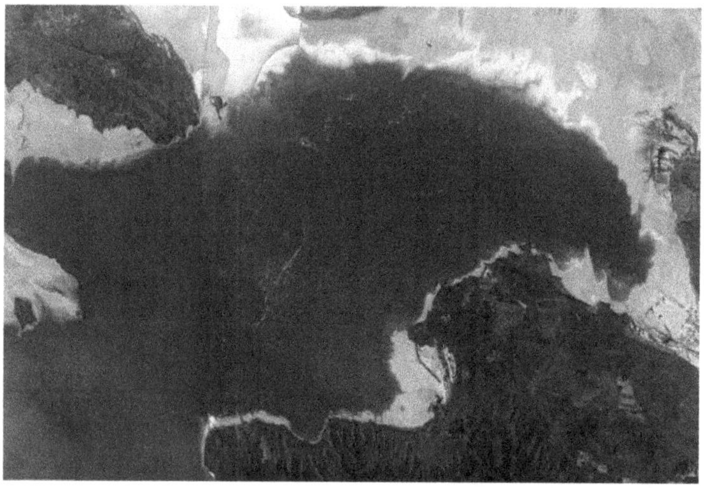

Figure 10.31 Photograph taken from space of the Great Salt Lake in Utah. The purple color is caused by high density of the alga Dunaliella and the archaean Halobacterium spp. (credit: NASA)

Dunaliella spp. counters the tremendous osmotic pressure of the environment with a high cytoplasmic concentration of glycerol and by actively pumping out salt ions. *Halobacterium* spp. accumulates large concentrations of K^+ and other ions in its cytoplasm. Its proteins are designed for high salt concentrations and lose activity at salt concentrations below 1–2 M. Although most **halotolerant** organisms, for example *Halomonas* spp. in salt marshes, do not need high concentrations of salt for growth, they will survive and divide in the presence of high salt. Not surprisingly, the staphylococci, micrococci, and corynebacteria that colonize our skin tolerate salt in their environment. Halotolerant pathogens are an important cause of food-borne illnesses because they survive and multiply in salty food. For example, the halotolerant bacteria *S. aureus, Bacillus cereus*, and *V. cholerae* produce dangerous enterotoxins and are major causes of food poisoning.

Microorganisms depend on available water to grow. Available moisture is measured as water activity (a_w), which is the ratio of the vapor pressure of the medium of interest to the vapor pressure of pure distilled water; therefore, the a_w of water is equal to 1.0. Bacteria require high a_w (0.97–0.99), whereas fungi can tolerate drier environments; for example, the range of a_w for growth of *Aspergillus* spp. is 0.8–0.75. Decreasing the water content of foods by drying, as in jerky, or through freeze-drying or by increasing osmotic pressure, as in brine and jams, are common methods of preventing spoilage.

Microorganisms that require high atmospheric pressure for growth are called **barophiles**. The bacteria that live at the bottom of the ocean must be able to withstand great pressures. Because it is difficult to retrieve intact specimens and reproduce such growth conditions in the laboratory, the characteristics of these microorganisms are largely unknown.

Light

Photoautotrophs, such as cyanobacteria or green sulfur bacteria, and photoheterotrophs, such as purple nonsulfur bacteria, depend on sufficient light intensity at the wavelengths absorbed by their pigments to grow and multiply. Energy from light is captured by pigments and converted into chemical energy that drives carbon fixation and other metabolic processes. The portion of the electromagnetic spectrum that is absorbed by these organisms is defined as photosynthetically active radiation (PAR). It lics within the visible light spectrum ranging from 400 to 700 nanometers (nm) and extends in the near infrared for some photosynthetic bacteria. A number of accessory pigments, such as fucoxanthin in brown algae and phycobilins in cyanobacteria, widen the useful range of wavelengths for photosynthesis and compensate for the low light levels available at greater depths of water. Other microorganisms, such as the archaea of the class Halobacteria, use light energy to drive their proton and sodium pumps. The light is absorbed by a pigment protein complex called bacteriorhodopsin, which is similar to the eye pigment rhodopsin. Photosynthetic bacteria are present not only in aquatic environments but also in soil and in symbiosis with fungi in lichens. The peculiar watermelon snow is caused by a microalga *Chlamydomonas nivalis*, a green alga rich in a secondary red carotenoid pigment (astaxanthin) which gives the pink hue to the snow where the alga grows.

MEDIA USED FOR BACTERIAL GROWTH

The study of microorganisms is greatly facilitated if we are able to culture them, that is, to keep reproducing populations alive under laboratory conditions. Culturing many microorganisms is challenging because of highly specific nutritional and environmental requirements and the diversity of these requirements among different species.

Nutritional Requirements

The number of available media to grow bacteria is considerable. Some media are considered general all- purpose media and support growth of a large variety of organisms. A prime example of an all-purpose medium is tryptic soy broth (TSB). Specialized media are used in the identification of bacteria and are supplemented with dyes, pH indicators, or antibiotics. One type, **enriched media,** contains growth factors, vitamins, and other essential nutrients to promote the growth of **fastidious organisms**, organisms that cannot make certain nutrients and require them to be added to the medium. When the complete chemical composition of a medium is known, it is called a **chemically defined medium.** For example, in EZ medium, all individual chemical components are identified and the exact amounts of each is known. In **complex media,** which contain extracts and digests of yeasts, meat, or plants, the precise chemical composition of the medium is not known. Amounts of individual components are undetermined and variable. Nutrient broth, tryptic soy broth, and brain heart infusion, are all examples of complex media.

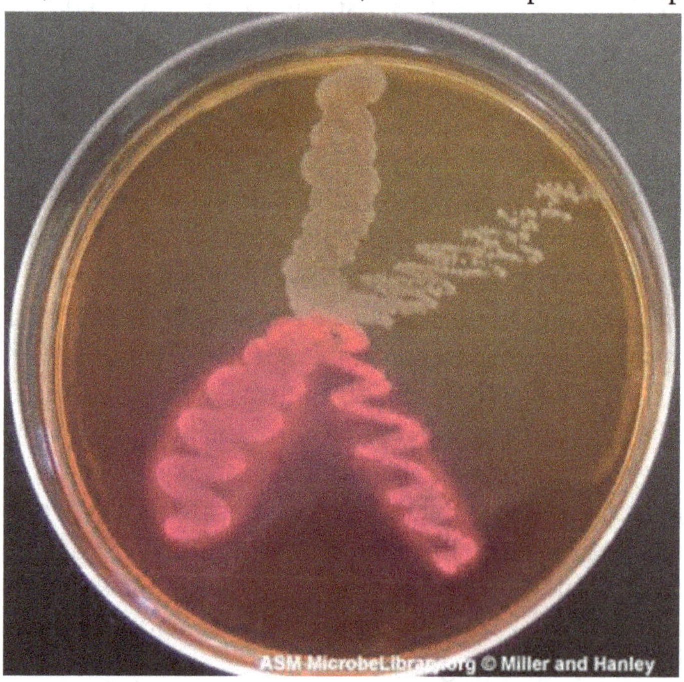

Figure 10.32 On this MacConkey agar plate, the lactose-fermenter *E. coli* colonies are bright pink. *Serratia marcescens*, which does not ferment lactose, forms a cream-colored streak on the tan medium. (credit: American Society for Microbiology)

Media that inhibit the growth of unwanted microorganisms and support the growth of the organism of interest by supplying nutrients and reducing competition are called **selective media**. An example of a selective medium is MacConkey agar. It contains bile salts and crystal violet, which interfere with the growth of many gram- positive bacteria and favour the growth of gram-negative bacteria, particularly the Enterobacteriaceae. These species are commonly named enterics, reside in the intestine, and are adapted to the presence of bile salts. The **enrichment culture**s foster the preferential growth of a desired microorganism that represents a fraction of the organisms present in an inoculum. For example, if we want to isolate bacteria that break down crude oil, hydrocarbonoclastic bacteria, sequential subculturing in a medium that supplies carbon only in the form of crude oil will enrich the cultures with oil-eating bacteria. The **differential media** make it easy to distinguish colonies of different bacteria by a change in the colour of the colonies or the colour of the medium. Colour changes are the result of end products created by interaction of bacterial enzymes with differential substrates in the medium or, in the case of hemolytic reactions, the lysis of red blood cells in the medium. In Figure 10.32, the differential fermentation of lactose can be observed on MacConkey agar. The lactose fermenters produce acid, which turns the medium and the colonies of strong fermenters hot pink. The medium is supplemented with the pH indicator neutral red, which turns to hot pink at low pH. Selective and differential media can be combined and play an important role in the identification of bacteria by biochemical methods.

CHAPTER-11

CONTROL OF MICROBIAL GROWTH AND ITS CONTROL

INTRODUCTION: AN OVERVIEW

In the ever-evolving landscape of microbiology, the control of microbial growth stands as a paramount concern with profound implications across various industries. From healthcare to food production and environmental management, the ability to regulate and manipulate the proliferation of microorganisms is pivotal for ensuring human health, safety, and the integrity of diverse ecosystems. This exploration into the control of microbial growth delves into the strategies and techniques employed to manage the expansion of bacteria, fungi, and other microorganisms. From antibiotics and disinfectants to sophisticated biotechnological approaches, this critical aspect of microbiology not only safeguards against infectious diseases but also plays a pivotal role in enhancing the quality and safety of numerous products and processes. As we navigate the intricacies of microbial growth control, we uncover the innovative measures that underscore the ongoing quest for mastery over these microscopic entities in our pursuit of a healthier, safer, and more sustainable future.

CONTROLLING MICROBIAL GROWTH

To prevent the spread of human disease, it is necessary to control the growth and abundance of microbes in or on various items frequently used by humans. Inanimate items, such as doorknobs, toys, or towels, which may harbor microbes and aid in disease transmission, are called **fomites**. Two factors heavily influence the level of cleanliness required for a particular fomite and, hence, the protocol chosen to achieve this level. The first factor is the application for which the item will be used. For example, invasive applications that require insertion into the human body require a much higher level of cleanliness than applications that do not. The second factor is the level of resistance to antimicrobial treatment by potential pathogens. For example, foods preserved by canning often become contaminated with the bacterium *Clostridium botulinum*, which produces the neurotoxin that causes botulism. Because *C. botulinum* can produce endospores that can survive harsh conditions, extreme temperatures and pressures must be used to eliminate the endospores. Other organisms may not require such extreme measures and can be controlled by a procedure such as washing clothes in a laundry machine.

Laboratory Biological Safety Levels

For researchers or laboratory personnel working with pathogens, the risks associated with specific pathogens determine the levels of cleanliness and control required. The Centers for Disease Control and Prevention (CDC) and the National Institutes of Health (NIH) have established four classification levels, called " biological safety levels" (BSLs). Various organizations around the world, including the World Health Organization (WHO) and the European Union (EU), use a similar classification scheme. According to the CDC, the BSL is determined by the agent's infectivity, ease of transmission, and potential disease severity, as well as the type of work being done with the agent.

Each BSL requires a different level of biocontainment to prevent contamination and spread of infectious agents to laboratory personnel and, ultimately, the community. For example, the lowest BSL, BSL-1, requires the fewest precautions because it applies to situations with the lowest risk for microbial infection.

BSL-1 agents are those that generally do not cause infection in healthy human adults. These include noninfectious bacteria, such as nonpathogenic strains of *Escherichia coli* and *Bacillus subtilis*, and viruses known to infect animals other than humans, such as baculoviruses (insect viruses). Because working with BSL-1 agents poses very little risk, few precautions are necessary. Laboratory workers use standard aseptic technique and may work with these agents at an open laboratory bench or table, wearing personal protective equipment (PPE) such as a laboratory coat, goggles, and gloves, as needed. Other than a sink for handwashing and doors to separate the laboratory from the rest of the building, no additional modifications are needed.

Agents classified as BSL-2 include those that pose moderate risk to laboratory workers and the community, and are typically "indigenous," meaning that they are commonly found in that geographical area. These include bacteria such as *Staphylococcus aureus* and *Salmonella* spp., and viruses like hepatitis, mumps, and measles viruses. BSL-2 laboratories require additional precautions beyond those of BSL-1, including restricted access; required PPE, including a face shield in some circumstances; and the use of biological safety cabinets for procedures that may disperse agents through the air (called "aerosolization"). BSL-2 laboratories are equipped with self-closing doors, an eyewash station, and an **autoclave**, which is a specialized device for sterilizing materials with pressurized steam before use or disposal. BSL-1 laboratories may also have an autoclave.

BSL-3 agents have the potential to cause lethal infections by inhalation. These may be either indigenous or "exotic," meaning that they are derived from a foreign location, and include pathogens such as *Mycobacterium tuberculosis*, *Bacillus anthracis*, West Nile virus, and human immunodeficiency virus (HIV). Because of the serious nature of the infections caused by BSL-3 agents, laboratories working with them require restricted access. Laboratory workers are under medical surveillance, possibly receiving vaccinations for the microbes with which they work. In addition to the standard PPE already mentioned, laboratory personnel in BSL-3 laboratories must also wear a respirator and work with microbes and infectious agents in a biological safety cabinet at all times. BSL-3 laboratories require a hands-free sink, an eyewash station near the exit, and two sets of self- closing and locking doors at the entrance. These laboratories are equipped with directional airflow, meaning that clean air is pulled through the laboratory from clean areas to potentially contaminated areas. This air cannot be recirculated, so a constant supply of clean air is required.

Figure 11.1 A protective suit like this one is an additional precaution for those who work in BSL-4 laboratories. This suit has its own air supply and maintains a positive pressure relative to the outside, so that if a leak occurs, air will flow out of the suit, not into it from the laboratory. (Credit: James Gathany, CDC, public domain)

BSL-4 agents are the most dangerous and often fatal. These microbes are typically exotic, are easily transmitted by inhalation, and cause infections for which there are no treatments or vaccinations. Examples include Ebola virus and Marburg virus, both of which cause hemorrhagic fevers, and smallpox virus. There are only a small number of laboratories in the United States and around the world appropriately equipped to work with these agents. In addition to BSL-3 precautions, laboratory workers in BSL-4 facilities must also change their clothing on entering the laboratory, shower on exiting, and decontaminate all material on exiting. While working in the laboratory, they must either wear a full-body protective suit with a designated air supply or conduct all work within a biological safety cabinet with a high-efficiency particulate air (HEPA)-filtered air supply and a doubly HEPA- filtered exhaust. If wearing a suit, the air pressure within the suit must be higher than that outside the suit, so that if a leak in the suit occurs, laboratory air that may be contaminated cannot be drawn into the suit (Figure 11.1). The laboratory itself must be located either in a separate building or in an isolated portion of a building and have its own air supply and exhaust system, as well as its own decontamination system. The BSLs are summarized in Figure 11.2.

Biosafety Levels			
Biological Safety Levels	Description	Examples	CDC Classification
BSL-4	Microbes are dangerous and exotic, posing a high risk of aerosol-transmitted infections, which are frequently fatal without treatment or vaccines. Few labs are at this level.	Ebola and Marburg viruses	
BSL-3	Microbes are indigenous or exotic and cause serious or potentially lethal diseases through respiratory transmission.	*Mycobacterium tuberculosis*	
BSL-2	Microbes are typically indigenous and are associated with diseases of varying severity. They pose moderate risk to workers and the environment.	*Staphylococcus aureus*	
BSL-1	Microbes are not known to cause disease in healthy hosts and pose minimal risk to workers and the environment.	Nonpathogenic strains of *Escherichia coli*	

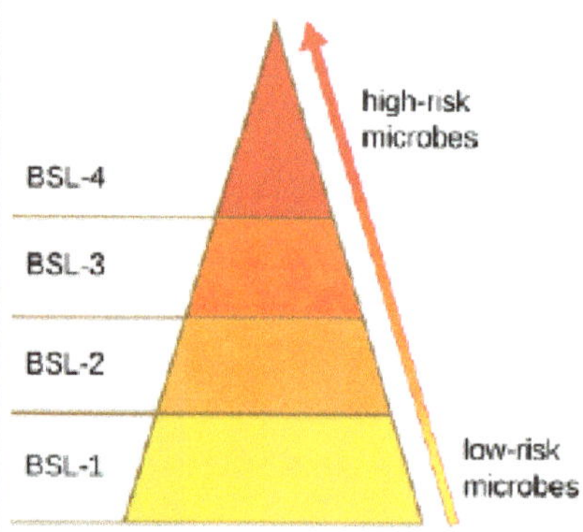

Figure 11.2 The CDC classifies infectious agents into four biosafety levels based on potential risk to laboratory personnel and the community. Each level requires a progressively greater level of precaution. (credit "pyramid": modification of work by Centers for Disease Control and Prevention)

Sterilization

The most extreme protocols for microbial control aim to achieve **sterilization**: the complete removal or killing of all vegetative cells, endospores, and viruses from the targeted item or environment. Sterilization protocols are generally reserved for laboratory, medical, manufacturing, and food industry settings, where it may be imperative for certain items to be completely free of potentially infectious agents. Sterilization can be accomplished through either physical means, such as exposure to high heat, pressure, or filtration through an appropriate filter, or by chemical means. Chemicals that can be used to achieve sterilization are called **sterilants**. Sterilants effectively kill all microbes and viruses, and, with appropriate exposure time, can also kill endospores.

For many clinical purposes, **aseptic technique** is necessary to prevent contamination of sterile surfaces. Aseptic technique involves a combination of protocols that collectively maintain sterility, or **asepsis**, thus preventing contamination of the patient with microbes and infectious agents. Failure to practice aseptic technique during many types of clinical procedures may introduce microbes to the patient's body and put the patient at risk for **sepsis**, a systemic inflammatory response to an infection that results in high fever, increased heart and respiratory rates, shock, and, possibly, death. Medical procedures that carry risk of contamination must be performed in a **sterile field**, a designated area that is kept free of all vegetative microbes, endospores, and viruses. Sterile fields are created according to protocols requiring the use of sterilized materials, such as packaging and drapings, and strict procedures for washing and application of sterilants. Other protocols are followed to maintain the sterile field while the medical procedure is being performed.

One food sterilization protocol, **commercial sterilization**, uses heat at a temperature low enough to preserve food quality but high enough to destroy common pathogens responsible for food poisoning, such as *C. botulinum*. Because *C. botulinum* and its endospores are commonly found in soil, they may easily contaminate crops during harvesting, and these endospores can later germinate within the anaerobic environment once foods are canned. Metal cans of food contaminated with *C. botulinum* will bulge due to the microbe's production of gases; contaminated jars of food typically bulge at the metal lid. To eliminate the risk for *C. botulinum* contamination, commercial food-canning protocols are designed with a large margin of error. They assume an impossibly large population of endospores (10^{12} per can) and aim to reduce this population to 1 endospore per can to ensure the safety of canned foods. For example, low- and medium-acid foods are heated to 121°C for a minimum of 2.52 minutes, which is the time it would take to reduce a population of 10^{12} endospores per can down to 1 endospore at this temperature.

Even so, commercial sterilization does not eliminate the presence of all microbes; rather, it targets those pathogens that cause spoilage and foodborne diseases, while allowing many non-pathogenic organisms to survive. Therefore, "sterilization" is somewhat of a misnomer in this context, and commercial sterilization may be more accurately described as "quasi-sterilization."

Other Methods of Control

Sterilization protocols require procedures that are not practical, or necessary, in many settings. Various other methods are used in clinical and nonclinical settings to reduce the microbial load on items. Although the terms for these methods are often used interchangeably, there are important distinctions (Figure 11.3).

The process of **disinfection** inactivates most microbes on the surface of a fomite by using antimicrobial chemicals or heat. Because some microbes remain, the disinfected item is not considered sterile. Ideally, **disinfectant**s should be fast acting, stable, easy to prepare, inexpensive, and easy to use. An example of a natural disinfectant is vinegar; its acidity kills most microbes. Chemical disinfectants, such as chlorine bleach or products containing chlorine, are used to clean nonliving surfaces such as laboratory benches, clinical surfaces, and bathroom sinks. Typical disinfection does not lead to sterilization because endospores tend to survive even when all vegetative cells have been killed.

Unlike disinfectants, **antiseptic**s are antimicrobial chemicals safe for use on living skin or tissues. Examples of antiseptics include hydrogen peroxide and isopropyl alcohol. The process of applying an antiseptic is called **antisepsis.** In addition to the characteristics of a good disinfectant, antiseptics must also be selectively effective against microorganisms and able to penetrate tissue deeply without causing tissue damage.

The type of protocol required to achieve the desired level of cleanliness depends on the particular item to be cleaned. For example, those used clinically are categorized as critical, semicritical, and noncritical. Critical items must be sterile because they will be used inside the body, often penetrating sterile tissues or the bloodstream; examples of **critical item**s include surgical instruments, catheters, and intravenous fluids. Gastrointestinal endoscopes and various types of equipment for respiratory therapies are examples of **semicritical items**; they may contact mucous membranes or nonintact skin but do not penetrate tissues. Semicritical items do not typically need to be sterilized but do require a high level of disinfection. Items that may contact but not penetrate intact skin are **noncritical items**; examples are bed linens, furniture, crutches, stethoscopes, and blood pressure cuffs. These articles need to be clean but not highly disinfected.

The act of hand washing is an example of **degerming**, in which microbial numbers are significantly reduced by gently scrubbing living tissue, most commonly skin, with a mild chemical (e.g., soap) to avoid the transmission of pathogenic microbes. Wiping the skin with an alcohol swab at an injection site is another example of degerming. These degerming methods remove most (but not all) microbes from the skin's surface.

The term **sanitization** refers to the cleansing of fomites to remove enough microbes to achieve levels deemed safe for public health. For example, commercial dishwashers used in the food service industry typically use very hot water and air for washing and drying; the high temperatures kill most microbes, sanitizing the dishes. Surfaces in hospital rooms are commonly sanitized using a chemical disinfectant to prevent disease transmission between patients. Figure 11.3 summarizes common protocols, definitions, applications, and agents used to control microbial growth.

Common Protocols for Control of Microbial Growth			
Protocol	**Definition**	**Common Application**	**Common Agents**
For Use on Fomites			
Disinfection	Reduces or destroys microbial load of an inanimate item through application of heat or antimicrobial chemicals	Cleaning surfaces like laboratory benches, clinical surfaces, and bathrooms	Chlorine bleach, phenols (e.g., Lysol), glutaraldehyde
Sanitization	Reduces microbial load of an inanimate item to safe public health levels through application of heat or antimicrobial chemicals	Commercial dishwashing of eating utensils, cleaning public restrooms	Detergents containing phosphates (e.g., Finish), industrial-strength cleaners containing quaternary ammonium compounds
Sterilization	Completely eliminates all vegetative cells, endospores, and viruses from an inanimate item	Preparation of surgical equipment and of needles used for injection	Pressurized steam (autoclave), chemicals, radiation
For Use on Living Tissue			
Antisepsis	Reduces microbial load on skin or tissue through application of an antimicrobial chemical	Cleaning skin broken due to injury; cleaning skin before surgery	Boric acid, isopropyl alcohol, hydrogen peroxide, iodine (betadine)
Degerming	Reduces microbial load on skin or tissue through gentle to firm scrubbing and the use of mild chemicals	Handwashing	Soap, alcohol swab

Figure 11.3

Measuring Microbial Control

Physical and chemical methods of microbial control that kill the targeted microorganism are identified by the suffix *-cide* (or *-cidal*). The prefix indicates the type of microbe or infectious agent killed by the treatment method: **bactericides** kill bacteria, **viricides** kill or inactivate viruses, and **fungicides** kill fungi. Other methods do not kill organisms but, instead, stop their growth, making their population static; such methods are identified by the suffix-*stat* (or -*static*). For example, **bacteriostatic** treatments inhibit the growth of bacteria, whereas **fungistatic** treatments inhibit the growth of fungi. Factors that determine whether a particular treatment is -*cidal* or -*static* include the types of microorganisms targeted, the concentration of the chemical used, and the nature of the treatment applied.

Although -*static* treatments do not actually kill infectious agents, they are often less toxic to humans and other animals, and may also better preserve the integrity of the item treated. Such treatments are typically sufficient to keep the microbial population of an item in check. The reduced toxicity of some of these -*static* chemicals also allows them to be impregnated safely into plastics to prevent the growth of microbes on these surfaces. Such plastics are used in products such as toys for children and cutting boards for food preparation. When used to treat an infection, -*static* treatments are typically sufficient in an otherwise healthy individual, preventing the pathogen from multiplying, thus allowing the individual's immune system to clear the infection.

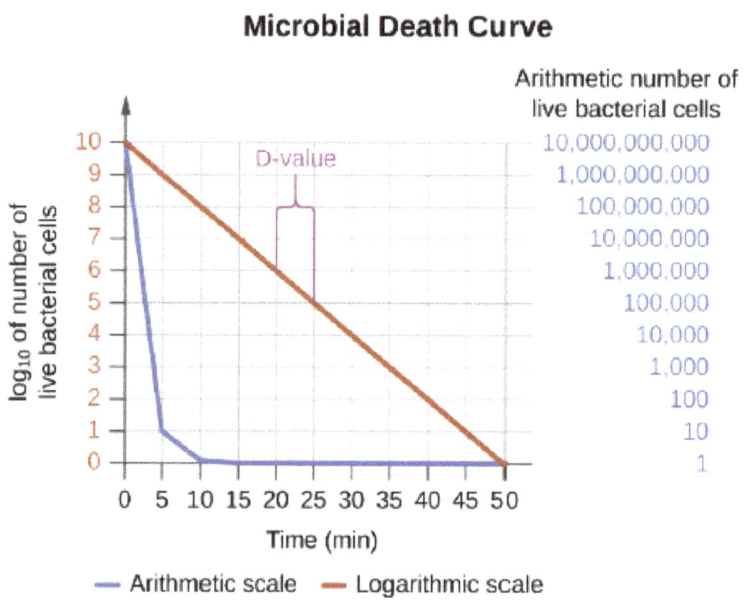

Time (min)	Number of cells
0	10^{10}
5	10^{9}
10	10^{8}
15	10^{7}
20	10^{6}
25	10^{5}
30	10^{4}
35	10^{3}
40	10^{2}
45	10^{1}
50	10^{0}

Figure 11.4 Microbial death is logarithmic and easily observed using a semilog plot instead of an arithmetic one. The decimal reduction time (D-value) is the time it takes to kill 90% of the population (a 1-log decrease in the total population) when exposed to a specific microbial control protocol, as indicated by the purple bracket.

The degree of microbial control can be evaluated using a **microbial death curve** to describe the progress and effectiveness of a particular protocol. When exposed to a particular microbial control protocol, a fixed percentage of the microbes within the population will die. Because the rate of killing remains constant even when the population size varies, the percentage killed is more useful information than the absolute number of microbes killed. Death curves are often plotted as semilog plots just like microbial growth curves because the reduction in microorganisms is typically logarithmic (Figure 11.4). The amount of time it takes for a specific protocol to produce a one order-of-magnitude decrease in the number of organisms, or the death of 90% of the population, is called the **decimal reduction time (DRT) or D-value.**

Several factors contribute to the effectiveness of a disinfecting agent or microbial control protocol. First, as demonstrated in Figure 11.4, the length of time of exposure is important. Longer exposure times kill more microbes. Because microbial death of a population exposed to a specific protocol is logarithmic, it takes longer to kill a high-population load than a low-population load exposed to the same protocol. A shorter treatment time (measured in multiples of the D-value) is needed when starting with a smaller number of organisms. Effectiveness also depends on the susceptibility of the agent to that disinfecting agent or protocol. The concentration of disinfecting agent or intensity of exposure is also important. For example, higher temperatures and higher concentrations of disinfectants kill microbes more quickly and effectively. Conditions that limit contact between the agent and the targeted cells cells—for example, the presence of bodily fluids, tissue, organic debris (e.g., mud or feces), or biofilms on surfaces—increase the cleaning time or intensity of the microbial control protocol required to reach the desired level of cleanliness. All these factors must be considered when choosing the appropriate protocol to control microbial growth in a given situation.

USING PHYSICAL METHODS TO CONTROL MICROORGANISMS

For thousands of years, humans have used various physical methods of microbial control for food preservation. Common control methods include the application of high temperatures, radiation, filtration, and desiccation (drying), among others. Many of these methods non-specifically kill cells by disrupting membranes, changing membrane permeability, or damaging proteins and nucleic acids by denaturation, degradation, or chemical modification. Various physical methods used for microbial control are described in this section.

Heat

Heating is one of the most common—and oldest—forms of microbial control. It is used in simple techniques like cooking and canning. Heat can kill microbes by altering their membranes and denaturing proteins. The **thermal death point (TDP)** of a microorganism is the lowest temperature at which all microbes are killed in a 10-minute exposure. Different microorganisms will respond differently to high temperatures, with some (e.g., endospore- formers such as *C. botulinum*) being more heat tolerant. A similar parameter, the **thermal death time (TDT)**, is the length of time needed to kill all microorganisms in a sample at a given temperature. These parameters are often used to describe sterilization procedures that use high heat, such as autoclaving. Boiling is one of the oldest methods of moist-heat control of microbes, and it is typically quite effective at killing vegetative cells and some viruses. However, boiling is less effective at killing endospores; some endospores are able to survive up to 20 hours of boiling. Additionally, boiling may be less effective at higher altitudes, where the boiling point of water is lower and the boiling time needed to kill microbes is therefore longer. For these reasons, boiling is not considered a useful sterilization technique in the laboratory or clinical setting.

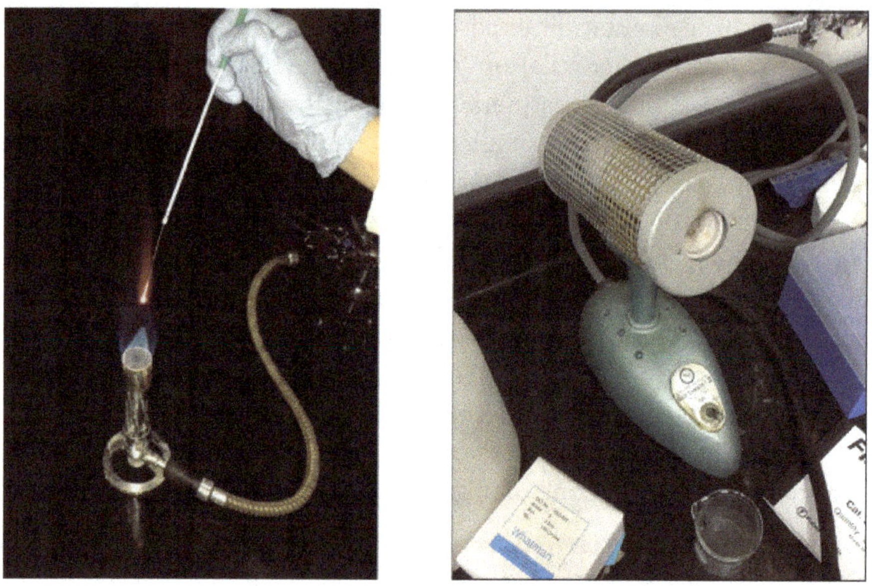

(a) (b)

Figure 11.5 (a) Sterilizing a loop, often referred to as "flaming a loop," is a common component of aseptic technique in the microbiology laboratory and is used to incinerate any microorganisms on the loop. (b) Alternatively, a bactericinerator may be used to reduce aerosolization of microbes and remove the presence of an open flame in the laboratory. These are examples of dry-heat sterilization by the direct application of high heat capable of incineration. (credit a: modification of work by Anh-Hue Tu; credit b: modification of work by Brian Forster)

Many different heating protocols can be used for sterilization in the laboratory or clinic, and these protocols can be broken down into two main categories: **dry-heat sterilization** and **moist-heat sterilization.** Aseptic technique in the laboratory typically involves some dry-heat sterilization protocols using direct application of high heat, such as sterilizing inoculating loops (Figure 11.5). Incineration at very high temperatures destroys all microorganisms. Dry heat can also be applied for relatively long periods of time (at least 2 hours) at temperatures up to 170 °C by using a dry-heat sterilizer, such as an oven. However, moist-heat sterilization is typically the more effective protocol because it penetrates cells better than dry heat does.

Autoclaves

Autoclaves rely on moist-heat sterilization. They are used to raise temperatures above the boiling point of water to sterilize items such as surgical equipment from vegetative cells, viruses, and especially endospores, which are known to survive boiling temperatures, without damaging the items. Charles Chamberland (1851–1908) designed the modern autoclave in 1879 while working in the laboratory of Louis Pasteur. The autoclave is still considered the most effective method of sterilization (Figure 11.6). Outside laboratory and clinical settings, large industrial autoclaves called retorts allow for moist-heat sterilization on a large scale.

In general, the air in the chamber of an autoclave is removed and replaced with increasing amounts of steam trapped within the enclosed chamber, resulting in increased interior pressure and temperatures above the boiling point of water. The two main types of autoclaves differ in the way that air is removed from the chamber. In gravity displacement autoclaves, steam is introduced into the chamber from the top or sides. Air, which is heavier than steam, sinks to the bottom of the chamber, where it is forced out through a vent. Complete displacement of air is difficult, especially in larger loads, so longer cycles may be required for such loads. In prevacuum sterilizers, air is removed completely using a high-speed vacuum before introducing steam into the chamber. Because air is more completely eliminated, the steam can more easily penetrate wrapped items. Many autoclaves are capable of both gravity and prevacuum cycles, using the former for the decontamination of waste and sterilization of media and unwrapped glassware, and the latter for sterilization of packaged instruments.

Standard operating temperatures for autoclaves are 121 °C or, in some cases, 132 °C, typically at a pressure of 15 to 20 pounds per square inch (psi). The length of exposure depends on the volume and nature of material being sterilized, but it is typically 20 minutes or more, with larger volumes requiring longer exposure times to ensure sufficient heat transfer to the materials being sterilized. The steam must directly contact the liquids or dry materials being sterilized, so containers are left loosely closed and instruments are loosely wrapped in paper or foil. The key to autoclaving is that the temperature must be high enough to kill endospores to achieve complete sterilization.

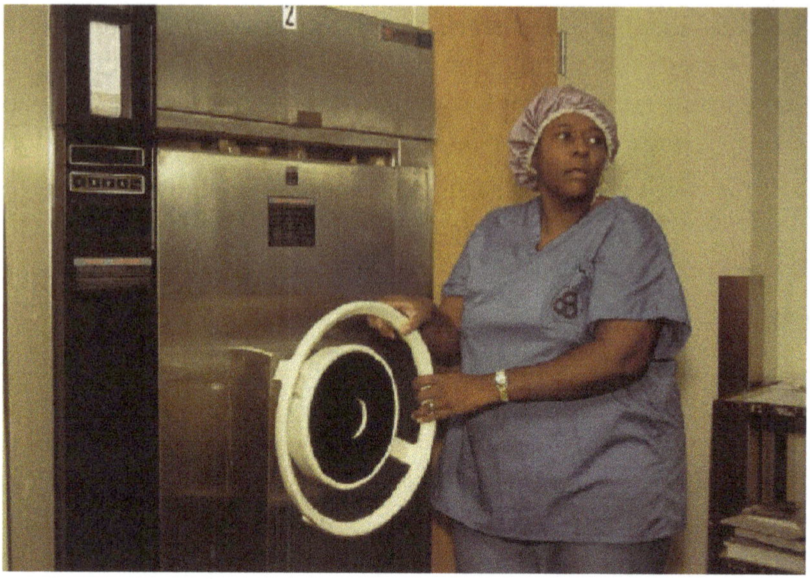

Figure 11.6 A technician sterilizes a sample using an autoclave. (Credit: Martha Cooper / Picryl; Public Domain.

Because sterilization is so important to safe medical and laboratory protocols, quality control is essential. Autoclaves may be equipped with recorders to document the pressures and temperatures achieved during each run. Additionally, internal indicators of various types should be autoclaved along with the materials to be sterilized to ensure that the proper sterilization temperature has been reached (Figure 11.7). One common type of indicator is the use of heat-sensitive autoclave tape, which has white stripes that turn black when the appropriate temperature is achieved during a successful autoclave run. This type of indicator is relatively inexpensive and can be used during every run. However, autoclave tape provides no indication of length of exposure, so it cannot be used as an indicator of sterility. Another type of indicator, a biological indicator spore test, uses either a strip of paper or a liquid suspension of the endospores of *Geobacillus stearothermophilus* to determine whether the endospores are killed by the process. The endospores of the obligate thermophilic bacterium *G. stearothermophilus* are the gold standard used for this purpose because of their extreme heat resistance. Biological spore indicators can also be used to test the effectiveness of other sterilization protocols, including ethylene oxide, dry heat, formaldehyde, gamma radiation, and hydrogen peroxide plasma sterilization using either *G. stearothermophilus*, *Bacillus atrophaeus*, *B. subtilis*, or *B. pumilus* spores. In the case of validating autoclave function, the endospores are incubated after autoclaving to ensure no viable endospores remain. Bacterial growth subsequent to endospore germination can be monitored by biological indicator spore tests that detect acid metabolites or fluorescence produced by enzymes derived from viable *G. stearothermophilus*. A third type of autoclave indicator is the Diack tube, a glass ampule containing a temperature-sensitive pellet that melts at the proper sterilization temperature. Spore strips or Diack tubes are used periodically to ensure the autoclave is functioning properly.

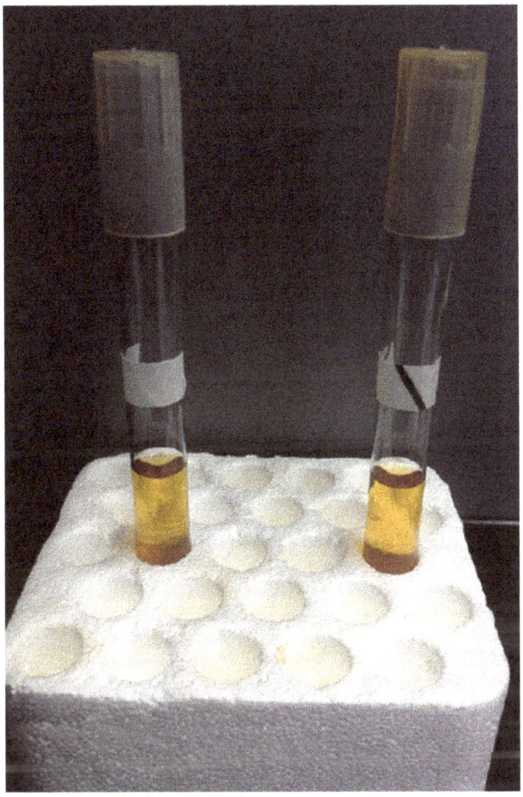

Figure 11.7 The white strips on autoclave tape (left tube) turn dark during a successful autoclave run (right tube). (credit: modification of work by Brian Forster)

Pasteurization

Although complete sterilization is ideal for many medical applications, it is not always practical for other applications and may also alter the quality of the product. Boiling and autoclaving are not ideal ways to control microbial growth in many foods because these methods may ruin the consistency and other organoleptic (sensory) qualities of the food. Pasteurization is a form of microbial control for food that uses heat but does not render the food sterile. Traditional **pasteurization** kills pathogens and reduces the number of spoilage-causing microbes while maintaining food quality. The process of pasteurization was first developed by Louis Pasteur in the 1860s as a method for preventing the spoilage of beer and wine. Today, pasteurization is most commonly used to kill heat-sensitive pathogens in milk and other food products (e.g., apple juice and honey) (Figure 11.8). However, because pasteurized food products are not sterile, they will eventually spoil.

The methods used for milk pasteurization balance the temperature and the length of time of treatment. One method, **high-temperature short-time (HTST) pasteurization**, exposes milk to a temperature of 72 °C for 15 seconds, which lowers bacterial numbers while preserving the quality of the milk. An alternative is **ultra-high- temperature (UHT) pasteurization**, in which the milk is exposed to a temperature of 138 °C for 2 or more seconds. UHT pasteurized milk can be stored for a long time in sealed containers without being refrigerated; however, the very high temperatures alter the proteins in the milk, causing slight changes in the taste and smell. Still, this method of pasteurization is advantageous in regions where access to refrigeration is limited.

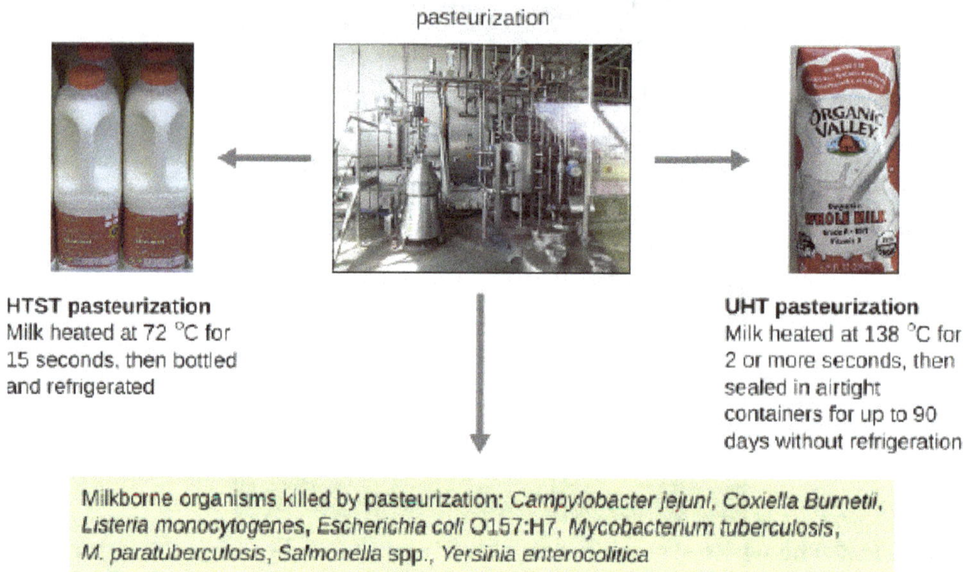

Figure 11.8 Two different methods of pasteurization, HTST and UHT, are commonly used to kill pathogens associated with milk spoilage. (credit left: modification of work by Mark Hillary; credit right: modification of work by Kerry Ceszyk)

Refrigeration and Freezing

Just as high temperatures are effective for controlling microbial growth, exposing microbes to low temperatures can also be an easy and effective method of microbial control, with the exception of psychrophiles, which prefer cold temperatures (see Temperature and Microbial Growth). Refrigerators used in home kitchens or in the laboratory maintain temperatures between 0 °C and 7 °C. This temperature range inhibits microbial metabolism, slowing the growth of microorganisms significantly and helping preserve refrigerated products such as foods or medical supplies. Certain types of laboratory cultures can be preserved by refrigeration for later use.

Freezing below -2 °C may stop microbial growth and even kill susceptible organisms. According to the US Department of Agriculture (USDA), the only safe ways that frozen foods can be thawed are in the refrigerator, immersed in cold water changed every 30 minutes, or in the microwave, keeping the food at temperatures not conducive for bacterial growth. In addition, halted bacterial growth can restart in thawed foods, so thawed foods should be treated like fresh perishables.

Bacterial cultures and medical specimens requiring long-term storage or transport are often frozen at ultra-low temperatures of -70 °C or lower. These ultra-low temperatures can be achieved by storing specimens on dry ice in an ultra-low freezer or in special liquid nitrogen tanks, which maintain temperatures lower than -196 °C

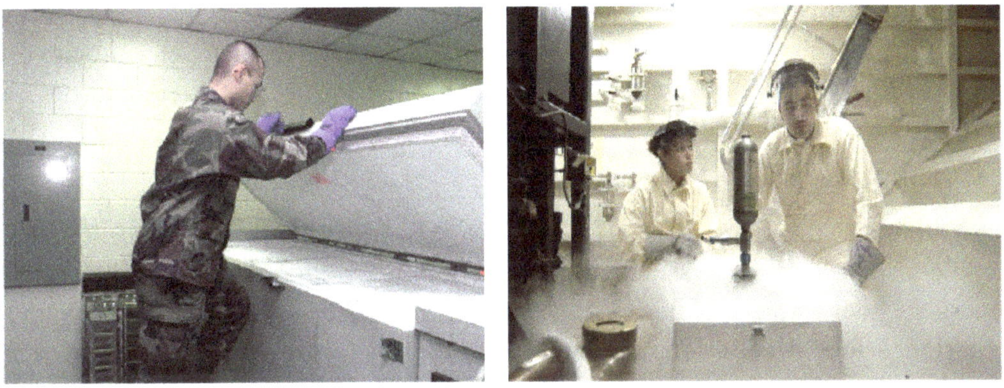

(a) (b)

Figure 11.9 Cultures and other medical specimens can be stored for long periods at ultra-low temperatures. (a) An ultra-low freezer maintains temperatures at or below -70 °C. (b) Two people stand in a room with a large nitrogen storage unit emitting a large amount of fog. (credit a: modification of work by "Expert Infantry"/Flickr; credit b: Credit: US Navy; Public Domain.)

Pressure

Exposure to high pressure kills many microbes. In the food industry, high-pressure processing (also called pascalization) is used to kill bacteria, yeast, molds, parasites, and viruses in foods while maintaining food quality and extending shelf life. The application of high pressure between 100 and 800 MPa (sea level atmospheric pressure is about 0.1 MPa) is sufficient to kill vegetative cells by protein denaturation, but endospores may survive these pressures.

In clinical settings, hyperbaric oxygen therapy is sometimes used to treat infections. In this form of therapy, a patient breathes pure oxygen at a pressure higher than normal atmospheric pressure, typically between 1 and 3 atmospheres (atm). This is achieved by placing the patient in a hyperbaric chamber or by supplying the pressurized oxygen through a breathing tube. Hyperbaric oxygen therapy helps increase oxygen saturation in tissues that become hypoxic due to infection and inflammation. This increased oxygen concentration enhances the body's immune response by increasing the activities of neutrophils and macrophages, white blood cells that fight infections. Increased oxygen levels also contribute to the formation of toxic free radicals that inhibit the growth of oxygen-sensitive or anaerobic bacteria like as *Clostridium perfringens*, a common cause of gas gangrene. In *C. perfringens* infections, hyperbaric oxygen therapy can also reduce secretion of a bacterial toxin that causes tissue destruction. Hyperbaric oxygen therapy also seems to enhance the effectiveness of antibiotic treatments. Unfortunately, some rare risks include oxygen toxicity and effects on delicate tissues, such as the eyes, middle ear, and lungs, which may be damaged by the increased air pressure.

High pressure processing is not commonly used for disinfection or sterilization of fomites. Although the application of pressure and steam in an autoclave is effective for killing endospores, it is the high temperature achieved, and not the pressure directly, that results in endospore death.

A Streak of Bad Potluck

One Monday in spring 2015, an Ohio woman began to experience blurred, double vision; difficulty swallowing; and drooping eyelids. She was rushed to the emergency department of her local hospital. During the examination, she began to experience abdominal cramping, nausea, paralysis, dry mouth, weakness of facial muscles, and difficulty speaking and breathing. Based on these symptoms, the hospital's incident command center was activated, and Ohio public health officials were notified of a possible caseof botulism. Meanwhile, other patients with similar symptoms began showing up at other local hospitals. Because of the suspicion of botulism, antitoxin was shipped overnight from the CDC to these medical facilities, to be administered to the affected patients. The first patient died of respiratory failure as a result of paralysis, and about half of the remaining victims required additional hospitalization following antitoxin administration, with at least two requiring ventilators for breathing.

Public health officials investigated each of the cases and determined that all of the patients had attended the same church potluck the day before. Moreover, they traced the source of the outbreak to a potato salad made with home-canned potatoes. More than likely, the potatoes were canned using boiling water, a method that allows endospores of *Clostridium botulinum* to survive. *C. botulinum* produces botulinum toxin, a neurotoxin that is often deadly once ingested. According to the CDC, the Ohio case was the largest botulism outbreak in the United States in nearly 40 years.

Killing *C. botulinum* endospores requires a minimum temperature of 116 °C (240 °F), well above the boiling point of water. This temperature can only be reached in a pressure canner, which is recommended for home canning of low-acid foods such as meat, fish, poultry, and vegetables (Figure 11.10). Additionally, the CDC recommends boiling home-canned foods for about 10 minutes before consumption. Since the botulinum toxin is heat labile (meaning that it is denatured by heat), 10 minutes of boiling will render non-functional any botulinum toxin that the food may contain.

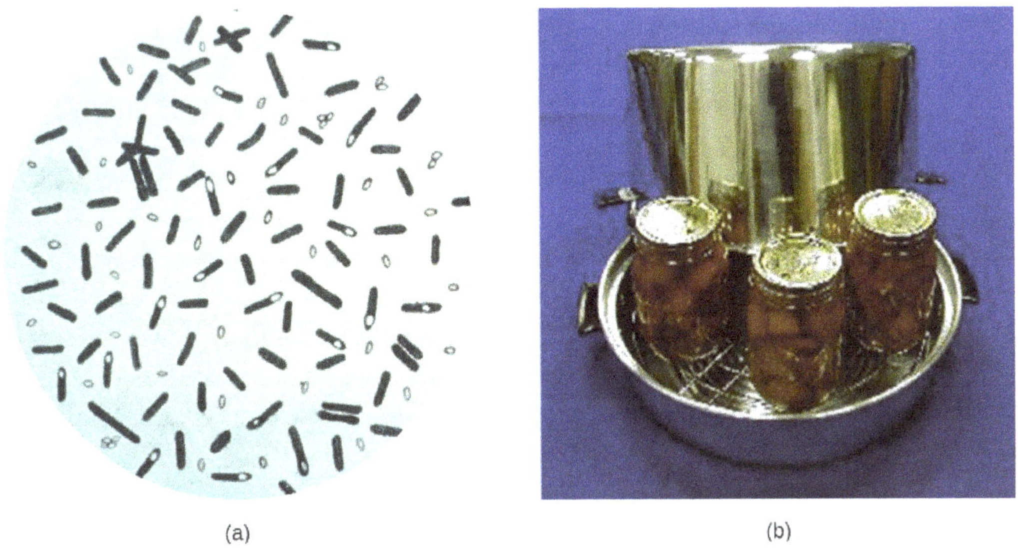

Figure 11.10 (a) *Clostridium botulinum* is the causative agent of botulism. (b) A pressure canner is recommended for home canning because endospores of *C. botulinum* can survive temperatures above the boiling point of water. (credit a: modification of work by Centers for Disease Control and Prevention; credit b: modification of work by National Center for Home Food Preservation)

Desiccation

Drying, also known as **desiccation** or dehydration is a method that has been used for millennia to preserve foods such as raisins, prunes, and jerky. It works because all cells, including microbes, require water for their metabolism and survival. Although drying controls microbial growth, it might not kill all microbes or their endospores, which may start to regrow when conditions are more favourable and water content is restored.

In some cases, foods are dried in the sun, relying on evaporation to achieve desiccation. Freeze-drying, or **lyophilization**, is another method of dessication in which an item is rapidly frozen ("snap-frozen") and placed under vacuum so that water is lost by sublimation. Lyophilization combines both exposure to cold temperatures and desiccation, making it quite effective for controlling microbial growth. In addition, lyophilization causes less damage to an item than conventional desiccation and better preserves the item's original qualities. Lyophilized items may be stored at room temperature if packaged appropriately to prevent moisture acquisition. Lyophilization is used for preservation in the food industry and is also used in the laboratory for the long-term storage and transportation of microbial cultures.

The water content of foods and materials, called the **water activity**, can be lowered without physical drying by the addition of solutes such as salts or sugars. At very high concentrations of salts or sugars, the amount of available water in microbial cells is reduced dramatically because water will be drawn from an area of low solute concentration (inside the cell) to an area of high solute concentration (outside the cell) (Figure 11.11). Many microorganisms do not survive these conditions of high osmotic pressure. Honey, for example, is 80% sucrose, an environment in which very few microorganisms are capable of growing, thereby eliminating the need for refrigeration. Salted meats and fish, like ham and cod, respectively, were critically important foods before the age of refrigeration. Fruits were preserved by adding sugar, making jams and jellies. However, certain microbes, such as molds and yeasts, tend to be more tolerant of desiccation and high osmotic pressures, and, thus, may still contaminate these types of foods.

(a) (b)

Figure 11.11 (a) The addition of a solute creates a hypertonic environment, drawing water out of cells. (b) Some foods can be dried directly, like raisins and jerky. Other foods are dried with the addition of salt, as in the case of salted fish, or sugar, as in the case of jam. (credit a: modification of work by "Bruce Blaus"/Wikimedia Commons; credit raisins: modification of work by Christian Schnettelker; credit jerky: modification of work by Larry Jacobsen; credit salted fish: modification of work by "The Photographer"/Wikimedia Commons; credit jam: modification of work by Kim Becker)

Radiation

Radiation in various forms, from high-energy radiation to sunlight, can be used to kill microbes or inhibit their growth. **Ionizing radiation** includes X-rays, gamma rays, and high-energy electron beams. Ionizing radiation is strong enough to pass into the cell, where it alters molecular structures and damages cell components. For example, ionizing radiation introduces double-strand breaks in DNA molecules. This may directly cause DNA mutations to occur, or mutations may be introduced when the cell attempts to repair the DNA damage. As these mutations accumulate, they eventually lead to cell death.

Both X-rays and gamma rays easily penetrate paper and plastic and can therefore be used to sterilize many packaged materials. In the laboratory, ionizing radiation is commonly used to sterilize materials that cannot be autoclaved, such as plastic Petri dishes and disposable plastic inoculating loops. For clinical use, ionizing radiation is used to sterilize gloves, intravenous tubing, and other latex and plastic items used for patient care. Ionizing radiation is also used for the sterilization of other types of delicate, heat-sensitive materials used clinically, including tissues for transplantation, pharmaceutical drugs, and medical equipment.

In Europe, gamma irradiation for food preservation is widely used, although it has been slow to catch on in the United States (see the Micro Connections box on this topic). Packaged dried spices are also often gamma-irradiated. Because of their ability to penetrate paper, plastic, thin sheets of wood and metal, and tissue, great care must be taken when using X-rays and gamma irradiation. These types of ionizing irradiation cannot penetrate thick layers of iron or lead, so these metals are commonly used to protect humans who may be potentially exposed.

Another type of radiation, **nonionizing radiation,** is commonly used for disinfection and uses less energy than ionizing radiation. It does not penetrate cells or packaging. Ultraviolet (UV) light is one example; it causes thymine dimers to form between adjacent thymines within a single strand of DNA (Figure 11.12). When DNA polymerase encounters the thymine dimer, it does not always incorporate the appropriate complementary nucleotides (two adenines), and this leads to formation of mutations that can ultimately kill microorganisms.

UV light can be used effectively by both consumers and laboratory personnel to control microbial growth. UV lamps are now commonly incorporated into water purification systems for use in homes. In addition, small portable UV lights are commonly used by campers to purify water from natural environments before drinking. Germicidal lamps are also used in surgical suites, biological safety cabinets, and transfer hoods, typically emitting UV light at a wavelength of 260 nm. Because UV light does not penetrate surfaces and will not pass through plastics or glass, cells must be exposed directly to the light source.

Sunlight has a very broad spectrum that includes UV and visible light. In some cases, sunlight can be effective against certain bacteria because of both the formation of thymine dimers by UV light and by the production of reactive oxygen products induced in low amounts by exposure to visible light.

Irradiated Food: Would You Eat That?

Of all the ways to prevent food spoilage and foodborne illness, gamma irradiation may be the most unappetizing. Although gamma irradiation is a proven method of eliminating potentially harmful microbes from food, the public has yet to buy in. Most of their concerns, however, stem from misinformation and a poor understanding of the basic principles of radiation.

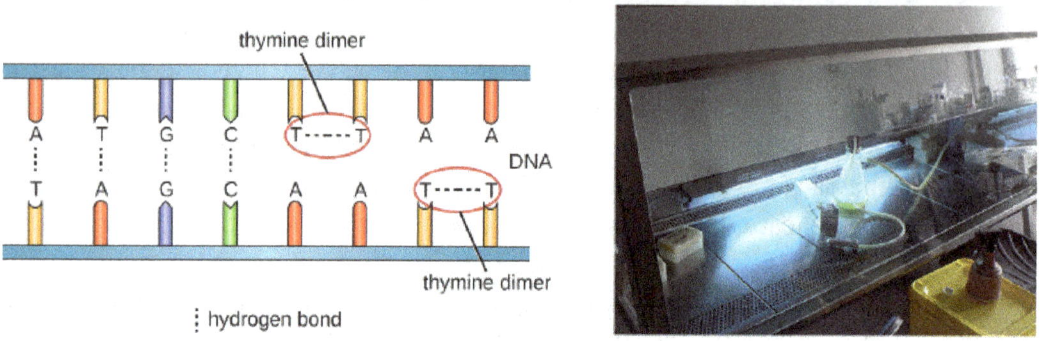

(a) (b)

Figure 11.12 (a) UV radiation causes the formation of thymine dimers in DNA, leading to lethal mutations in the exposed microbes. (b) Germicidal lamps that emit UV light are commonly used in the laboratory to disinfect equipment.

The most common method of irradiation is to expose food to cobalt-60 or cesium-137 by passing it through a radiation chamber on a conveyor belt. The food does not directly contact the radioactive material and does not become radioactive itself. Thus, there is no risk for exposure to radioactive material through eating gamma-irradiated foods. Additionally, irradiated foods are not significantly altered in terms of nutritional quality, aside from the loss of certain vitamins, which is also exacerbated by extended storage. Alterations in taste or smell may occur in irradiated foods with high fat content, such as fatty meats and dairy products, but this effect can be minimized by using lower doses of radiation at colder temperatures.

In the United States, the CDC, Environmental Protection Agency (EPA), and the Food and Drug Administration (FDA) have deemed irradiation safe and effective for various types of meats, poultry, shellfish, fresh fruits and vegetables, eggs with shells, and spices and seasonings. Gamma irradiation of foods has also been approved for use in many other countries, including France, the Netherlands, Portugal, Israel, Russia, China, Thailand, Belgium, Australia, and South Africa. To help ameliorate consumer concern and assist with education efforts, irradiated foods are now clearly labeled and marked with the international irradiation symbol, called the "radura" (Figure 11.13). Consumer acceptance seems to be rising, as indicated by several recent studies.

Sonication

The use of high-frequency ultrasound waves to disrupt cell structures is called **sonication.** Application of ultrasound waves causes rapid changes in pressure within the intracellular liquid; this leads to cavitation, the formation of bubbles inside the cell, which can disrupt cell structures and eventually cause the cell to lyse or collapse. Sonication is useful in the laboratory for efficiently lysing cells to release their contents for further research; outside the laboratory, sonication is used for cleaning surgical instruments, lenses, and a variety of other objects such as coins, tools, and musical instruments.

(a) (b)

Figure 11.13 (a) Foods are exposed to gamma radiation by passage on a conveyor belt through a radiation chamber. (b) Gamma-irradiated foods must be clearly labeled and display the irradiation symbol, known as the "radura." (credit a, b: modification of work by U.S. Department of Agriculture)

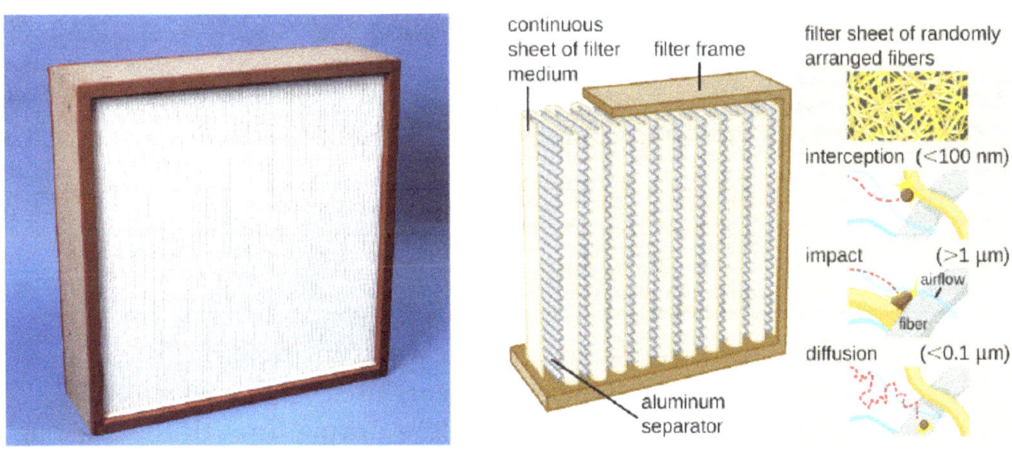

(a) (b)

Figure 11.14 (a) HEPA filters like this one remove microbes, endospores, and viruses as air flows through them. (b) A schematic of a HEPA filter. (credit a: modification of work by CSIRO; credit b: modification of work by "LadyofHats"/ Mariana Ruiz Villareal)

Filtration

Filtration is a method of physically separating microbes from samples. Air is commonly filtered through **high- efficiency particulate air (HEPA) filters** (Figure 11.14). HEPA filters have effective pore sizes of 0.3 μm, small enough to capture bacterial cells, endospores, and many viruses, as air passes through these filters, nearly sterilizing the air on the other side of the filter. HEPA filters have a variety of applications and are used widely in clinical settings, in cars and airplanes, and even in the home. For example, they may be found in vacuum cleaners, heating and air-conditioning systems, and air purifiers.

Biological Safety Cabinets

Biological safety cabinets are a good example of the use of HEPA filters. HEPA filters in biological safety cabinets (BSCs) are used to remove particulates in the air either entering the cabinet (air intake), leaving the cabinet (air exhaust), or treating both the intake and exhaust. Use of an air-intake HEPA filter prevents environmental contaminants from entering the BSC, creating a clean area for handling biological materials. Use of an air- exhaust HEPA filter prevents laboratory pathogens from contaminating the laboratory, thus maintaining a safe work area for laboratory personnel.

There are three classes of BSCs: I, II, and III. Each class is designed to provide a different level of protection for laboratory personnel and the environment; BSC II and III are also designed to protect the materials or devices in the cabinet. Table 11.1 summarizes the level of safety provided by each class of BSC for each BSL.

Table 11.1 Biological Risks and BSCs

Biological Risk Assessed	BSC Class	Protection of Personnel	Protection of Environment	Protection of Product
BSL-1, BSL-2, BSL-3	I	Yes	Yes	No
BSL-1, BSL-2, BSL-3	II	Yes	Yes	Yes
BSL-4	III; II when used in suit room with suit	Yes	Yes	Yes

Class I BSCs protect laboratory workers and the environment from a low to moderate risk for exposure to biological agents used in the laboratory. Air is drawn into the cabinet and then filtered before exiting through the building's exhaust system. Class II BSCs use directional air flow and partial barrier systems to contain infectious agents. Class III BSCs are designed for working with highly infectious agents like those used in BSL-4 laboratories. They are gas tight, and materials entering or exiting the cabinet must be passed through a double- door system, allowing the intervening space to be decontaminated between uses. All air is passed through one or two HEPA filters and an air incineration system before being exhausted directly to the outdoors (not through the building's exhaust system). Personnel can manipulate materials inside the Class III cabinet by using long rubber gloves sealed to the cabinet.

Filtration in Hospitals

HEPA filters are also commonly used in hospitals and surgical suites to prevent contamination and the spread of airborne microbes through ventilation systems. HEPA filtration systems may be designed for entire buildings or for individual rooms. For example, burn units, operating rooms, or isolation units may require special HEPA- filtration systems to remove opportunistic pathogens from the environment because patients in these rooms are particularly vulnerable to infection.

Membrane Filters

Filtration can also be used to remove microbes from liquid samples using **membrane filtration**. Membrane filters for liquids function similarly to HEPA filters for air. Typically, membrane filters that are used to remove bacteria have an effective pore size of 0.2 µm, smaller than the average size of a bacterium (1 µm), but filters with smaller pore sizes are available for more specific needs. Membrane filtration is useful for removing bacteria from various types of heat-sensitive solutions used in the laboratory, such as antibiotic solutions and vitamin solutions. Large volumes of culture media may also be filter sterilized rather than autoclaved to protect heat-sensitive components. Often when filtering small volumes, syringe filters are used, but vacuum filters are typically used for filtering larger volumes (Figure 11.15).

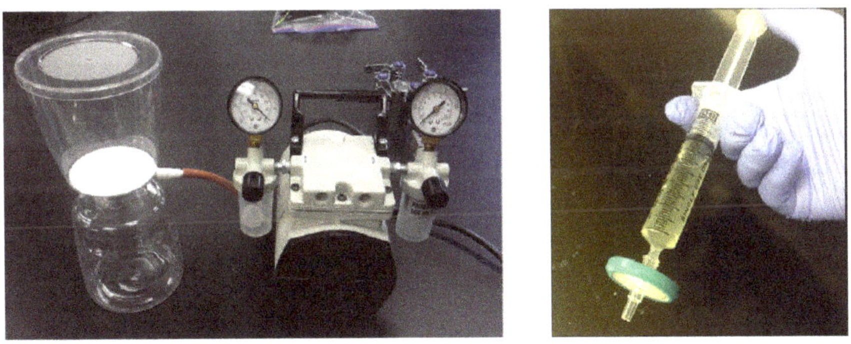

(a) (b)

Figure 11.15 Membrane filters come in a variety of sizes, depending on the volume of solution being filtered. (a) Larger volumes are filtered in units like these. The solution is drawn through the filter by connecting the unit to a vacuum. (b) Smaller volumes are often filtered using syringe filters, which are units that fit on the end of a syringe. In this case, the solution is pushed through by depressing the syringe's plunger. (credit a, b: modification of work by Brian Forster)

Figure 11.16 and **Figure 11.17** summarize the physical methods of control discussed in this section.

Physical Methods of Control			
Method	**Conditions**	**Mode of Action**	**Example Uses**
Heat			
Boiling	100 °C at sea level	Denatures proteins and alters membranes	Cooking, personal use, preparing certain laboratory media
Dry-heat oven	170 °C for 2 hours	Denatures proteins and alters membranes, dehydration, desiccation	Sterilization of heat-stable medical and laboratory equipment and glassware
Incineration	Exposure to flame	Destroy by burning	Flaming loop, microincinerator
Autoclave	Typical settings: 121 °C for 15 minutes at 15 pounds per square inch (psi)	Denatures proteins and alters membranes	Sterilization of microbiological media, heat-stable medical and laboratory equipment, and other heat-stable items
Pasteurization	Can vary. One type is 72 °C for 15 seconds (HTST)	Denatures proteins and alters membranes	Prevents spoilage of milk, apple juice, honey, and other ingestible liquids
Cold			
Refrigeration	0 °C to 7 °C	Inhibits metabolism (slows or arrests cell division)	Preservation of food or laboratory materials (solutions, cultures)
Freezing	Below −2 °C	Stops metabolism, may kill microbes	Long-term storage of food, laboratory cultures, or medical specimens
Pressure			
High-pressure processing	100–800 MPa	Denatures proteins and can cause cell lysis	Preservation of food
Hyberbaric oxygen therapy	Air pressure three times higher than normal	Inhibits metabolism and growth of anaerobic microbes	Treatment of certain infections (e.g., gas gangrene)
Desiccation			
Simple desiccation	Drying	Inhibits metabolism	Dried fruits, jerky
Reduce water activity	Addition of salt or water	Inhibits metabolism and can cause lysis	Salted meats and fish, honey, jams and jellies
Lyophilization	Rapid freezing under vacuum	Inhibits metabolism	Preservation of food, laboratory cultures, or reagents
Radiation			
Ionizing radiation	Exposure to X-rays or gamma rays	Alters molecular structures, introduces double-strand breaks into DNA	Sterilization of spices and heat-sensitive laboratory and medical items; used for food sterilization in Europe but not widely accepted in US
Nonionizing radiation	Exposure to ultraviolet light	Introduces thymine dimers, leading to mutations	Disinfection of surfaces in laboratories and rooms in health-care environment, and disinfection of water and air

Figure 11.15

Physical Methods of Control (continued)			
Method	Conditions	Mode of Action	Example Uses
Sonication			
Sonication	Exposure to ultrasonic waves	Cavitation (formation of empty space) disrupts cells, lysing them	Laboratory research to lyse cells; cleaning jewelry, lenses, and equipment
Filtration			
HEPA filtration	Use of high-efficiency particulate air (HEPA) filter with 0.3 μm pore size	Physically removes microbes from air	Laboratory biological safety cabinets, operating rooms, isolation units, heating and air conditioning systems, vacuum cleaners
Membrane filtration	Use of membrane filter with 0.2-μm or smaller pore size	Physically removes microbes from liquid solutions	Removal of bacteria from heat-sensitive solutions like vitamins, antibiotics, and media with heat-sensitive components

Figure 11.16

USING CHEMICALS TO CONTROL MICROORGANISMS

Phenolics

In the 1800s, scientists began experimenting with a variety of chemicals for disinfection. In the 1860s, British surgeon Joseph Lister (1827–1912) began using carbolic acid, known as phenol, as a disinfectant for the treatment of surgical wounds (see Foundations of Modern Cell Theory (https://legacy.cnx.org/content/m58791/latest/)). In 1879, Lister's work inspired the American chemist Joseph Lawrence (1836–1909) to develop Listerine, an alcohol-based mixture of several related compounds that is still used today as an oral antiseptic. Today, carbolic acid is no longer used as a surgical disinfectant because it is a skin irritant, but the chemical compounds found in antiseptic mouthwashes and throat lozenges are called **phenolics**.

Chemically, phenol consists of a benzene ring with an –OH group, and phenolics are compounds that have this group as part of their chemical structure (Figure 11.18). Phenolics such as thymol and eucalyptol occur naturally in plants. Other phenolics can be derived from creosote, a component of coal tar. Phenolics tend to be stable, persistent on surfaces, and less toxic than phenol. They inhibit microbial growth by denaturing proteins and disrupting membranes.

Since Lister's time, several phenolic compounds have been used to control microbial growth. Phenolics like cresols (methylated phenols) and o-phenylphenol were active ingredients in various formulations of Lysol since its invention in 1889. o-Phenylphenol was also commonly used in agriculture to control bacterial and fungal growth on harvested crops, especially citrus fruits, but its use in the United States is now far more limited. The bisphenol hexachlorophene, a disinfectant, is the active ingredient in pHisoHex, a topical cleansing detergent widely used for handwashing in hospital settings. pHisoHex is particularly effective against gram-positive bacteria, including those causing staphylococcal and streptococcal skin infections. Ph isoHex was formerly used for bathing infants, but this practice has been discontinued because it has been shown that exposure to hexachlorophene can lead to neurological problems.

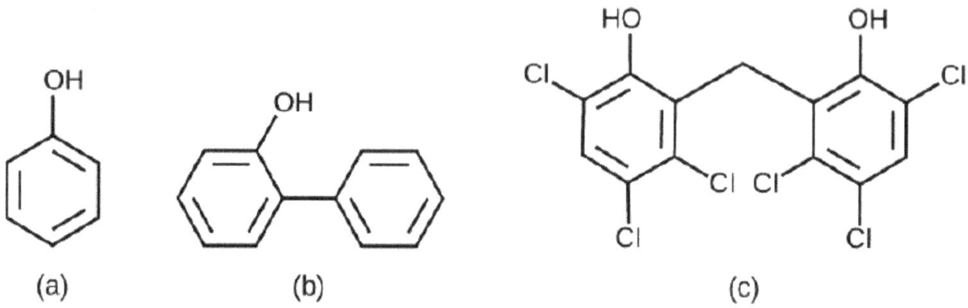

Figure 11.18 Phenol and phenolic compounds have been used to control microbial growth. (a) Chemical structure of phenol, also known as carbolic acid. (b) o-Phenylphenol, a type of phenolic, has been used as a disinfectant as well as to control bacterial and fungal growth on harvested citrus fruits. (c) Hexachlorophene, another phenol, known as a bisphenol (two rings), is the active ingredient in pHisoHex.

Triclosan is another bisphenol compound that has seen widespread application in antibacterial products over the last several decades. Initially used in toothpastes, triclosan has also been used in hand soaps and impregnated into a wide variety of other products, including cutting boards, knives, shower curtains, clothing, and concrete, to make them antimicrobial. However, in 2016 the FDA banned the marketing of over-the-counter antiseptic products containing triclosan and 18 other chemicals. This ruling was based on the lack of evidence of safety or efficacy, as well as concerns about the health risks of long-term exposure (See Micro Connections below). In 2019 the FDA issued an updated ban ruling to included 28 chemicals. Rulings on benzalkonium chloride, ethyl alcohol, and isopropyl alcohol have been deferred to allow for the submission of additional safety and efficacy data.

Triclosan: Antibacterial Overkill?

Hand soaps and other cleaning products are often marketed as "antibacterial," suggesting that they provide a level of cleanliness superior to that of conventional soaps and cleansers. But are the antibacterial ingredients in these products really safe and effective?

About 75% of antibacterial liquid hand soaps and 30% of bar soaps contain the chemical triclosan, a phenolic, (Figure 11.**19**).[9] Triclosan blocks an enzyme in the bacterial fatty acid-biosynthesis pathway that is not found in the comparable human pathway. Although the use of triclosan in the home increased dramatically during the 1990s, more than 40 years of research by the FDA have turned up no conclusive evidence that washing with triclosan-containing products provides increased health benefits compared with washing with traditional soap. Although some studies indicate that fewer bacteria may remain on a person's hands after washing with triclosan-based soap, compared with traditional soap, no evidence points to any reduction in the transmission of bacteria that cause respiratory and gastrointestinal illness. In short, soaps with triclosan may remove or kill a few more germs but not enough to reduce the spread of disease.

Perhaps more disturbing, some clear risks associated with triclosan-based soaps have come to light. The widespread use of triclosan has led to an increase in triclosan-resistant bacterial strains, including those of clinical importance, such as *Salmonella enterica*; this resistance may render triclosan useless as an antibacterial in the long run. Bacteria can easily gain resistance to triclosan through a change to a single gene encoding the targeted enzyme in the bacterial fatty acid-synthesis pathway. Other disinfectants with a less specific mode of action are much less prone to engendering resistance because it would take much more than a single genetic change.

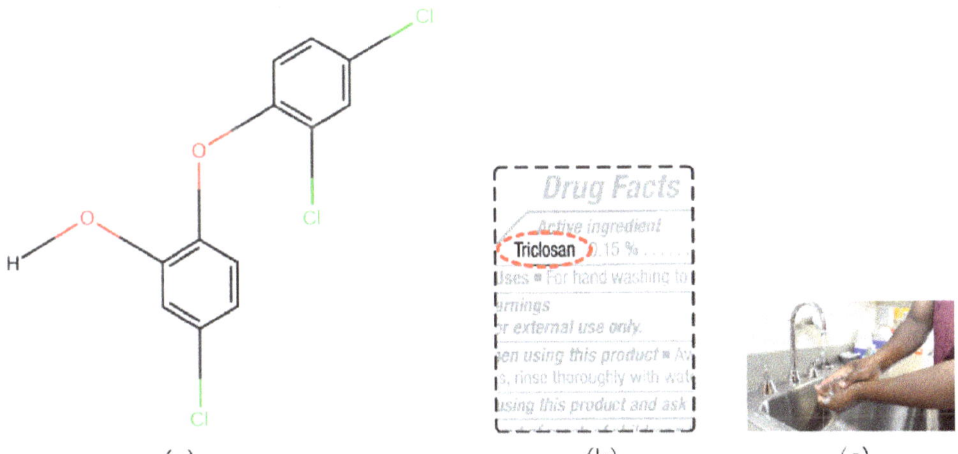

(a) (b) (c)

Figure 11.19 Triclosan is a common ingredient in antibacterial soaps despite evidence that it poses environmental and health risks and offers no significant health benefit compared to conventional soaps. (credit b: modification of work by FDA; c: Michelle Gigante/US Air Force; Public Domain.)

Use of triclosan over the last several decades has also led to a build-up of the chemical in the environment. Triclosan in hand soap is directly introduced into wastewater and sewage systems as a result of the hand washing process. There, its antibacterial properties can inhibit or kill bacteria responsible for the decomposition of sewage, causing septic systems to clog and back up. Eventually, triclosan in wastewater finds its way into surface waters, streams, lakes, sediments, and soils, disrupting natural populations of bacteria that carry out important environmental functions, such as inhibiting algae. Triclosan also finds its way into the bodies of amphibians and fish, where it can act as an endocrine disruptor. Detectable levels of triclosan have also been found in various human bodily fluids, including breast milk, plasma, and urine. In fact, a study conducted by the CDC found detectable levels of triclosan in the urine of 75% of 2,517 people tested in 2003–2004.This finding is even more troubling given the evidence that triclosan may affect immune function in humans.

In December 2013, the FDA gave soap manufacturers until 2016 to prove that antibacterial soaps provide a significant benefit over traditional soaps; if unable to do so, manufacturers will be forced to remove these products from the market.

Heavy Metals

Some of the first chemical disinfectants and antiseptics to be used were heavy metals. Heavy metals kill microbes by binding to proteins, thus inhibiting enzymatic activity (Figure 11.20). Heavy metals are oligodynamic, meaning that very small concentrations show significant antimicrobial activity. Ions of heavy metals bind to sulfur-containing amino acids strongly and bioaccumulate within cells, allowing these metals to reach high localized concentrations. This causes proteins to denature.

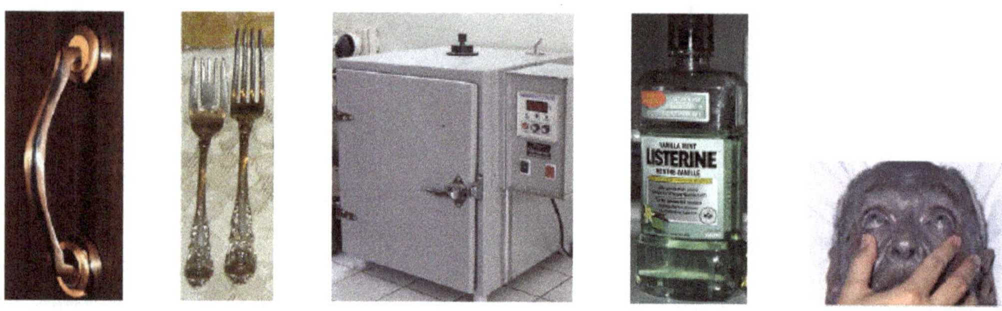

(a) (b) (c) (d) (e)

Figure 11.20 Heavy metals denature proteins, impairing cell function and, thus, giving them strong antimicrobial properties. (a) Copper in fixtures like this door handle kills microbes that otherwise might accumulate on frequently touched surfaces. (b) Eating utensils contain small amounts of silver to inhibit microbial growth. (c) Copper commonly lines incubators to minimize contamination of cell cultures stored inside. (d) Antiseptic mouthwashes commonly contain zinc chloride. (e) This patient is suffering from argyria, an irreversible condition caused by bioaccumulation of silver in the body. (credit b: modification of work by "Shoshanah"/Flickr; credit e: modification of work by Herbert L. Fred and Hendrik A. van Dijk)

Heavy metals are not selectively toxic to microbial cells. They may bioaccumulate in human or animal cells, as well, and excessive concentrations can have toxic effects on humans. If too much silver accumulates in the body, for example, it can result in a condition called argyria, in which the skin turns irreversibly blue-gray. One way to reduce the potential toxicity of heavy metals is by carefully controlling the duration of exposure and concentration of the heavy metal.

Mercury

Mercury is an example of a heavy metal that has been used for many years to control microbial growth. It was used for many centuries to treat syphilis. Mercury compounds like mercuric chloride are mainly bacteriostatic and have a very broad spectrum of activity. Various forms of mercury bind to sulfur-containing amino acids within proteins, inhibiting their functions.

In recent decades, the use of such compounds has diminished because of mercury's toxicity. It is toxic to the central nervous, digestive, and renal systems at high concentrations, and has negative environmental effects, including bioaccumulation in fish. Topical antiseptics such as mercurochrome, which contains mercury in low concentrations, and merthiolate, a **tincture** (a solution of mercury dissolved in alcohol) were once commonly used. However, because of concerns about using mercury compounds, these antiseptics are no longer sold in the United States.

Silver

Silver has long been used as an antiseptic. In ancient times, drinking water was stored in silver jugs. Silvadene cream is commonly used to treat topical wounds and is particularly helpful in preventing infection in burn wounds. Silver nitrate drops were once routinely applied to the eyes of newborns to protect against ophthalmia neonatorum, eye infections that can occur due to exposure to pathogens in the birth canal, but antibiotic creams are more now commonly used. Silver is often combined with antibiotics, making the antibiotics thousands of times more effective. Silver is also commonly incorporated into catheters and bandages, rendering them antimicrobial; however, there is evidence that heavy metals may also enhance selection for antibiotic resistance.

Copper, Nickel, and Zinc

Several other heavy metals also exhibit antimicrobial activity. Copper sulfate is a common algicide used to control algal growth in swimming pools and fish tanks. The use of metallic copper to minimize microbial growth is also becoming more widespread. Copper linings in incubators help reduce contamination of cell cultures. The use of copper pots for water storage in underdeveloped countries is being investigated as a way to combat diarrheal diseases. Copper coatings are also becoming popular for frequently handled objects such as doorknobs, cabinet hardware, and other fixtures in health-care facilities in an attempt to reduce the spread of microbes.

Nickel and zinc coatings are now being used in a similar way. Other forms of zinc, including zinc chloride and zinc oxide, are also used commercially. Zinc chloride is quite safe for humans and is commonly found in mouthwashes, substantially increasing their length of effectiveness. Zinc oxide is found in a variety of products, including topical antiseptic creams such as calamine lotion, diaper ointments, baby powder, and dandruff shampoos.

Halogens

Other chemicals commonly used for disinfection are the halogens iodine, chlorine, and fluorine. Iodine works by oxidizing cellular components, including sulfur-containing amino acids, nucleotides, and fatty acids, and destabilizing the macromolecules that contain these molecules. It is often used as a topical tincture, but it may cause staining or skin irritation. An **iodophor** is a compound of iodine complexed with an organic molecule, thereby increasing iodine's stability and, in turn, its efficacy. One common iodophor is povidone-iodine, which includes a wetting agent that releases iodine relatively slowly. Betadine is a brand of povidone-iodine commonly used as a hand scrub by medical personnel before surgery and for topical antisepsis of a patient's skin before incision (Figure 11.21).

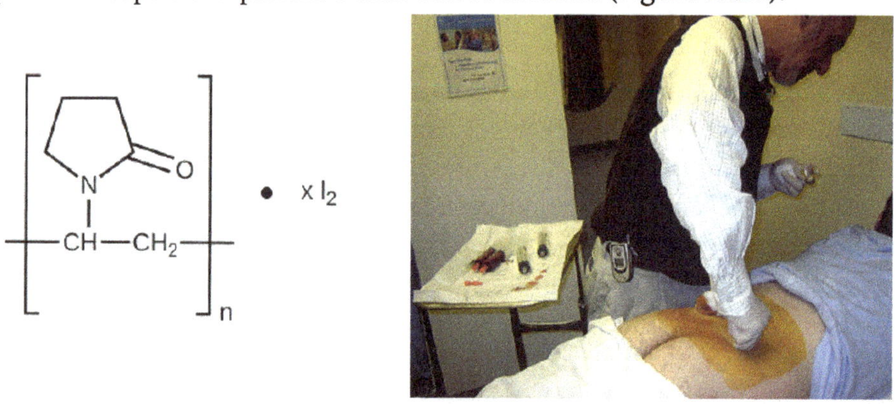

(a) (b)

Figure 11.21 (a) Betadine is a solution of the iodophor povidone-iodine. (b) It is commonly used as a topical antiseptic on a patient's skin before incision during surgery. (credit b: modification of work by Andrew Ratto)

Chlorine is another halogen commonly used for disinfection. When chlorine gas is mixed with water, it produces a strong oxidant called hypochlorous acid, which is uncharged and enters cells easily. Chlorine gas is commonly used in municipal drinking water and wastewater treatment plants, with the resulting hypochlorous acid producing the actual antimicrobial effect. Those working at water treatment facilities need to take great care to minimize personal exposure to chlorine gas. Sodium hypochlorite is the chemical component of common household bleach, and it is also used for a wide variety of disinfecting purposes. Hypochlorite salts, including sodium and calcium hypochlorites, are used to disinfect swimming pools. Chlorine gas, sodium hypochlorite, and calcium hypochlorite are also commonly used disinfectants in the food processing and restaurant industries to reduce the spread of foodborne diseases. Workers in these industries also need to take care to use these products correctly to ensure their own safety as well as the safety of consumers. A recent joint statement published by the Food and Agriculture Organization (FAO) of the United Nations and WHO indicated that none of the many beneficial uses of chlorine products in food processing to reduce the spread of foodborne illness posed risks to consumers.

Another class of chlorinated compounds called chloramines are widely used as disinfectants. Chloramines are relatively stable, releasing chlorine over long periods time. Chloramines are derivatives of ammonia by substitution of one, two, or all three hydrogen atoms with chlorine atoms (Figure 11.22).

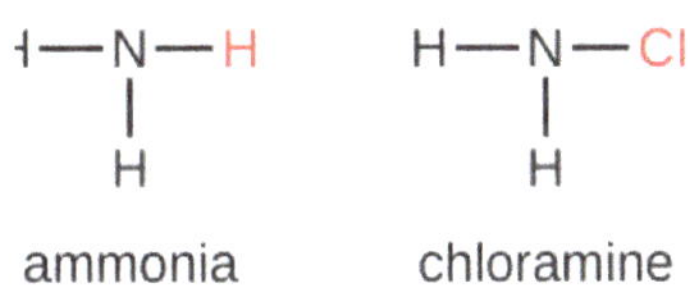

Figure 11.22 Monochloroamine, one of the chloramines, is derived from ammonia by the replacement of one hydrogen atom with a chlorine atom.

Chloramines and other cholorine compounds may be used for disinfection of drinking water, and chloramine tablets are frequently used by the military for this purpose. After a natural disaster or other event that compromises the public water supply, the CDC recommends disinfecting tap water by adding small amounts of regular household bleach. Recent research suggests that sodium dichloroisocyanurate (NaDCC) may also be a good alternative for drinking water disinfection. Currently, NaDCC tablets are available for general use and for use by the military, campers, or those with emergency needs; for these uses, NaDCC is preferable to chloramine tablets. Chlorine dioxide, a gaseous agent used for fumigation and sterilization of enclosed areas, is also commonly used for the disinfection of water.

Although chlorinated compounds are relatively effective disinfectants, they have their disadvantages. Some may irritate the skin, nose, or eyes of some individuals, and they may not completely eliminate certain hardy organisms from contaminated drinking water. The protozoan parasite *Cryptosporidium*, for example, has a protective outer shell that makes it resistant to chlorinated disinfectants.

Thus, boiling of drinking water in emergency situations is recommended when possible.

The halogen fluorine is also known to have antimicrobial properties that contribute to the prevention of dental caries (cavities). Fluoride is the main active ingredient of toothpaste and is also commonly added to tap water to help communities maintain oral health. Chemically, fluoride can become incorporated into the hydroxyapatite of tooth enamel, making it more resistant to corrosive acids produced by the fermentation of oral microbes. Fluoride also enhances the uptake of calcium and phosphate ions in tooth enamel, promoting remineralization. In addition to strengthening enamel, fluoride also seems to be bacteriostatic. It accumulates in plaque-forming bacteria, interfering with their metabolism and reducing their production of the acids that contribute to tooth decay.

Alcohols

Alcohols make up another group of chemicals commonly used as disinfectants and antiseptics. They work by rapidly denaturing proteins, which inhibits cell metabolism, and by disrupting membranes, which leads to cell lysis. Once denatured, the proteins may potentially refold if enough water is present in the solution. Alcohols are typically used at concentrations of about 70% aqueous solution and, in fact, work better in aqueous solutions than 100% alcohol solutions. This is because alcohols coagulate proteins. In higher alcohol concentrations, rapid coagulation of surface proteins prevents effective penetration of cells. The most commonly used alcohols for disinfection are ethyl alcohol (ethanol) and isopropyl alcohol (isopropanol, rubbing alcohol) (Figure 11.23).

Alcohols tend to be bactericidal and fungicidal, but may also be viricidal for enveloped viruses only. Although alcohols are not sporicidal, they do inhibit the processes of sporulation and germination. Alcohols are volatile and dry quickly, but they may also cause skin irritation because they dehydrate the skin at the site of application. One common clinical use of alcohols is swabbing the skin for degerming before needle injection. Alcohols also are the active ingredients in instant hand sanitizers, which have gained popularity in recent years. The alcohol in these hand sanitizers works both by denaturing proteins and by disrupting the microbial cell membrane, but will not work effectively in the presence of visible dirt.

Last, alcohols are used to make tinctures with other antiseptics, such as the iodine tinctures discussed previously in this chapter. All in all, alcohols are inexpensive and quite effective for the disinfection of a broad range of vegetative microbes. However, one disadvantage of alcohols is their high volatility, limiting their effectiveness to immediately after application.

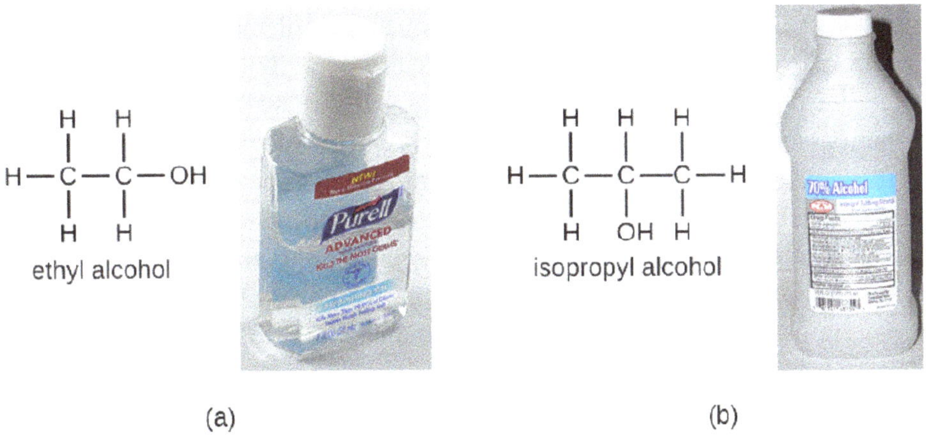

Figure 11.23 (a) Ethyl alcohol, the intoxicating ingredient found in alcoholic drinks, is also used commonly as a disinfectant. (b) Isopropyl alcohol, also called rubbing alcohol, has a related molecular structure and is another commonly used disinfectant. (credit a photo: modification of work by D Coetzee; credit b photo: modification of work by Craig Spurrier)

Surfactants

Surface-active agents, or **surfactants,** are a group of chemical compounds that lower the surface tension of water. Surfactants are the major ingredients in soaps and detergents. Soaps are salts of long-chain fatty acids and have both polar and nonpolar regions, allowing them to interact with polar and nonpolar regions in other molecules (Figure 11.24). They can interact with nonpolar oils and grease to create emulsions in water, loosening and lifting away dirt and microbes from surfaces and skin. Soaps do not kill or inhibit microbial growth and so are not considered antiseptics or disinfectants. However, proper use of soaps mechanically carries away microorganisms, effectively degerming a surface. Some soaps contain added bacteriostatic agents such as triclocarban or cloflucarban, compounds structurally related to triclosan, that introduce antiseptic or disinfectant properties to the soaps.

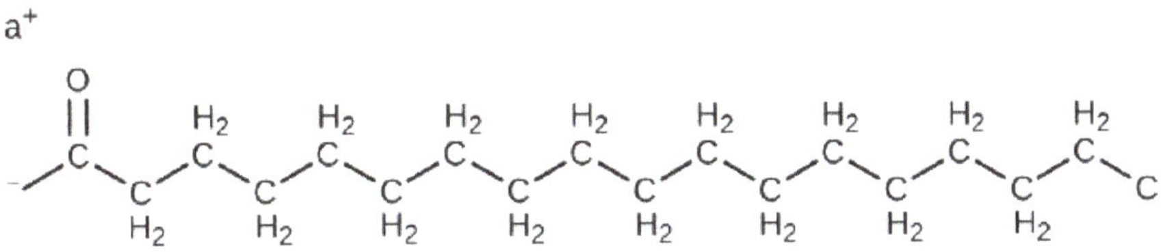

Figure 11.24 Soaps are the salts (sodium salt in the illustration) of fatty acids and have the ability to emulsify lipids, fats, and oils by interacting with water through their hydrophilic heads and with the lipid at their hydrophobic tails.

Soaps, however, often form films that are difficult to rinse away, especially in hard water, which contains high concentrations of calcium and magnesium mineral salts. Detergents contain synthetic surfactant molecules with both polar and nonpolar regions that have strong cleansing activity but are more soluble, even in hard water, and, therefore, leave behind no soapy deposits. Anionic detergents, such as those used for laundry, have a negatively charged anion at one end attached to a long hydrophobic chain, whereas cationic detergents have a positively charged cation instead. Cationic detergents include an important class of disinfectants and antiseptics called the **quaternary ammonium salts (quats)**, named for the characteristic quaternary nitrogen atom that confers the positive charge (Figure 11.25). Overall, quats have properties similar to phospholipids, having hydrophilic and hydrophobic ends. As such, quats have the ability to insert into the bacterial phospholipid bilayer and disrupt membrane integrity. The cationic charge of quats appears to confer their antimicrobial properties, which are diminished when neutralized. Quats have several useful properties. They are stable, nontoxic, inexpensive, colorless, odorless, and tasteless. They tend to be bactericidal by disrupting membranes. They are also active against fungi, protozoans, and enveloped viruses, but endospores are unaffected. In clinical settings, they may be used as antiseptics or to disinfect surfaces. Mixtures of quats are also commonly found in household cleaners and disinfectants, including many current formulations of Lysol brand products, which contain benzalkonium chlorides as the active ingredients. Benzalkonium chlorides, along with the quat cetylpyrimidine chloride, are also found in products such as skin antiseptics, oral rinses, and mouthwashes.

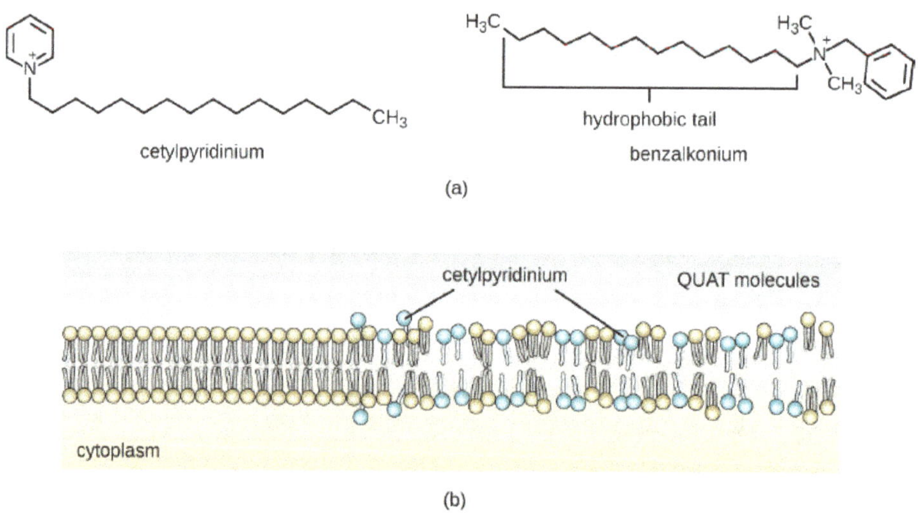

Figure 11.25 (a) Two common quats are benzylalkonium chloride and cetylpyrimidine chloride. Note the hydrophobic nonpolar carbon chain at one end and the nitrogen-containing cationic component at the other end. (b) Quats are able to infiltrate the phospholipid plasma membranes of bacterial cells and disrupt their integrity, leading to death of the cell.

Hand Washing the Right Way

Hand washing is critical for public health and should be emphasized in a clinical setting. For the general public, the CDC recommends hand washing before, during, and after food handling; before eating; before and after interacting with someone who is ill; before and after treating a wound; after using the toilet or changing diapers; after coughing, sneezing, or blowing the nose; after handling garbage; and after interacting with an animal, its feed, or its waste. Figure 11.26 illustrates the five steps of proper hand washing recommended by the CDC.

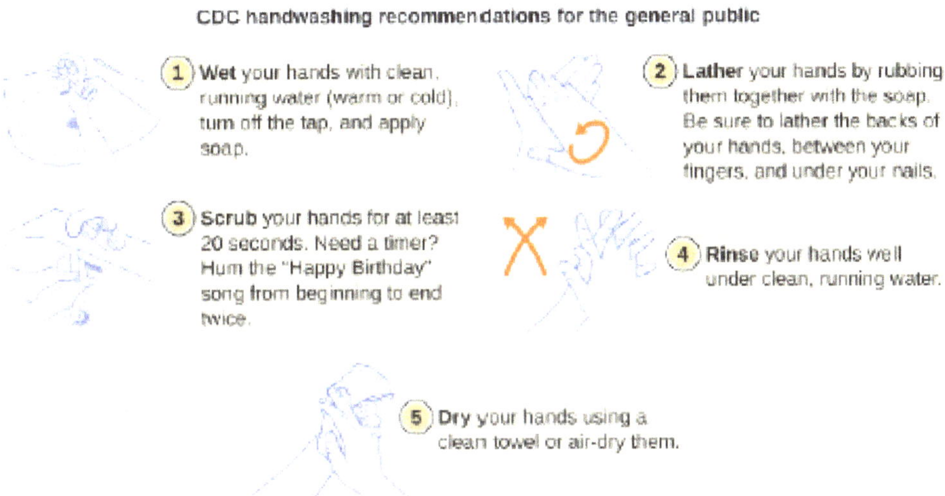

(a)

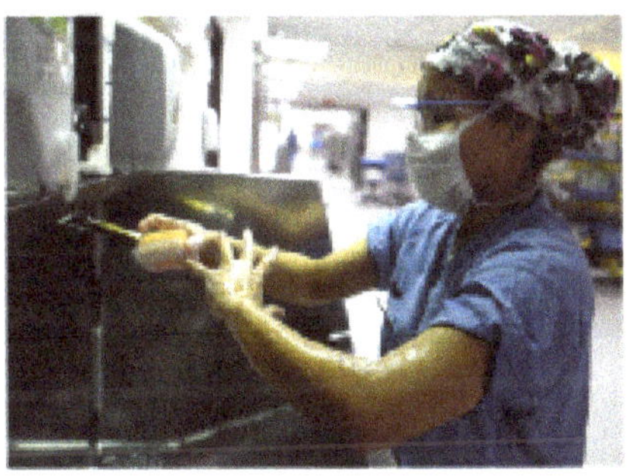

(b)

Figure 11.26 (a) The CDC recommends five steps as part of typical handwashing for the general public. (b) Surgical scrubbing is more extensive, requiring scrubbing starting from the fingertips, extending to the hands and forearms, and then up beyond the elbows, as shown here. (credit a: modification of work by World Health Organization; credit b: b: Staff Sgt. Kevin Iinuma / US Air Force; Public Domain)

Hand washing is even more important for health-care workers, who should wash their hands thoroughly between every patient contact, after the removal of gloves, after contact with bodily fluids and potentially infectious fomites, and before and after assisting a surgeon with invasive procedures. Even with the use of proper surgical attire, including gloves, scrubbing for surgery is more involved than routine hand washing. The goal of surgical scrubbing is to reduce the normal microbiota on the skin's surface to prevent the introduction of these microbes into a patient's surgical wounds.

There is no single widely accepted protocol for surgical scrubbing. Protocols for length of time spent scrubbing may depend on the antimicrobial used; health-care workers should always check the manufacturer's recommendations. According to the Association of Surgical Technologists (AST), surgical scrubs may be performed with or without the use of brushes (Figure 11.26).

Bisbiguanides

Bisbiguanides were first synthesized in the 20th century and are cationic (positively charged) molecules known for their antiseptic properties (Figure 11.27). One important **bisbiguanide** antiseptic is chlorhexidine. It has broad-spectrum activity against yeasts, gram-positive bacteria, and gram-negative bacteria, with the exception of *Pseudomonas aeruginosa*, which may develop resistance on repeated exposure.Chlorhexidine disrupts cell membranes and is bacteriostatic at lower concentrations or bactericidal at higher concentrations, in which it actually causes the cells' cytoplasmic contents to congeal. It also has activity against enveloped viruses. However, chlorhexidine is poorly effective against *Mycobacterium tuberculosis* and nonenveloped viruses, and it is not sporicidal. Chlorhexidine is typically used in the clinical setting as a surgical scrub and for other handwashing needs for medical personnel, as well as for topical antisepsis for patients before surgery or needle injection. It is more persistent than iodophors, providing long-lasting antimicrobial activity. Chlorhexidine solutions may also be used as oral rinses after oral procedures or to treat gingivitis. Another bisbiguanide, alexidine, is gaining popularity as a surgical scrub and an oral rinse because it acts faster than chlorhexidine.

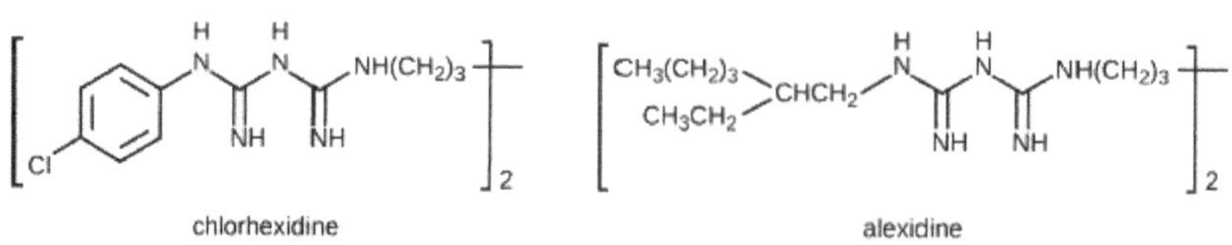

Figure 11.27 The bisbiguanides chlorhexadine and alexidine are cationic antiseptic compounds commonly used as surgical scrubs.

Alkylating Agents

The **alkylating agents** are a group of strong disinfecting chemicals that act by replacing a hydrogen atom within a molecule with an alkyl group (C_nH_{2n+1}), thereby inactivating enzymes and nucleic acids (Figure 11.28). The alkylating agent formaldehyde (CH_2OH) is commonly used in solution at a concentration of 37% (known as formalin) or as a gaseous disinfectant and biocide. It is a strong, broad-spectrum disinfectant and biocide that has the ability to kill bacteria, viruses, fungi, and endospores, leading to sterilization at low temperatures, which is sometimes a convenient alternative to the more labour-intensive heat sterilization methods. It also cross- links proteins and has been widely used as a chemical fixative. Because of this, it is used for the storage of tissue specimens and as an embalming fluid. It also has been used to inactivate infectious agents in vaccine preparation. Formaldehyde is very irritating to living tissues and is also carcinogenic; therefore, it is not used as an antiseptic.

Glutaraldehyde is structurally similar to formaldehyde but has two reactive aldehyde groups, allowing it to act more quickly than formaldehyde. It is commonly used as a 2% solution for sterilization and is marketed under the brand name Cidex. It is used to disinfect a variety of surfaces and surgical and medical equipment. However, similar to formaldehyde, glutaraldehyde irritates the skin and is not used as an antiseptic.

A new type of disinfectant gaining popularity for the disinfection of medical equipment is o-phthalaldehyde (OPA), which is found in some newer formulations of Cidex and similar products, replacing glutaraldehyde. o-Phthalaldehyde also has two reactive aldehyde groups, but they are linked by an aromatic bridge. o-Phthalaldehyde is thought to work similarly to glutaraldehyde and formaldehyde, but is much less irritating to skin and nasal passages, produces a minimal odor, does not require processing before use, and is more effective against mycobacteria.

Ethylene oxide is a type of alkylating agent that is used for gaseous sterilization. It is highly penetrating and can sterilize items within plastic bags such as catheters, disposable items in laboratories and clinical settings (like packaged Petri dishes), and other pieces of equipment. Ethylene oxide exposure is a form of cold sterilization, making it useful for the sterilization of heat-sensitive items. Great care needs to be taken with the use of ethylene oxide, however; it is carcinogenic, like the other alkylating agents, and is also highly explosive. With careful use and proper aeration of the products after treatment, ethylene oxide is highly effective, and ethylene oxide sterilizers are commonly found in medical settings for sterilizing packaged materials.

β-Propionolactone is an alkylating agent with a different chemical structure than the others already discussed. Like other alkylating agents, β-propionolactone binds to DNA, thereby inactivating it (Figure 11.28). It is a clear liquid with a strong odor and has the ability to kill endospores. As such, it has been used in either liquid form or as a vapor for the sterilization of medical instruments and tissue grafts, and it is a common component of vaccines, used to maintain their sterility. It has also been used for the sterilization of nutrient broth, as well as blood plasma, milk, and water. It is quickly metabolized by animals and humans to lactic acid. It is also an irritant, however, and may lead to permanent damage of the eyes, kidneys, or liver. Additionally, it has been shown to be carcinogenic in animals; thus, precautions are necessary to minimize human exposure to β-propionolactone.

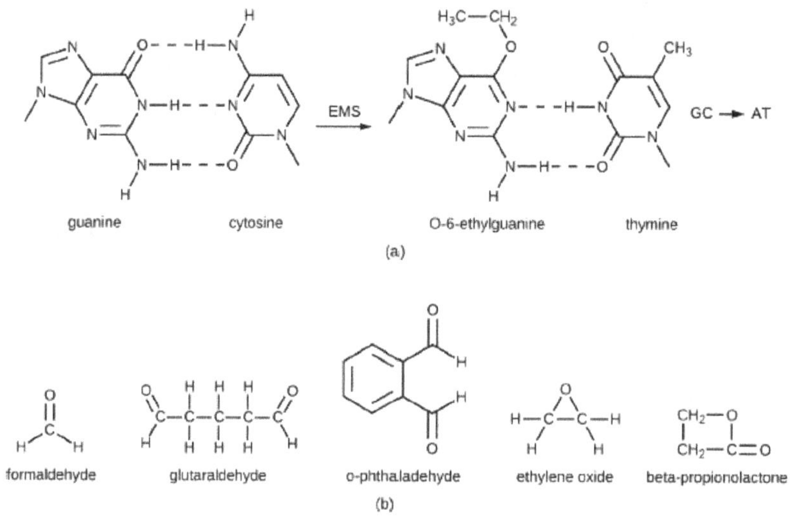

Figure 11.28 (a) Alkylating agents replace hydrogen atoms with alkyl groups. Here, guanine is alkylated, resulting in its hydrogen bonding with thymine, instead of cytosine. (b) The chemical structures of several alkylating agents.

Diehard Prions

Prions, the acellular, misfolded proteins responsible for incurable and fatal diseases such as kuru and Creutzfeldt-Jakob disease (see Viroids, Virusoids, and Prions (https://legacy.cnx.org/content/m58811/ latest/)), are notoriously difficult to destroy. Prions are extremely resistant to heat, chemicals, and radiation. They are also extremely infectious and deadly; thus, handling and disposing of prion-infected items requires extensive training and extreme caution.

Typical methods of disinfection can reduce but not eliminate the infectivity of prions. Autoclaving is not completely effective, nor are chemicals such as phenol, alcohols, formalin, and β-propiolactone. Even when fixed in formalin, affected brain and spinal cord tissues remain infectious.

Personnel who handle contaminated specimens or equipment or work with infected patients must wear a protective coat, face protection, and cut-resistant gloves. Any contact with skin must be immediately washed with detergent and warm water without scrubbing. The skin should then be washed with 1 N NaOH or a 1:10 dilution of bleach for 1 minute. Contaminated waste must be incinerated or autoclaved in a strong basic solution, and instruments must be cleaned and soaked in a strong basic solution.

Peroxygens

Peroxygens are strong oxidizing agents that can be used as disinfectants or antiseptics. The most widely used **peroxygen** is hydrogen peroxide (H_2O_2), which is often used in solution to disinfect surfaces and may also be used as a gaseous agent. Hydrogen peroxide solutions are inexpensive skin antiseptics that break down into water and oxygen gas, both of which are environmentally safe. This decomposition is accelerated in the presence of light, so hydrogen peroxide solutions typically are sold in brown or opaque bottles. One disadvantage of using hydrogen peroxide as an antiseptic is that it also causes damage to skin that may delay healing or lead to scarring. Contact lens cleaners often include hydrogen peroxide as a disinfectant.

Hydrogen peroxide works by producing free radicals that damage cellular macromolecules. Hydrogen peroxide has broad-spectrum activity, working against gram-positive and gram-negative bacteria (with slightly greater efficacy against gram-positive bacteria), fungi, viruses, and endospores. However, bacteria that produce the oxygen-detoxifying enzymes catalase or peroxidase may have inherent tolerance to low hydrogen peroxide concentrations (Figure 11.29). To kill endospores, the length of exposure or concentration of solutions of hydrogen peroxide must be increased. Gaseous hydrogen peroxide has greater efficacy and can be used as a sterilant for rooms or equipment.

$$2H_2O_2 \xrightarrow{\text{catalase}} 2H_2O + O_2$$

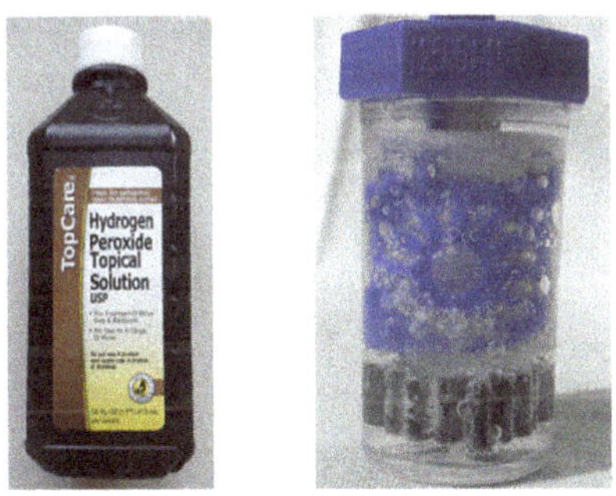

Figure 11.29 Catalase enzymatically converts highly reactive hydrogen peroxide (H_2O_2) into water and oxygen. Hydrogen peroxide can be used to clean wounds. Hydrogen peroxide is used to sterilize items such as contact lenses. (credit photos: modification of work by Kerry Ceszyk)

Plasma, a hot, ionized gas, described as the fourth state of matter, is useful for sterilizing equipment because it penetrates surfaces and kills vegetative cells and endospores. Hydrogen peroxide and peracetic acid, another commonly used peroxygen, each may be introduced as a plasma. Peracetic acid can be used as a liquid or plasma sterilant insofar as it readily kills endospores, is more effective than hydrogen peroxide even at rather low concentrations, and is immune to inactivation by catalases and peroxidases. It also breaks down to environmentally innocuous compounds; in this case, acetic acid and oxygen.

Other examples of peroxygens include benzoyl peroxide and carbamide peroxide. Benzoyl peroxide is a peroxygen that used in acne medication solutions. It kills the bacterium *Propionibacterium acnes*, which is associated with acne. Carbamide peroxide, an ingredient used in toothpaste, is a peroxygen that combats oral biofilms that cause tooth discoloration and halitosis (bad breath). Last, ozone gas is a peroxygen with disinfectant qualities and is used to clean air or water supplies. Overall, peroxygens are highly effective and commonly used, with no associated environmental hazard.

- How do peroxides kill cells?

Supercritical Fluids

Within the last 15 years, the use of **supercritical fluids**, especially supercritical carbon dioxide ($scCO_2$), has gained popularity for certain sterilizing applications. When carbon dioxide is brought to approximately 10 times atmospheric pressure, it reaches a supercritical state that has physical properties between those of liquids and gases. Materials put into a chamber in which carbon dioxide is pressurized in this way can be sterilized because of the ability of $scCO_2$ to penetrate surfaces.

Supercritical carbon dioxide works by penetrating cells and forming carbonic acid, thereby lowering the cell pH considerably. This technique is effective against vegetative cells and is also used in combination with peracetic acid to kill endospores. Its efficacy can also be augmented with increased temperature or by rapid cycles of pressurization and depressurization, which more likely produce cell lysis.

Benefits of $scCO_2$ include the nonreactive, nontoxic, and non-flammable properties of carbon dioxide, and this protocol is effective at low temperatures. Unlike other methods, such as heat and irradiation, that can degrade the object being sterilized, the use of $scCO_2$ preserves the object's integrity and is commonly used for treating foods (including spices and juices) and medical devices such as endoscopes. It is also gaining popularity for disinfecting tissues such as skin, bones, tendons, and ligaments prior to transplantation. $scCO_2$ can also be used for pest control because it can kill insect eggs and larvae within products.

Chemical Food Preservatives

Chemical preservatives are used to inhibit microbial growth and minimize spoilage in some foods. Commonly used chemical preservatives include sorbic acid, benzoic acid, and propionic acid, and their more soluble salts potassium sorbate, sodium benzoate, and calcium propionate, all of which are used to control the growth of molds in acidic foods. Each of these preservatives is nontoxic and readily metabolized by humans. They are also flavourless, so they do not compromise the flavour of the foods they preserve.

Sorbic and benzoic acids exhibit increased efficacy as the pH decreases. Sorbic acid is thought to work by inhibiting various cellular enzymes, including those in the citric acid cycle, as well as catalases and peroxidases. It is added as a preservative in a wide variety of foods, including dairy, bread, fruit, and vegetable products. Benzoic acid is found naturally in many types of fruits and berries, spices, and fermented products. It is thought to work by decreasing intracellular pH, interfering with mechanisms such as oxidative phosphorylation and the uptake of molecules such as amino acids into cells. Foods preserved with benzoic acid or sodium benzoate include fruit juices, jams, ice creams, pastries, soft drinks, chewing gum, and pickles.

Propionic acid is thought to both inhibit enzymes and decrease intracellular pH, working similarly to benzoic acid. However, propionic acid is a more effective preservative at a higher pH than either sorbic acid or benzoic acid. Propionic acid is naturally produced by some cheeses during their ripening and is added to other types of cheese and baked goods to prevent mold contamination. It is also added to raw dough to prevent contamination by the bacterium *Bacillus mesentericus*, which causes bread to become ropy.

Other commonly used chemical preservatives include sulfur dioxide and nitrites. Sulfur dioxide prevents browning of foods and is used for the preservation of dried fruits; it has been used in winemaking since ancient times. Sulfur dioxide gas dissolves in water readily, forming sulfites. Although sulfites can be metabolized by the body, some people have sulfite allergies, including asthmatic reactions. Additionally, sulfites degrade thiamine, an important nutrient in some foods. The mode of action of sulfites is not entirely clear, but they may interfere with the disulfide bond (see m58816 (https://legacy.cnx.org/content/m58816/ latest/#OSC_Microbio_07_04_tertiary)) formation in proteins, inhibiting enzymatic activity. Alternatively, they may reduce the intracellular pH of the cell, interfering with proton motive force-driven mechanisms.

Nitrites are added to processed meats to maintain colour and stop the germination of *Clostridium botulinum* endospores. Nitrites are reduced to nitric oxide, which reacts with heme groups and iron-sulfur groups. When nitric oxide reacts with the heme group within the myoglobin of meats, a red product forms, giving meat its red colour. Alternatively, it is thought that when nitric acid reacts with the iron-sulfur enzyme ferredoxin within bacteria, this electron transport-chain carrier is destroyed, preventing ATP synthesis. Nitrosamines, however, are carcinogenic and can be produced through exposure of nitrite-preserved meats (e.g., hot dogs, lunch meat, breakfast sausage, bacon, meat in canned soups) to heat during cooking.

Natural Chemical Food Preservatives

The discovery of natural antimicrobial substances produced by other microbes has added to the arsenal of preservatives used in food. Nisin is an antimicrobial peptide produced by the bacterium *Lactococcus lactis* and is particularly effective against gram-positive organisms. Nisin works by disrupting cell wall production, leaving cells more prone to lysis. It is used to preserve cheeses, meats, and beverages.

Natamycin is an antifungal macrolide antibiotic produced by the bacterium *Streptomyces natalensis*. It was approved by the FDA in 1982 and is used to prevent fungal growth in various types of dairy products, including cottage cheese, sliced cheese, and shredded cheese. Natamycin is also used for meat preservation in countries outside the United States.

- ⦿ What are the advantages and drawbacks of using sulfites and nitrites as food preservatives?

Chemical Disinfectants

Chemical	Mode of Action	Examples of Use
Phenolics		
Cresols o-Phenylphenol Hexachlorophene Triclosan	Denature proteins and disrupt membranes	Disinfectant in Lysol Prevent contamination of crops (citrus) Antibacterial soap pHisoHex for handwashing in hospitals
Metals		
Mercury Silver Copper Nickel Zinc	Bind to proteins and inhibit enzyme activity	Topical antiseptic Treatment of wounds and burns Prevention of eye infections in newborns Antibacterial in catheters and bandages Mouthwash Algicide for pools and fish tanks Containers for long-term water storage

Chemical	Mode of Action	Example Uses
Halogens		
Iodine Chlorine Fluorine	Oxidation and destabilization of cellular macromolecules	Topical antiseptic Hand scrub for medical personnel Water disinfectant Water treatment plants Household bleach Food processing Prevention of dental carries
Alcohols		
Ethanol Isopropanol	Denature proteins and disrupt membranes	Disinfectant Antiseptic
Surfactants		
Quaternary ammonium salts	Lowers surface tension of water to help with washing away of microbes, and disruption of cell membranes	Soaps and detergent Disinfectant Antiseptic Mouthwash
Bisbiguanides		
Chlorhexidine Alexidine	Disruption of cell membranes	Oral rinse Hand scrub for medical personnel
Alkylating Agents		
Formaldehyde Glutaraldehyde o-Phthalaldehyde Ethylene oxide β- Propionolactone	Inactivation of enzymes and nucleic acid	Disinfectant Tissue specimen storage Embalming Sterilization of medical equipment Vaccine component for sterility
Peroxygens		
Hydrogen peroxide Peracetic acid Benzoyl peroxide Carbamide peroxide Ozone gas	Oxidation and destabilization of cellular macromolecules	Antiseptic Disinfectant Acne medication Toothpaste ingredient

[illegible]	[illegible]	[illegible]
Supercritical Gases		
Carbon dioxide	Penetrates cells, forms carbonic acid, lowers intracellular pH	Food preservation Disinfection of medical devices Disinfection of transplant tissues
Chemical Food Preservatives		
Sorbic acid Benzoic acid Propionic acid Potassium sorbate Sodium benzoate Calcium propionate Sulfur dioxide Nitrites	Decrease pH and inhibit enzymatic function	Preservation of food products
Natural Food Preservatives		
Nisin Natamycin	Inhibition of cell wall synthesis (Nisin)	Preservation of dairy products, meats, and beverages

CHAPTER-12

NUCLEIC ACIDS: STRUCTURE & FUNCTIONS

INTRODUCTION: AN OVERVIEW

Nucleic acids, the fundamental building blocks of genetic information, exhibit an intricate structure that underlies their paramount importance in the biological realm. Deoxyribonucleic acid (DNA) and ribonucleic acid (RNA) serve as the carriers of genetic instructions, orchestrating the complex processes essential for life. The double-helix structure of DNA, with its elegant arrangement of nucleotide base pairs, encodes the blueprint for cellular functions and inheritance. RNA, in its diverse forms, plays crucial roles in translating this genetic information into the synthesis of proteins and participating in various cellular activities. Together, the structure and functions of nucleic acids form the basis of heredity, molecular biology, and the remarkable diversity of life, illustrating the profound impact these molecules have on the fundamental processes that drive living organisms.

STRUCTURE AND FUNCTION OF DNA

In Microbial Metabolism (https://legacy.cnx.org/content/m58818/latest/), we discussed the microbial catabolism of three classes of macromolecules: proteins, lipids and carbohydrates. In this chapter, we will discuss the genetic role of a fourth class of molecules: nucleic acids. Like other macromolecules, **nucleic acid**s are composed of monomers, called **nucleotide**s, which are polymerized to form large strands. Each nucleic acid strand contains certain nucleotides that appear in a certain order within the strand, called its **base sequence**. The base sequence of **deoxyribonucleic acid (DNA)** is responsible for carrying and retaining the hereditary information in a cell. In Mechanisms of Microbial Genetics (https://legacy.cnx.org/content/m58839/latest/), we will discuss in detail the ways in which DNA uses its own base sequence to direct its own synthesis, as well as the synthesis of RNA and proteins, which, in turn, gives rise to products with diverse structure and function. In this section, we will discuss the basic structure and function of DNA.

DNA Nucleotides

The building blocks of nucleic acids are nucleotides. Nucleotides that compose DNA are called **deoxyribonucleotides.** The three components of a deoxyribonucleotide are a five-carbon sugar called deoxyribose, a phosphate group, and a **nitrogenous base**, a nitrogen-containing ring structure that is responsible for complementary base pairing between nucleic acid strands (Figure 12.1). The carbon atoms of the five-carbon deoxyribose are numbered 1′, 2′, 3′, 4′, and 5′ (1′ is read as "one prime"). A nucleoside comprises the five-carbon sugar and nitrogenous base.

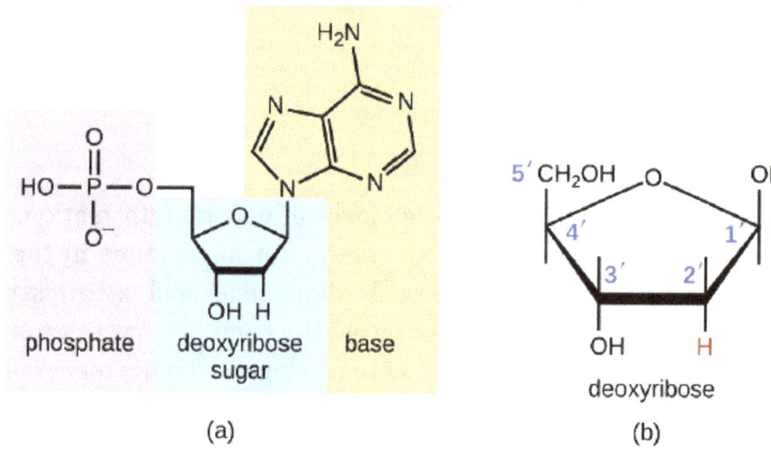

Figure 12.1 (a) Each deoxyribonucleotide is made up of a sugar called deoxyribose, a phosphate group, and a nitrogenous base—in this case, adenine. (b) The five carbons within deoxyribose are designated as 1', 2', 3', 4', and 5'.

The deoxyribonucleotide is named according to the nitrogenous bases (Figure 12.2). The nitrogenous bases **adenine** (A) and **guanine** (G) are the **purines;** they have a double-ring structure with a six-carbon ring fused to a five-carbon ring. The **pyrimidines, cytosine** (C) and **thymine** (T), are smaller nitrogenous bases that have only a six-carbon ring structure.

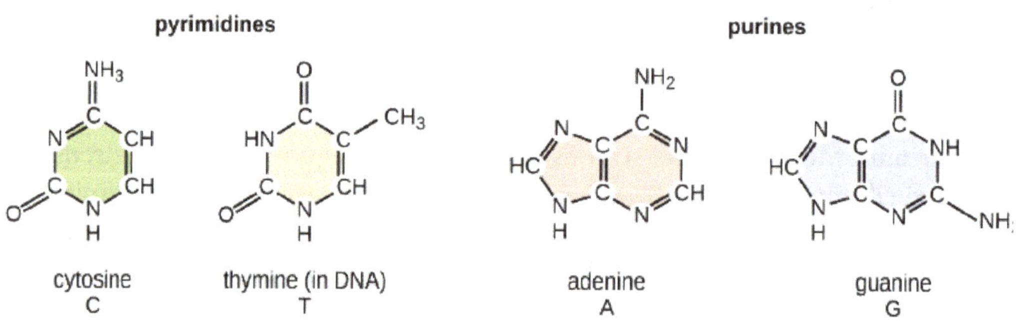

Figure 12.2 Nitrogenous bases within DNA are categorized into the two-ringed purines adenine and guanine and the single-ringed pyrimidines cytosine and thymine. Thymine is unique to DNA.

Individual nucleoside triphosphates combine with each other by covalent bonds known as 5′-3′ **phosphodiester bonds**, or linkages whereby the phosphate group attached to the 5′ carbon of the sugar of one nucleotide bonds to the hydroxyl group of the 3′ carbon of the sugar of the next nucleotide. Phosphodiester bonding between nucleotides forms the **sugar-phosphate backbone**, the alternating sugar-phosphate structure composing the framework of a nucleic acid strand (Figure 12.3). During the polymerization process, deoxynucleotide triphosphates (dNTP) are used. To construct the sugar-phosphate backbone, the two terminal phosphates are released from the dNTP as a pyrophosphate. The resulting strand of nucleic acid has a free phosphate group at the 5′ carbon end and a free hydroxyl group at the 3′ carbon end. The two unused phosphate groups from the nucleotide triphosphate are released as pyrophosphate during phosphodiester bond formation. Pyrophosphate is subsequently hydrolyzed, releasing the energy used to drive nucleotide polymerization.

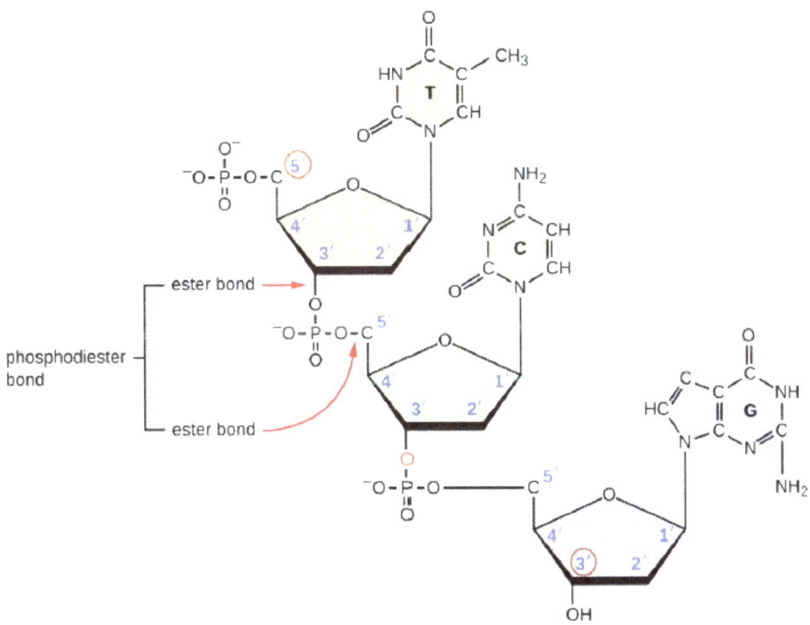

Figure 12.3 Phosphodiester bonds form between the phosphate group attached to the 5′ carbon of one nucleotide and the hydroxyl group of the 3′ carbon in the next nucleotide, bringing about polymerization of nucleotides in to nucleic acid strands. Note the 5′ and 3′ ends of this nucleic acid strand.

⦿ What is meant by the 5′ and 3' ends of a nucleic acid strand?

Discovering the Double Helix

By the early 1950s, considerable evidence had accumulated indicating that DNA was the genetic material of cells, and now the race was on to discover its three-dimensional structure. Around this time, Austrian biochemist Erwin Chargaff

(1905–2002) examined the content of DNA in different species and discovered that adenine, thymine, guanine, and cytosine were not found in equal quantities, and that it varied from species to species, but not between individuals of the same species. He found that the amount of adenine was very close to equaling the amount of thymine, and the amount of cytosine was very close to equaling the amount of guanine, or A = T and G = C. These relationships are also known as Chargaff's rules.

Other scientists were also actively exploring this field during the mid-20th century. In 1952, American scientist Linus Pauling (1901–1994) was the world's leading structural chemist and odds-on favorite to solve the structure of DNA. Pauling had earlier discovered the structure of protein α helices, using X-ray diffraction, and, based upon X-ray d DNA. At the same time, British researchers Rosalind Franklin (1920–1958) and her graduate student R.G. Gosling were also using X-ray diffraction to understand the structure of DNA (iffraction images of DNA made in his laboratory, he proposed a triple-stranded model of Figure 12.4). It was Franklin's scientific expertise that resulted in the production of more well-defined X-ray diffraction images of DNA that would clearly show the overall double-helix structure of DNA.

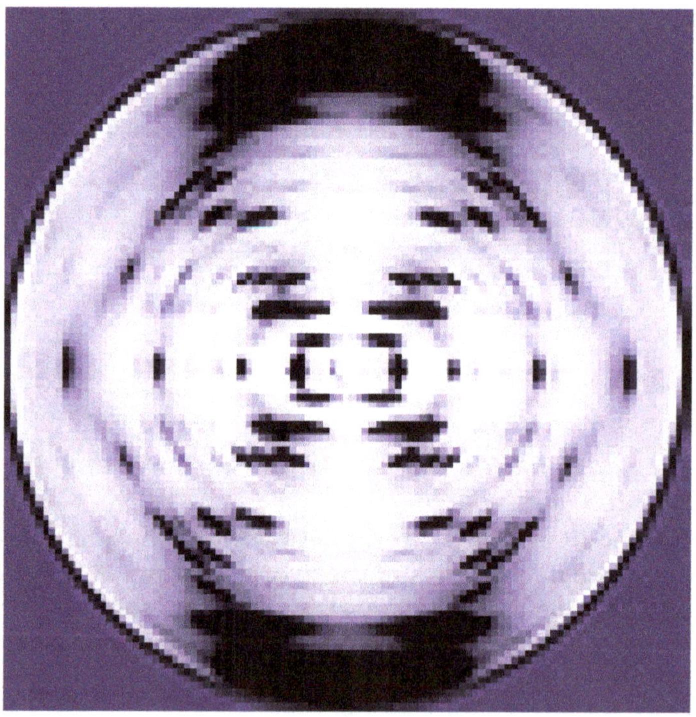

Figure 12.4 The X-ray diffraction pattern of DNA shows its helical nature. (credit: National Institutes of Health)

James Watson (1928–), an American scientist, and Francis Crick (1916–2004), a British scientist, were working together in the 1950s to discover DNA's structure.

They used Chargaff's rules and Franklin and Wilkins' X- ray diffraction images of DNA fibers to piece together the purine-pyrimidine pairing of the double helical DNA molecule (Figure 12.5). In April 1953, Watson and Crick published their model of the DNA double helix in *Nature.* The same issue additionally included papers by Wilkins and colleagues, as well as by Franklin and Gosling, each describing different aspects of the molecular structure of DNA. In 1962, James Watson, Francis Crick, and Maurice Wilkins were awarded the Nobel Prize in Physiology and Medicine. Unfortunately, by then Franklin had died, and Nobel prizes at the time were not awarded posthumously. Work continued, however, on learning about the structure of DNA. In 1973, Alexander Rich (1924–2015) and colleagues were able to analyze DNA crystals to confirm and further elucidate DNA structure.

Figure 12.5 In 1953, James Watson and Francis Crick built this model of the structure of DNA, shown here on display at the Science Museum in London.

DNA Structure

Watson and Crick proposed that DNA is made up of two strands that are twisted around each other to form a right-handed helix. The two DNA strands are **antiparallel**, such that the 3′ end of one strand faces the 5′ end of the other (Figure 12.6). The 3′ end of each strand has a free hydroxyl group, while the 5′ end of each strand has a free phosphate group. The sugar and phosphate of the polymerized nucleotides form the backbone of the structure, whereas the nitrogenous bases are stacked inside. These nitrogenous bases on the interior of the molecule interact with each other, base pairing.

Analysis of the diffraction patterns of DNA has determined that there are approximately 10 bases per turn in DNA. The asymmetrical spacing of the sugar-phosphate backbones generates major grooves (where the backbone is far apart) and minor grooves (where the backbone is close together) (Figure 12.6). These grooves are locations where proteins can bind to DNA. The binding of these proteins can alter the structure of DNA, regulate replication, or regulate transcription of DNA into RNA.

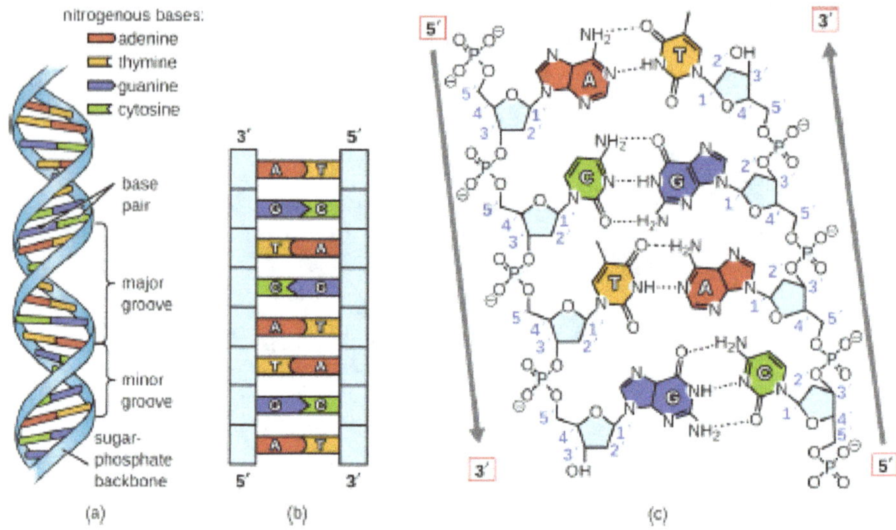

Figure 12.6 Watson and Crick proposed the double helix model for DNA. (a) The sugar-phosphate backbones are on the outside of the double helix and purines and pyrimidines form the "rungs" of the DNA helix ladder. (b) The two DNA strands are antiparallel to each other. (c) The direction of each strand is identified by numbering the carbons (1 through in each sugar molecule. The 5′ end is the one where carbon #5 is not bound to another nucleotide; the 3′end is the one where carbon #3 is not bound to another nucleotide.

Base pairing takes place between a purine and pyrimidine. In DNA, adenine (A) and thymine (T) are **complementary base pairs**, and cytosine (C) and guanine (G) are also complementary base pairs, explaining Chargaff's rules (Figure 12.7). The base pairs are stabilized by hydrogen bonds; adenine and thymine form two hydrogen bonds between them, whereas cytosine and guanine form three hydrogen bonds between them.

In the laboratory, exposing the two DNA strands of the double helix to high temperatures or to certain chemicals can break the hydrogen bonds between complementary bases, thus separating the strands into two separate single strands of DNA (single-stranded DNA [ssDNA]). This process is called DNA denaturation and is analogous to protein denaturation, as described in Proteins (https://legacy.cnx.org/content/m58816/latest/).

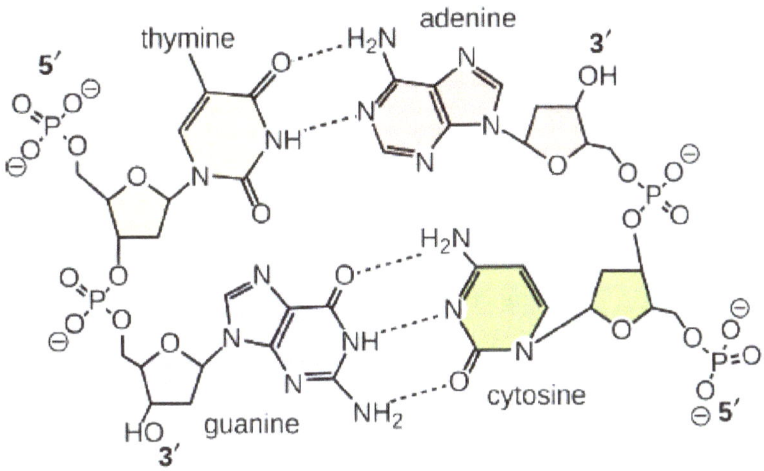

Figure 12.7 Hydrogen bonds form between complementary nitrogenous bases on the interior of DNA.

The ssDNA strands can also be put back together as double-stranded DNA (dsDNA), through reannealing or renaturing by cooling or removing the chemical denaturants, allowing these hydrogen bonds to reform. The ability to artificially manipulate DNA in this way is the basis for several important techniques in biotechnology (Figure 12.8). Because of the additional hydrogen bonding between the C = G base pair, DNA with a high GC content is more difficult to denature than DNA with a lower GC content.

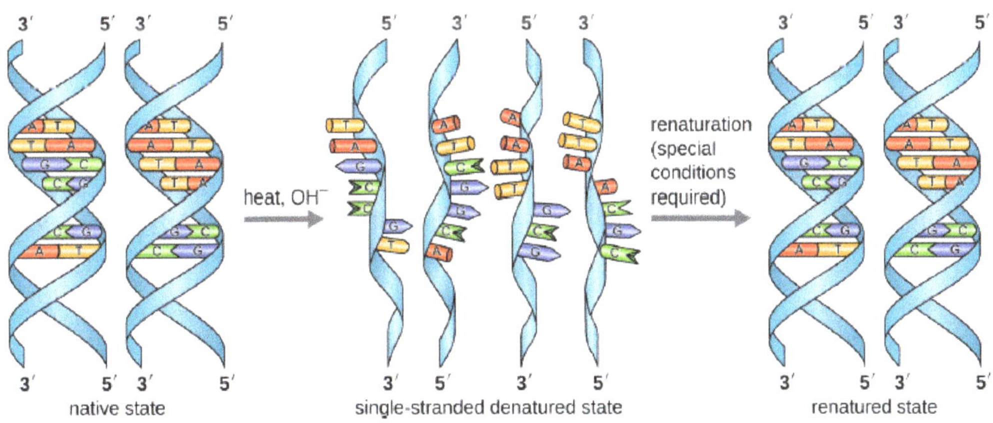

Figure 12.8 In the laboratory, the double helix can be denatured to single-stranded DNA through exposure to heat or chemicals, and then renatured through cooling or removal of chemical denaturants to allow the DNA strands to reanneal. (credit: modification of work by Hernández-Lemus E, Nicasio-Collazo LA, Castañeda-Priego R)

- What are the two complementary base pairs of DNA and how are they bonded together?

DNA Function

DNA stores the information needed to build and control the cell. The transmission of this information from mother to daughter cells is called **vertical gene transfer** and it occurs through the process of DNA replication. DNA is replicated when a cell makes a duplicate copy of its DNA, then the cell divides, resulting in the correct distribution of one DNA copy to each resulting cell. DNA can also be enzymatically degraded and used as a source of nucleosides and nucleotides for the cell. Unlike other macromolecules, DNA does not serve a structural role in cells.

Paving the Way for Women in Science and Health Professions

Historically, women have been underrepresented in the sciences and in medicine, and often their pioneering contributions have gone relatively unnoticed. For example, although Rosalind Franklin performed the X-ray diffraction studies demonstrating the double helical structure of DNA, it is Watson and Crick who became famous for this discovery, building on her data. There still remains great controversy over whether their acquisition of her data was appropriate and whether personality conflicts and gender bias contributed to the delayed recognition of her significant contributions. Similarly, Barbara McClintock did pioneering work in maize (corn) genetics from the 1930s through 1950s, discovering transposons (jumping genes), but she was not recognized until much later, receiving a Nobel Prize in Physiology or Medicine in 1983 (Figure 12.9).

Today, women still remain underrepresented in many fields of science and medicine. While more than half of the undergraduate degrees in science are awarded to women, only 46% of doctoral degrees in science are awarded to women. In academia, the number of women at each level of career advancement continues to decrease, with women holding less than one-third of the positions of Ph.D.-level scientists in tenure-track positions, and less than one-quarter of the full professorships at 4-year colleges and universities.Even in the health professions, like nearly all other fields, women are often underrepresented in many medical careers and earn significantly less than their male counterparts, as shown in a 2013 study published by the *Journal of the American Medical Association.*

Why do such disparities continue to exist and how do we break these cycles? The situation is complex and likely results from the combination of various factors, including how society conditions the behaviours of girls from a young age and supports their interests, both professionally and personally. Some have suggested that women do not belong in the laboratory, including Nobel Prize winner Tim Hunt, whose 2015 public comments suggesting that women are too emotional for science were met with widespread condemnation.

Perhaps girls should be supported more from a young age in the areas of science and math (Figure 12.9). Science, technology, engineering, and mathematics (STEM) programs sponsored by the American Association of University Women (AAUW) and National Aeronautics and Space Administration (NASA) are excellent examples of programs that offer such support. Contributions by women in science should be made known more widely to the public, and marketing targeted to young girls should include more images of historically and professionally successful female scientists and medical professionals, encouraging all bright young minds, including girls and women, to pursue careers in science and medicine.

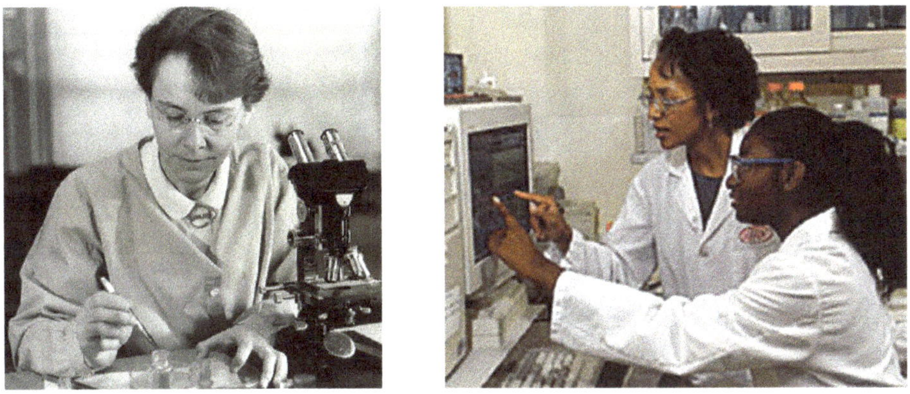

(a) (b)

Figure 12.9 (a) Barbara McClintock's work on maize genetics in the 1930s through 1950s resulted in the discovery of transposons, but its significance was not recognized at the time. (b) Efforts to appropriately mentor and to provide continued societal support for women in science and medicine may someday help alleviate some of the issues preventing gender equality at all levels in science and medicine. (credit a: modification of work by Smithsonian Institution; credit b: modification of work by Haynie SL, Hinkle AS, Jones NL, Martin CA, Olsiewski PJ, Roberts MF)

ETEC produces several plasmid-encoded virulence factors that make it pathogenic compared with typical *E coli*. These include the secreted toxins heat-labile enterotoxin (LT) and heat-stabile enterotoxin (ST), as well as colonization factor (CF). Both LT and ST cause the excretion of chloride ions from intestinal cells to the intestinal lumen, causing a consequent loss of water from intestinal cells, resulting in diarrhea. CF encodes a bacterial protein that aids in allowing the bacterium to adhere to the lining of the small intestine.

ROSALIND FRANKLIN

A short biography of Rosalind Franklin and her contributions to the discovery of the structure of DNA

Rosalind Franklin and Photo 51

Rosalind Elsie Franklin (25 July 1920 – 16 April 1958) was an English chemist and X-ray crystallographer who made contributions to the understanding of the ***molecular***

structures of DNA (deoxyribonucleic acid), RNA (ribonucleic acid), viruses, coal, and graphite. Although her works on coal and viruses were appreciated in her lifetime, her contributions to the discovery of the structure of DNA were largely recognized posthumously.

Figure 12.10 Rosalind Elsie Franklin (25 July 1920 – 16 April 1958). Image Credit (left): By Jewish Chronicle Archive/Heritage-Imageshttp://www.britannica.com/EBchecked/topic-art/217394/99712/Rosalind-Franklin, Fair use, https://en.wikipedia.org/w/index.php?curid=24959067(https://en.wikipedia.org/w/index.php?curid=24959067)

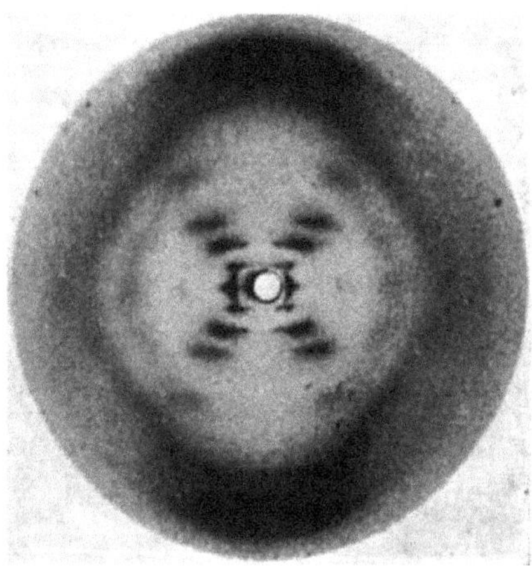

Figure 12.11 Rosalind Franklin's Photo 51, credited for the elucidation of the alpha helix structure theory of DNA. By Source (WP:NFCC#4), Fair use, https://en.wikipedia.org/w/index.php?curid=38068629 (https://en.wikipedia.org/ w/index.php?curid=38068629)

Born to a prominent British Jewish family, Franklin studied the Natural Sciences Tripos at Newnham College, Cambridge, from which she graduated in 1941. Earning a research fellowship, she joined the University of Cambridge physical chemistry laboratory under Ronald George Wreyford Norrish, where she earned a Ph.D. in 1945. She went to Paris in 1947 as a chercheur (post-doctoral researcher) under Jacques Mering (https://en.wikipedia.org/wiki/Jacques_Mering) at the Laboratoire Central des Services Chimiques de l'Etat, where she became an accomplished X-ray crystallographer. She became a research associate at King's College London in 1951 and worked on X-ray diffraction studies, which would *facilitate* the double helix (https://en.wikipedia.org/wiki/Double_helix*) theory of the DNA*. She died in 1958 at the age of 37 of ovarian cancer.

Franklin is best known for her work on the X-ray diffraction images of DNA, particularly Photo 51 (https://en.wikipedia.org/wiki/Photo_51) , while at King's College London, which led to the discovery of the DNA double helix for which James Watson (https://en.wikipedia.org/wiki/James_Watson) , Francis Crick (https://en.wikipedia.org/wiki/Francis_Crick) and Maurice Wilkins (https://en.wikipedia.org/wiki/ Maurice_Wilkins) shared the Nobel Prize in Physiology or Medicine in 1962. Watson suggested that Franklin would have ideally been awarded a Nobel Prize in Chemistry, along with Wilkins, but, although there was not yet a rule against posthumous awards, the Nobel Committee generally does not make posthumous nominations.

After finishing her work on DNA, Franklin led pioneering work at Birkbeck on the molecular structures of viruses. Her team member Aaron Klug (https://en.wikipedia.org/wiki/Aaron_Klug) continued her research, winning the Nobel Prize in Chemistry in 1982. Read more about Rosalind Franklin's legacy at Wikipedia (https://en.wikipedia.org/wiki/Rosalind_Franklin) .

Image and Content Credit: https://en.wikipedia.org/wiki/Rosalind_Franklin (https://en.wikipedia.org/wiki/ Rosalind_Franklin) , https://en.wikipedia.org/wiki/Photo_51 (https://en.wikipedia.org/wiki/Photo_51)

DNA REPLICATION

When a cell divides, it is important that each daughter cell receives an identical copy of the DNA. This is accomplished by the process of DNA replication. The replication of DNA occurs during the synthesis phase, or S phase, of the cell cycle, before the cell enters mitosis or meiosis.

The elucidation of the structure of the double helix provided a hint as to how DNA is copied. Recall that adenine nucleotides pair with thymine nucleotides, and cytosine with guanine. This means that the two strands are complementary to each other. For example, a strand of DNA with a nucleotide sequence of AGTCATGA will have a complementary strand with the sequence TCAGTACT (Figure 12.12).

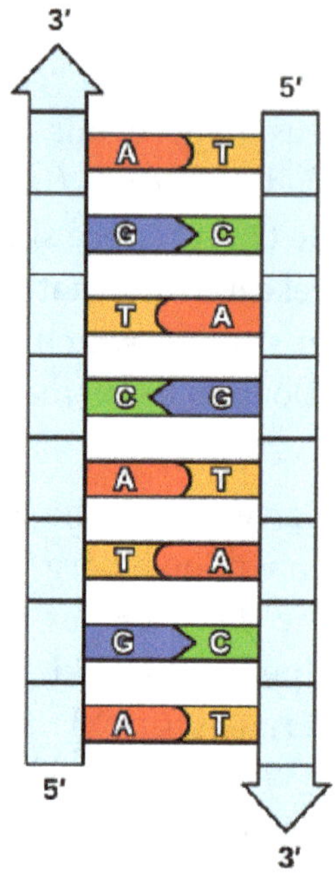

Figure 12.12 The two strands of DNA are complementary, meaning the sequence of bases in one strand can be used to create the correct sequence of bases in the other strand.

Because of the complementarity of the two strands, having one strand means that it is possible to recreate the other strand. This model for replication suggests that the two strands of the double helix separate during replication, and each strand serves as a template from which the new complementary strand is copied (Figure 12.13).

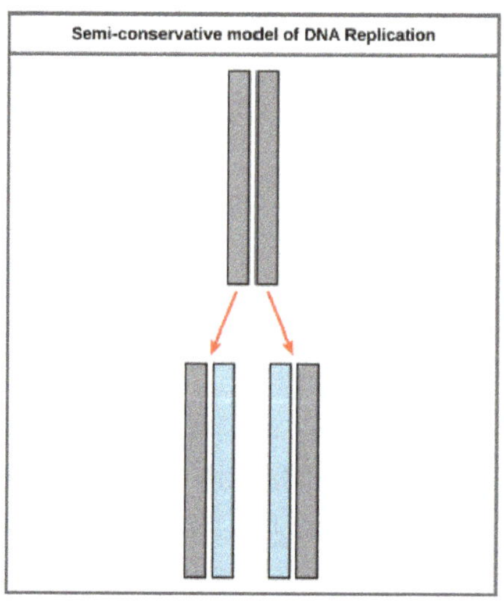

Figure 12.13 The semiconservative model of DNA replication is shown. Gray indicates the original DNA strands, and blue indicates newly synthesized DNA.

During DNA replication, each of the two strands that make up the double helix serves as a template from which new strands are copied. The new strand will be complementary to the parental or "old" strand. Each new double strand consists of one parental strand and one new daughter strand. This is known as **semiconservative replication**. When two DNA copies are formed, they have an identical sequence of nucleotide bases and are divided equally into two daughter cells.

DNA Replication in Eukaryotes

Because eukaryotic genomes are very complex, DNA replication is a very complicated process that involves several enzymes and other proteins. It occurs in three main stages: initiation, elongation, and termination.

Recall that eukaryotic DNA is bound to proteins known as histones to form structures called nucleosomes. During initiation, the DNA is made accessible to the proteins and enzymes involved in the replication process. How does the replication machinery know where on the DNA double helix to begin? It turns out that there are specific nucleotide sequences called origins of replication at which replication begins. Certain proteins bind to the origin of replication while an enzyme called **helicase** unwinds and opens up the DNA helix. As the DNA opens up, Y-shaped structures called **replication forks** are formed (Figure 12.14). Two replication forks are formed at the origin of replication, and these get extended in both directions as replication proceeds. There are multiple origins of replication on the eukaryotic chromosome, such that replication can occur simultaneously from several places in the genome.

During elongation, an enzyme called **DNA polymerase** adds DNA nucleotides to the 3' end of the template. Because DNA polymerase can only add new nucleotides at the end of a backbone, a **primer** sequence, which provides this starting point, is added with complementary RNA nucleotides. This primer is removed later, and the nucleotides are replaced with DNA nucleotides. One strand, which is complementary to the parental DNA strand, is synthesized continuously toward the replication fork so the polymerase can add nucleotides in this direction. This continuously synthesized strand is known as the **leading strand**. Because DNA polymerase can only synthesize DNA in a 5' to 3' direction, the other new strand is put together in short pieces called **Okazaki fragments**. The Okazaki fragments each require a primer made of RNA to start the synthesis. The strand with the Okazaki fragments is known as the **lagging strand**. As synthesis proceeds, an enzyme removes the RNA primer, which is then replaced with DNA nucleotides, and the gaps between fragments are sealed by an enzyme called **DNA ligase**.

The process of DNA replication can be summarized as follows:

- DNA unwinds at the origin of replication.
- New bases are added to the complementary parental strands. One new strand is made continuously, while the other strand is made in pieces.
- Primers are removed, new DNA nucleotides are put in place of the primers and the backbone is sealed by DNA ligase.

Visual Connection

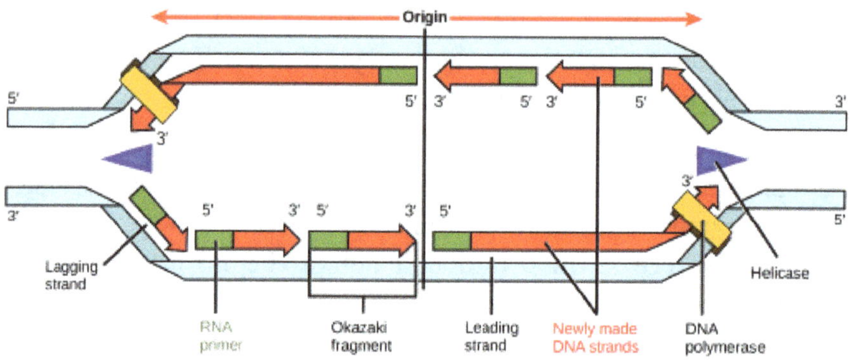

Figure 12.14 A replication fork is formed by the opening of the origin of replication, and helicase separates the DNA strands. An RNA primer is synthesized, and is elongated by the DNA polymerase. On the leading strand, DNA is synthesized continuously, whereas on the lagging strand, DNA is synthesized in short stretches. The DNA fragments are joined by DNA ligase (not shown).

You isolate a cell strain in which the joining together of Okazaki fragments is impaired and suspect that a mutation has occurred in an enzyme found at the replication fork. Which enzyme is most likely to be mutated?

Telomere Replication

Because eukaryotic chromosomes are linear, DNA replication comes to the end of a line in eukaryotic chromosomes. As you have learned, the DNA polymerase enzyme can add nucleotides in only one direction. In the leading strand, synthesis continues until the end of the chromosome is reached; however, on the lagging strand there is no place for a primer to be made for the DNA fragment to be copied at the end of the chromosome. This presents a problem for the cell because the ends remain unpaired, and over time these ends get progressively shorter as cells continue to divide. The ends of the linear chromosomes are known as **telomeres**, which have repetitive sequences that do not code for a particular gene. As a consequence, it is telomeres that are shortened with each round of DNA replication instead of genes. For example, in humans, a six base-pair sequence, TTAGGG, is repeated 100 to 1000 times. The discovery of the enzyme **telomerase** (Figure 12.15) helped in the understanding of how chromosome ends are maintained. The telomerase attaches to the end of the chromosome, and complementary bases to the RNA template are added on the end of the DNA strand. Once the lagging strand template is sufficiently elongated, DNA polymerase can now add nucleotides that are complementary to the ends of the chromosomes. Thus, the ends of the chromosomes are replicated.

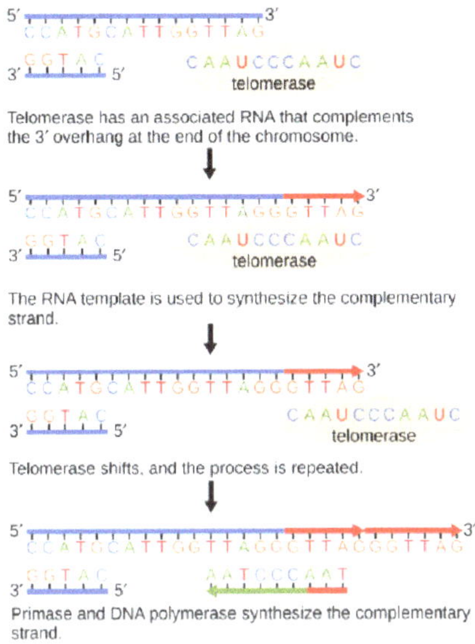

Figure 12.15 The ends of linear chromosomes are maintained by the action of the telomerase enzyme.

Telomerase is typically found to be active in germ cells, adult stem cells, and some cancer cells. For her discovery of telomerase and its action, Elizabeth Blackburn (Figure 12.16) received the Nobel Prize for Medicine and Physiology in 2009. Later research using HeLa cells (obtained from Henrietta Lacks) confirmed that telomerase is present in human cells. And in 2001, researchers including Diane L. Wright found that telomerase is necessary for cells in human embryos to rapidly proliferate.

Figure 12.16 Elizabeth Blackburn, 2009 Nobel Laureate, was the scientist who discovered how telomerase works. (credit: U.S. Embassy, Stockholm, Sweden)

Telomerase is not active in adult somatic cells. Adult somatic cells that undergo cell division continue to have their telomeres shortened. This essentially means that telomere shortening is associated with aging. In 2010, scientists found that telomerase can reverse some age-related conditions in mice, and this may have potential in regenerative medicine. Telomerase-deficient mice were used in these studies; these mice have tissue atrophy, stem-cell depletion, organ system failure, and impaired tissue injury responses. Telomerase reactivation in these mice caused extension of telomeres, reduced DNA damage, reversed neurodegeneration, and improved functioning of the testes, spleen, and intestines. Thus, telomere reactivation may have potential for treating age- related diseases in humans.

DNA Replication in Prokaryotes

Recall that the prokaryotic chromosome is a circular molecule with a less extensive coiling structure than eukaryotic chromosomes. The eukaryotic chromosome is linear and highly coiled around proteins. While there are many similarities in the DNA replication process, these structural differences necessitate some differences in the DNA replication process in these two life forms.

DNA replication has been extremely well-studied in prokaryotes, primarily because of the small size of the genome and large number of variants available. *Escherichia coli* has 4.6 million base pairs in a single circular chromosome, and all of it gets replicated in approximately 42 minutes, starting from a single origin of replication and proceeding around the chromosome in both directions. This means that approximately 1000 nucleotides are added per second. The process is much more rapid than in eukaryotes. Table 12.1 summarizes the differences between prokaryotic and eukaryotic replications.

Table 12.1 Differences between Prokaryotic and Eukaryotic Replications

Property	Prokaryotes	Eukaryotes
Origin of replication	Single	Multiple
Rate of replication	1000 nucleotides/s	50 to 100 nucleotides/s
Chromosome structure	circular	linear
Telomerase	Not present	Present

DNA Repair

DNA polymerase can make mistakes while adding nucleotides. It edits the DNA by proofreading every newly added base. Incorrect bases are removed and replaced by the correct base, and then polymerization continues (Figure 12.17a). Most mistakes are corrected during replication, although when this does not happen, the mismatch repair mechanism is employed. Mismatch repair enzymes recognize the wrongly incorporated base and excise it from the DNA, replacing it with the correct base (Figure 12.17b). In yet another type of repair, nucleotide excision repair, the DNA double strand is unwound and separated, the incorrect bases are removed along with a few bases on the 5' and 3' end, and these are replaced by copying the template with the help of DNA polymerase (Figure 12.17c). Nucleotide excision repair is particularly important in correcting thymine dimers, which are primarily caused by ultraviolet light. In a thymine dimer, two thymine nucleotides adjacent to each other on one strand are covalently bonded to each other rather than their complementary bases. If the dimer is not removed and repaired it will lead to a mutation. Individuals with flaws in their nucleotide excision repair genes show extreme sensitivity to sunlight and develop skin cancers early in life.

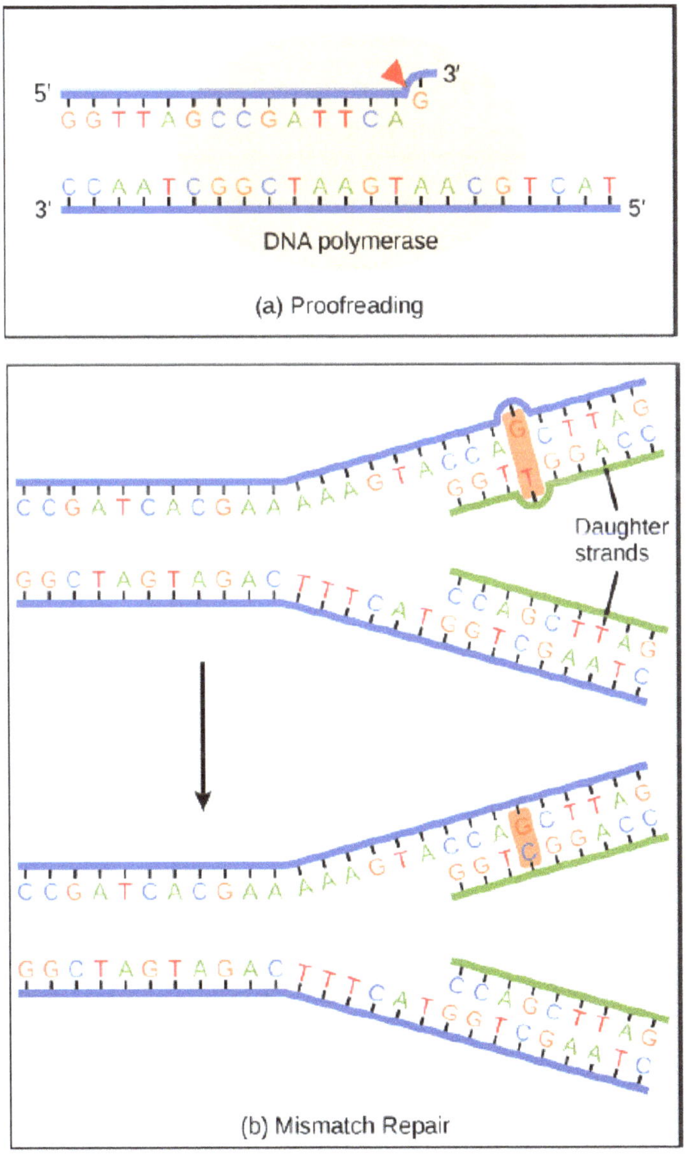

(a) Proofreading

(b) Mismatch Repair

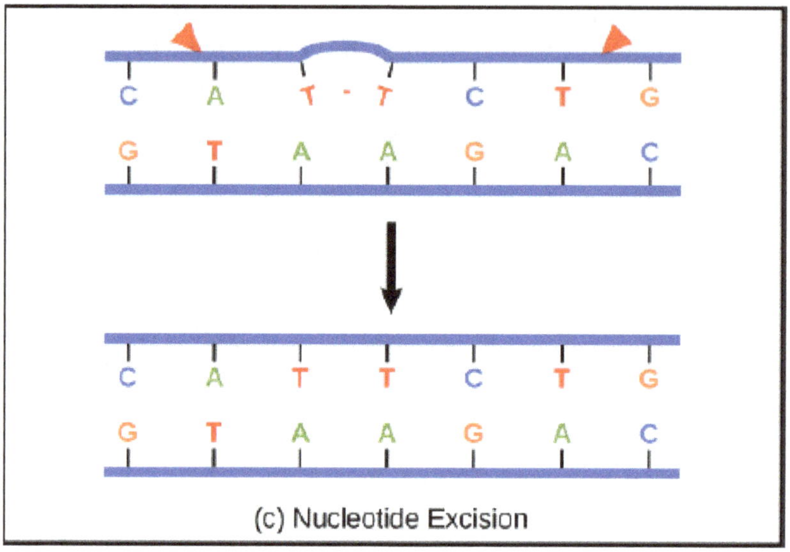

Figure 12.17 Proofreading by DNA polymerase (a) corrects errors during replication. In mismatch repair (b), the incorrectly added base is detected after replication. The mismatch repair proteins detect this base and remove it from the newly synthesized strand by nuclease action. The gap is now filled with the correctly paired base. Nucleotide excision (c) repairs thymine dimers. When exposed to UV, thymines lying adjacent to each other can form thymine dimers. In normal cells, they are excised and replaced.

Most mistakes are corrected; if they are not, they may result in a **mutation**—defined as a permanent change in the DNA sequence. Mutations in repair genes may lead to serious consequences like cancer.

STRUCTURE AND FUNCTION OF RNA

Structurally speaking, **ribonucleic acid (RNA)**, is quite similar to DNA. However, whereas DNA molecules are typically long and double stranded, RNA molecules are much shorter and are typically single stranded. RNA molecules perform a variety of roles in the cell but are mainly involved in the process of protein synthesis (translation) and its regulation.

RNA Structure

RNA is typically single stranded and is made of **ribonucleotides** that are linked by phosphodiester bonds. A ribonucleotide in the RNA chain contains ribose (the pentose sugar), one of the four nitrogenous bases (A, U, G, and C), and a phosphate group. The subtle structural difference between the sugars gives DNA added stability, making DNA more suitable for storage of genetic information, whereas the relative instability of RNA makes it more suitable for its more short-term functions. The RNA-specific pyrimidine **uracil** forms a complementary base pair with adenine and is used instead of the thymine used in DNA. Even though RNA is single stranded, most types of RNA molecules show extensive intramolecular base pairing between complementary sequences within the RNA strand, creating a predictable three-dimensional structure essential for their function (Figure 12.18 and Figure 12.19).

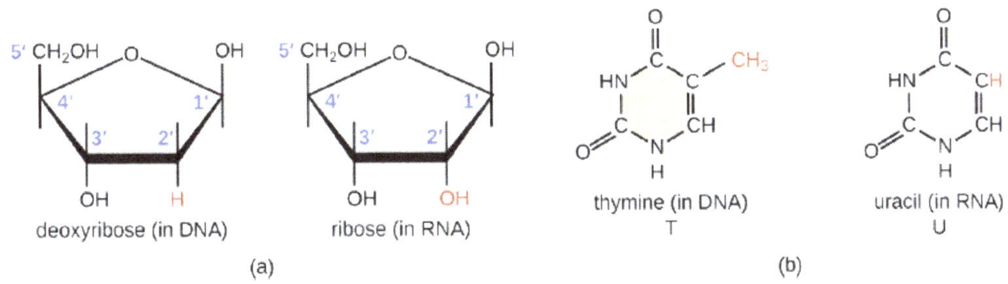

Figure 12.18 (a) Ribonucleotides contain the pentose sugar ribose instead of the deoxyribose found in deoxyribonucleotides. (b) RNA contains the pyrimidine uracil in place of thymine found in DNA.

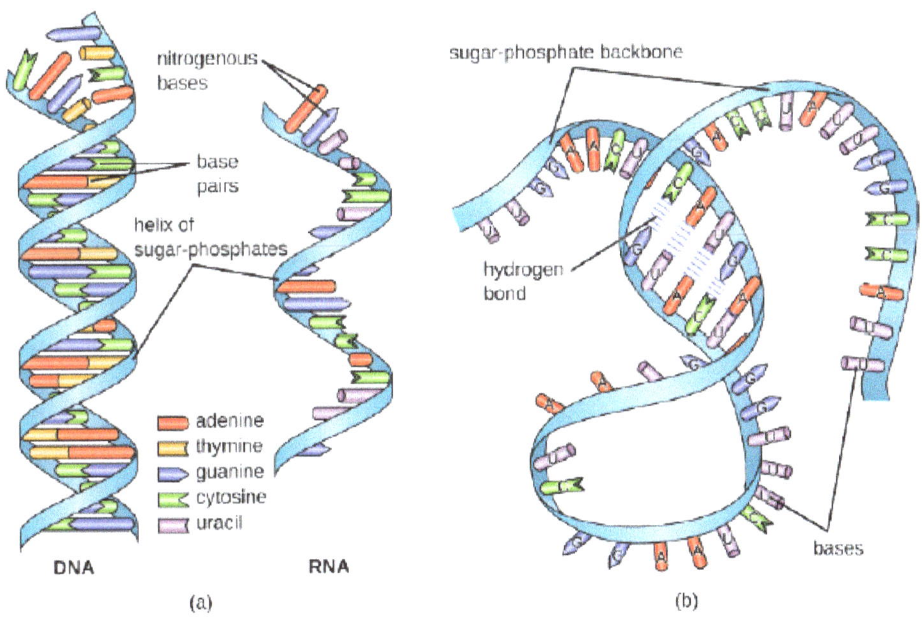

Figure 12.19 (a) DNA is typically double stranded, whereas RNA is typically single stranded. (b) Although it is single stranded, RNA can fold upon itself, with the folds stabilized by short areas of complementary base pairing within the molecule, forming a three-dimensional structure.

Functions of RNA in Protein Synthesis

Cells access the information stored in DNA by creating RNA to direct the synthesis of proteins through the process of translation. Proteins within a cell have many functions, including building cellular structures and serving as enzyme catalysts for cellular chemical reactions that give cells their specific characteristics. The three main types of RNA directly involved in protein synthesis are **messenger RNA (mRNA), ribosomal RNA (rRNA)**, and **transfer RNA (tRNA).**

In 1961, French scientists François Jacob and Jacques Monod hypothesized the existence of an intermediary between DNA and its protein products, which they called messenger RNA. Evidence supporting their hypothesis was gathered soon afterwards showing that information from DNA is transmitted to the ribosome for protein synthesis using mRNA. If DNA serves as the complete library of cellular information, mRNA serves as a photocopy of specific information needed at a particular point in time that serves as the instructions to make a protein.

The mRNA carries the message from the DNA, which controls all of the cellular activities in a cell. If a cell requires a certain protein to be synthesized, the gene for this product is "turned on" and the mRNA is synthesized through the process of transcription (see RNA Transcription (https://legacy.cnx.org/content/m58842/latest/)). The mRNA then interacts with ribosomes and other cellular machinery (Figure 12.20) to direct the synthesis of the protein it encodes during the process of translation (see Protein Synthesis (https://legacy.cnx.org/ content/m58843/latest/)). mRNA is relatively unstable and short-lived in the cell, especially in prokaryotic cells, ensuring that proteins are only made when needed.

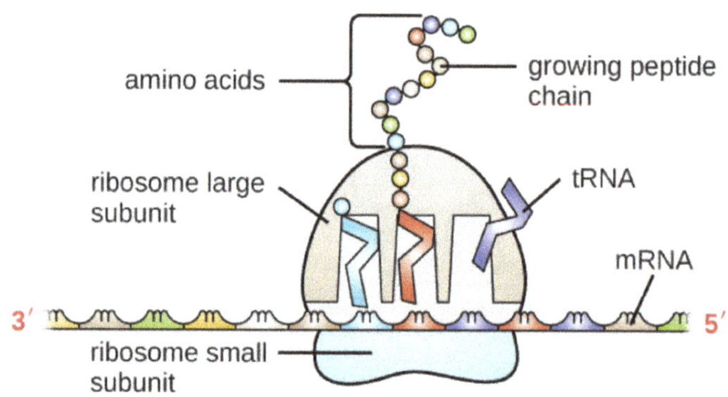

Figure 12.20 A generalized illustration of how mRNA and tRNA are used in protein synthesis within a cell.

rRNA and tRNA are stable types of RNA. In prokaryotes and eukaryotes, tRNA and rRNA are encoded in the DNA, then copied into long RNA molecules that are cut to release smaller fragments containing the individual mature RNA species. In eukaryotes, synthesis, cutting, and assembly of rRNA into ribosomes takes place in the nucleolus region of the nucleus, but these activities occur in the cytoplasm of prokaryotes. Neither of these types of RNA carries instructions to direct the synthesis of a polypeptide, but they play other important roles in protein synthesis.

Ribosomes are composed of rRNA and protein. As its name suggests, rRNA is a major constituent of ribosomes, composing up to about 60% of the ribosome by mass and providing the location where the mRNA binds. The rRNA ensures the proper alignment of the mRNA, tRNA, and the ribosomes; the rRNA of the ribosome also has an enzymatic activity (peptidyl transferase) and catalyzes the formation of the peptide bonds between two aligned amino acids during protein synthesis. Although rRNA had long been thought to serve primarily a structural role, its catalytic role within the ribosome was proven in 2000. Scientists in the laboratories of Thomas Steitz (1940–) and Peter Moore (1939–) at Yale University were able to crystallize the ribosome structure from *Haloarcula marismortui*, a halophilic archaeon isolated from the Dead Sea. Because of the importance of this work, Steitz shared the 2009 Nobel Prize in Chemistry with other scientists who made significant contributions to the understanding of ribosome structure.

Transfer RNA is the third main type of RNA and one of the smallest, usually only 70–90 nucleotides long. It carries the correct amino acid to the site of protein synthesis in the ribosome. It is the base pairing between the tRNA and mRNA that allows for the correct amino acid to be inserted in the polypeptide chain being synthesized (Figure 12.21). Any mutations in the tRNA or rRNA can result in global problems for the cell because both are necessary for proper protein synthesis (Table 12.2).

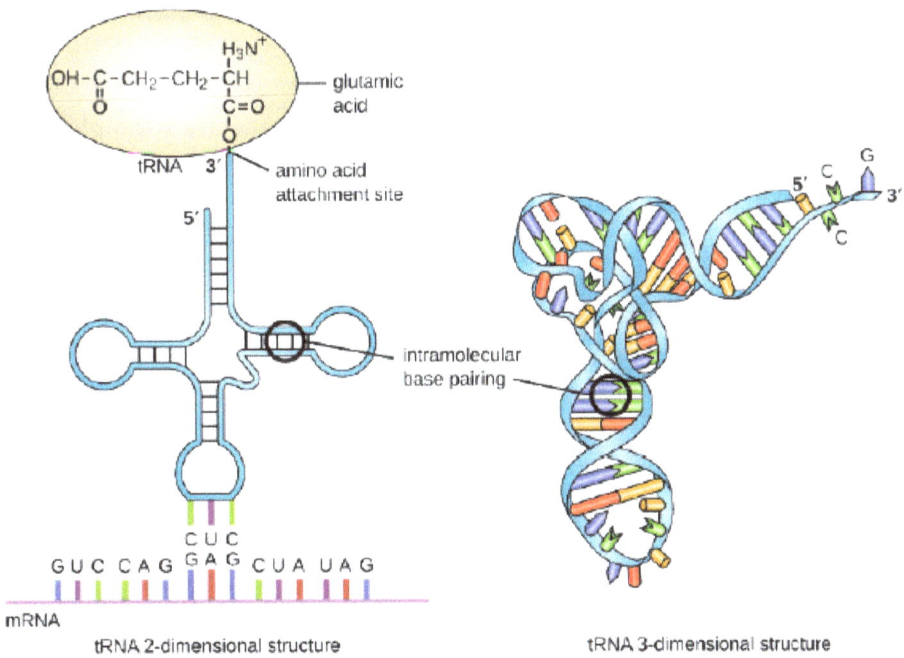

Figure 12.21 A tRNA molecule is a single-stranded molecule that exhibits significant intracellular base pairing, giving it its characteristic three-dimensional shape.

Table 12.2 Structure and Function of RNA

	mRNA	rRNA	tRNA
Structure	Short, unstable, single- stranded RNA corresponding to a gene encoded within DNA	Longer, stable RNA molecules composing 60% of ribosome's mass	Short (70-90 nucleotides), stable RNA with extensive intramolecular base pairing; contains an amino acid binding site and an mRNA binding site
Function	Serves as intermediary between DNA and protein; used by ribosome to direct synthesis of protein it encodes	Ensures the proper alignment of mRNA, tRNA, and ribosome during protein synthesis; catalyzes peptide bond formation between amino acids	Carries the correct amino acid to the site of protein synthesis in the ribosome

- What are the functions of the three major types of RNA molecules involved in protein synthesis?

RNA as Hereditary Information

Although RNA does not serve as the hereditary information in most cells, RNA does hold this function for many viruses that do not contain DNA. Thus, RNA clearly does have the additional capacity to serve as genetic information. Although RNA is typically single stranded within cells, there is significant diversity in viruses. Rhinoviruses, which cause the common cold; influenza viruses; and the Ebola virus are single-stranded RNA viruses. Rotaviruses, which cause severe gastroenteritis in children and other immunocompromised individuals, are examples of double-stranded RNA viruses. Because double-stranded RNA is uncommon in eukaryotic cells, its presence serves as an indicator of viral infection.

TRANSCRIPTION

In both prokaryotes and eukaryotes, the second function of DNA (the first was replication) is to provide the information needed to construct the proteins necessary so that the cell can perform all of its functions. To do this, the DNA is "read" or transcribed into an **mRNA** molecule. The mRNA then provides the code to form a protein by a process called translation. Through the processes of transcription and translation, a protein is built with a specific sequence of amino acids that was originally encoded in the DNA. This module discusses the details of transcription.

The Central Dogma: DNA Encodes RNA; RNA Encodes Protein

The flow of genetic information in cells from DNA to mRNA to protein is described by the central dogma (Figure 12.22), which states that genes specify the sequences of mRNAs, which in turn specify the sequences of proteins.

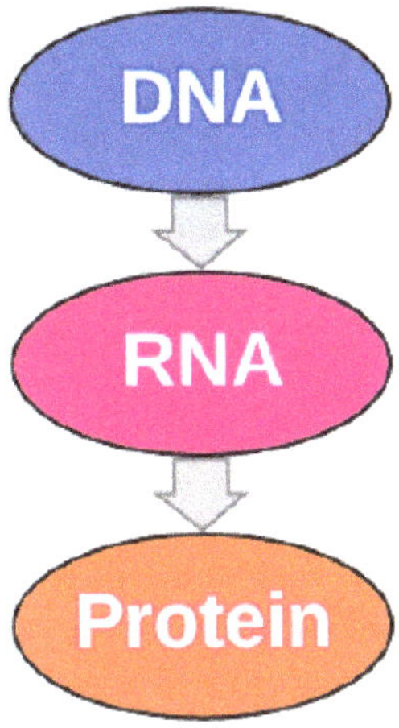

Figure 12.22 The central dogma states that DNA encodes RNA, which in turn encodes protein.

The copying of DNA to mRNA is relatively straightforward, with one nucleotide being added to the mRNA strand for every complementary nucleotide read in the DNA strand. The translation to protein is more complex because groups of three mRNA nucleotides correspond to one amino acid of the protein sequence. However, as we shall see in the next module, the translation to protein is still systematic, such that nucleotides 1 to 3 correspond to amino acid 1, nucleotides 4 to 6 correspond to amino acid 2, and so on.

Transcription: from DNA to mRNA

Both prokaryotes and eukaryotes perform fundamentally the same process of transcription, with the important difference of the membrane-bound nucleus in eukaryotes. With the genes bound in the nucleus, transcription occurs in the nucleus of the cell and the mRNA transcript must be transported to the cytoplasm. The prokaryotes, which include bacteria and archaea, lack membrane-bound nuclei and other organelles, and transcription occurs in the cytoplasm of the cell. In both prokaryotes and eukaryotes, transcription occurs in three main stages: initiation, elongation, and termination.

Initiation

Transcription requires the DNA double helix to partially unwind in the region of mRNA synthesis. The region of unwinding is called a **transcription bubble**. The DNA sequence onto which the proteins and enzymes involved in transcription bind to initiate the process is called a **promoter**. In most cases, promoters exist upstream of the genes they regulate. The specific sequence of a promoter is very important because it determines whether the corresponding gene is transcribed all of the time, some of the time, or hardly at all (Figure 12.23).

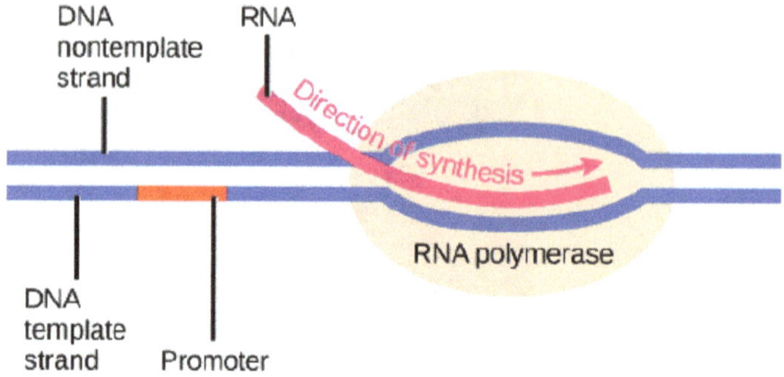

Figure 12.23 The initiation of transcription begins when DNA is unwound, forming a transcription bubble. Enzymes and other proteins involved in transcription bind at the promoter.

Elongation

Transcription always proceeds from one of the two DNA strands, which is called the **template strand.** The mRNA product is complementary to the template strand and is almost identical to the other DNA strand, called the **nontemplate strand**, with the exception that RNA contains a uracil (U) in place of the thymine (T) found in DNA. During elongation, an enzyme called **RNA polymerase** proceeds along the DNA template adding nucleotides by base pairing with the DNA template in a manner similar to DNA replication, with the difference that an RNA strand is being synthesized that does not remain bound to the DNA template. As elongation proceeds, the DNA is continuously unwound ahead of the core enzyme and rewound behind it (Figure 12.24).

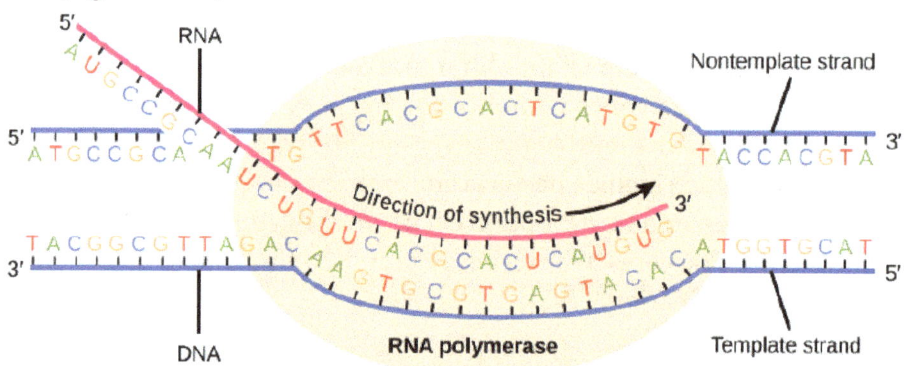

Figure 12.24 During elongation, RNA polymerase tracks along the DNA template, synthesizes mRNA in the 5' to 3' direction, and unwinds then rewinds the DNA as it is read.

Termination

Once a gene is transcribed, the prokaryotic polymerase needs to be instructed to dissociate from the DNA template and liberate the newly made mRNA. Depending on the gene being transcribed, there are two kinds of termination signals, but both involve repeated nucleotide sequences in the DNA template that result in RNA polymerase stalling, leaving the DNA template, and freeing the mRNA transcript.

On termination, the process of transcription is complete. In a prokaryotic cell, by the time termination occurs, the transcript would already have been used to partially synthesize numerous copies of the encoded protein because these processes can occur concurrently using multiple ribosomes (polyribosomes) (Figure 12.25). In contrast, the presence of a nucleus in eukaryotic cells precludes simultaneous transcription and translation.

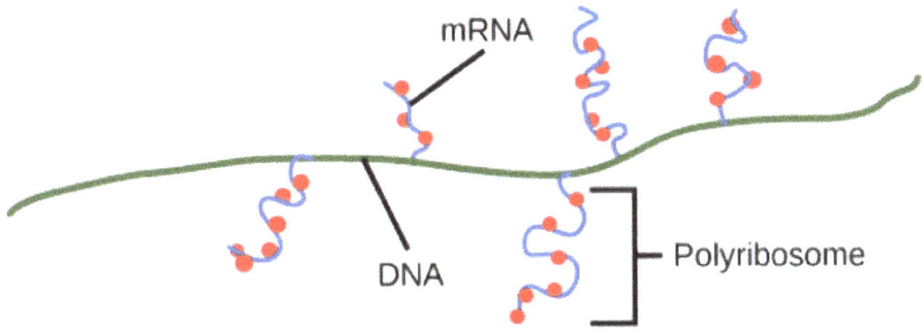

Figure 12.25 Multiple polymerases can transcribe a single bacterial gene while numerous ribosomes concurrently translate the mRNA transcripts into polypeptides. In this way, a specific protein can rapidly reach a high concentration in the bacterial cell.

Eukaryotic RNA Processing

The newly transcribed eukaryotic mRNAs must undergo several processing steps before they can be transferred from the nucleus to the cytoplasm and translated into a protein. The additional steps involved in eukaryotic mRNA maturation create a molecule that is much more stable than a prokaryotic mRNA. For example, eukaryotic mRNAs last for several hours, whereas the typical prokaryotic mRNA lasts no more than five seconds.

The mRNA transcript is first coated in RNA-stabilizing proteins to prevent it from degrading while it is processed and exported out of the nucleus. This occurs while the pre-mRNA still is being synthesized by adding a special nucleotide "cap" to the 5' end of the growing transcript. In addition to preventing degradation, factors involved in protein synthesis recognize the cap to help initiate translation by ribosomes.

Once elongation is complete, an enzyme then adds a string of approximately 200 adenine residues to the 3' end, called the poly-A tail. This modification further protects the pre-mRNA from degradation and signals to cellular factors that the transcript needs to be exported to the cytoplasm.

Eukaryotic genes are composed of protein-coding sequences called **exons** (*ex*-on signifies that they are *ex*pressed) and *int*ervening sequences called **introns** (*int*-ron denotes their *int*ervening role). Introns are removed from the pre-mRNA during processing. Intron sequences in mRNA do not encode functional proteins. It is essential that all of a pre-mRNA's introns be completely and precisely removed before protein synthesis so that the exons join together to code for the correct amino acids. If the process errs by even a single nucleotide, the sequence of the rejoined exons would be shifted, and the resulting protein would be nonfunctional. The process of removing introns and reconnecting exons is called **splicing** (Figure 12.26). Introns are removed and degraded while the pre-mRNA is still in the nucleus.

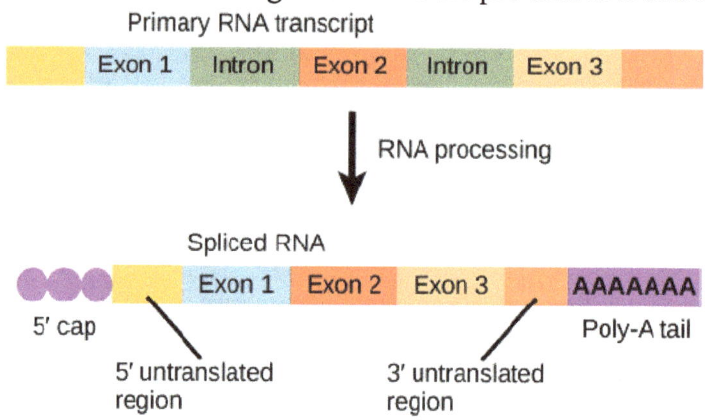

Figure 12.26 Eukaryotic mRNA contains introns that must be spliced out. A 5' cap and 3' tail are also added.

TRANSLATION

The synthesis of proteins is one of a cell's most energy-consuming metabolic processes. In turn, proteins account for more mass than any other component of living organisms (with the exception of water), and proteins perform a wide variety of the functions of a cell. The process of translation, or protein synthesis, involves decoding an mRNA message into a polypeptide product. Amino acids are covalently strung together in lengths ranging from approximately 50 amino acids to more than 1,000.

The Protein Synthesis Machinery

In addition to the mRNA template, many other molecules contribute to the process of translation. The composition of each component may vary across species; for instance, ribosomes may consist of different numbers of ribosomal RNAs (**rRNA**) and polypeptides depending on the organism. However, the general structures and functions of the protein synthesis machinery are comparable from bacteria to human cells. Translation requires the input of an mRNA template, ribosomes, tRNAs, and various enzymatic factors (Figure 12.27).

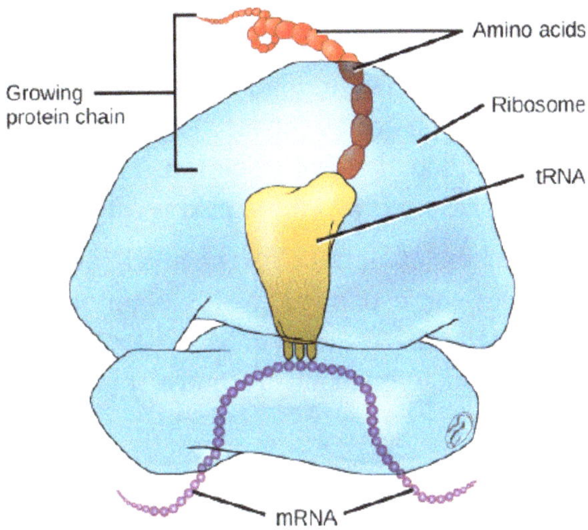

Figure 12.27 The protein synthesis machinery includes the large and small subunits of the ribosome, mRNA, and tRNA. (credit: modification of work by NIGMS, NIH)

In *E. coli*, there are 200,000 ribosomes present in every cell at any given time. A ribosome is a complex macromolecule composed of structural and catalytic rRNAs, and many distinct polypeptides. In eukaryotes, the nucleolus is completely specialized for the synthesis and assembly of rRNAs.

Ribosomes are located in the cytoplasm in prokaryotes and in the cytoplasm and endoplasmic reticulum of eukaryotes. Ribosomes are made up of a large and a small subunit that come together for translation. The small subunit is responsible for binding the mRNA template, whereas the large subunit sequentially binds **tRNAs**, a type of RNA molecule that brings amino acids to the growing chain of the polypeptide. Each mRNA molecule is simultaneously translated by many ribosomes, all synthesizing protein in the same direction.

Depending on the species, 40 to 60 types of tRNA exist in the cytoplasm. Serving as adaptors, specific tRNAs bind to sequences on the mRNA template and add the corresponding amino acid to the polypeptide chain. Therefore, tRNAs are the molecules that actually "translate" the language of RNA into the language of proteins.

For each tRNA to function, it must have its specific amino acid bonded to it. In the process of tRNA "charging," each tRNA molecule is bonded to its correct amino acid.

The Genetic Code

To summarize what we know to this point, the cellular process of transcription generates messenger RNA (mRNA), a mobile molecular copy of one or more genes with an alphabet of A, C, G, and uracil (U). Translation of the mRNA template converts nucleotide-based genetic information into a protein product. Protein sequences consist of 20 commonly occurring amino acids; therefore, it can be said that the protein alphabet consists of 20 letters. Each amino acid is defined by a three-nucleotide sequence called the triplet **codon.** The relationship between a nucleotide codon and its corresponding amino acid is called the **genetic code.**

Given the different numbers of "letters" in the mRNA and protein "alphabets," combinations of nucleotides corresponded to single amino acids. Using a three-nucleotide code means that there are a total of 64 (4 × 4 × 4) possible combinations; therefore, a given amino acid is encoded by more than one nucleotide triplet (Figure 12.28).

Second letter

First letter	U	C	A	G	Third letter
U	UUU Phe UUC Phe UUA Leu UUG Leu	UCU Ser UCC Ser UCA Ser UCG Ser	UAU Tyr UAC Tyr UAA Stop UAG Stop	UGU Cys UGC Cys UGA Stop UGG Trp	U C A G
C	CUU Leu CUC Leu CUA Leu CUG Leu	CCU Pro CCC Pro CCA Pro CCG Pro	CAU His CAC His CAA Gln CAG Gln	CGU Arg CGC Arg CGA Arg CGG Arg	U C A G
A	AUU Ile AUC Ile AUA Ile AUG Met	ACU Thr ACC Thr ACA Thr ACG Thr	AAU Asn AAC Asn AAA Lys AAG Lys	AGU Ser AGC Ser AGA Arg AGG Arg	U C A G
G	GUU Val GUC Val GUA Val GUG Val	GCU Ala GCC Ala GCA Ala GCG Ala	GAU Asp GAC Asp GAA Glu GAG Glu	GGU Gly GGC Gly GGA Gly GGG Gly	U C A G

Figure 12.28 This figure shows the genetic code for translating each nucleotide triplet, or codon, in mRNA into an amino acid or a termination signal in a nascent protein. (credit: modification of work by NIH)

Three of the 64 codons terminate protein synthesis and release the polypeptide from the translation machinery. These triplets are called **stop codons.** Another codon, AUG, also has a special function. In addition to specifying the amino acid methionine, it also serves as the **start codon** to initiate translation. The reading frame for translation is set by the AUG start codon near the 5' end of the mRNA. The genetic code is universal. With a few exceptions, virtually all species use the same genetic code for protein synthesis, which is powerful evidence that all life on Earth shares a common origin.

The Mechanism of Protein Synthesis

Just as with mRNA synthesis, protein synthesis can be divided into three phases: initiation, elongation, and termination. The process of translation is similar in prokaryotes and eukaryotes. Here we will explore how translation occurs in *E. coli*, a representative prokaryote, and specify any differences between prokaryotic and eukaryotic translation.

Protein synthesis begins with the formation of an initiation complex. In *E. coli*, this complex involves the small ribosome subunit, the mRNA template, three initiation factors, and a special initiator tRNA. The initiator tRNA interacts with the AUG start codon, and links to a special form of the amino acid methionine that is typically removed from the polypeptide after translation is complete.

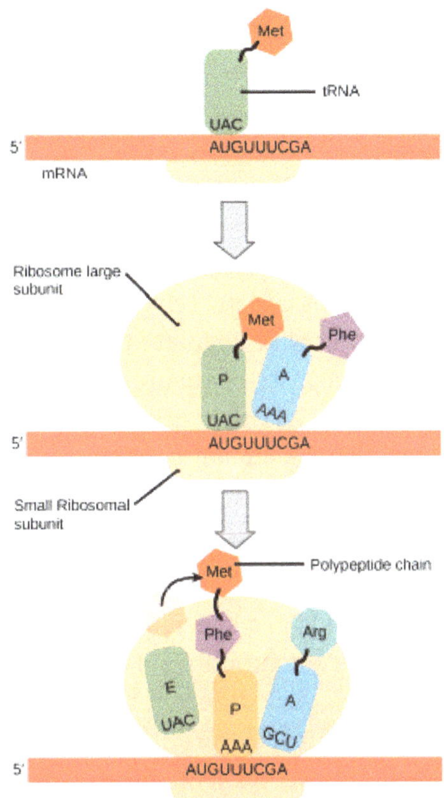

Figure 12.29 Translation begins when a tRNA anticodon recognizes a codon on the mRNA. The large ribosomal subunit joins the small subunit, and a second tRNA is recruited. As the mRNA moves relative to the ribosome, the polypeptide chain is formed. Entry of a release factor into the A site terminates translation and the components dissociate.

In prokaryotes and eukaryotes, the basics of polypeptide elongation are the same, so we will review elongation from the perspective of *E. coli*. The large ribosomal subunit of *E. coli* consists of three compartments: the A site binds incoming charged tRNAs (tRNAs with their attached specific amino acids). The P site binds charged tRNAs carrying amino acids that have formed bonds with the growing polypeptide chain but have not yet dissociated from their corresponding tRNA. The E site releases dissociated tRNAs so they can be recharged with free amino acids. The ribosome shifts one codon at a time, catalyzing each process that occurs in the three sites. With each step, a charged tRNA enters the complex, the polypeptide becomes one amino acid longer, and an uncharged tRNA departs. The energy for each bond between amino acids is derived from GTP, a molecule similar to ATP (Figure 12.29). Amazingly, the *E. coli* translation apparatus takes only 0.05 seconds to add each amino acid, meaning that a -200amino acid polypeptide could be translated in just 10 seconds.

Termination of translation occurs when a stop codon (UAA, UAG, or UGA) is encountered. When the ribosome encounters the stop codon, the growing polypeptide is released and the ribosome subunits dissociate and leave the mRNA. After many ribosomes have completed translation, the mRNA is degraded so the nucleotides can be reused in another transcription reaction.

HOW GENES ARE REGULATED

For a cell to function properly, necessary proteins must be synthesized at the proper time. All organisms and cells control or regulate the transcription and translation of their DNA into protein. The process of turning on a gene to produce RNA and protein is called **gene expression**. Whether in a simple unicellular organism or in a complex multicellular organism, each cell controls when and how its genes are expressed. For this to occur, there must be a mechanism to control when a gene is expressed to make RNA and protein, how much of the protein is made, and when it is time to stop making that protein because it is no longer needed.

Cells in multicellular organisms are specialized; cells in different tissues look very different and perform different functions. For example, a muscle cell is very different from a liver cell, which is very different from a skin cell. These differences are a consequence of the expression of different sets of genes in each of these cells. All cells have certain basic functions they must perform for themselves, such as converting the energy in sugar molecules into energy in ATP. Each cell also has many genes that are not expressed, and expresses many that are not expressed by other cells, such that it can carry out its specialized functions. In addition, cells will turn on or off certain genes at different times in response to changes in the environment or at different times during the development of the organism. Unicellular organisms, both eukaryotic and prokaryotic, also turn on and off genes in response to the demands of their environment so that they can respond to special conditions.

The control of gene expression is extremely complex. Malfunctions in this process are detrimental to the cell and can lead to the development of many diseases, including cancer.

Prokaryotic Versus Eukaryotic Gene Expression

To understand how gene expression is regulated, we must first understand how a gene becomes a functional protein in a cell. The process occurs in both prokaryotic and eukaryotic cells, just in slightly different fashions.

Because prokaryotic organisms lack a cell nucleus, the processes of transcription and translation occur almost simultaneously. When the protein is no longer needed, transcription stops. As a result, the primary method to control what type and how much protein is expressed in a prokaryotic cell is through the regulation of DNA transcription into RNA. All the subsequent steps happen automatically. When more protein is required, more transcription occurs. Therefore, in prokaryotic cells, the control of gene expression is almost entirely at the transcriptional level.

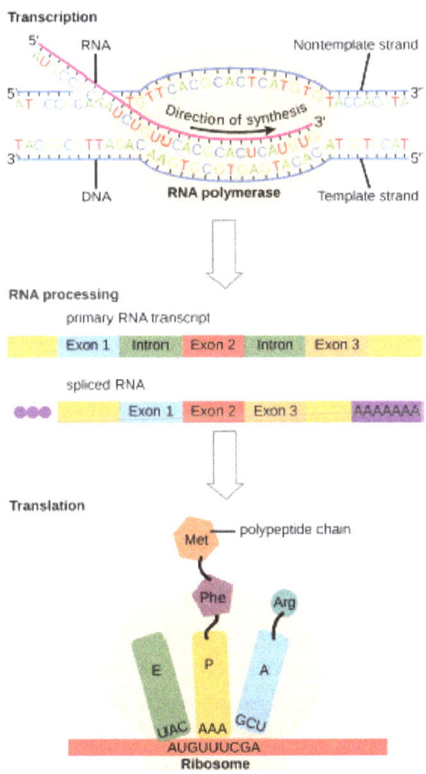

Figure 12.30 Eukaryotic gene expression is regulated during transcription and RNA processing, which take place in the nucleus, as well as during protein translation, which takes place in the cytoplasm. Further regulation may occur through post-translational modifications of proteins.

The first example of such control was discovered using *E. coli* in the 1950s and 1960s by French researchers and is called the *lac* operon. The *lac* operon is a stretch of DNA with three adjacent genes that code for proteins that participate in the absorption and metabolism of lactose, a food source for *E. coli*. When lactose is not present in the bacterium's environment, the *lac* genes are transcribed in small amounts. When lactose is present, the genes are transcribed and the bacterium is able to use the lactose as a food source. The operon also contains a promoter sequence to which the RNA polymerase binds to begin transcription; between the promoter and the three genes is a region called the operator. When there is no lactose present, a protein known as a repressor binds to the operator and prevents RNA polymerase from binding to the promoter, except in rare cases. Thus very little of the protein products of the three genes is made. When lactose is present, an end product of lactose metabolism binds to the repressor protein and prevents it from binding to the operator. This allows RNA polymerase to bind to the promoter and freely transcribe the three genes, allowing the organism to metabolize the lactose.

Eukaryotic cells, in contrast, have intracellular organelles and are much more complex. Recall that in eukaryotic cells, the DNA is contained inside the cell's nucleus and it is transcribed into mRNA there. The newly synthesized mRNA is then transported out of the nucleus into the cytoplasm, where ribosomes translate the mRNA into protein. The processes of transcription and translation are physically separated by the nuclear membrane; transcription occurs only within the nucleus, and translation only occurs outside the nucleus in the cytoplasm. The regulation of gene expression can occur at all stages of the process (Figure 12.30). Regulation may occur when the DNA is uncoiled and loosened from nucleosomes to bind transcription factors (**epigenetic** level), when the RNA is transcribed (transcriptional level), when RNA is processed and exported to the cytoplasm after it is transcribed (**post-transcriptional** level), when the RNA is translated into protein (translational level), or after the protein has been made (**post-translational** level).

The differences in the regulation of gene expression between prokaryotes and eukaryotes are summarized in Table 12.3.

Table 12.3 Differences in the Regulation of Gene Expression of Prokaryotic and Eukaryotic Organisms

Prokaryotic organisms	Eukaryotic organisms
Lack nucleus	Contain nucleus
RNA transcription and protein translation occur almost simultaneously	RNA transcription occurs prior to protein translation, and it takes place in the nucleus. RNA translation to protein occurs in the cytoplasm. RNA post-processing includes addition of a 5' cap, poly-A tail, and excision of introns and splicing of exons.
Gene expression is regulated primarily at the transcriptional level	Gene expression is regulated at many levels (epigenetic, transcriptional, post-transcriptional, translational, and post-translational)

Alternative RNA Splicing

In the 1970s, genes were first observed that exhibited **alternative RNA splicing.** Alternative RNA splicing is a mechanism that allows different protein products to be produced from one gene when different combinations of introns (and sometimes exons) are removed from the transcript (Figure 12.31). This alternative splicing can be haphazard, but more often it is controlled and acts as a mechanism of gene regulation, with the frequency of different splicing alternatives controlled by the cell as a way to control the production of different protein products in different cells, or at different stages of development. Alternative splicing is now understood to be a common mechanism of gene regulation in eukaryotes; according to one estimate, 70% of genes in humans are expressed as multiple proteins through alternative splicing.

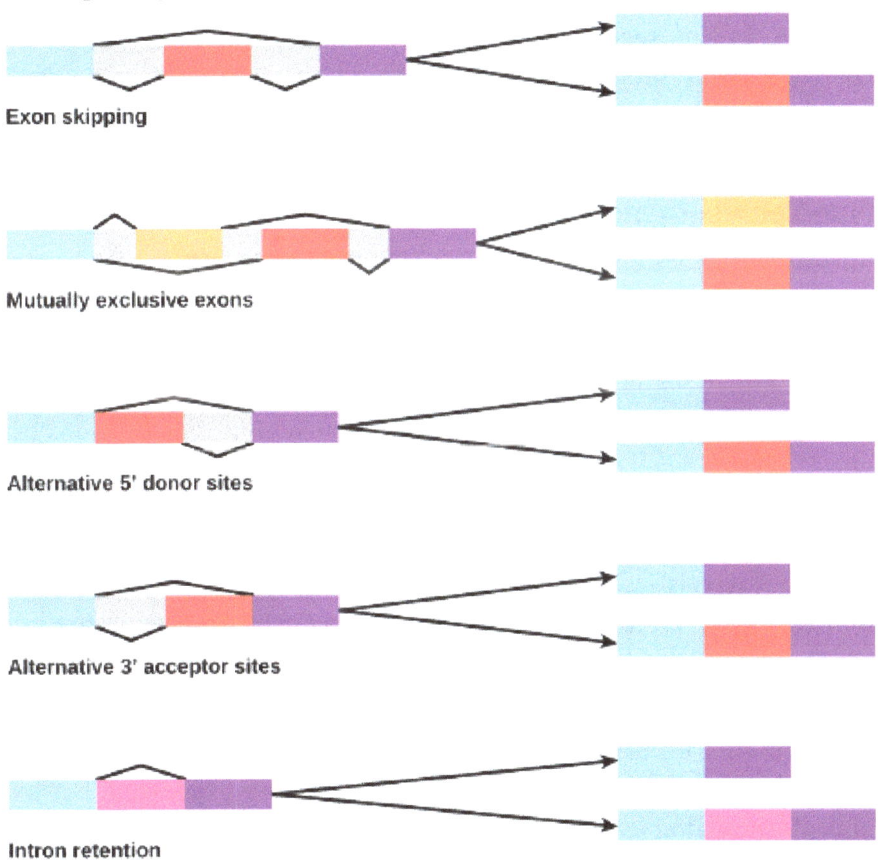

Figure 12.31 There are five basic modes of alternative splicing. Segments of pre-mRNA with exons shown in blue, red, orange, and pink can be spliced to produce a variety of new mature mRNA segments.

How could alternative splicing evolve? Introns have a beginning and ending recognition sequence, and it is easy to imagine the failure of the splicing mechanism to identify the end of an intron and find the end of the next intron, thus removing two introns and the intervening exon. In fact, there are mechanisms in place to prevent such exon skipping, but mutations are likely to lead to their failure. Such "mistakes" would more than likely produce a non-functional protein. Indeed, the cause of many genetic diseases is alternative splicing rather than mutations in a sequence. However, alternative splicing would create a protein variant without the loss of the original protein, opening up possibilities for adaptation of the new variant to new functions. Gene duplication has played an important role in the evolution of new functions in a similar way—by providing genes that may evolve without eliminating the original functional protein.

STRUCTURE AND FUNCTION OF CELLULAR GENOMES

Thus far, we have discussed the structure and function of individual pieces of DNA and RNA. In this section, we will discuss how all of an organism's genetic material—collectively referred to as its **genome**—is organized inside of the cell. Since an organism's genetics to a large extent dictate its characteristics, it should not be surprising that organisms differ in the arrangement of their DNA and RNA.

Genotype Versus Phenotype

All cellular activities are encoded within a cell's DNA. The sequence of bases within a DNA molecule represents the genetic information of the cell. Segments of DNA molecules are called **genes**, and individual genes contain the instructional code necessary for synthesizing various proteins, enzymes, or stable RNA molecules.

The full collection of genes that a cell contains within its genome is called its **genotype**. However, a cell does not express all of its genes simultaneously. Instead, it turns on (expresses) or turns off certain genes when necessary. The set of genes being expressed at any given point in time determines the cell's activities and its observable characteristics, referred to as its **phenotype**. Genes that are always expressed are known as constitutive genes; some constitutive genes are known as housekeeping genes because they are necessary for the basic functions of the cell.

While the genotype of a cell remains constant, the phenotype may change in response to environmental signals (e.g., changes in temperature or nutrient availability) that affect which nonconstitutive genes are expressed. For example, the oral bacterium *Streptococcus mutans* produces a sticky slime layer that allows it to adhere to teeth, forming dental plaque; however, the genes that control the production of the slime layer are only expressed in the presence of sucrose (table sugar). Thus, while the genotype of *S. mutans* is constant, its phenotype changes depending on the presence and absence of sugar in its environment. Temperature can also regulate gene expression. For example, the gram-negative bacterium *Serratia marcescens*, a pathogen frequently associated with hospital-acquired infections, produces a red pigment at 28°C but not at 37 °C, the normal internal temperature of the human body (Figure 12.32).

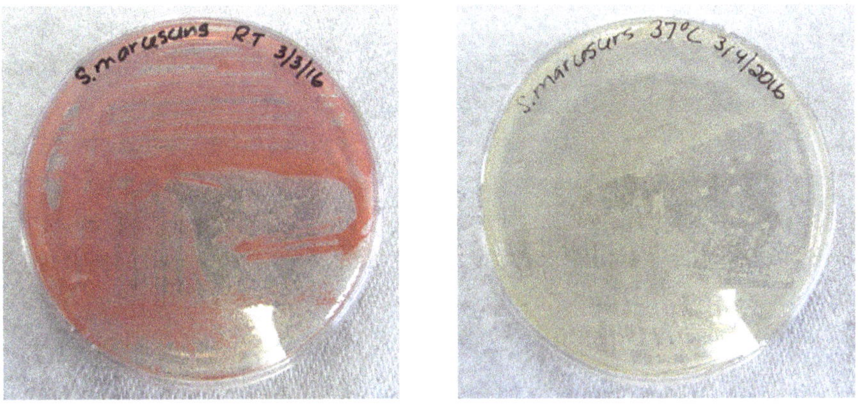

Figure 12.32 Both plates contain strains of *Serratia marcescens* that have the gene for red pigment. However, this gene is expressed at 28 °C (left) but not at 37 °C (right). (credit: modification of work by Ann Auman)

Organization of Genetic Material

The vast majority of an organism's genome is organized into the cell's **chromosomes**, which are discrete DNA structures within cells that control cellular activity. Recall that while eukaryotic chromosomes are housed in the membrane-bound nucleus, most prokaryotes contain a single, circular chromosome that is found in an area of the cytoplasm called the nucleoid (see Unique Characteristics of Prokaryotic Cells (https://legacy.cnx.org/ content/m58792/latest/)). A chromosome may contain several thousand genes.

Organization of Eukaryotic Chromosome

Chromosome structure differs somewhat between eukaryotic and prokaryotic cells. Eukaryotic chromosomes are typically linear, and eukaryotic cells contain multiple distinct chromosomes. Many eukaryotic cells contain two copies of each chromosome and, therefore, are **diploid**.

The length of a chromosome greatly exceeds the length of the cell, so a chromosome needs to be packaged into a very small space to fit within the cell. For example, the combined length of all of the 3 billion base pairs of DNA of the human genome would measure approximately 2 meters if completely stretched out, and some eukaryotic genomes are many times larger than the human genome. DNA **supercoiling** refers to the process by which DNA is twisted to fit inside the cell. Supercoiling may result in DNA that is either underwound (less than one turn of the helix per 10 base pairs) or overwound (more than one turn per 10 base pairs) from its normal relaxed state. Proteins known to be involved in supercoiling include **topoisomerases**; these enzymes help maintain the structure of supercoiled chromosomes, preventing overwinding of DNA during certain cellular processes like DNA replication.

During **DNA packaging**, DNA-binding proteins called **histones** perform various levels of DNA wrapping and attachment to scaffolding proteins. The combination of DNA with these attached proteins is referred to as **chromatin.** In eukaryotes, the packaging of DNA by histones may be influenced by environmental factors that affect the presence of methyl groups on certain cytosine nucleotides of DNA. The influence of environmental factors on DNA packaging is called epigenetics. Epigenetics is another mechanism for regulating gene expression without altering the sequence of nucleotides. Epigenetic changes can be maintained through multiple rounds of cell division and, therefore, can be heritable.

Organization of Prokaryotic Chromosomes

Chromosomes in bacteria and archaea are usually circular, and a prokaryotic cell typically contains only a single chromosome within the nucleoid. Because the chromosome contains only one copy of each gene, prokaryotes are **haploid.** As in eukaryotic cells, DNA supercoiling is necessary for the genome to fit within the prokaryotic cell. The DNA in the bacterial chromosome is arranged in several supercoiled domains. As with eukaryotes, topoisomerases are involved in supercoiling DNA. DNA gyrase is a type of topoisomerase, found in bacteria and some archaea, that helps prevent the overwinding of DNA. (Some antibiotics kill bacteria by targeting DNA gyrase.) In addition, histone-like proteins bind DNA and aid in DNA packaging. Other proteins bind to the origin of replication, the location in the chromosome where DNA replication initiates. Because different regions of DNA are packaged differently, some regions of chromosomal DNA are more accessible to enzymes and thus may be used more readily as templates for gene expression. Interestingly, several bacteria, including *Helicobacter pylori* and *Shigella flexneri*, have been shown to induce epigenetic changes in their hosts upon infection, leading to chromatin remodeling that may cause long-term effects on host immunity.

Noncoding DNA

In addition to genes, a genome also contains many regions of noncoding DNA that do not encode proteins or stable RNA products. Noncoding DNA is commonly found in areas prior to the start of coding sequences of genes as well as in intergenic regions (i.e., DNA sequences located between genes) (Figure 12.33).

Prokaryotes appear to use their genomes very efficiently, with only an average of 12% of the genome being taken up by noncoding sequences. In contrast, noncoding DNA can represent about 98% of the genome in eukaryotes, as seen in humans, but the percentage of noncoding DNA varies between species.

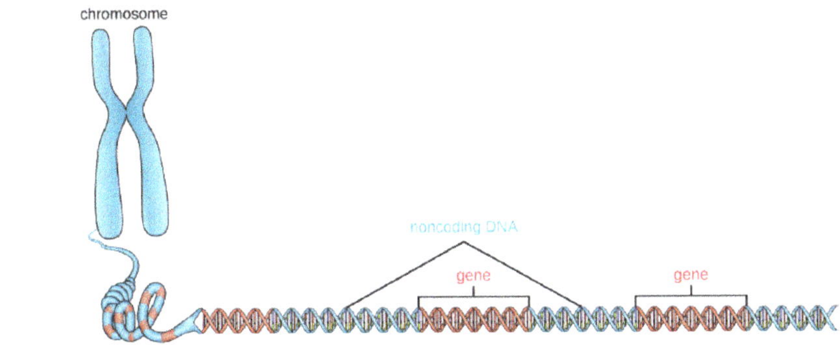

Figure 12.33 Chromosomes typically have a significant amount of noncoding DNA, often found in intergenic regions.

These noncoding DNA regions were once referred to as "junk DNA"; however, this terminology is no longer widely accepted because scientists have since found roles for some of these regions, many of which contribute to the regulation of transcription or translation through the production of small noncoding RNA molecules, DNA packaging, and chromosomal stability. Although scientists may not fully understand the roles of all noncoding regions of DNA, it is generally believed that they do have purposes within the cell.

Extrachromosomal DNA

Although most DNA is contained within a cell's chromosomes, many cells have additional molecules of DNA outside the chromosomes, called **extrachromosomal DNA**, that are also part of its genome. The genomes of eukaryotic cells would also include the chromosomes from any organelles such as mitochondria and/or chloroplasts that these cells maintain (Figure 12.34).

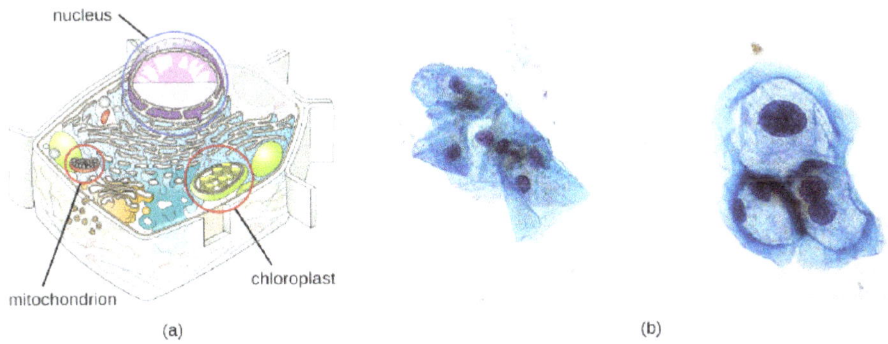

Figure 12.34 The genome of a eukaryotic cell consists of the chromosome housed in the nucleus, and extra chromosomal DNA found in the mitochondria (all cells) and chloroplasts (plants and algae). The cells shown in represent cells obtained from a pap smear. The cells on the left are normal squamous cells whereas the cells on the right are infected with human papillomavirus and show enlarged nuclei with increased staining (hyperchromasia).

The maintenance of circular chromosomes in these organelles is a vestige of their prokaryotic origins and supports the endosymbiotic theory (see Foundations of Modern Cell Theory (https://legacy.cnx.org/content/m58791/latest/)). In some cases, genomes of certain DNA viruses can also be maintained independently in host cells during latent viral infection. In these cases, these viruses are another form of extrachromosomal DNA. For example, the human papillomavirus (HPV) may be maintained in infected cells in this way.

Bacteria can exchange these plasmids with other bacteria in a process known as horizontal gene transfer (HGT). The exchange of genetic material on plasmids sometimes provides microbes with new genes beneficial for growth and survival under special conditions. In some cases, genes obtained from plasmids may have clinical implications, encoding virulence factors that give a microbe the ability to cause disease or make a microbe resistant to certain antibiotics. Plasmids are also used heavily in genetic engineering and biotechnology as a way to move genes from one cell to another.

Lethal Plasmids

Maria, a 20-year-old anthropology student from Texas, recently became ill in the African nation of Botswana, where she was conducting research as part of a study-abroad program. Maria's research was focused on traditional African methods of tanning hides for the production of leather. Over a period of three weeks, she visited a tannery daily for several hours to observe and participate in the tanning process. One day, after returning from the tannery, Maria developed a fever, chills, and a headache, along with chest pain, muscle aches, nausea, and other flu-like symptoms. Initially, she was not concerned, but when her fever spiked and she began to cough up blood, her African host family became alarmed and rushed her to the hospital, where her condition continued to worsen.

After learning about her recent work at the tannery, the physician suspected that Maria had been exposed to anthrax. He ordered a chest X-ray, a blood sample, and a spinal tap, and immediately started her on a course of intravenous penicillin. Unfortunately, lab tests confirmed the physician's presumptive diagnosis. Maria's chest X-ray exhibited pleural effusion, the accumulation of fluid in the space between the pleural membranes, and a Gram stain of her blood revealed the presence of gram-positive, rod-shaped bacteria in short chains, consistent with *Bacillus anthracis*. Blood and bacteria were also shown to be present in her cerebrospinal fluid, indicating that the infection had progressed to meningitis. Despite supportive treatment and aggressive antibiotic therapy, Maria slipped into an unresponsive state and died three days later.

Anthrax is a disease caused by the introduction of endospores from the gram-positive bacterium *B. anthracis* into the body. Once infected, patients typically develop meningitis, often with fatal results. In Maria's case, she inhaled the endospores while handling the hides of animals that had been infected.

The genome of *B. anthracis* illustrates how small structural differences can lead to major differences in virulence. In 2003, the genomes of *B. anthracis* and *Bacillus cereus*, a similar but less pathogenic bacterium of the same genus, were sequenced and compared.[18] Researchers discovered that the 16S rRNA gene sequences of these bacteria are more than 99% identical, meaning that they are actually members of the same species despite their traditional classification as separate species. Although their chromosomal sequences also revealed a great deal of similarity, several virulence factors of *B. anthracis* were found to be encoded on two large plasmids not found in *B. cereus*. The plasmid pX01 encodes a three-part toxin that suppresses the host immune system, whereas the plasmid pX02 encodes a capsular polysaccharide that further protects the bacterium from the host immune system (Figure 12.35). Since *B. cereus* lacks these plasmids, it does not produce these virulence factors, and although it is still pathogenic, it is typically associated with mild cases of diarrhea from which the body can quickly recover. Unfortunately for Maria, the presence of these toxin-encoding plasmids in *B. anthracis* gives it its lethal virulence.

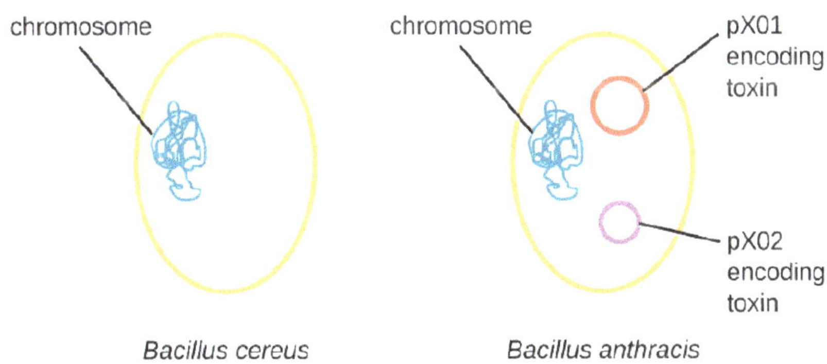

Figure 12.35 Genome sequencing of *Bacillus anthracis* and its close relative *B. cereus* reveals that the pathogenicity of *B. anthracis* is due to the maintenance of two plasmids, pX01 and pX02, which encode virulence factors.

Resolution

Within 24 hours, the results of the diagnostic test analysis of Alex's stool sample revealed that it was positive for heat-labile enterotoxin (LT), heat-stabile enterotoxin (ST), and colonization factor (CF), confirming the hospital physician's suspicion of ETEC. During a follow-up with Alex's family physician, this physician noted that Alex's symptoms were not resolving quickly and he was experiencing discomfort that was preventing him from returning to classes. The family physician prescribed Alex a course of ciprofloxacin to resolve his symptoms. Fortunately, the ciprofloxacin resolved Alex's symptoms within a few days.

Alex likely got his infection from ingesting contaminated food or water. Emerging industrialized countries like Mexico are still developing sanitation practices that prevent the contamination of water with fecal material. Travelers in such countries should avoid the ingestion of undercooked foods, especially meats, seafood, vegetables, and unpasteurized dairy products. They should also avoid use of water that has not been treated; this includes drinking water, ice cubes, and even water used for brushing teeth. Using bottled water for these purposes is a good alternative. Good hygiene (handwashing) can also aid the prevention of an ETEC infection. Alex had not been careful about his food or water consumption, which led to his illness.

Alex's symptoms were very similar to those of cholera, caused by the gram-negative bacterium *Vibrio cholerae*, which also produces a toxin similar to ST and LT. At some point in the evolutionary history of ETEC, a nonpathogenic strain of *E. coli* similar to those typically found in the gut may have acquired the genes encoding the ST and LT toxins from *V. cholerae*. The fact that the genes encoding those toxins are encoded on extrachromosomal plasmids in ETEC supports the idea that these genes were acquired by *E. coli* and are likely maintained in bacterial populations through horizontal gene transfer.

Viral Genomes

Viral genomes exhibit significant diversity in structure. Some viruses have genomes that consist of DNA as their genetic material. This DNA may be single stranded, as exemplified by human parvoviruses, or double stranded, as seen in the herpesviruses and poxviruses. Additionally, although all cellular life uses DNA as its genetic material, some viral genomes are made of either single-stranded or double-stranded RNA molecules, as we have discussed. Viral genomes are typically smaller than most bacterial genomes, encoding only a few genes, because they rely on their hosts to carry out many of the functions required for their replication. The diversity of viral genome structures and their implications for viral replication life cycles are discussed in more detail in The Viral Life Cycle (https://legacy.cnx.org/content/m58809/latest/) .

Genome Size Matters

There is great variation in size of genomes among different organisms. Most eukaryotes maintain multiple chromosomes; humans, for example have 23 pairs, giving them 46 chromosomes. Despite being large at 3 billion base pairs, the human genome is far from the largest genome. Plants often maintain very large genomes, up to 150 billion base pairs, and commonly are polyploid, having multiple copies of each chromosome. The size of bacterial genomes also varies considerably, although they tend to be smaller than eukaryotic genomes (Figure 12.36). Some bacterial genomes may be as small as only 112,000 base pairs. Often, the size of a bacterium's genome directly relates to how much the bacterium depends on its host for survival. When a bacterium relies on the host cell to carry out certain functions, it loses the genes encoding the abilities to carry out those functions itself. These types of bacterial endosymbionts are reminiscent of the prokaryotic origins of mitochondria and chloroplasts.

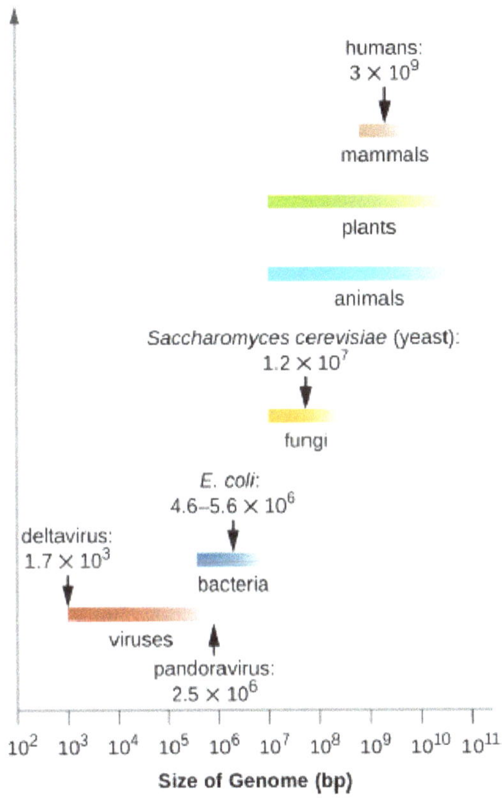

Figure 12.36 There is great variability as well as overlap among the genome sizes of various groups of organisms and viruses.

From a clinical perspective, obligate and facultative intracellular pathogens also tend to have small genomes (some around 1 million base pairs). Because host cells can supply most of their nutrients, they tend to have a reduced number of genes encoding metabolic functions, making their cultivation in the laboratory difficult if not impossible Due to their small sizes, the genomes of organisms like *Mycoplasma genitalium* (580,000 base pairs), *Chlamydia trachomatis* (1.0 million), *Rickettsia prowazekii* (1.1 million), and *Treponema pallidum* (1.1 million) were some of the earlier bacterial genomes sequenced. Respectively, these pathogens cause urethritis and pelvic inflammation, chlamydia, typhus, and syphilis.

Whereas obligate intracellular pathogens have unusually small genomes, other bacteria with a great variety of metabolic and enzymatic capabilities have unusually large bacterial genomes. *Pseudomonas aeruginosa*, for example, is a bacterium commonly found in the environment and is able to grow on a wide range of substrates. Its genome contains 6.3 million base pairs, giving it a high metabolic ability and the ability to produce virulence factors that cause several types of opportunistic infections.

Interestingly, there has been significant variability in genome size in viruses as well, ranging from 3,500 base pairs to 2.5 million base pairs, significantly exceeding the size of many bacterial genomes. The great variation observed in viral genome sizes further contributes to the great diversity of viral genome characteristics already discussed.

MAPPING GENOMES

Genomics is the study of entire genomes, including the complete set of genes, their nucleotide sequence and organization, and their interactions within a species and with other species. **Genome mapping** is the process of finding the locations of genes on each chromosome. The maps created by genome mapping are comparable to the maps that we use to navigate streets. A **genetic map** is an illustration that lists genes and their location on a chromosome. Genetic maps provide the big picture (similar to a map of interstate highways) and use genetic markers (similar to landmarks). A **genetic marker** is a gene or sequence on a chromosome that co-segregates (shows genetic linkage) with a specific trait. Early geneticists called this linkage analysis. Physical maps present the intimate details of smaller regions of the chromosomes (similar to a detailed road map). A **physical map** is a representation of the physical distance, in nucleotides, between genes or genetic markers. Both genetic linkage maps and physical maps are required to build a complete picture of the genome. Having a complete map of the genome makes it easier for researchers to study individual genes. Human genome maps help researchers in their efforts to identify human disease-causing genes related to illnesses like cancer, heart disease, and cystic fibrosis. Genome mapping can be used in a variety of other applications, such as using live microbes to clean up pollutants or even prevent pollution. Research involving plant genome mapping may lead to producing higher crop yields or developing plants that better adapt to climate change.

Genetic Maps

The study of genetic maps begins with **linkage analysis**, a procedure that analyzes the recombination frequency between genes to determine if they are linked or show independent assortment. The term *linkage* was used before the discovery of DNA.

Early geneticists relied on the observation of phenotypic changes to understand the genotype of an organism. Shortly after Gregor Mendel (the father of modern genetics) proposed that traits were determined by what are now known as genes, other researchers observed that different traits were often inherited together, and thereby deduced that the genes were physically linked by being located on the same chromosome. The mapping of genes relative to each other based on linkage analysis led to the development of the first genetic maps.

Observations that certain traits were always linked and certain others were not linked came from studying the offspring of crosses between parents with different traits. For example, in experiments performed on the garden pea, it was discovered that the colour of the flower and shape of the plant's pollen were linked traits, and therefore the genes encoding these traits were in close proximity on the same chromosome. The exchange of DNA between homologous pairs of chromosomes is called **genetic recombination**, which occurs by the crossing over of DNA between homologous strands of DNA, such as nonsister chromatids. Linkage analysis involves studying the recombination frequency between any two genes. The greater the distance between two genes, the higher the chance that a recombination event will occur between them, and the higher the recombination frequency between them. Two possibilities for recombination between two nonsister chromatids during meiosis are shown in Figure 12.37. If the recombination frequency between two genes is less than 50 percent, they are said to be linked.

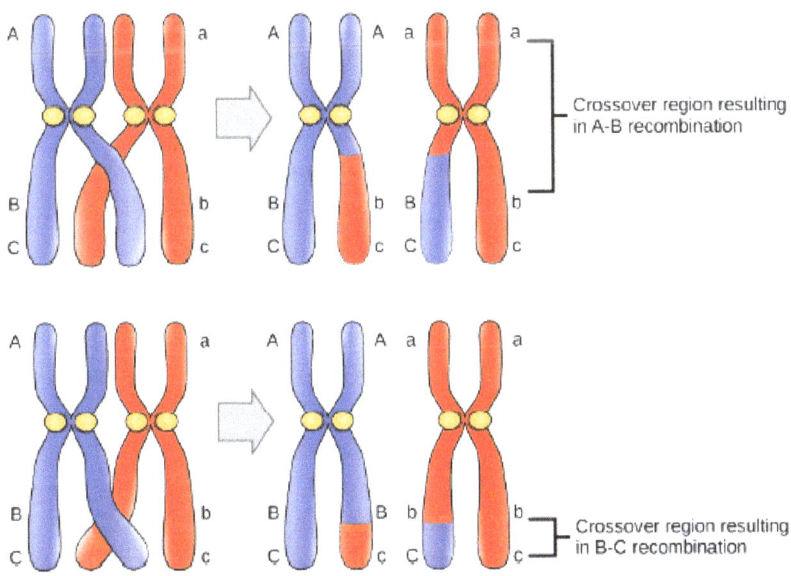

Figure 12.37 Crossover may occur at different locations on the chromosome. Recombination between genes *A* and *B* is more frequent than recombination between genes *B* and *C* because genes *A* and *B* are farther apart; a crossover is therefore more likely to occur between them.

The generation of genetic maps requires markers, just as a road map requires landmarks (such as rivers and mountains). Early genetic maps were based on the use of known genes as markers. More sophisticated markers, including those based on non-coding DNA, are now used to compare the genomes of individuals in a population. Although individuals of a given species are genetically similar, they are not identical; every individual has a unique set of traits. These minor differences in the genome between individuals in a population are useful for the purposes of genetic mapping. In general, a good genetic marker is a region on the chromosome that shows variability or polymorphism (multiple forms) in the population.

Some genetic markers used in generating genetic maps are **restriction fragment length polymorphisms** (RFLP), variable number of tandem repeats (VNTRs), **microsatellite polymorphisms**, and the **single nucleotide polymorphisms** (SNPs). RFLPs (sometimes pronounced "rif-lips") are detected when the DNA of an individual is cut with a restriction endonuclease that recognizes specific sequences in the DNA to generate a series of DNA fragments, which are then analyzed by gel electrophoresis. The DNA of every individual will give rise to a unique pattern of bands when cut with a particular set of restriction endonucleases; this is sometimes referred to as an individual's DNA "fingerprint." Certain regions of the chromosome that are subject to polymorphism will lead to the generation of the unique banding pattern. VNTRs are repeated sets of nucleotides present in the non-coding regions of DNA. Non-coding, or "junk," DNA has no known biological function; however, research shows that much of this DNA is actually transcribed. While its function is uncertain, it is certainly active, and it may be involved in the regulation of coding genes. The number of repeats may vary in individual organisms of a population. Microsatellite polymorphisms are similar to VNTRs, but the repeat unit is very small. SNPs are variations in a single nucleotide.

Because genetic maps rely completely on the natural process of recombination, mapping is affected by natural increases or decreases in the level of recombination in any given area of the genome. Some parts of the genome are recombination hotspots, whereas others do not show a propensity for recombination. For this reason, it is important to look at mapping information developed by multiple methods.

Physical Maps

A physical map provides detail of the actual physical distance between genetic markers, as well as the number of nucleotides. There are three methods used to create a physical map: cytogenetic mapping, radiation hybrid mapping, and sequence mapping. **Cytogenetic mapping** uses information obtained by microscopic analysis of stained sections of the chromosome (Figure 12.38). It is possible to determine the approximate distance between genetic markers using cytogenetic mapping, but not the exact distance (number of base pairs). **Radiation hybrid mapping** uses radiation, such as x-rays, to break the DNA into fragments. The amount of radiation can be adjusted to create smaller or larger fragments. This technique overcomes the limitation of genetic mapping and is not affected by increased or decreased recombination frequency. **Sequence mapping** resulted from DNA sequencing technology that allowed for the creation of detailed physical maps with distances measured in terms of the number of base pairs. The creation of **genomic libraries** and **complementary DNA (cDNA) libraries** (collections of cloned sequences or all DNA from a genome) has sped up the process of physical mapping. A genetic site used to generate a physical map with sequencing technology (a sequence- tagged site, or STS) is a unique sequence in the genome with a known exact chromosomal location. An **expressed sequence tag (EST)** and a single sequence length polymorphism (SSLP) are common STSs. An EST is a short STS that is identified with cDNA libraries, while SSLPs are obtained from known genetic markers and provide a link between genetic maps and physical maps.

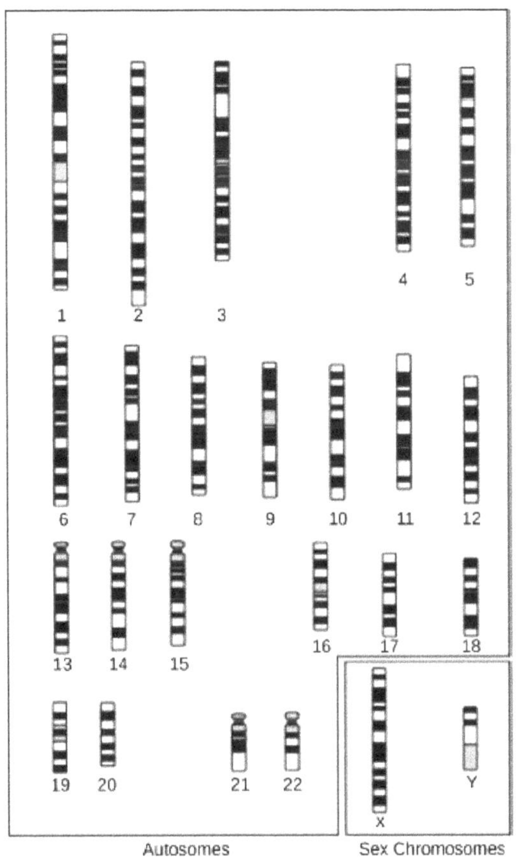

Figure 12.38 A cytogenetic map shows the appearance of a chromosome after it is stained and examined under a microscope. (credit: National Human Genome Research Institute)

Integration of Genetic and Physical Maps

Genetic maps provide the outline and physical maps provide the details. It is easy to understand why both types of genome mapping techniques are important to show the big picture. Information obtained from each technique is used in combination to study the genome. Genomic mapping is being used with different model organisms that are used for research. Genome mapping is still an ongoing process, and as more advanced techniques are developed, more advances are expected. Genome mapping is similar to completing a complicated puzzle using every piece of available data. Mapping information generated in laboratories all over the world is entered into central databases, such as GenBank at the National Center for Biotechnology Information (NCBI). Efforts are being made to make the information more easily accessible to researchers and the general public. Just as we use global positioning systems instead of paper maps to navigate through roadways, NCBI has created a genome viewer tool to simplify the data-mining process.

APPLYING GENOMICS

The introduction of DNA sequencing and whole genome sequencing projects, particularly the Human Genome project, has expanded the applicability of DNA sequence information. Genomics is now being used in a wide variety of fields, such as metagenomics, pharmacogenomics, and mitochondrial genomics. The most commonly known application of genomics is to understand and find cures for diseases.

Predicting Disease Risk at the Individual Level

Predicting the risk of disease involves screening currently healthy individuals by genome analysis at the individual level. Intervention with lifestyle changes and drugs can be recommended before disease onset. However, this approach is most applicable when the problem resides within a single gene defect. Such defects only account for approximately 5 percent of diseases in developed countries. Most of the common diseases, such as heart disease, are multi-factored or **polygenic**, which is a phenotypic characteristic that involves two or more genes, and also involve environmental factors such as diet. In April 2010, scientists at Stanford University published the genome analysis of a healthy individual (Stephen Quake, a scientist at Stanford University, who had his genome sequenced); the analysis predicted his propensity to acquire various diseases. A risk assessment was performed to analyze Quake's percentage of risk for 55 different medical conditions. A rare genetic mutation was found, which showed him to be at risk for sudden heart attack. He was also predicted to have a 23 percent risk of developing prostate cancer and a 1.4 percent risk of developing Alzheimer's. The scientists used databases and several publications to analyze the genomic data. Even though genomic sequencing is becoming more affordable and analytical tools are becoming more reliable, ethical issues surrounding genomic analysis at a population level remain to be addressed.

Art Connection

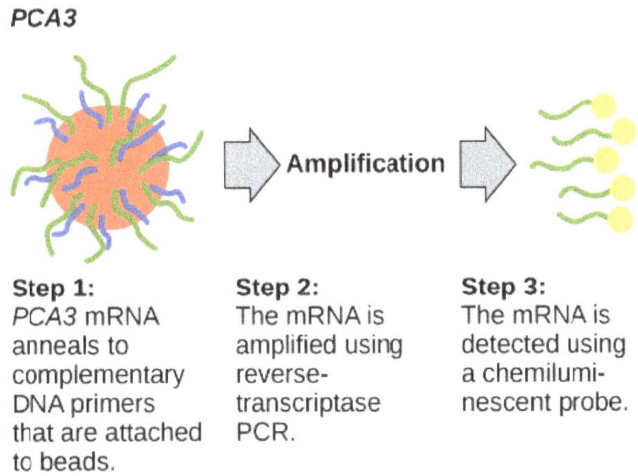

Figure 12.39 PCA3 is a gene that is expressed in prostate epithelial cells and over expressed in cancerous cells. A high concentration of *PCA3* in urine is indicative of prostate cancer. The *PCA3* test is considered to be a better indicator of cancer than the more well know PSA test, which measures the level of PSA (prostate-specific antigen) in the blood.

In 2011, the United States Preventative Services Task Force recommended against using the PSA test to screen healthy men for prostate cancer. Their recommendation is based on evidence that screening does not reduce the risk of death from prostate cancer. Prostate cancer often develops very slowly and does not cause problems, while the cancer treatment can have severe side effects. The *PCA3* test is considered to be more accurate, but screening may still result in men who would not have been harmed by the cancer itself suffering side effects from treatment. What do you think? Should all healthy men be screened for prostate cancer using the *PCA3* or PSA test? Should people in general be screened to find out if they have a genetic risk for cancer or other diseases?

Pharmacogenomics and Toxicogenomics

Pharmacogenomics, also called toxicogenomics, involves evaluating the effectiveness and safety of drugs on the basis of information from an individual's genomic sequence. Genomic responses to drugs can be studied using experimental animals (such as laboratory rats or mice) or live cells in the laboratory before embarking on studies with humans. Studying changes in gene expression could provide information about the transcription profile in the presence of the drug, which can be used as an early indicator of the potential for toxic effects. For example, genes involved in cellular growth and controlled cell death, when disturbed, could lead to the growth of cancerous cells. Genome-wide studies can also help to find new genes involved in drug toxicity. Personal genome sequence information can be used to prescribe medications that will be most effective and least toxic on the basis of the individual patient's genotype. The gene signatures may not be completely accurate, but can be tested further before pathologic symptoms arise.

Microbial Genomics: Metagenomics

Traditionally, microbiology has been taught with the view that microorganisms are best studied under **pure culture** conditions, which involves isolating a single type of cell and culturing it in the laboratory. Because microorganisms can go through several generations in a matter of hours, their gene expression profiles adapt to the new laboratory environment very quickly. In addition, the vast majority of bacterial species resist being cultured in isolation. Most microorganisms do not live as isolated entities, but in microbial communities known as biofilms. For all of these reasons, pure culture is not always the best way to study microorganisms. **Metagenomics** is the study of the collective genomes of multiple species that grow and interact in an environmental niche. Metagenomics can be used to identify new species more rapidly and to analyze the effect of pollutants on the environment (Figure 12.40).

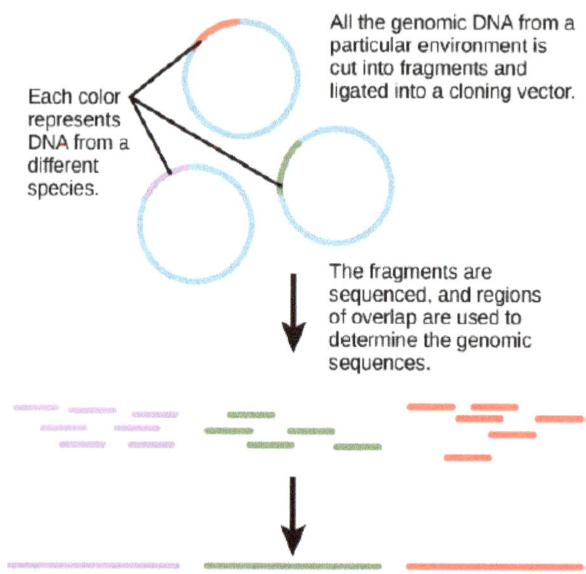

Figure 12.40 Metagenomics involves isolating DNA from multiple species within an environmental niche.

Microbial Genomics: Creation of New Biofuels

Knowledge of the genomics of microorganisms is being used to find better ways to harness biofuels from algae and cyanobacteria. The primary sources of fuel today are coal, oil, wood, and other plant products, such as ethanol. Although plants are renewable resources, there is still a need to find more alternative renewable sources of energy to meet our population's energy demands. The microbial world is one of the largest resources for genes that encode new enzymes and produce new organic compounds, and it remains largely untapped. Microorganisms are used to create products, such as enzymes that are used in research, antibiotics, and other anti-microbial mechanisms. Microbial genomics is helping to develop diagnostic tools, improved vaccines, new disease treatments, and advanced environmental cleanup techniques.

Mitochondrial Genomics

Mitochondria are intracellular organelles that contain their own DNA. Mitochondrial DNA mutates at a rapid rate and is often used to study evolutionary relationships. Another feature that makes studying the mitochondrial genome interesting is that the mitochondrial DNA in most multicellular organisms is passed on from the mother during the process of fertilization. For this reason, mitochondrial genomics is often used to trace genealogy.

Information and clues obtained from DNA samples found at crime scenes have been used as evidence in court cases, and genetic markers have been used in forensic analysis. Genomic analysis has also become useful in this field. In 2001, the first use of genomics in forensics was published. It was a collaborative attempt between academic research institutions and the FBI to solve the mysterious cases of anthrax communicated via the US Postal Service. Using microbial genomics, researchers determined that a specific strain of anthrax was used in all the mailings.

Genomics in Agriculture

Genomics can reduce the trials and failures involved in scientific research to a certain extent, which could improve the quality and quantity of crop yields in agriculture. Linking traits to genes or gene signatures helps to improve crop breeding to generate hybrids with the most desirable qualities. Scientists use genomic data to identify desirable traits, and then transfer those traits to a different organism. Scientists are discovering how genomics can improve the quality and quantity of agricultural production. For example, scientists could use desirable traits to create a useful product or enhance an existing product, such as making a drought-sensitive crop more tolerant of the dry season.

DNA Replication

DNA replicates by a semi-conservative method in which each of the two parental DNA strands act as a template for new DNA to be synthesized. After replication, each DNA has one parental or "old" strand, and one daughter or "new" strand.

Replication in eukaryotes starts at multiple origins of replication, while replication in prokaryotes starts from a single origin of replication. The DNA is opened with enzymes, resulting in the formation of the replication fork. Primase synthesizes an RNA primer to initiate synthesis by DNA polymerase, which can add nucleotides in only one direction. One strand is synthesized continuously in the direction of the replication fork; this is called the leading strand. The other strand is synthesized in a direction away from the replication fork, in short stretches of DNA known as Okazaki fragments. This strand is known as the lagging strand. Once replication is completed, the RNA primers are replaced by DNA nucleotides and the DNA is sealed with DNA ligase.

The ends of eukaryotic chromosomes pose a problem, as polymerase is unable to extend them without a primer. Telomerase, an enzyme with an inbuilt RNA template, extends the ends by copying the RNA template and extending one end of the chromosome. DNA polymerase can then extend the DNA using the primer. In this way, the ends of the chromosomes are protected. Cells have mechanisms for repairing DNA when it becomes damaged or errors are made in replication. These mechanisms include mismatch repair to replace nucleotides that are paired with a non-complementary base and nucleotide excision repair, which removes bases that are damaged such as thymine dimers.

Structure and Function of RNA

- **Ribonucleic acid (RNA)** is typically single stranded and contains ribose as its pentose sugar and the pyrimidine uracil instead of thymine. An RNA strand can undergo significant intramolecular base pairing to take on a three-dimensional structure.
- There are three main types of RNA, all involved in protein synthesis.
- Messenger RNA (**mRNA**) serves as the intermediary between DNA and the synthesis of protein products during translation.
- Ribosomal RNA (**rRNA**) is a type of stable RNA that is a major constituent of ribosomes. It ensures the proper alignment of the mRNA and the ribosomes during protein synthesis and catalyzes the formation of the peptide bonds between two aligned amino acids during protein synthesis.
- Transfer RNA (**tRNA**) is a small type of stable RNA that carries an amino acid to the corresponding site of protein synthesis in the ribosome. It is the base pairing between the tRNA and mRNA that allows for the correct amino acid to be inserted in the polypeptide chain being synthesized.
- Although RNA is not used for long-term genetic information in cells, many viruses do use RNA as their genetic material.

Transcription

In prokaryotes, mRNA synthesis is initiated at a promoter sequence on the DNA template. Elongation synthesizes new mRNA. Termination liberates the mRNA and occurs by mechanisms that stall the RNA polymerase and cause it to fall off the DNA template. Newly transcribed eukaryotic mRNAs are modified with a cap and a poly-A tail. These structures protect the mature mRNA from degradation and help export it from the nucleus. Eukaryotic mRNAs also undergo splicing, in which introns are removed and exons are reconnected with single-nucleotide accuracy. Only finished mRNAs are exported from the nucleus to the cytoplasm.

Translation

The central dogma describes the flow of genetic information in the cell from genes to mRNA to proteins. Genes are used to make mRNA by the process of transcription; mRNA is used to synthesize proteins by the process of translation. The genetic code is the correspondence between the three-nucleotide mRNA codon and an amino acid. The genetic code is "translated" by the tRNA molecules, which associate a specific codon with a specific amino acid. The genetic code is degenerate because 64 triplet codons in mRNA specify only 20 amino acids and three stop codons. This means that more than one codon corresponds to an amino acid. Almost every species on the planet uses the same genetic code.

The players in translation include the mRNA template, ribosomes, tRNAs, and various enzymatic factors. The small ribosomal subunit binds to the mRNA template. Translation begins at the initiating AUG on the mRNA. The formation of bonds occurs between sequential amino acids specified by the mRNA template according to the genetic code. The ribosome accepts charged tRNAs, and as it steps along the mRNA, it catalyzes bonding between the new amino acid and the end of the growing polypeptide. The entire mRNA is translated in three- nucleotide "steps" of the ribosome. When a stop codon is encountered, a release factor binds and dissociates the components and frees the new protein.

How Genes Are Regulated

While all somatic cells within an organism contain the same DNA, not all cells within that organism express the same proteins. Prokaryotic organisms express the entire DNA they encode in every cell, but not necessarily all at the same time. Proteins are expressed only when they are needed. Eukaryotic organisms express a subset of the DNA that is encoded in any given cell. In each cell type, the type and amount of protein is regulated by controlling gene expression. To express a protein, the DNA is first transcribed into RNA, which is then translated into proteins. In prokaryotic cells, these processes occur almost simultaneously. In eukaryotic cells, transcription occurs in the nucleus and is separate from the translation that occurs in the cytoplasm. Gene expression in prokaryotes is regulated only at the transcriptional level, whereas in eukaryotic cells, gene expression is regulated at the epigenetic, transcriptional, post-transcriptional, translational, and post-translational levels.

Structure and Function of Cellular Genomes

The entire genetic content of a cell is its **genome.**

- **Genes** code for proteins, or stable RNA molecules, each of which carries out a specific function in the cell.
- Although the **genotype** that a cell possesses remains constant, expression of genes is dependent on environmental conditions.
- A **phenotype** is the observable characteristics of a cell (or organism) at a given point in time and results from the complement of genes currently being used.
- The majority of genetic material is organized into **chromosomes** that contain the DNA that controls cellular activities.
- Prokaryotes are typically haploid, usually having a single circular chromosome found in the nucleoid. Eukaryotes are diploid; DNA is organized into multiple linear chromosomes found in the nucleus.
- Supercoiling and DNA packaging using DNA binding proteins allows lengthy molecules to fit inside a cell. Eukaryotes and archaea use histone proteins, and bacteria use different proteins with similar function.
- Prokaryotic and eukaryotic genomes both contain **noncoding DNA**, the function of which is not well understood. Some noncoding DNA appears to participate in the formation of small noncoding RNA molecules that influence gene expression; some appears to play a role in maintaining chromosomal structure and in DNA packaging.
- **Extrachromosomal DNA** in eukaryotes includes the chromosomes found within organelles of prokaryotic origin (mitochondria and chloroplasts) that evolved by endosymbiosis. Some viruses may also maintain themselves extrachromosomally.
- Extrachromosomal DNA in prokaryotes is commonly maintained as **plasmids** that encode a few nonessential genes that may be helpful under specific conditions. Plasmids can be spread through a bacterial community by horizontal gene transfer.
- Viral genomes show extensive variation and may be composed of either RNA or DNA, and may be either double or single stranded.

Mapping Genomes

Genome mapping is similar to solving a big, complicated puzzle with pieces of information coming from laboratories all over the world. Genetic maps provide an outline for the location of genes within a genome, and they estimate the distance between genes and genetic markers on the basis of recombination frequencies during meiosis. Physical maps provide detailed information about the physical distance between the genes. The most detailed information is available through sequence mapping. Information from all mapping and sequencing sources is combined to study an entire genome.

Applying Genomics

Imagination is the only barrier to the applicability of genomics. Genomics is being applied to most fields of biology; it is being used for personalized medicine, prediction of disease risks at an individual level, the study of drug interactions before the conduct of clinical trials, and the study of microorganisms in the environment as opposed to the laboratory. It is also being applied to developments such as the generation of new biofuels, genealogical assessment using mitochondria, advances in forensic science, and improvements in agriculture.

CHAPTER-13

PROTEIN: STRUCTURE AND FUNCTIONS

INTRODUCTION: AN OVERVIEW

Since the rediscovery of Mendel's work in 1900, the definition of the gene has progressed from an abstract unit of heredity to a tangible molecular entity capable of replication, expression, and mutation. Genes are composed of DNA and are linearly arranged on chromosomes. Genes specify the sequences of amino acids, which are the building blocks of proteins. In turn, proteins are responsible for orchestrating nearly every function of the cell. Both genes and the proteins they encode are absolutely essential to life as we know it.

Proteins are one of the most abundant organic molecules in living systems and have the most diverse range of functions of all macromolecules. Proteins may be structural, regulatory, contractile, or protective; they may serve in transport, storage, or membranes; or they may be toxins or enzymes. Each cell in a living system may contain thousands of proteins, each with a unique function. Their structures, like their functions, vary greatly. They are all, however, polymers of amino acids, arranged in a linear sequence.

TYPES AND FUNCTIONS OF PROTEINS

Enzymes, which are produced by living cells, are catalysts in biochemical reactions (like digestion) and are usually complex or conjugated proteins. Each enzyme is specific for the substrate (a reactant that binds to an enzyme) it acts on. The enzyme may help in breakdown, rearrangement, or synthesis reactions. Enzymes that break down their substrates are called catabolic enzymes, enzymes that build more complex molecules from their substrates are called anabolic enzymes, and enzymes that affect the rate of reaction are called catalytic enzymes. It should be noted that all enzymes increase the rate of reaction and, therefore, are considered to be organic catalysts. An example of an enzyme is salivary amylase, which hydrolyzes its substrate amylose, a component of starch.

Hormones are chemical-signaling molecules, usually small proteins or steroids, secreted by endocrine cells that act to control or regulate specific physiological processes, including growth, development, metabolism, and reproduction. For example, insulin is a protein hormone that helps to regulate the blood glucose level. The primary types and functions of proteins are listed in Table 13.1.

Table 13.1 Protein Types and Functions

Type	Examples	Functions
Digestive Enzymes	Amylase, lipase, pepsin, trypsin	Help in digestion of food by catabolizing nutrients into monomeric units
Transport	Hemoglobin, albumin	Carry substances in the blood or lymph throughout the body
Structural	Actin, tubulin, keratin	Construct different structures, like the cytoskeleton
Hormones	Insulin, thyroxine	Coordinate the activity of different body systems
Defense	Immunoglobulins	Protect the body from foreign pathogens
Contractile	Actin, myosin	Effect muscle contraction
Storage	Legume storage proteins, egg white (albumin)	Provide nourishment in early development of the embryo and the seedling

Proteins have different shapes and molecular weights; some proteins are globular in shape whereas others are fibrous in nature. For example, hemoglobin is a globular protein, but collagen, found in our skin, is a fibrous protein. Protein shape is critical to its function, and this shape is maintained by many different types of chemical bonds. Changes in temperature, pH, and exposure to chemicals may lead to permanent changes in the shape of the protein, leading to loss of function, known as **denaturation.** All proteins are made up of different arrangements of the same 20 types of amino acids.

Amino Acids

Amino acids are the monomers that make up proteins. Each amino acid has the same fundamental structure, which consists of a central carbon atom, also known as the alpha (α) carbon, bonded to an amino group (NH_2), a carboxyl group (COOH), and to a hydrogen atom. Every amino acid also has another atom or group of atoms bonded to the central atom known as the R group (Figure 13.1).

The name "amino acid" is derived from the fact that they contain both amino group and carboxyl-acid-group in their basic structure. As mentioned, there are 20 amino acids present in proteins. Nine of these are considered essential amino acids in humans because the human body cannot produce them and they are obtained from the diet. For each amino acid, the R group (or side chain) is different (Figure 13.2).

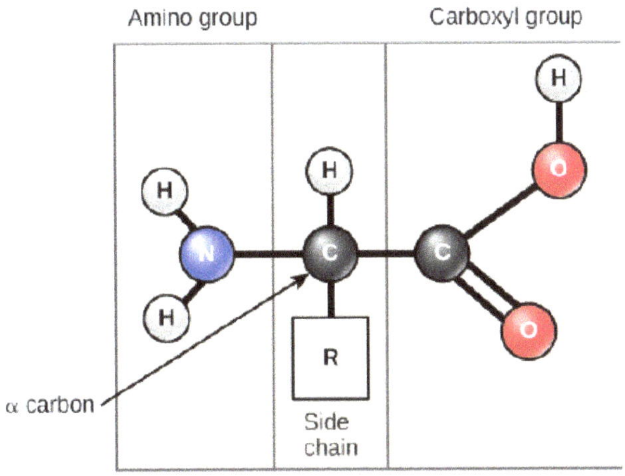

Figure 13.1 Amino acids have a central asymmetric carbon to which an amino group, a carboxyl group, a hydrogen atom, and a side chain (R group) are attached.

Art Connection

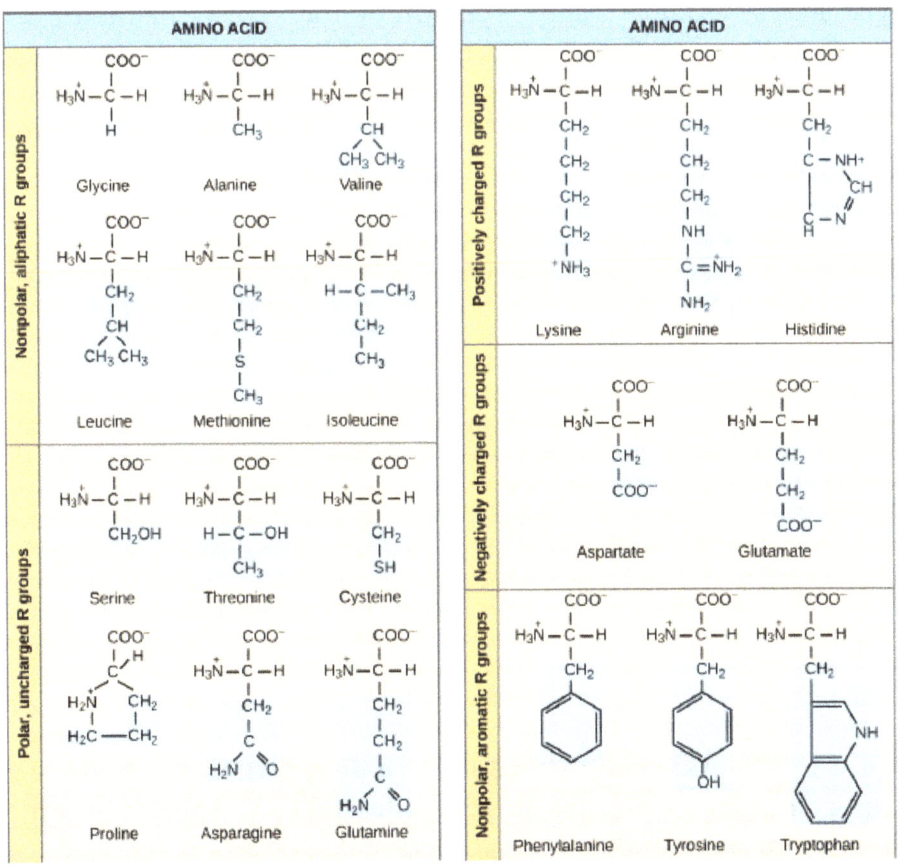

Figure 13.2 There are 20 common amino acids commonly found in proteins, each with a different R group (variant group) that determines its chemical nature.

Which categories of amino acid would you expect to find on the surface of a soluble protein, and which would you expect to find in the interior? What distribution of amino acids would you expect to find in a protein embedded in a lipid bilayer?

The chemical nature of the side chain determines the nature of the amino acid (that is, whether it is acidic, basic, polar, or nonpolar). For example, the amino acid glycine has a hydrogen atom as the R group. Amino acids such as valine, methionine, and alanine are nonpolar or hydrophobic in nature, while amino acids such as serine, threonine, and cysteine are polar and have hydrophilic side chains. The side chains of lysine and arginine are positively charged, and therefore these amino acids are also known as basic amino acids. Proline has an R group that is linked to the amino group, forming a ring-like structure. Proline is an exception to the standard structure of an amino acid since its amino group is not separate from the side chain (Figure 13.2).

Amino acids are represented by a single upper case letter or a three-letter abbreviation. For example, valine is known by the letter V or the three-letter symbol val. Just as some fatty acids are essential to a diet, some amino acids are necessary as well. They are known as essential amino acids, and in humans they include isoleucine, leucine, and cysteine. Essential amino acids refer to those necessary for construction of proteins in the body, although not produced by the body; which amino acids are essential varies from organism to organism.

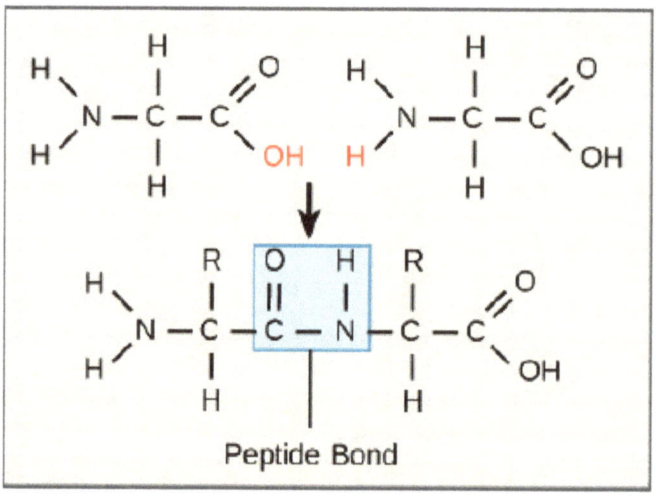

Figure 13.3 Peptide bond formation is a dehydration synthesis reaction. The carboxyl group of one amino acid is linked to the amino group of the incoming amino acid. In the process, a molecule of water is released.

The sequence and the number of amino acids ultimately determine the protein's shape, size, and function. Each amino acid is attached to another amino acid by a covalent bond, known as a **peptide bond**, which is formed by a dehydration reaction. The carboxyl group of one amino acid and the amino group of the incoming amino acid combine, releasing a molecule of water. The resulting bond is the peptide bond (Figure 13.3).

The products formed by such linkages are called peptides. As more amino acids join to this growing chain, the resulting chain is known as a polypeptide. Each polypeptide has a free amino group at one end. This end is called the N terminal, or the amino terminal, and the other end has a free carboxyl group, also known as the C or carboxyl terminal. While the terms polypeptide and protein are sometimes used interchangeably, a polypeptide is technically a polymer of amino acids, whereas the term protein is used for a polypeptide or polypeptides that have combined together, often have bound non-peptide prosthetic groups, have a distinct shape, and have a unique function. After protein synthesis (translation), most proteins are modified. These are known as post- translational modifications. They may undergo cleavage, phosphorylation, or may require the addition of other chemical groups. Only after these modifications is the protein completely functional.

The Evolutionary Significance of Cytochrome C

Cytochrome c is an important component of the electron transport chain, a part of cellular respiration, and it is normally found in the cellular organelle, the mitochondrion. This protein has a heme prosthetic group, and the central ion of the heme gets alternately reduced and oxidized during electron transfer. Because this essential protein's role in producing cellular energy is crucial, it has changed very little over millions of years. Protein sequencing has shown that there is a considerable amount of cytochrome c amino acid sequence homology among different species; in other words, evolutionary kinship can be assessed by measuring the similarities or differences among various species' DNA or protein sequences.

Scientists have determined that human cytochrome c contains 104 amino acids. For each cytochrome c molecule from different organisms that has been sequenced to date, 37 of these amino acids appear in the same position in all samples of cytochrome c. This indicates that there may have been a common ancestor. On comparing the human and chimpanzee protein sequences, no sequence difference was found. When human and rhesus monkey sequences were compared, the single difference found was in one amino acid. In another comparison, human to yeast sequencing shows a difference in the 44th position.

Protein Structure

As discussed earlier, the shape of a protein is critical to its function. For example, an enzyme can bind to a specific substrate at a site known as the active site. If this active site is altered because of local changes or changes in overall protein structure, the enzyme may be unable to bind to the substrate. To understand how the protein gets its final shape or conformation, we need to understand the four levels of protein structure: primary, secondary, tertiary, and quaternary.

Primary Structure

The unique sequence of amino acids in a polypeptide chain is its **primary structure**. For example, the pancreatic hormone insulin has two polypeptide chains, A and B, and they are linked together by disulfide bonds. The N terminal amino acid of the A chain is glycine, whereas the C terminal amino acid is asparagine (Figure 13.4). The sequences of amino acids in the A and B chains are unique to insulin.

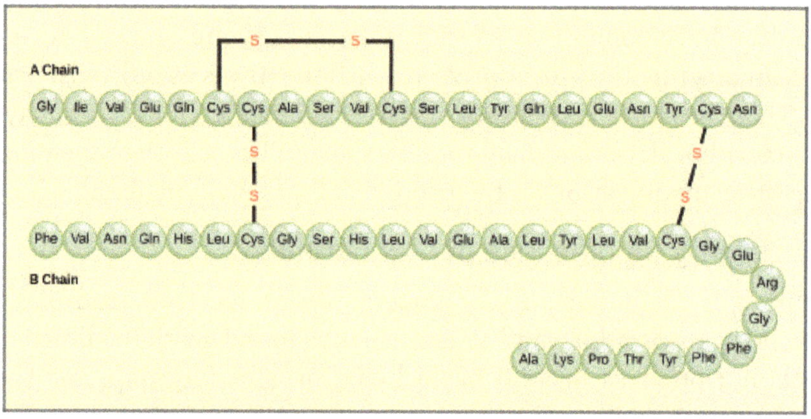

Figure 13.4 Bovine serum insulin is a protein hormone made of two peptide chains, A (21 amino acids long) and B (30 amino acids long). In each chain, primary structure is indicated by three-letter abbreviations that represent the names of the amino acids in the order they are present. The amino acid cysteine (cys) has a sulfhydryl (SH) group as a side chain. Two sulfhydryl groups can react in the presence of oxygen to form a disulfide (S-S) bond. Two disulfide bonds connect the A and B chains together, and a third helps the A chain fold into the correct shape. Note that all disulfide bonds are the same length, but are drawn different sizes for clarity.

The unique sequence for every protein is ultimately determined by the gene encoding the protein. A change in nucleotide sequence of the gene's coding region may lead to a different amino acid being added to the growing polypeptide chain, causing a change in protein structure and function. In sickle cell anemia, the hemoglobin β chain (a small portion of which is shown in Figure 13.5) has a single amino acid substitution, causing a change in protein structure and function. Specifically, the amino acid glutamic acid is substituted by valine in the β chain. What is most remarkable to consider is that a hemoglobin molecule is made up of two alpha chains and two beta chains that each consist of about 150 amino acids. The molecule, therefore, has about 600 amino acids. The structural difference between a normal hemoglobin molecule and a sickle cell molecule—which dramatically decreases life expectancy—is a single amino acid of the 600. What is even more remarkable is that those 600 amino acids are encoded by three nucleotides each, and the mutation is caused by a single base change (point mutation), 1 in 1800 bases.

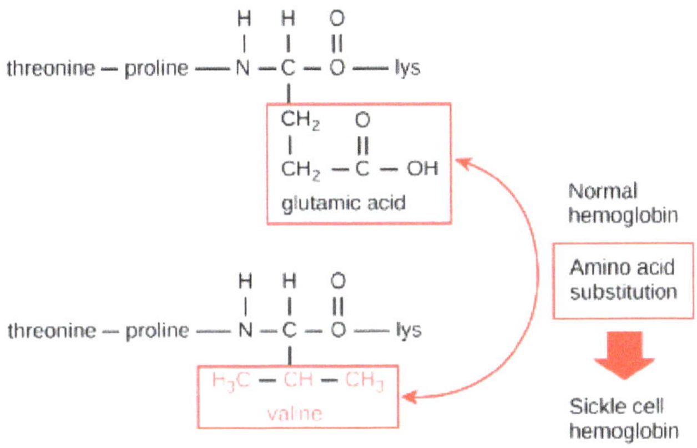

Figure 13.5 The beta chain of hemoglobin is 147 residues in length, yet a single amino acid substitution leads to sickle cell anemia. In normal hemoglobin, the amino acid at position seven is glutamate. In sickle cell hemoglobin, this glutamate is replaced by a valine.

Because of this change of one amino acid in the chain, hemoglobin molecules form long fibers that distort the biconcave, or disc-shaped, red blood cells and assume a crescent or "sickle" shape, which clogs arteries (Figure 13.6). This can lead to myriad serious health problems such as breathlessness, dizziness, headaches, and abdominal pain for those affected by this disease.

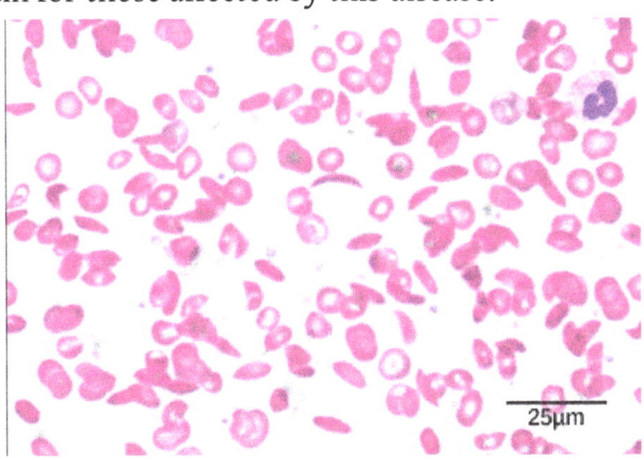

Figure 13.6 In this blood smear, visualized at 535x magnification using bright field microscopy, sickle cells are crescent shaped, while normal cells are disc-shaped. (credit: modification of work by Ed Uthman; scale-bar data from Matt Russell)

Secondary Structure

The local folding of the polypeptide in some regions gives rise to the **secondary structure** of the protein. The most common are the ***α*-helix** and ***β*-pleated sheet** structures (Figure 13.7). Both structures are the *α*-helix structure—the helix held in shape by hydrogen bonds. The hydrogen bonds form between the oxygen atom in the carbonyl group in one amino acid and another amino acid that is four amino acids farther along the chain.

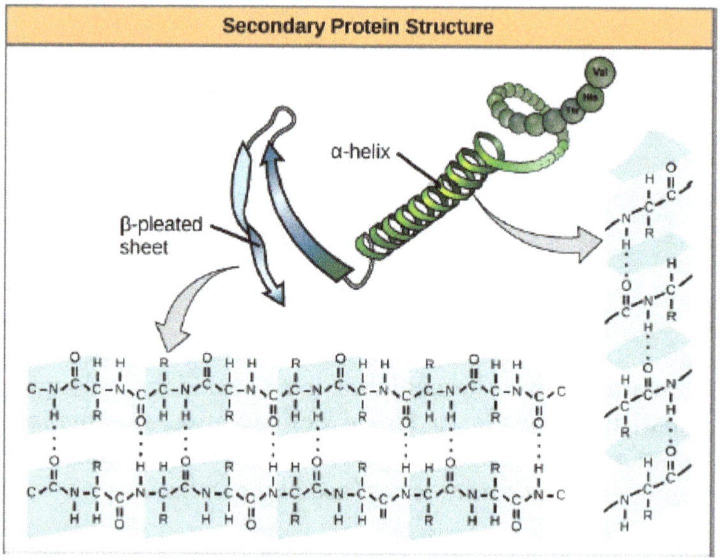

Figure 13.7 The α-helix and β-pleated sheet are secondary structures of proteins that form because of hydrogen bonding between carbonyl and amino groups in the peptide backbone. Certain amino acids have a propensity to form an *α*-helix, while others have a propensity to form a *β*-pleated sheet.

Every helical turn in an alpha helix has 3.6 amino acid residues. The R groups (the variant groups) of the polypeptide protrude out from the *α*-helix chain. In the *β*-pleated sheet, the "pleats" are formed by hydrogen bonding between atoms on the backbone of the polypeptide chain. The R groups are attached to the carbons and extend above and below the folds of the pleat. The pleated segments align parallel or antiparallel to each other, and hydrogen bonds form between the partially positive nitrogen atom in the amino group and the partially negative oxygen atom in the carbonyl group of the peptide backbone. The *α*-helix and *β*-pleated sheet structures are found in most globular and fibrous proteins and they play an important structural role.

Tertiary Structure

The unique three-dimensional structure of a polypeptide is its tertiary structure (Figure 13.8). This structure is in part due to chemical interactions at work on the polypeptide chain. Primarily, the interactions among R groups creates the complex three-dimensional tertiary structure of a protein. The nature of the R groups found in the amino acids involved can counteract the formation of the hydrogen bonds described for standard secondary structures. For example, R groups with like charges are repelled by each other and those with unlike charges are attracted to each other (ionic bonds). When protein folding takes place, the hydrophobic R groups of nonpolar amino acids lay in the interior of the protein, whereas the hydrophilic R groups lay on the outside. The former types of interactions are also known as hydrophobic interactions. Interaction between cysteine side chains forms disulfide linkages in the presence of oxygen, the only covalent bond forming during protein folding.

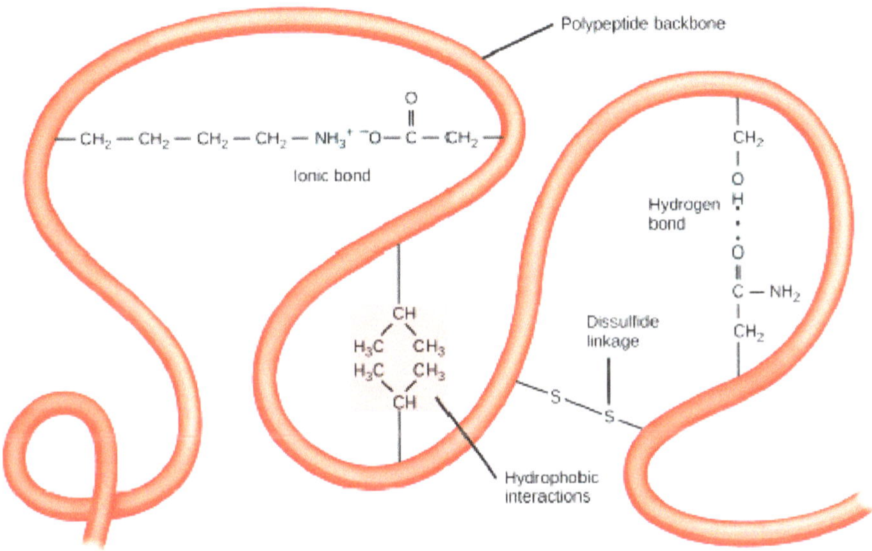

Figure 13.8 The tertiary structure of proteins is determined by a variety of chemical interactions. These include hydrophobic interactions, ionic bonding, hydrogen bonding and disulfide linkages.

All of these interactions, weak and strong, determine the final three-dimensional shape of the protein. When a protein loses its three-dimensional shape, it may no longer be functional.

Quaternary Structure

In nature, some proteins are formed from several polypeptides, also known as subunits, and the interaction of these subunits forms the **quaternary structure.** Weak interactions between the subunits help to stabilize the overall structure. For example, insulin (a globular protein) has a combination of hydrogen bonds and disulfide bonds that cause it to be mostly clumped into a ball shape. Insulin starts out as a single polypeptide and loses some internal sequences in the presence of post-translational modification after the formation of the disulfide linkages that hold the remaining chains together. Silk (a fibrous protein), however, has a β-pleated sheet structure that is the result of hydrogen bonding between different chains.

The four levels of protein structure (primary, secondary, tertiary, and quaternary) are illustrated in Figure 13.9.

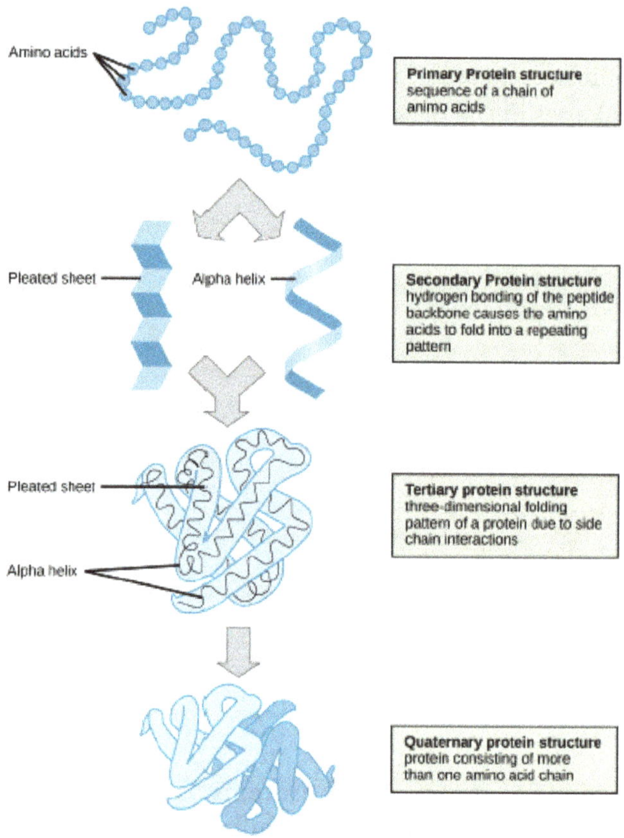

Figure 13.9 The four levels of protein structure can be observed in these illustrations. (credit: modification of work by National Human Genome Research Institute)

Denaturation and Protein Folding

Each protein has its own unique sequence and shape that are held together by chemical interactions. If the protein is subject to changes in temperature, pH, or exposure to chemicals, the protein structure may change, losing its shape without losing its primary sequence in what is known as denaturation. Denaturation is often reversible because the primary structure of the polypeptide is conserved in the process if the denaturing agent is removed, allowing the protein to resume its function. Sometimes denaturation is irreversible, leading to loss of function.

One example of irreversible protein denaturation is when an egg is fried. The albumin protein in the liquid egg white is denatured when placed in a hot pan. Not all proteins are denatured at high temperatures; for instance, bacteria that survive in hot springs have proteins that function at temperatures close to boiling. The stomach is also very acidic, has a low pH, and denatures proteins as part of the digestion process; however, the digestive enzymes of the stomach retain their activity under these conditions.

Protein folding is critical to its function. It was originally thought that the proteins themselves were responsible for the folding process. Only recently was it found that often they receive assistance in the folding process from protein helpers known as **chaperones** (or chaperonins) that associate with the target protein during the folding process.

ENZYMES

A substance that helps a chemical reaction to occur is a catalyst, and the special molecules that catalyze biochemical reactions are enzymes. Almost all enzymes are proteins, comprised of amino acid chains, and they perform the critical task of lowering the activation energies of chemical reactions inside the cell. Enzymes do this by binding to the reactant molecules, and holding them in such a way as to make the chemical bond- breaking and bond-forming processes take place more readily. It is important to remember that enzymes do not change the reaction's ΔG. In other words, they do not change whether a reaction is exergonic (spontaneous) or endergonic. This is because they do not change the reactants' or products' free energy. They only reduce the activation energy required to reach the transition state (Figure 13.10).

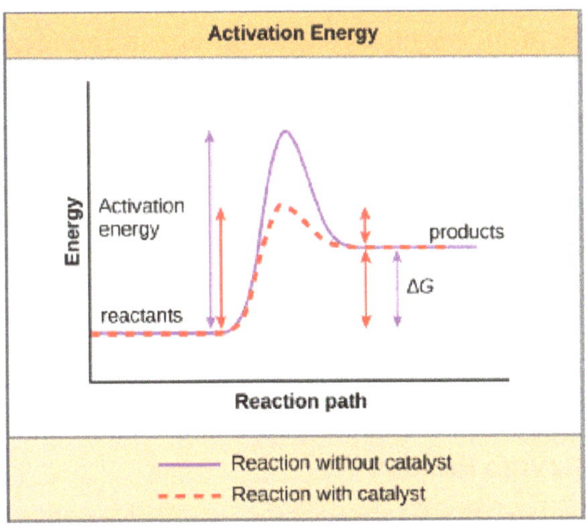

Figure 13.10 Enzymes lower the reaction's activation energy but do not change the reaction's free energy.

Enzyme Active Site and Substrate Specificity

The chemical reactants to which an enzyme binds are the enzyme's **substrates**. There may be one or more substrates, depending on the particular chemical reaction. In some reactions, a single-reactant substrate breaks down into multiple products. In others, two substrates may come together to create one larger molecule. Two reactants might also enter a reaction, both become modified, and leave the reaction as two products. The location within the enzyme where the substrate binds is the enzyme's **active site**. This is where the "action" happens. Since enzymes are proteins, there is a unique combination of amino acid residues (also side chains, or R groups) within the active site. Different properties characterize each residue. These can be large or small, weakly acidic or basic, hydrophilic or hydrophobic, positively or negatively charged, or neutral. The unique combination of amino acid residues, their positions, sequences, structures, and properties, creates a very specific chemical environment within the active site. This specific environment is suited to bind, albeit briefly, to a specific chemical substrate (or substrates). Due to this jigsaw puzzle-like match between an enzyme and its substrates (which adapts to find the best fit between the transition state and the active site), enzymes are known for their specificity. The "best fit" results from the shape and the amino acid functional group's attraction to the substrate. There is a specifically matched enzyme for each substrate and, thus, for each chemical reaction; however, there is flexibility as well.

The fact that active sites are so perfectly suited to provide specific environmental conditions also means that they are subject to local environmental influences. It is true that increasing the environmental temperature generally increases reaction rates, enzyme-catalyzed or otherwise. However, increasing or decreasing the temperature outside of an optimal range can affect chemical bonds within the active site in such a way that they are less well suited to bind substrates. High temperatures will eventually cause enzymes, like other biological molecules, to **denature**, a process that changes the substance's natural properties. Likewise, the local environment's pH can also affect enzyme function. Active site amino acid residues have their own acidic or basic properties that are optimal for catalysis. These residues are sensitive to changes in pH that can impair the way substrate molecules bind. Enzymes are suited to function best within a certain pH range, and, as with temperature, extreme environmental pH values (acidic or basic) can cause enzymes to denature.

Induced Fit and Enzyme Function

For many years, scientists thought that enzyme-substrate binding took place in a simple "lock-and-key" fashion. This model asserted that the enzyme and substrate fit together perfectly in one instantaneous step. However, current research supports a more refined view scientists call **induced fit** (Figure 13.11). This model expands upon the lock-and-key model by describing a more dynamic interaction between enzyme and substrate. As the enzyme and substrate come together, their interaction causes a mild shift in the enzyme's structure that confirms an ideal binding arrangement between the enzyme and the substrate's transition state. This ideal binding maximizes the enzyme's ability to catalyze its reaction.

When an enzyme binds its substrate, it forms an enzyme-substrate complex. This complex lowers the reaction's activation energy and promotes its rapid progression in one of many ways. On a basic level, enzymes promote chemical reactions that involve more than one substrate by bringing the substrates together in an optimal orientation. The appropriate region (atoms and bonds) of one molecule is juxtaposed to the other molecule's appropriate region with which it must react. Another way in which enzymes promote substrate reaction is by creating an optimal environment within the active site for the reaction to occur. Certain chemical reactions might proceed best in a slightly acidic or non-polar environment. The chemical properties that emerge from the particular arrangement of amino acid residues within an active site create the perfect environment for an enzyme's specific substrates to react.

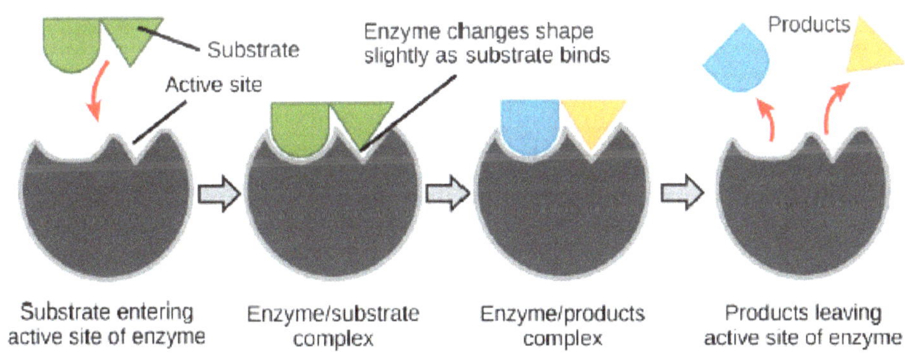

Figure 13.11 According to the induced-fit model, both enzyme and substrate undergo dynamic conformational changes upon binding. The enzyme contorts the substrate into its transition state, thereby increasing the reaction's rate.

You have learned that the activation energy required for many reactions includes the energy involved in manipulating or slightly contorting chemical bonds so that they can easily break and allow others to reform. Enzymatic action can aid this process. The enzyme-substrate complex can lower the activation energy by contorting substrate molecules in such a way as to facilitate bond-breaking, helping to reach the transition state. Finally, enzymes can also lower activation energies by taking part in the chemical reaction itself. The amino acid residues can provide certain ions or chemical groups that actually form covalent bonds with substrate molecules as a necessary step of the reaction process. In these cases, it is important to remember that the enzyme will always return to its original state at the reaction's completion. One of enzymes' hallmark properties is that they remain ultimately unchanged by the reactions they catalyze. After an enzyme catalyzes a reaction, it releases its product(s).

Metabolism Control Through Enzyme Regulation

It would seem ideal to have a scenario in which all the encoded enzymes in an organism's genome existed in abundant supply and functioned optimally under all cellular conditions, in all cells, at all times. In reality, this is far from the case. A variety of mechanisms ensure that this does not happen. Cellular needs and conditions vary from cell to cell, and change within individual cells over time. The required enzymes and energetic demands of stomach cells are different from those of fat storage cells, skin cells, blood cells, and nerve cells. Furthermore, a digestive cell works much harder to process and break down nutrients during the time that closely follows a meal compared with many hours after a meal. As these cellular demands and conditions vary, so do the amounts and functionality of different enzymes.

Since the rates of biochemical reactions are controlled by activation energy, and enzymes lower and determine activation energies for chemical reactions, the relative amounts and functioning of the variety of enzymes within a cell ultimately determine which reactions will proceed and at which rates. This determination is tightly controlled. In certain cellular environments, environmental factors like pH and temperature partly control enzyme activity. There are other mechanisms through which cells control enzyme activity and determine the rates at which various biochemical reactions will occur.

Molecular Regulation of Enzymes

Enzymes can be regulated in ways that either promote or reduce their activity. There are many different kinds of molecules that inhibit or promote enzyme function, and various mechanisms exist for doing so. For example, in some cases of enzyme inhibition, an inhibitor molecule is similar enough to a substrate that it can bind to the active site and simply block the substrate from binding. When this happens, the enzyme is inhibited through **competitive inhibition**, because an inhibitor molecule competes with the substrate for active site binding (Figure 13.12). On the other hand, in **noncompetitive inhibition**, an inhibitor molecule binds to the enzyme in a location other than the active site, called an allosteric site, but still manages to prevent substrate binding to the active site. Some inhibitor molecules bind to enzymes in a location where their binding induces a conformational change that reduces the enzyme activity as it no longer effectively catalyzes the conversion of the substrate to product.

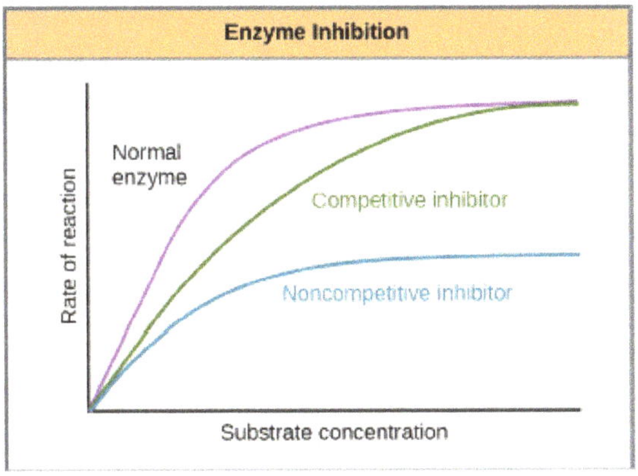

Figure 13.12 Competitive and noncompetitive inhibition affect the reaction's rate differently. Competitive inhibitors affect the initial rate but do not affect the maximal rate; whereas, noncompetitive inhibitors affect the maximal rate.

Some inhibitor molecules bind to enzymes in a location where their binding induces a conformational change that reduces the enzyme's affinity for its substrate. This type of inhibition is an **allosteric inhibition** (Figure 13.13). More than one polypeptide comprise most allosterically regulated enzymes, meaning that they have more than one protein subunit. When an allosteric inhibitor binds to an enzyme, all active sites on the protein subunits change slightly such that they bind their substrates with less efficiency. There are allosteric activators as well as inhibitors. Allosteric activators bind to locations on an enzyme away from the active site, inducing a conformational change that increases the affinity of the enzyme's active site(s) for its substrate(s).

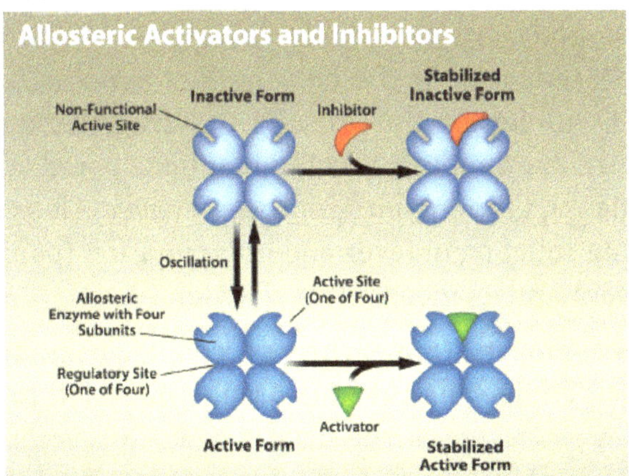

Figure 13.13 Allosteric inhibitors modify the enzyme's active site so that substrate binding is reduced or prevented. In contrast, allosteric activators modify the enzyme's active site so that the affinity for the substrate increases. Credit: Rao, A., Hawkins, A., Fletcher, S. and Tag, A. Department of Biology, Texas A&M University.

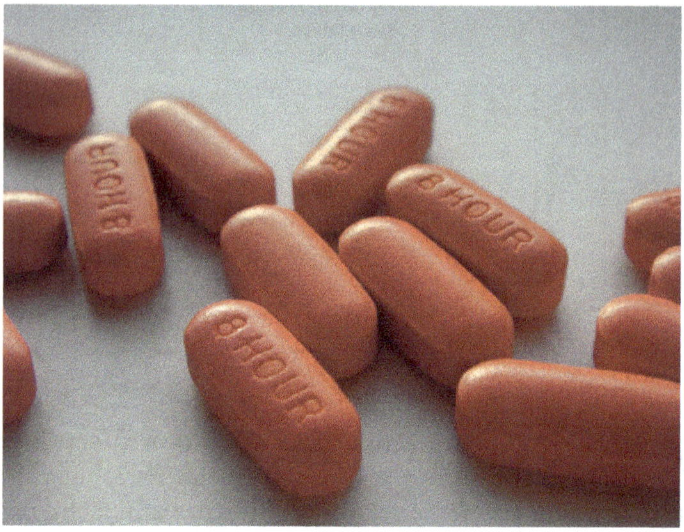

Figure 13.14 Have you ever wondered how pharmaceutical drugs are developed? (credit: Deborah Austin)

Drug Discovery by Looking for Inhibitors of Key Enzymes in Specific Pathways

Enzymes are key components of metabolic pathways. Understanding how enzymes work and how they can be regulated is a key principle behind developing many pharmaceutical drugs (**Figure 13.14)** on the market today. Biologists working in this field collaborate with other scientists, usually chemists, to design drugs.

Consider statins for example—which is a class of drugs that reduces cholesterol levels. These compounds are essentially inhibitors of the enzyme HMG-CoA reductase. HMG-CoA reductase is the enzyme that synthesizes cholesterol from lipids in the body. By inhibiting this enzyme, the drug reduces cholesterol levels synthesized in the body. Similarly, acetaminophen, popularly marketed under the brand name Tylenol, is an inhibitor of the enzyme cyclooxygenase. While it is effective in providing relief from fever and inflammation (pain), scientists still do not completely understand its mechanism of action.

How are drugs developed? One of the first challenges in drug development is identifying the specific molecule that the drug is intended to target. In the case of statins, HMG-CoA reductase is the drug target. Researchers identify targets through painstaking research in the laboratory. Identifying the target alone is not sufficient. Scientists also need to know how the target acts inside the cell and which reactions go awry in the case of disease. Once researchers identify the target and the pathway, then the actual drug design process begins. During this stage, chemists and biologists work together to design and synthesize molecules that can either block or activate a particular reaction. However, this is only the beginning:

both if and when a drug prototype is successful in performing its function, then it must undergo many tests from *in vitro* experiments to clinical trials before it can obtain FDA approval to be on the market.

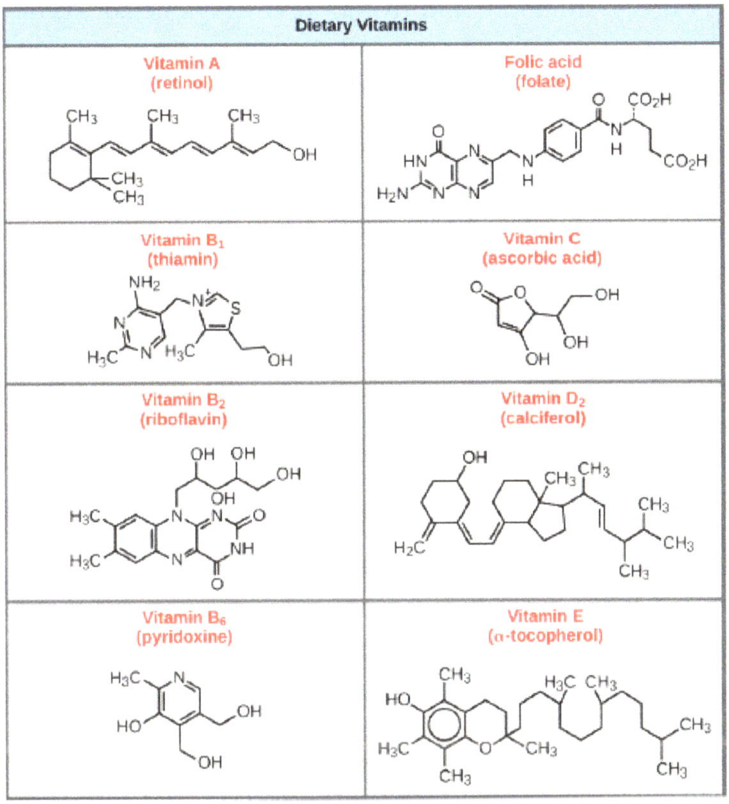

Figure 13.15 Vitamins are important coenzymes or precursors of coenzymes, and are required for enzymes to function properly. Multivitamin capsules usually contain mixtures of all the vitamins at different percentages.

Many enzymes don't work optimally, or even at all, unless bound to other specific non-protein helper molecules, either temporarily through ionic or hydrogen bonds or permanently through stronger covalent bonds. Two types of helper molecules are **cofactors** and **coenzymes.** Binding to these molecules promotes optimal conformation and function for their respective enzymes. Cofactors are inorganic ions such as iron (Fe++) and magnesium (Mg++). One example of an enzyme that requires a metal ion as a cofactor is the enzyme that builds DNA molecules, DNA polymerase, which requires a bound zinc ion (Zn++) to function. Coenzymes are organic helper molecules, with a basic atomic structure comprised of carbon and hydrogen, which are required for enzyme action. The most common sources of coenzymes are dietary vitamins (Figure 13.15). Some vitamins are precursors to coenzymes and others act directly as coenzymes. Vitamin C is a coenzyme for multiple enzymes that take part in building the important connective tissue component, collagen. An important step in breaking down glucose to yield energy is catalysis by a multi-enzyme complex scientists call pyruvate dehydrogenase. Pyruvate dehydrogenase is a complex of several enzymes that actually requires one cofactor (a magnesium ion) and five different organic coenzymes to catalyze its specific chemical reaction. Therefore, enzyme function is, in part, regulated by an abundance of various cofactors and coenzymes, which the diets of most organisms supply.

Enzyme Compartmentalization

In eukaryotic cells, molecules such as enzymes are usually compartmentalized into different organelles. This allows for yet another level of regulation of enzyme activity. Enzymes required only for certain cellular processes are sometimes housed separately along with their substrates, allowing for more efficient chemical reactions. Examples of this sort of enzyme regulation based on location and proximity include the enzymes involved in the latter stages of cellular respiration, which take place exclusively in the mitochondria, and the enzymes involved in digesting cellular debris and foreign materials, located within lysosomes.

Feedback Inhibition in Metabolic Pathways

Molecules can regulate enzyme function in many ways. However, a major question remains: What are these molecules and from where do they come? Some are cofactors and coenzymes, ions, and organic molecules, as you have learned. What other molecules in the cell provide enzymatic regulation, such as allosteric modulation, and competitive and noncompetitive inhibition? The answer is that a wide variety of molecules can perform these roles. Some include pharmaceutical and non-pharmaceutical drugs, toxins, and poisons from the environment. Perhaps the most relevant sources of enzyme regulatory molecules, with respect to cellular metabolism, are cellular metabolic reaction products themselves. In a most efficient and elegant way, cells have evolved to use their own reactions' products for feedback inhibition of enzyme activity. **Feedback inhibition** involves using a reaction product to regulate its own further production (Figure 13.16). The cell responds to the abundance of specific products by slowing down production during anabolic or catabolic reactions. Such reaction products may inhibit the enzymes that catalyzed their production through the mechanisms that we described above.

Producing both amino acids and nucleotides is controlled through feedback inhibition. Additionally, ATP is an allosteric regulator of some of the enzymes involved in sugar's catabolic breakdown, the process that produces ATP. In this way, when ATP is abundant, the cell can prevent its further production. Remember that ATP is an unstable molecule that can spontaneously dissociate into ADP and inorganic phosphate.

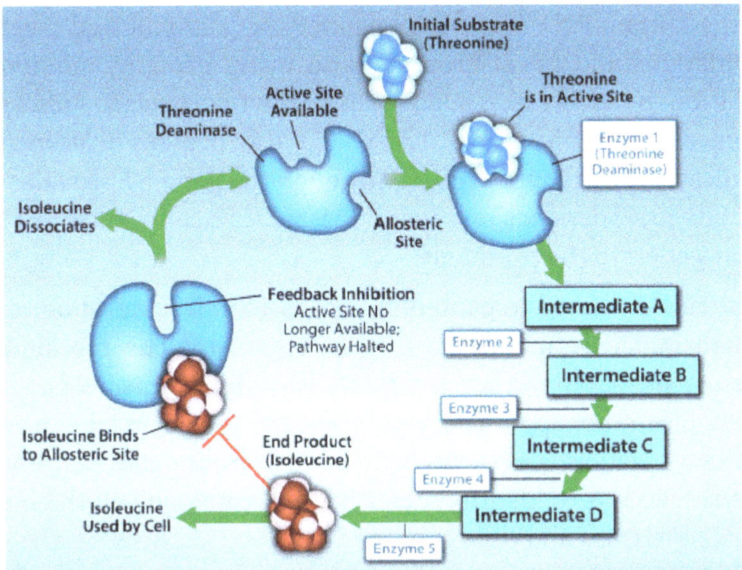

Figure 13.16 Metabolic pathways are a series of reactions that multiple enzymes catalyze.

If too much ATP were present in a cell, much of it would go to waste. Alternatively, ADP serves as a positive allosteric regulator (an allosteric activator) for some of the same enzymes that ATP inhibits. Thus, when relative ADP levels are high compared to ATP, the cell is triggered to produce more ATP through sugar catabolism.

APPLYING PROTEOMICS

Proteins are the final products of genes, which help perform the function that the gene encodes. Amino acids comprise proteins and play important roles in the cell. All enzymes (except ribozymes) are proteins that act as catalysts to affect the rate of reactions. Proteins are also regulatory molecules, and some are hormones. Transport proteins, such as hemoglobin, help transport oxygen to various organs. Antibodies that defend against foreign particles are also proteins. In the diseased state, protein function can be impaired because of changes at the genetic level or because of direct impact on a specific protein.

A **proteome** is the entire set of proteins that a cell type produces. We can study proteoms using the knowledge of genomes because genes code for mRNAs, and the mRNAs encode proteins. Although mRNA analysis is a step in the right direction, not all mRNAs are translated into proteins. **Proteomics** is the study of proteomes' function. Proteomics complements genomics and is useful when scientists want to test their hypotheses that they based on genes. Even though all multicellular organisms' cells have the same set of genes, the set of proteins produced in different tissues is different and dependent on gene expression. Thus, the genome is constant, but the proteome varies and is dynamic within an organism. In addition, RNAs can be alternately spliced (cut and pasted to create novel combinations and novel proteins) and many proteins modify themselves after translation by processes such as proteolytic cleavage, phosphorylation, glycosylation, and ubiquitination. There are also protein-protein interactions, which complicate studying proteomes. Although the genome provides a blueprint, the final architecture depends on several factors that can change the progression of events that generate the proteome.

Metabolomics is related to genomics and proteomics. **Metabolomics** involves studying small molecule metabolites in an organism. The **metabolome** is the complete set of metabolites that are related to an organism's genetic makeup. Metabolomics offers an opportunity to compare genetic makeup and physical characteristics, as well as genetic makeup and environmental factors. The goal of metabolome research is to identify, quantify, and catalogue all the metabolites in living organisms' tissues and fluids.

Basic Techniques in Protein Analysis

The ultimate goal of proteomics is to identify or compare the proteins expressed from a given genome under specific conditions, study the interactions between the proteins, and use the information to predict cell behaviour or develop drug targets. Just as scientists analyze the genome using the basic DNA sequencing technique, proteomics requires techniques for protein analysis. The basic technique for protein analysis, analogous to DNA sequencing, is mass spectrometry. Mass spectrometry identifies and determines a molecule's characteristics. Advances in spectrometry have allowed researchers to analyze very small protein samples. X-ray crystallography, for example, enables scientists to determine a protein crystal's three-dimensional structure at atomic resolution. Another protein imaging technique, nuclear magnetic resonance (NMR), uses atoms' magnetic properties to determine the protein's three-dimensional structure in aqueous solution. Scientists have also used protein microarrays to study protein interactions. Large-scale adaptations of the basic two-hybrid screen (Figure 13.17) have provided the basis for protein microarrays. Scientists use computer software to analyze the vast amount of data for proteomic analysis.

Genomic- and proteomic-scale analyses are part of **systems biology**, which is the study of whole biological systems (genomes and proteomes) based on interactions within the system. The European Bioinformatics Institute and the Human Proteome Organization (HUPO) are developing and establishing effective tools to sort through the enormous pile of systems biology data. Because proteins are the direct products of genes and reflect activity at the genomic level, it is natural to use proteomes to compare the protein profiles of different cells to identify proteins and genes involved in disease processes. Most pharmaceutical drug trials target proteins. Researchers use information that they obtain from proteomics to identify novel drugs and to understand their mechanisms of action.

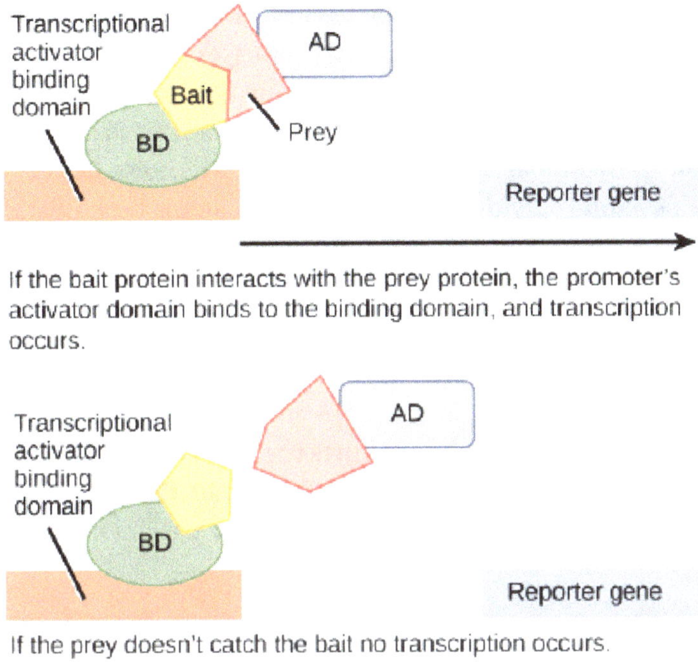

Figure 13.17 Scientists use two-hybrid screening to determine whether two proteins interact. In this method, a transcription factor splits into a DNA-binding domain (BD) and an activator domain (AD). The binding domain is able to bind the promoter in the activator domain's absence, but it does not turn on transcription. The bait protein attaches to the BD, and the prey protein attaches to the AD. Transcription occurs only if the prey "catches" the bait.

Scientists are challenged when implementing proteomic analysis because it is difficult to detect small protein quantities. Although mass spectrometry is good for detecting small protein amounts, variations in protein expression in diseased states can be difficult to discern. Proteins are naturally unstable molecules, which makes proteomic analysis much more difficult than genomic analysis.

Cancer Proteomics

Researchers are studying patients' genomes and proteomes to understand the genetic basis of diseases. The most prominent disease researchers are studying with proteomic approaches is cancer. These approaches improve screening and early cancer detection. Researchers are able to identify proteins whose expression indicates the disease process. An individual protein is a **biomarker**; whereas, a set of proteins with altered expression levels is a **protein signature**. For a biomarker or protein signature to be useful as a candidate for early cancer screening and detection, they must secrete in body fluids, such as sweat, blood, or urine, such that health professionals can perform large-scale screenings in a noninvasive fashion. The current problem with using biomarkers for early cancer detection is the high rate of false-negative results. A **false negative** is an incorrect test result that should have been positive. In other words, many cancer cases go undetected, which makes biomarkers unreliable. Some examples of protein biomarkers in cancer detection are CA-125 for ovarian cancer and PSA for prostate cancer. Protein signatures may be more reliable than biomarkers to detect cancer cells. Researchers are also using proteomics to develop individualized treatment plans, which involves predicting whether or not an individual will respond to specific drugs and the side effects that the individual may experience. Researchers also use proteomics to predict the possibility of disease recurrence.

The National Cancer Institute has developed programs to improve cancer detection and treatment. The Clinical Proteomic Technologies for Cancer and the Early Detection Research Network are efforts to identify protein signatures specific to different cancer types. The Biomedical Proteomics Program identifies protein signatures and designs effective therapies for cancer patients.

CHAPTER-14

NUCLEIC ACIDS AND PROTEINS: LABORATORY TECHNIQUES

INTRODUCTION: AN OVERVIEW

In the dynamic realm of molecular biology, the manipulation and analysis of nucleic acids and proteins represent a cornerstone of scientific inquiry. The intricate dance between DNA, RNA, and proteins orchestrates the molecular processes governing life itself. Laboratory techniques tailored to these vital biomolecules enable scientists to dissect and understand the mechanisms underlying genetic information, cellular functions, and the intricate networks that regulate biological systems. This exploration into laboratory techniques for nucleic acids and proteins unveils the sophisticated methodologies employed in the quest for unlocking the mysteries of genetic expression, heredity, and cellular function. From DNA sequencing and PCR amplification to protein purification and mass spectrometry, these techniques empower researchers to unravel the complexities of life at the molecular level, driving advancements in fields ranging from medicine to biotechnology.

Watson and Crick's identification of the structure of DNA in 1953 was the seminal event in the field of genetic engineering. Since the 1970s, there has been a veritable explosion in scientists' ability to manipulate DNA in ways that have revolutionized the fields of biology, medicine, diagnostics, forensics, and industrial manufacturing. Many of the molecular tools discovered in recent decades have been produced using prokaryotic microbes. In this chapter, we will explore some of those tools, especially as they relate to applications in medicine and health care.

As an example, the thermal cycler is used to perform a diagnostic technique called the polymerase chain reaction (PCR), which relies on DNA polymerase enzymes from thermophilic bacteria. Other molecular tools, such as restriction enzymes and plasmids obtained from microorganisms, allow scientists to insert genes from humans or other organisms into microorganisms. The microorganisms are then grown on an industrial scale to synthesize products such as insulin, vaccines, and biodegradable polymers. These are just a few of the numerous applications of microbial genetics that we will explore in this chapter.

TOOLS OF GENETIC ENGINEERING

Kayla, a 24-year-old electrical engineer and running enthusiast, just moved from Arizona to New Hampshire to take a new job. On her weekends off, she loves to explore her new surroundings, going for long runs in the pine forests. In July she spent a week hiking through the mountains. In early August, Kayla developed a low fever, headache, and mild muscle aches, and she felt a bit fatigued. Not thinking much of it, she took some ibuprofen to combat her symptoms and vowed to get more rest.

The science of using living systems to benefit humankind is called **biotechnology**. Technically speaking, the domestication of plants and animals through farming and breeding practices is a type of biotechnology. However, in a contemporary sense, we associate biotechnology with the direct alteration of an organism's genetics to achieve desirable traits through the process of **genetic engineering**. Genetic engineering involves the use of **recombinant DNA technology**, the process by which a DNA sequence is manipulated *in vitro,* thus creating **recombinant DNA molecules** that have new combinations of genetic material. The recombinant DNA is then introduced into a host organism. If the DNA that is introduced comes from a different species, the host organism is now considered to be **transgenic**.

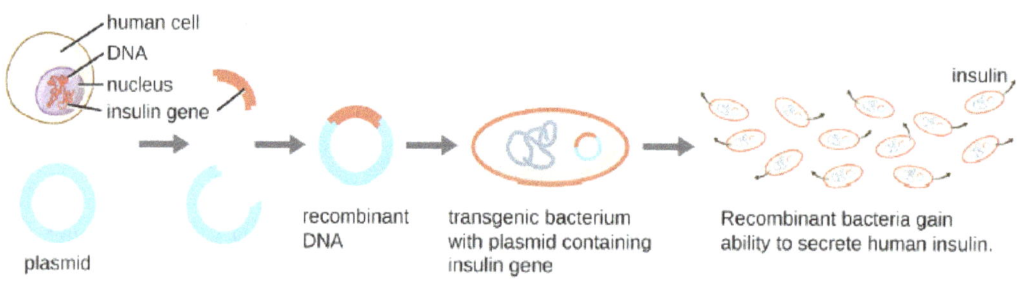

Figure 14.1 Recombinant DNA technology is the artificial recombination of DNA from two organisms. In this example, the human insulin gene is inserted into a bacterial plasmid. This recombinant plasmid can then be used to transform bacteria, which gain the ability to produce the insulin protein.

Molecular Cloning

Herbert Boyer and Stanley Cohen first demonstrated the complete **molecular cloning** process in 1973 when they successfully cloned genes from the African clawed frog (*Xenopus laevis*) into a bacterial plasmid that was then introduced into the bacterial host *Escherichia coli*. Molecular cloning is a set of methods used to construct recombinant DNA and incorporate it into a host organism; it makes use of a number of molecular tools that are derived from microorganisms.

Restriction Enzymes and Ligases

In recombinant DNA technology, DNA molecules are manipulated using naturally occurring enzymes derived mainly from bacteria and viruses. The creation of recombinant DNA molecules is possible due to the use of naturally occurring **restriction endonucleases** (**restriction enzymes**), bacterial enzymes produced as a protection mechanism to cut and destroy foreign cytoplasmic DNA that is most commonly a result of bacteriophage infection. Stewart Linn and Werner Arber discovered restriction enzymes in their 1960s studies of how *E. coli* limits bacteriophage replication on infection. Today, we use restriction enzymes extensively for cutting DNA fragments that can then be spliced into another DNA molecule to form recombinant molecules. Each restriction enzyme cuts DNA at a characteristic **recognition site**, a specific, usually palindromic, DNA sequence typically between four to six base pairs in length. A palindrome is a sequence of letters that reads the same forward as backward. (The word "level" is an example of a palindrome.) Palindromic DNA sequences contain the same base sequences in the 5′ to 3′ direction on one strand as in the 5′ to 3′ direction on the complementary strand. A restriction enzyme recognizes the DNA palindrome and cuts each backbone at identical positions in the palindrome. Some restriction enzymes cut to produce molecules that have complementary overhangs (**sticky ends**) while others cut without generating such overhangs, instead producing **blunt ends** (Figure 14.2).

Molecules with complementary sticky ends can easily **anneal**, or form hydrogen bonds between complementary bases, at their sticky ends. The annealing step allows **hybridization** of the single-stranded overhangs. Hybridization refers to the joining together of two complementary single strands of DNA. Blunt ends can also attach together, but less efficiently than sticky ends due to the lack of complementary overhangs facilitating the process. In either case, **ligation** by DNA ligase can then rejoin the two sugar-phosphate backbones of the DNA through covalent bonding, making the molecule a continuous double strand. In 1972, Paul Berg, a Stanford biochemist, was the first to produce a recombinant DNA molecule using this technique, combining the SV40 monkey virus with *E. coli* bacteriophage lambda to create a hybrid.

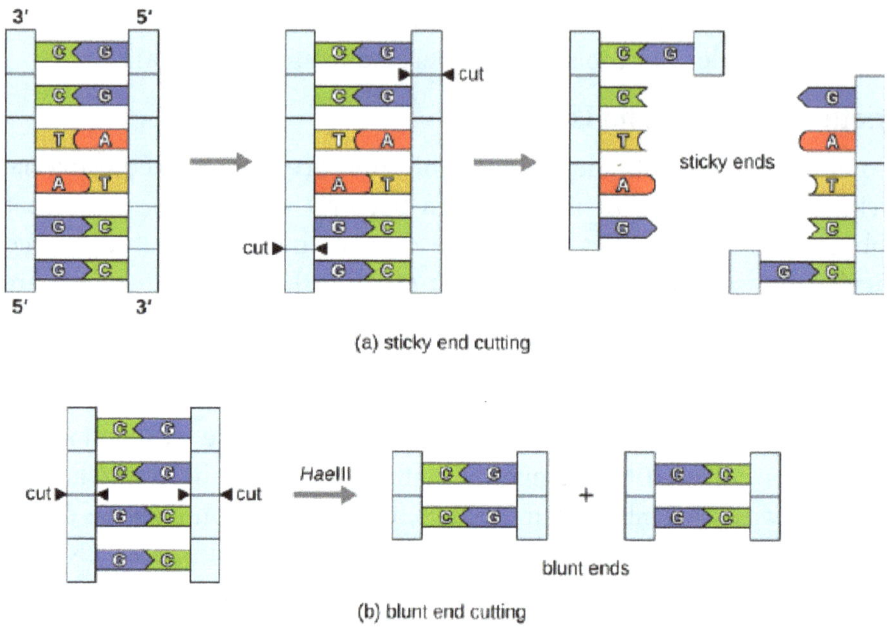

Figure 14.2 (a) In this six-nucleotide restriction enzyme site, recognized by the enzyme BamHI, notice that the sequence reads the same in the 5' to 3' direction on both strands. This is known as a palindrome. The cutting of the DNA by the restriction enzyme at the sites (indicated by the black arrows) produces DNA fragments with sticky ends. Another piece of DNA cut with the same restriction enzyme could attach to one of these sticky ends, forming a recombinant DNA molecule. (b) This four-nucleotide recognition site also exhibits a palindromic sequence. The cutting of the DNA by the restriction enzyme HaeIII at the indicated sites produces DNA fragments with blunt ends. Any other piece of blunt DNA could attach to one of the blunt ends produced, forming a recombinant DNA molecule.

Plasmids

After restriction digestion, genes of interest are commonly inserted into plasmids, small pieces of typically circular, double-stranded DNA that replicate independently of the bacterial chromosome (see Unique Characteristics of Prokaryotic Cells (https://legacy.cnx.org/content/m58792/latest/)). In recombinant DNA technology, plasmids are often used as **vectors**, DNA molecules that carry DNA fragments from one organism to another. Plasmids used as vectors can be genetically engineered by researchers and scientific supply companies to have specialized properties, as illustrated by the commonly used plasmid vector pUC19 (Figure 14.3). Some plasmid vectors contain genes that confer antibiotic resistance; these resistance genes allow researchers to easily find plasmid-containing colonies by plating them on media containing the corresponding antibiotic. The antibiotic kills all host cells that do not harbor the desired plasmid vector, but those that contain the vector are able to survive and grow.

Plasmid vectors used for cloning typically have a **polylinker site**, or **multiple cloning site (MCS).** A polylinker site is a short sequence containing multiple unique restriction enzyme recognition sites that are used for inserting DNA into the plasmid after restriction digestion of both the DNA and the plasmid. Having these multiple restriction enzyme recognition sites within the polylinker site makes the plasmid vector versatile, so it can be used for many different cloning experiments involving different restriction enzymes.

This polylinker site is often found within a **reporter gene**, another gene sequence artificially engineered into the plasmid that encodes a protein that allows for visualization of DNA insertion. The reporter gene allows a researcher to distinguish host cells that contain recombinant plasmids with cloned DNA fragments from host cells that only contain the non-recombinant plasmid vector. The most common reporter gene used in plasmid vectors is the bacterial *lacZ* gene encoding beta-galactosidase, an enzyme that naturally degrades lactose but can also degrade a colourless synthetic analog X-gal, thereby producing blue colonies on X-gal–containing media. The *lacZ* reporter gene is disabled when the recombinant DNA is spliced into the plasmid. Because the LacZ protein is not produced when the gene is disabled, X-gal is not degraded and white colonies are produced, which can then be isolated. This **blue-white screening** method is described later and shown in Figure 14.4. In addition to these features, some plasmids come pre-digested and with an enzyme linked to the linearized plasmid to aid in ligation after the insertion of foreign DNA fragments.

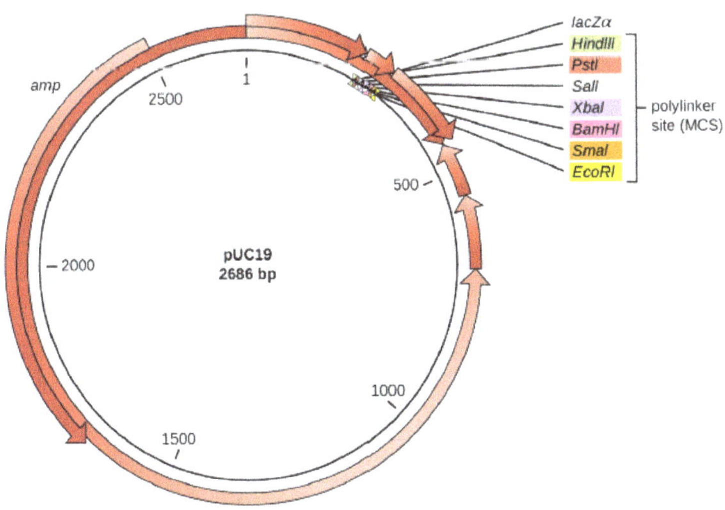

Figure 14.3 The artificially constructed plasmid vector pUC19 is commonly used for cloning foreign DNA. Arrows indicate the directions in which the genes are transcribed. Note the polylinker site, containing multiple unique restriction enzyme recognition sites, found within the *lacZ* reporter gene. Also note the ampicillin (*amp*) resistance gene encoded on the plasmid.

Molecular Cloning using Transformation

The most commonly used mechanism for introducing engineered plasmids into a bacterial cell is transformation, a process in which bacteria take up free DNA from their surroundings. In nature, free DNA typically comes from other lysed bacterial cells; in the laboratory, free DNA in the form of recombinant plasmids is introduced to the cell's surroundings.

Some bacteria, such as *Bacillus* spp., are naturally competent, meaning they are able to take up foreign DNA. However, not all bacteria are naturally competent. In most cases, bacteria must be made artificially competent in the laboratory by increasing the permeability of the cell membrane. This can be achieved through chemical treatments that neutralize charges on the cell membrane or by exposing the bacteria to an electric field that creates microscopic pores in the cell membrane. These methods yield chemically competent or electro competent bacteria, respectively.

Following the transformation protocol, bacterial cells are plated onto an antibiotic-containing medium to inhibit the growth of the many host cells that were not transformed by the plasmid conferring antibiotic resistance. A technique called **blue-white screening** is then used for *lacZ*-encoding plasmid vectors such as pUC19. Blue colonies have a functional beta-galactosidase enzyme because the *lacZ* gene is uninterrupted, with no foreign DNA inserted into the polylinker site. These colonies typically result from the digested, linearized plasmid religating to itself. However, white colonies lack a functional beta-galactosidase enzyme, indicating the insertion of foreign DNA within the polylinker site of the plasmid vector, thus disrupting the *lacZ* gene. Thus, white colonies resulting from this blue-white screening contain plasmids with an insert and can be further screened to characterize the foreign DNA. To be sure the correct DNA was incorporated into the plasmid, the DNA insert can then be sequenced.

Molecular Cloning Using Conjugation or Transduction

Alternatively, bacteriophages can be used to introduce recombinant DNA into host bacterial cells through a manipulation of the transduction process . In the laboratory, DNA fragments of interest can be engineered into **phagemids**, which are plasmids that have phage sequences that allow them to be packaged into bacteriophages. Bacterial cells can then be infected with these bacteriophages so that the recombinant phagemids can be introduced into the bacterial cells. Depending on the type of phage, the recombinant DNA may be integrated into the host bacterial genome (lysogeny), or it may exist as a plasmid in the host's cytoplasm.

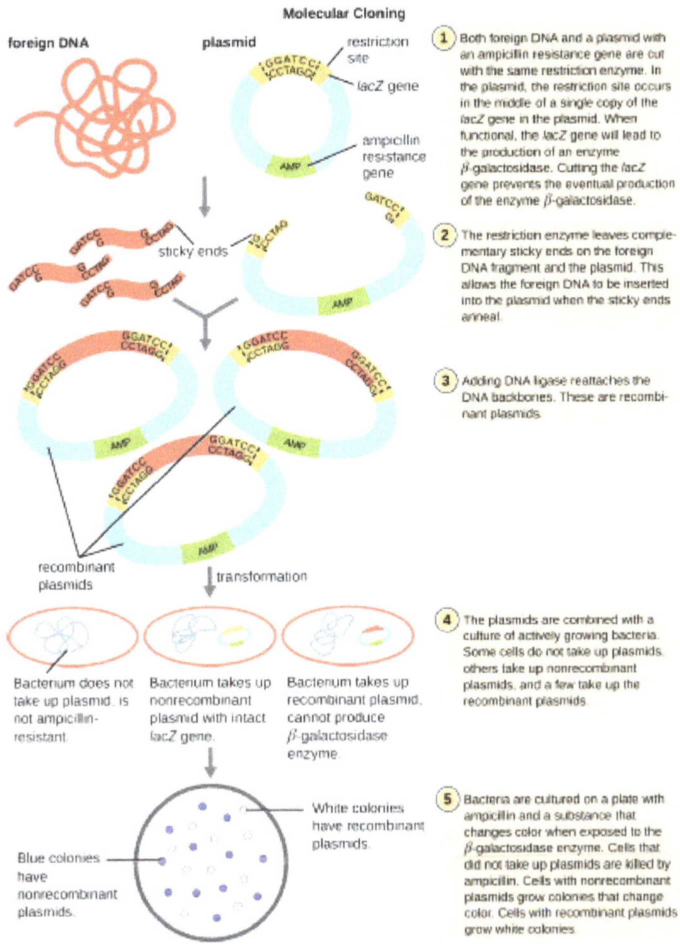

Figure 14.4 The steps involved in molecular cloning using bacterial transformation are outlined in this graphic flowchart.

Creating a Genomic Library

Molecular cloning may also be used to generate a **genomic library**. The library is a complete (or nearly complete) copy of an organism's genome contained as recombinant DNA plasmids engineered into unique clones of bacteria. Having such a library allows a researcher to create large quantities of each fragment by growing the bacterial host for that fragment. These fragments can be used to determine the sequence of the DNA and the function of any genes present.

One method for generating a genomic library is to ligate individual restriction enzyme-digested genomic fragments into plasmid vectors cut with the same restriction enzyme (Figure 14.5). After transformation into a bacterial host, each transformed bacterial cell takes up a single recombinant plasmid and grows into a colony of cells. All of the cells in this colony are identical **clones** and carry the same recombinant plasmid. The resulting library is a collection of colonies, each of which contains a fragment of the original organism's genome, that are each separate and distinct and can each be used for further study. This makes it possible for researchers to screen these different clones to discover the one containing a gene of interest from the original organism's genome.

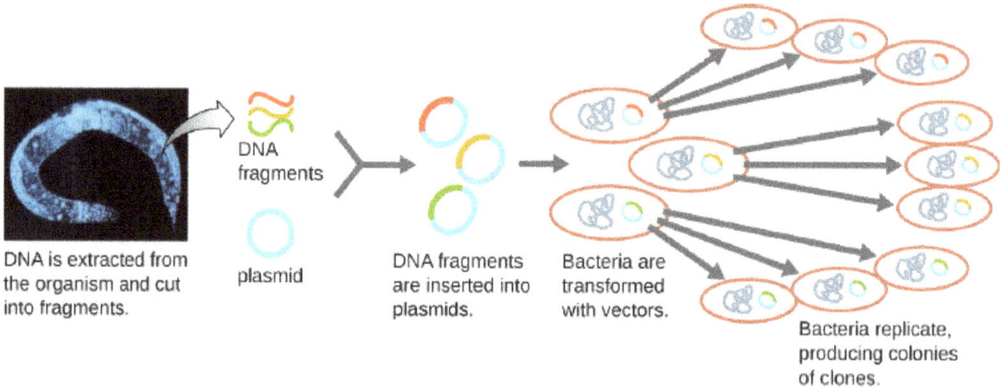

Figure 14.5 The generation of a genomic library facilitates the discovery of the genomic DNA fragment that contains a gene of interest.

To construct a genomic library using larger fragments of genomic DNA, an E. coli bacteriophage, such as lambda, can be used as a host (Figure 14.6). Genomic DNA can be sheared or enzymatically digested and ligated into a pre-digested bacteriophage lambda DNA vector. Then, these recombinant phage DNA molecules can be packaged into phage particles and used to infect *E. coli* host cells on a plate. During infection within each cell, each recombinant phage will make many copies of itself and lyse the *E. coli* lawn, forming a plaque. Thus, each plaque from a phage library represents a unique recombinant phage containing a distinct genomic DNA fragment. Plaques can then be screened further to look for genes of interest. One advantage to producing a library using phages instead of plasmids is that a phage particle holds a much larger insert of foreign DNA compared with a plasmid vector, thus requiring a much smaller number of cultures to fully represent the entire genome of the original organism.

To focus on the expressed genes in an organism or even a tissue, researchers construct libraries using the organism's messenger RNA (mRNA) rather than its genomic DNA. Whereas all cells in a single organism will have the same genomic DNA, different tissues express different genes, producing different complements of mRNA. For example, all human cells' genomic DNA contains the gene for insulin, but only cells in the pancreas express mRNA directing the production of insulin. Because mRNA cannot be cloned directly, in the laboratory mRNA must be used as a template by the retroviral enzyme reverse transcriptase to make complementary DNA (cDNA).

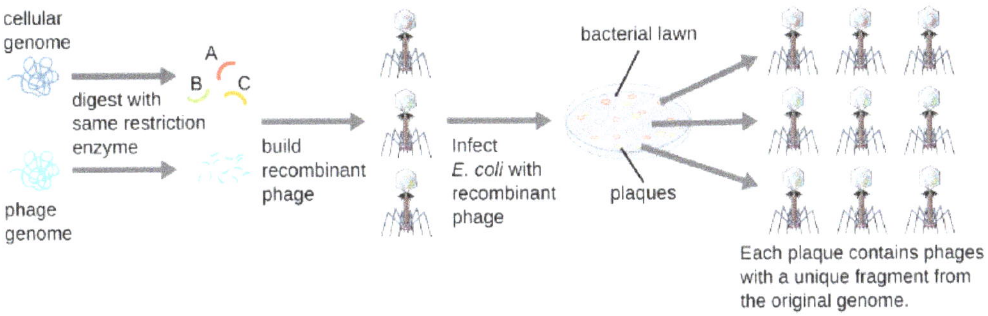

Figure 14.6 Recombinant phage DNA molecules are made by ligating digested phage particles with fragmented genomic DNA molecules. These recombinant phage DNA molecules are packaged into phage particles and allowed to infect a bacterial lawn. Each plaque represents a unique recombinant DNA molecule that can be further screened for genes of interest.

A cell's full complement of mRNA can be reverse-transcribed into cDNA molecules, which can be used as a template for DNA polymerase to make double-stranded DNA copies; these fragments can subsequently be ligated into either plasmid vectors or bacteriophage to produce a cDNA library. The benefit of a cDNA library is that it contains DNA from only the expressed genes in the cell. This means that the introns, control sequences such as promoters, and DNA not destined to be translated into proteins are not represented in the library. The focus on translated sequences means that the library cannot be used to study the sequence and structure of the genome in its entirety. The construction of a cDNA genomic library is shown in Figure 14.7.

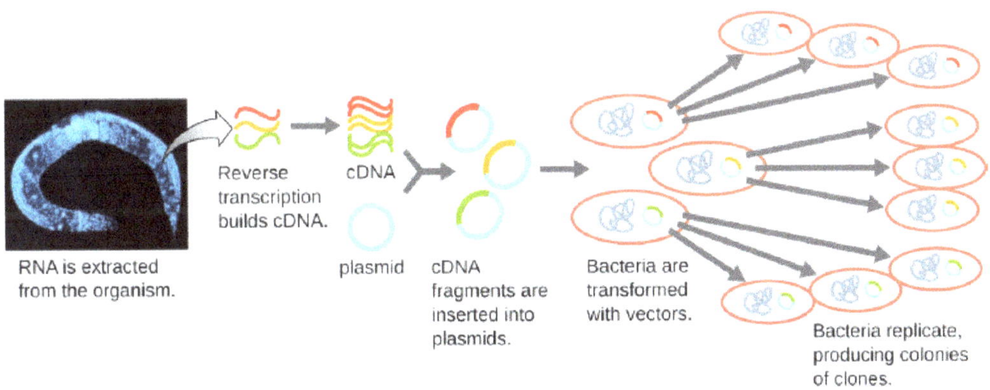

Figure 14.7 Complementary DNA (cDNA) is made from mRNA by the retroviral enzyme reverse transcriptase, converted into double-stranded copies, and inserted into either plasmid vectors or bacteriophage, producing a cDNA library. (credit "micrograph": modification of work by National Institutes of Health)

Introducing Recombinant Molecules into Eukaryotic Hosts

The use of bacterial hosts for genetic engineering laid the foundation for recombinant DNA technology; however, researchers have also had great interest in genetically engineering eukaryotic cells, particularly those of plants and animals. The introduction of recombinant DNA molecules into eukaryotic hosts is called **transfection**. Genetically engineered plants, called transgenic plants, are of significant interest for agricultural and pharmaceutical purposes. The first transgenic plant sold commercially was the Flavr Savr delayed-ripening tomato, which came to market in 1994. Genetically engineered livestock have also been successfully produced, resulting, for example, in pigs with increased nutritional value[1] and goats that secrete pharmaceutical products in their milk.[2]

Electroporation

Compared to bacterial cells, eukaryotic cells tend to be less amenable as hosts for recombinant DNA molecules. Because eukaryotes are typically neither competent to take up foreign DNA nor able to maintain plasmids, transfection of eukaryotic hosts is far more challenging and requires more intrusive techniques for success. One method used for transfecting cells in cell culture is called **electroporation.** A brief electric pulse induces the formation of transient pores in the phospholipid bilayers of cells through which the gene can be introduced. At the same time, the electric pulse generates a short-lived positive charge on one side of the cell's interior and a negative charge on the opposite side; the charge difference draws negatively charged DNA molecules into the cell (Figure 14.8).

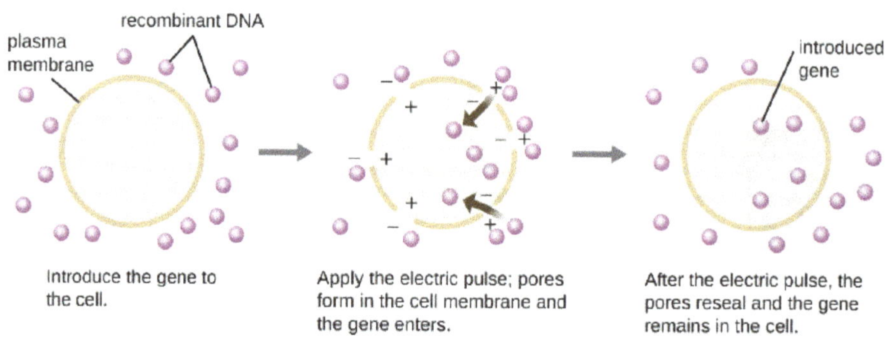

Figure 14.8 Electroporation is one laboratory technique used to introduce DNA into eukaryotic cells.

Microinjection

An alternative method of transfection is called **microinjection.** Because eukaryotic cells are typically larger than those of prokaryotes, DNA fragments can sometimes be directly injected into the cytoplasm using a glass micropipette, as shown in Figure 14.9.

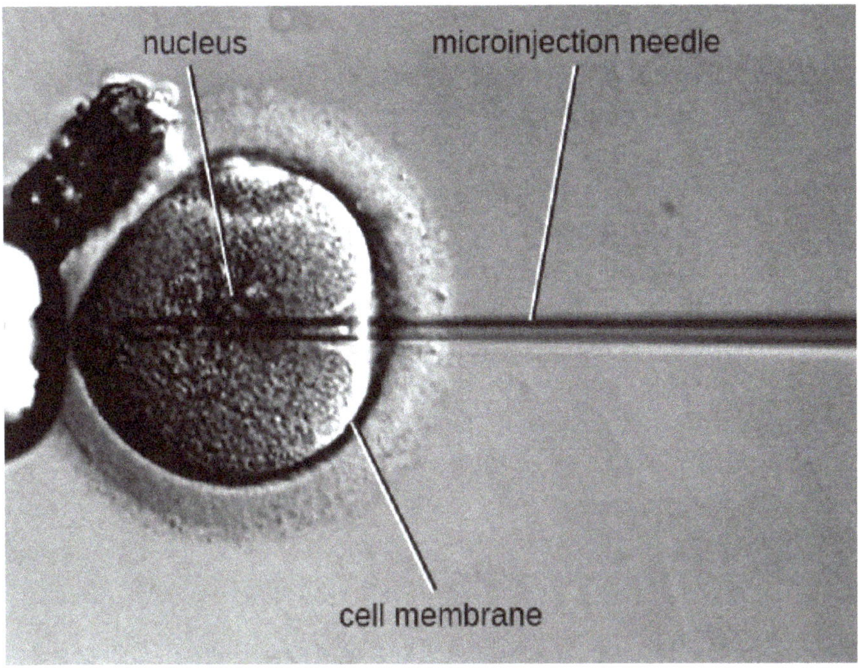

Figure 14.9 Microinjection is another technique for introducing DNA into eukaryotic cells. A microinjection needle containing recombinant DNA is able to penetrate both the cell membrane and nuclear envelope.

Gene Guns

Transfecting plant cells can be even more difficult than animal cells because of their thick cell walls. One approach involves treating plant cells with enzymes to remove their cell walls, producing protoplasts. Then, a **gene gun** is used to shoot gold or tungsten particles coated with recombinant DNA molecules into the plant protoplasts at high speeds. Recipient protoplast cells can then recover and be used to generate new transgenic plants (Figure 14.10).

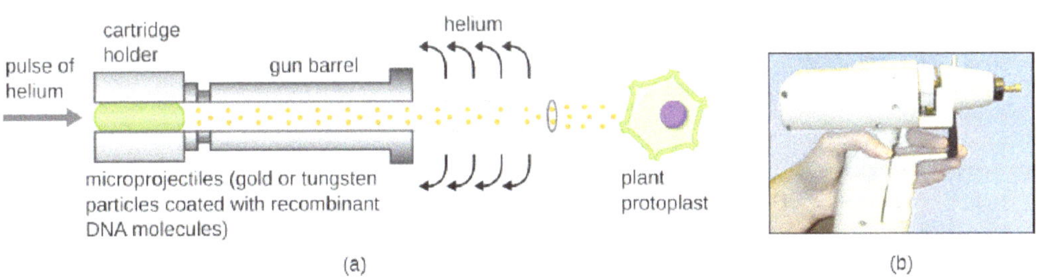

Figure 14.10 Heavy-metal particles coated with recombinant DNA are shot into plant protoplasts using a gene gun. The resulting transformed cells are allowed to recover and can be used to generate recombinant plants. (a) A schematic of a gene gun. (b) A photograph of a gene gun. (credit a, b: modification of work by JA O'Brien, SC Lummis)

Shuttle Vectors

Another method of transfecting plants involves **shuttle vectors**, plasmids that can move between bacterial and eukaryotic cells. The **tumor-inducing (Ti) plasmids** originating from the bacterium *Agrobacterium tumefaciens* are commonly used as shuttle vectors for incorporating genes into plants (Figure 14.11). In nature, the Ti plasmids of *A. tumefaciens* cause plants to develop tumors when they are transferred from bacterial cells to plant cells. Researchers have been able to manipulate these naturally occurring plasmids to remove their tumor- causing genes and insert desirable DNA fragments. The resulting recombinant Ti plasmids can be transferred into the plant genome through the natural transfer of Ti plasmids from the bacterium to the plant host. Once inside the plant host cell, the gene of interest recombines into the plant cell's genome.

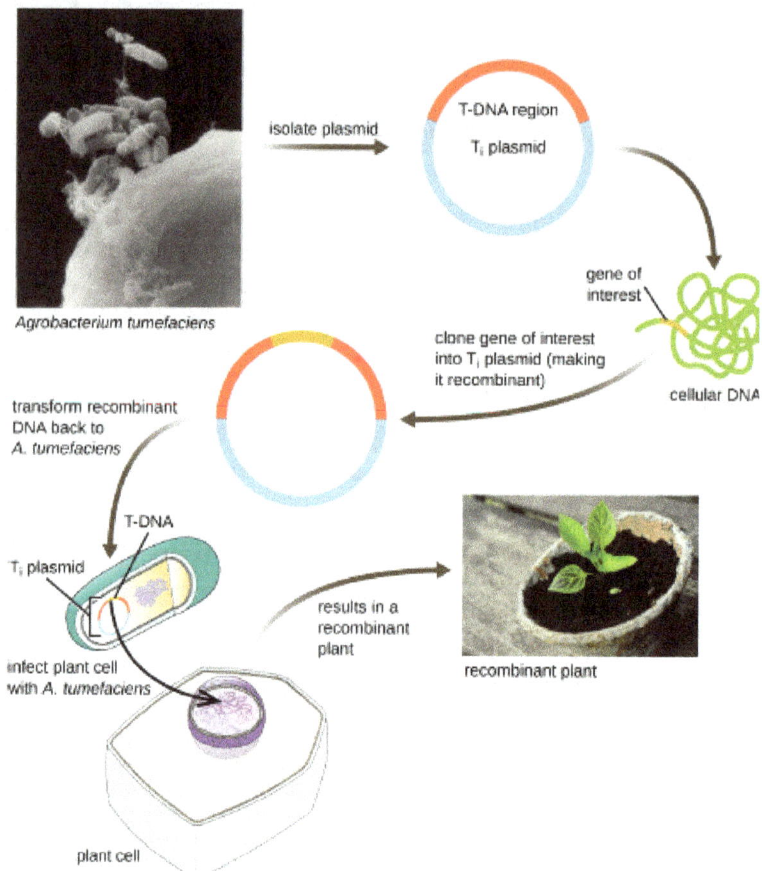

Figure 14.11 The Ti plasmid of *Agrobacterium tumefaciens* is a useful shuttle vector for the uptake of genes of interest into plant cells. The gene of interest is cloned into the Ti plasmid, which is then introduced into plant cells. The gene of interest then recombines into the plant cell's genome, allowing for the production of transgenic plants.

Viral Vectors

Viral vectors can also be used to transfect eukaryotic cells. In fact, this method is often used in gene therapy (see Gene Therapy) to introduce healthy genes into human patients suffering from diseases that result from genetic mutations. Viral genes can be deleted and replaced with the gene to be delivered to the patient; the virus then infects the host cell and delivers the foreign DNA into the genome of the targeted cell. Adenoviruses are often used for this purpose because they can be grown to high titer and can infect both nondividing and dividing host cells. However, use of viral vectors for gene therapy can pose some risks for patients, as discussed in **Gene Therapy.**

PHARMACEUTICAL APPLICATIONS OF GENETIC ENGINEERING

Genomics, Transcriptomics, and Proteomics

The study and comparison of entire genomes, including the complete set of genes and their nucleotide sequence and organization, is called **genomics**. This field has great potential for future medical advances through the study of the human genome as well as the genomes of infectious organisms. Analysis of microbial genomes has contributed to the development of new antibiotics, diagnostic tools, vaccines, medical treatments, and environmental cleanup techniques.

The field of **transcriptomics** is the science of the entire collection of mRNA molecules produced by cells. Scientists compare gene expression patterns between infected and uninfected host cells, gaining important information about the cellular responses to infectious disease. Additionally, transcriptomics can be used to monitor the gene expression of virulence factors in microorganisms, aiding scientists in better understanding pathogenic processes from this viewpoint.

When genomics and transcriptomics are applied to entire microbial communities, we use the terms **metagenomics** and **meta-transcriptomics**, respectively. Metagenomics and meta-transcriptomics allow researchers to study genes and gene expression from a collection of multiple species, many of which may not be easily cultured or cultured at all in the laboratory. A DNA microarray (discussed in the previous section) can be used in metagenomics studies.

Another up-and-coming clinical application of genomics and transcriptomics is **pharmacogenomics**, also called **toxico-genomics**, which involves evaluating the effectiveness and safety of drugs on the basis of information from an individual's genomic sequence. Genomic responses to drugs can be studied using experimental animals (such as laboratory rats or mice) or live cells in the laboratory before embarking on studies with humans. Changes in gene expression in the presence of a drug can sometimes be an early indicator of the potential for toxic effects. Personal genome sequence information may someday be used to prescribe medications that will be most effective and least toxic on the basis of the individual patient's genotype.

The study of **proteomics** is an extension of genomics that allows scientists to study the entire complement of proteins in an organism, called the proteome. Even though all cells of a multicellular organism have the same set of genes, cells in various tissues produce different sets of proteins. Thus, the genome is constant, but the proteome varies and is dynamic within an organism. Proteomics may be used to study which proteins are expressed under various conditions within a single cell type or to compare protein expression patterns between different organisms.

The most prominent disease being studied with proteomic approaches is cancer, but this area of study is also being applied to infectious diseases. Research is currently underway to examine the feasibility of using proteomic approaches to diagnose various types of hepatitis, tuberculosis, and HIV infection, which are rather difficult to diagnose using currently available techniques.

A recent and developing proteomic analysis relies on identifying proteins called **biomarkers,** whose expression is affected by the disease process. Biomarkers are currently being used to detect various forms of cancer as well as infections caused by pathogens such as Yersinia pestis and Vaccinia virus.

Other "-omic" sciences related to genomics and proteomics include metabolomics, glycomics, and lipidomics, which focus on the complete set of small-molecule metabolites, sugars, and lipids, respectively, found within a cell. Through these various global approaches, scientists continue to collect, compile, and analyze large amounts of genetic information. This emerging field of **bioinformatics** can be used, among many other applications, for clues to treating diseases and understanding the workings of cells.

Additionally, researchers can use reverse genetics, a technique related to classic mutational analysis, to determine the function of specific genes. Classic methods of studying gene function involved searching for the genes responsible for a given phenotype. Reverse genetics uses the opposite approach, starting with a specific DNA sequence and attempting to determine what phenotype it produces. Alternatively, scientists can attach known genes (called reporter genes) that encode easily observable characteristics to genes of interest, and the location of expression of such genes of interest can be easily monitored. This gives the researcher important information about what the gene product might be doing or where it is located in the organism. Common reporter genes include bacterial *lacZ*, which encodes beta-galactosidase and whose activity can be monitored by changes in colony colour in the presence of X-gal as previously described, and the gene encoding the jellyfish protein green fluorescent protein (GFP) whose activity can be visualized in colonies under ultraviolet light exposure (Figure 14.12).

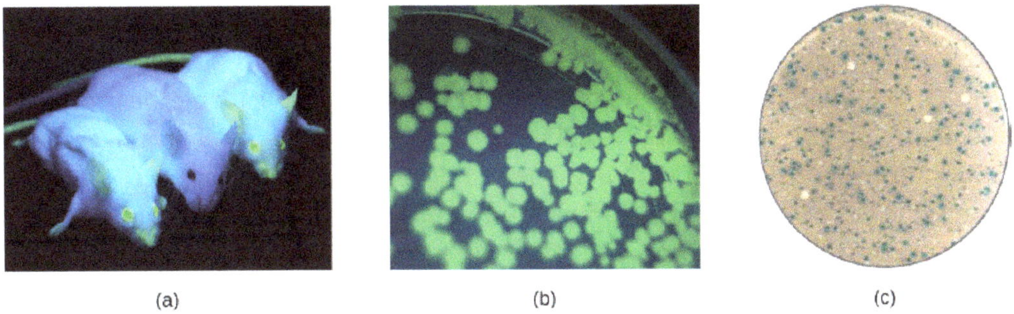

Figure 14.12 (a) The gene encoding green fluorescence protein is a commonly used reporter gene for monitoring gene expression patterns in organisms. Under ultraviolet light, GFP fluoresces.

Recombinant DNA Technology and Pharmaceutical Production

Genetic engineering has provided a way to create new pharmaceutical products **called recombinant DNA pharmaceuticals**. Such products include antibiotic drugs, vaccines, and hormones used to treat various diseases. Table 14.1 lists examples of recombinant DNA products and their uses.

For example, the naturally occurring antibiotic synthesis pathways of various *Streptomyces* spp., long known for their antibiotic production capabilities, can be modified to improve yields or to create new antibiotics through the introduction of genes encoding additional enzymes. More than 200 new antibiotics have been generated through the targeted inactivation of genes and the novel combination of antibiotic synthesis genes in antibiotic-producing *Streptomyces* hosts.

Genetic engineering is also used to manufacture subunit vaccines, which are safer than other vaccines because they contain only a single antigenic molecule and lack any part of the genome of the pathogen (see Vaccines). For example, a vaccine for hepatitis B is created by inserting a gene encoding a hepatitis B surface protein into a yeast; the yeast then produces this protein, which the human immune system recognizes as an antigen. The hepatitis B antigen is purified from yeast cultures and administered to patients as a vaccine. Even though the vaccine does not contain the hepatitis B virus, the presence of the antigenic protein stimulates the immune system to produce antibodies that will protect the patient against the virus in the event of exposure.

Genetic engineering has also been important in the production of other therapeutic proteins, such as insulin, interferons, and human growth hormone, to treat a variety of human medical conditions. For example, at one time, it was possible to treat diabetes only by giving patients pig insulin, which caused allergic reactions due to small differences between the proteins expressed in human and pig insulin. However, since 1978, recombinant DNA technology has been used to produce large-scale quantities of human insulin using *E. coli* in a relatively inexpensive process that yields a more consistently effective pharmaceutical product. Scientists have also genetically engineered *E. coli* capable of producing human growth hormone (HGH), which is used to treat growth disorders in children and certain other disorders in adults. The HGH gene was cloned from a cDNA library and inserted into *E. coli* cells by cloning it into a bacterial vector. Eventually, genetic engineering will be used to produce DNA vaccines and various gene therapies, as well as customized medicines for fighting cancer and other diseases.

Table 14.1 Some Genetically Engineered Pharmaceutical Products and Applications

[illegible]	[illegible]
Atrial natriuretic peptide	Treatment of heart disease (e.g., congestive heart failure), kidney disease, high blood pressure
D Nase	Treatment of viscous lung secretions in cystic fibrosis
Erythropoietin	Treatment of severe anemia with kidney damage
Factor VIII	Treatment of hemophilia
Hepatitis B vaccine	Prevention of hepatitis B infection
Human growth hormone	Treatment of growth hormone deficiency, Turner's syndrome, burns
Human insulin	Treatment of diabetes
Interferons	Treatment of multiple sclerosis, various cancers (e.g., melanoma), viral infections (e.g., Hepatitis B and C)
Tetracenomycins	Used as antibiotics
Tissue plasminogen activator	Treatment of pulmonary embolism in ischemic stroke, myocardial infarction

RNA Interference Technology

In **Structure and Function of RNA**, we described the function of mRNA, rRNA, and tRNA. In addition to these types of RNA, cells also produce several types of small noncoding RNA molecules that are involved in the regulation of gene expression. These include **antisense RNA** molecules, which are complementary to regions of specific mRNA molecules found in both prokaryotes and eukaryotic cells. Non-coding RNA molecules play a major role in **RNA interference (RNAi)**, a natural regulatory mechanism by which mRNA molecules are prevented from guiding the synthesis of proteins. RNA interference of specific genes results from the base pairing of short, single-stranded antisense RNA molecules to regions within complementary mRNA molecules, preventing protein synthesis. Cells use RNA interference to protect themselves from viral invasion, which may introduce double-stranded RNA molecules as part of the viral replication process (Figure 14.13).

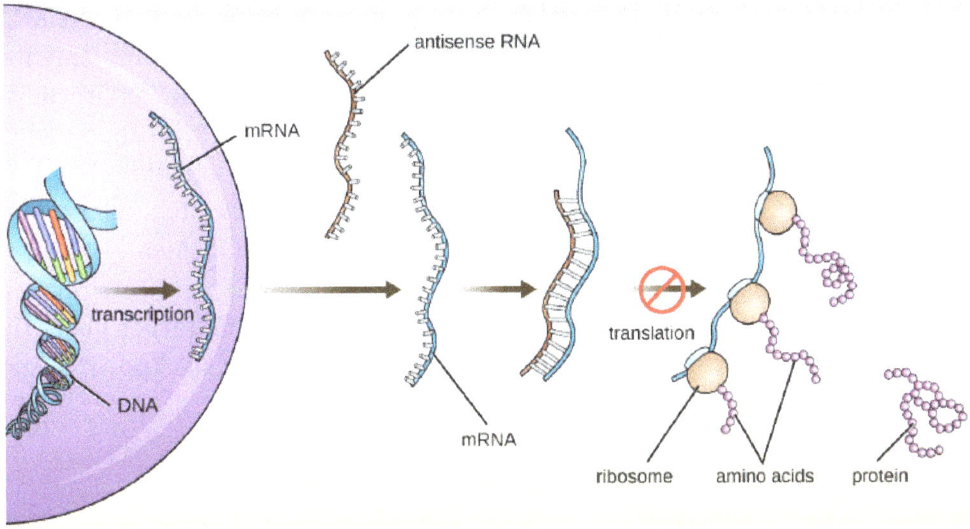

Figure 14.13 Cells like the eukaryotic cell shown in this diagram commonly make small antisense RNA molecules with sequences complementary to specific mRNA molecules. When an antisense RNA molecule is bound to an mRNA molecule, the mRNA can no longer be used to direct protein synthesis. (credit: modification of work by Robinson R)

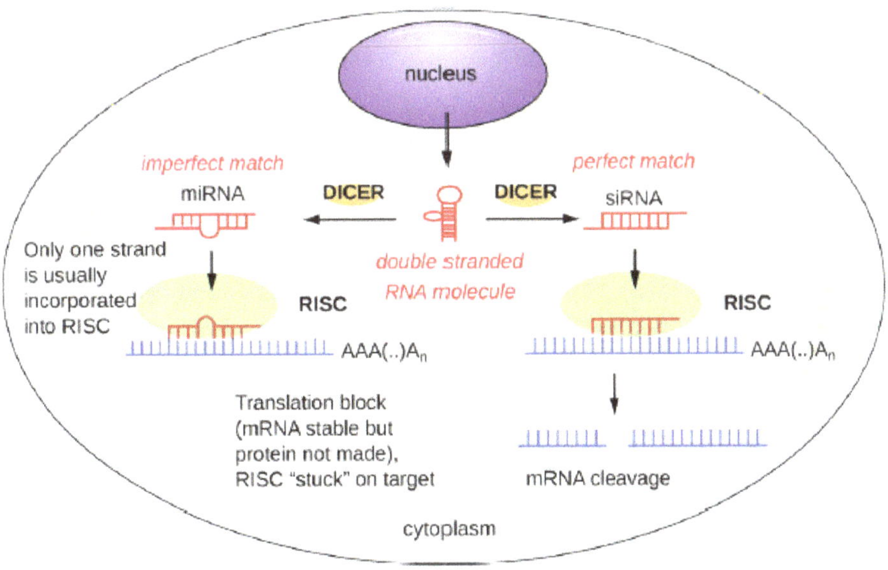

Figure 14.14 The process of using siRNA or miRNA in a eukaryotic cell to silence genes involved in the pathogenesis of various diseases. (credit: modification of work by National Center for Biotechnology Information)

Researchers are currently developing techniques to mimic the natural process of RNA interference as a way to treat viral infections in eukaryotic cells. RNA interference technology involves using small interfering RNAs (siRNAs) or microRNAs (miRNAs) (Figure 14.14). siRNAs are completely complementary to the mRNA transcript of a specific gene of interest while miRNAs are mostly complementary. These double-stranded RNAs are bound to DICER, an endonuclease that cleaves the RNA into short molecules (approximately 20 nucleotides long). The RNAs are then bound to RNA-induced silencing complex (RISC), a ribonucleoprotein. The siRNA- RISC complex binds to mRNA and cleaves it. For miRNA, only one of the two strands binds to RISC. The miRNA-RISC complex then binds to mRNA, inhibiting translation. If the miRNA is completely complementary to the target gene, then the mRNA can be cleaved. Taken together, these mechanisms are known as **gene silencing.**

CELLULAR DISRUPTION TECHNIQUES

The first step of downstream processing development is determining the best method to break open the cells to release the recombinant protein. There are many methods for cell disruption to choose from with the goal of determining the gentlest method that gives the highest yield of functioning protein. If the extraction method is too harsh, it risks denaturing and destroying the target protein. If the extraction method is too mild, the cells will remain intact and may lose precious recombinant protein stuck inside the cell. This module discusses various cell disruption techniques.

Cell Disruption Techniques

According to an Evaluate Pharma report, the worldwide pharmaceutical industry has now topped a trillion dollars in sales, with most of this growth propelled by new FDA approvals of recombinant protein products such as monoclonal antibody therapeutics, biologics, and biosimilars (generic protein products). In the U.S. alone there were over $250 billion in annual sales and upwards of $150 billion invested in biopharmaceutical R&D (EvaluatePharma, 2016). A critical part of research and development in making recombinant protein products is deciding what organism to use in the manufacture of the recombinant protein and what purification steps to employ in separating the target protein from the cellular contaminants.

Expression Vectors

As mentioned previously, a gene of interest may be inserted into a vector, and the recombinant plasmid is placed into a cell where the gene can be expressed. For instance, one might desire to clone the gene coding for human growth hormone or insulin or other medically important proteins and have a bacterium or yeast make large quantities of it very cheaply.

To clone a gene so that it can be expressed, one needs to set up the proper conditions for the human protein to be made in the bacterial cells. This typically involves the use of specially designed plasmids. These plasmids have been engineered to 1) replicate in high numbers; 2) carry markers that allow researchers to identify cells carrying them (antibiotic resistance, for example) and 3) contain sequences (such as a promoter and Shine- Dalgarno sequence) necessary for expression of the desired protein, with convenient sites for insertion of the gene of interest in the appropriate place relative to the control sequences. A plasmid which has all these features is referred to as an expression vector.

In addition to plasmids that can be used for expression in bacterial cells, expression vectors are also available that allow protein expression in a variety of eukaryotic cells. Many sophisticated variations on such vectors have been created that have made it easy to produce and purify large amounts of any protein of interest for which the gene has been cloned. A handy feature in some expression vectors is a sequence encoding an affinity tag either up or downstream of the gene being expressed. This sequence allows a short affinity tag (such as a run of histidine residues) to be fused onto the encoded protein. The tag can be used to readily purify the protein, as described in the section on affinity chromatography.

Recombinant Protein Purification

The optimal protein purification strategy for a biopharmaceutical depends on the properties of the protein, the starting concentration of the extract, and the types of contaminating materials present. Most proteins produced commercially rely on fermentation by microbial or animal cell culture. The process of harvesting and purifying a protein in an industrial setting is referred to as "downstream processing" and includes all steps of production downstream of the fermentation step. Since downstream processing of a protein can often exceed all other costs of production combined, it must be a carefully designed strategy, often requiring extensive development by scientists and engineers.

Downstream Processing

- Since most proteins are not secreted from cells, the first phase of downstream processing consists of cell disruption, followed by removal of cell debris. Cell disruption can be performed using a relatively mild treatment with chemicals or with a more severe physical disruption by sonication. Clearing the lysate of insoluble debris can be carried out using centrifugation or by filtration.
- Since processing of large volumes is expensive, the first purification step usually includes concentrating the protein extract to a smaller volume. Protein can be concentrated via precipitation or by ultrafiltration.

- Once the protein solution volume is reduced to a more manageable size, purification can proceed. Column chromatography offers the highest resolution. Some examples include size exclusion, ion-exchange, and hydrophobic interaction chromatography.
- For biopharmaceutical products that require higher levels of purity, more sophisticated techniques can be used to polish the purified protein and may include immunoaffinity techniques in combination with high- performance liquid chromatography (HPLC).
- When the protein is purified sufficiently, it is dried by lyophilization, freeze-dried, or formulated into a solution that stabilizes its activity and integrity.

Cell Disruption Methods

The first step of downstream processing development is determining the best method to break open the cells to release the recombinant protein. There are many methods for cell disruption to choose from with the goal of determining ***the gentlest method that gives the highest yield of functioning protein.*** If the extraction method is too harsh, it risks denaturing and destroying the target protein. If the extraction method is too mild, the cells will remain intact and may lose precious recombinant protein stuck inside the cell.

Part of picking the right method is realizing different types of cells have diverse cellular structures and will, therefore, require appropriate cell disruption methods. Some organisms have cell walls or capsules surrounding the plasma membrane, and others only a plasma membrane. For example, different cell walls require different enzymes to break them; others may need the assistance of sheer force. However, some proteins are susceptible to the shear forces required to break down cell walls; others are sensitive to damage by exposure to air, or heat from the extraction method, or nucleases present in the cellular extract. So how can the target protein stay intact while disrupting as many cells as possible? This is typically done experimentally, through processes of trial and error development. A summary of common cell disruption methods is provided in the table 14.2.

Table 14.2 Summary of common cell disruption techniques

[illegible]	[illegible]	[illegible]
	Cell lysis	Osmotic shock: rapid immersion in a hypotonic solution
	Enzyme Digest/cell lysis	Digestion of cell wall; contents release following osmotic shock

Relative Hardness	Cell Disruption Method	Description
	Potter-Elvehjem homogenizer	Cells forced through a narrow gap, disruption by shear forces
Moderate	Freeze-thaw	Slow freeze-thaw cycles break cell walls by ice crystal formation
Harsh	Waring blender	Shear forces of rotating blades break cells
	Organic extraction	Mixing with an immiscible organic solvent to dissolve membranes
	Grinding	Abrasive grinding with glass beads or a mortar and pestle
Vigorous	French press cell	Cells forced through a small orifice at high pressure
	Bead mill	Rapid vibrations with glass beads grind cell walls
(most harsh)	Ultra-sonication	High-pressure sound waves cause cell rupture by cavitation

Osmotic shock and enzymes: One way to lyse cells is by lowering the ionic strength of the cellular medium. This can cause cells to swell and burst. Mild surfactants may be used to disrupt membranes. Most bacteria, yeast, and plant tissues are resistant to osmotic shocks, because of the presence of cell walls, and stronger disruption techniques are usually required. Enzymes may be useful in helping to degrade the cell walls. Lysozyme, for example, is very useful for breaking down bacterial walls. Other enzymes commonly employed include cellulase (plants), proteases, mannoses, and others.

Mechanical disruption: Mechanical agitation may be employed in the form of beads that are shaken with a mixture of cells. In this method, cells are bombarded with tiny, glass beads that break the cells open. **Sonication** (2050 kHz sound waves) provides an alternative type of agitation that can be effective. The method is noisy, however, and generates heat that can be problematic for heat-sensitive compounds.

Pressure disruption: Another means of disrupting cells involves using a "cell bomb." In this method, cells are placed under very high pressure (up to 25,000 psi) and then the pressure is rapidly released. The rapid pressure change causes dissolved gases in cells to be released as bubbles, which, in turn, break open cells.

Cryogenic Pulverizing: Cryogenic pulverizing, or grinding, is often employed for samples having a tough extracellular matrix, such as connective tissue, seed, and cartilage. In this technique, tissues are frozen using liquid nitrogen and then impact pulverization (typically, grinding, using a mortar and pestle or a powerful electric grinder) is performed. The powder so obtained is then suspended in the appropriate buffer.

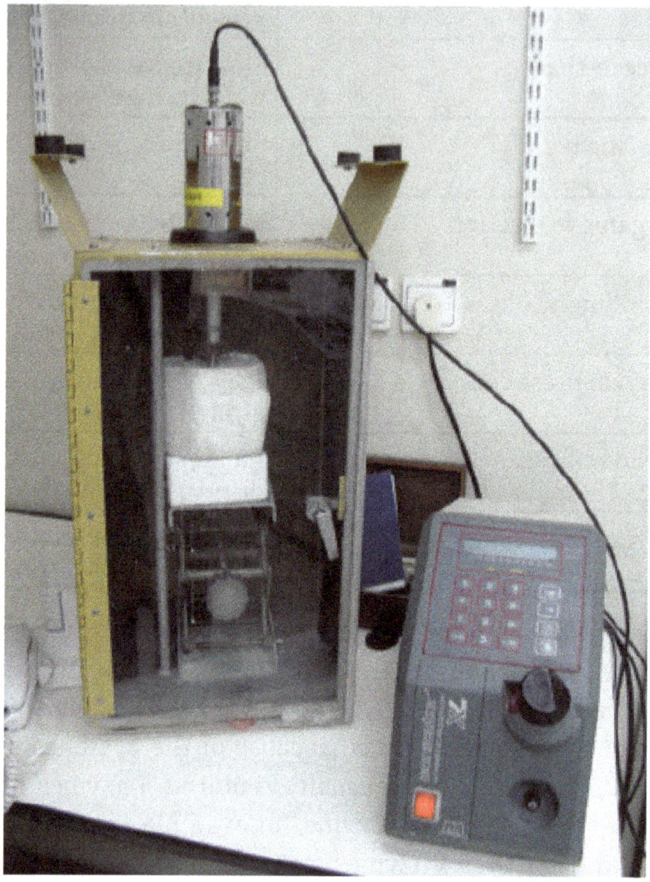

Figure 14.15 A sonication apparatus is set up in a sound containment box. A long metal probe is inserted into a cell sample sitting in ice. Sonication waves are transmitted through a cable attached to the metal probe controlled by an outside controller. Image credit: Dr. Kevin Ahern (http://biochem.science.oregonstate.edu/content/kevin- ahern) and Dr. Indira Rajagopal (http://biochem.science.oregonstate.edu/content/indira-rajagopal) https://biochem.science.oregonstate.edu/content/biochemistry-free-and-easy

Protease Inhibitors

Proteins must be protected from degradation by proteases during isolation steps by working ***quickly*** and working ***cold*** (keeping all reagents on ice), while also being mindful of ***buffering*** condition. Additionally, protease inhibitors can be employed to reduce proteolytic damage to proteins. For example, PMSF is a broad inhibitor and is successfully used in animal, plant and yeast tissues. Other inhibitors include EDTA, aprotinin, chromostatin, pepstatin, and benzamidine.

Clarification

Once cells have been disrupted, the cell lysate must be clarified, removing cell debris and unbroken cells either by filtration or by centrifugation. Cellular materials are highly compressible, they rapidly clog pores of filters, so centrifugation is usually the method used to remove particulate matter when the volumes are small.

Fractionation: Using low-speed centrifugation, one can remove cell debris, leaving a supernatant containing the contents of the cell. By using successively higher centrifugation speeds (and resulting g-forces), it is possible to separate different cellular components, like nuclei, mitochondria from the cytoplasm. These may then be separately lysed to release molecules that are specific to a cellular compartment. The soluble fraction of any lysate can, then, be further separated into its constituents using various methods. The most common for biopharmaceuticals is **column chromatography**.

COLUMN CHROMATOGRAPHY

Even when produced by a recombinant host at high expression levels, recombinant protein extracts are complex mixtures and frequently require multiple purification steps to remove the contaminants. The protein must be separated from hundreds of other proteins with similar properties, as well as all the lipids, nucleic acids, and carbohydrate-containing biomolecules of the cell. One of the most effective and powerful tools of protein purification is column chromatography; a technique used to separate the components of a mixture based on differences in their size, charge or other characteristics.

Even when produced by a recombinant host at high expression levels, recombinant protein extracts are complex mixtures and frequently require multiple purification steps to remove the contaminants. The protein must be separated from hundreds of other proteins with similar properties, as well as all the lipids, nucleic acids, and carbohydrate-containing biomolecules of the cell. One of the most effective and powerful tools of protein purification is column chromatography.

We will consider several chromatographic approaches in this section. Chromatography is used to separate the components of a mixture based on differences in their size, charge or other characteristics. During chromatography, the mobile phase (buffer or another solvent) moves through the stationary phase (usually a solid matrix) carrying the components of the mixture. Separation of the components is achieved because of the different affinities the components have for the mobile and stationary phases. Molecules in a chromatography method move at different rates, for reasons that vary, depending on the type of chromatography used.

Purification Strategy

There are many types of column chromatography, based on the chemistry of the stationary phase. There is no single best way to purify all proteins. Finding an optimal protein purification strategy requires trial and error because the best method for purification depends on the properties of the protein being isolated as well as the other contaminants in the lysate. In choosing the best types of chromatography to use, there are at least five factors that must be evaluated and include resolution (how well the method removes target away from contaminants), capacity, speed, percent recovery, and cost. In the next two labs, students will separate their target recombinant protein using first a gravity HIC column, then using an ion-exchange FPLC process.

We will consider several different kinds of chromatography to illustrate this process. 1. Ion exchange chromatography 2. Gel exclusion chromatography 3. Affinity chromatography 4. HPLC These variations on chromatography are performed with the stationary phase (matrix) held within columns (Figure). Matrix supports are composed of tiny beads suspended in buffer (Figure) and are designed to exploit the chemistry or size differences of the components of the samples and thus provide a means of separation. Columns are "packed" or filled into a column, and a buffer or solvent carries the mixture of compounds to be separated. Molecules in the sample interact differentially with the support and, consequently, travel through it at different speeds, thus enabling separation.

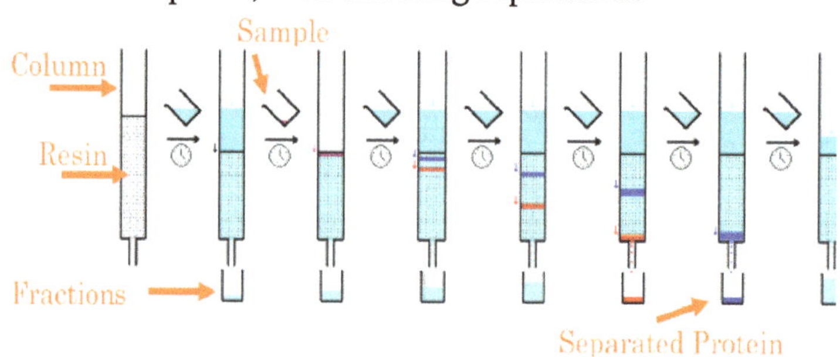

Figure 14.16 Schematic diagram demonstrating the separation of molecules using column chromatography and identifying important elements: column, resin, fraction, sample, and separated molecules. Image Credit: Edited by J. O'Grady, original, by quantockgoblin - Own work, Public Domain, https:// upload.wikimedia.org/wikipedia/ commons/3/3f/ Column_chromatography_ sequence.png (https://upload.wikimedia.org/wikipedia/commons/3/ 3f/ Column_chromatography_sequence.png)

Size Exclusion Chromatography (SEC)

Size exclusion chromatography (also called molecular exclusion chromatography, gel exclusion chromatography, or gel filtration chromatography) is a low-resolution separation method that employs beads with tiny "tunnels" in them that each has a precise opening. In SEC, separation on the column is a function of size alone: larger molecules are excluded from the pores of the chromatography beads and elute from the column more quickly than smaller molecules (Figure).

In a size exclusion column, the resolving power is a function of the length of the column on which the proteins are separated on, as well as the pore size of the beads, the flow rate of the elution buffer, and the relative sizes of proteins being separated. Since the permeation of the beads by smaller proteins is a diffusion dependent process, the flow rates should be slow enough to allow for this to happen. In general, the resolution of SEC on small columns is not great, but this step can be useful for changing the buffer in which the protein is dissolved in, to the buffer used to elute the proteins from the column. Significant drawbacks to SEC are that the volume of the protein sample applied to the column must not exceed 5% of the column volume, and the protein eluted from the column is diluted by a factor of 10-fold. This means that an SEC step must usually be followed by a step that will concentrate the protein fraction.

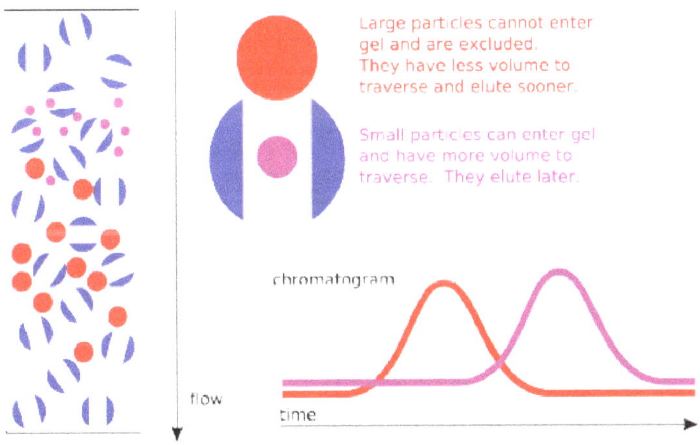

Figure 14.17 Cartoon image is depicting size exclusion chromatography. Left is a cartoon column showing blue beads and large molecules being excluded from them, while pink smaller molecules are entering the beads and have more volume to traverse, delaying their exit from the column.

Adsorption Chromatography

The other forms of chromatography are a form of "adsorption" chromatography, in that the protein that is applied to the column matrix will be adsorbed to the solid phase matrix, allowing separation of proteins that have no affinity to the stationary phase. The proteins left in the column can then be eluted by changing the buffering conditions of the mobile phase. A gradual change in buffering conditions can elute proteins one at a time, allowing their separation from each other.

Adsorption is a surface phenomenon. In the example of column chromatography, protein is adsorbed (collected) on the surface of a solid phase matrix. This is very different from ***absorption***, which describes a substance that diffuses into a liquid or solid to form a solution. For example, in spectrophotometry, we are measuring light being absorbed by a sample.

Adsorption chromatography can be used to concentrate proteins since a large volume of protein can be applied to the column without affecting the amount of protein adsorbed to the column matrix. Careful selection of elution conditions can remove the protein from the column in a much smaller volume. Since adsorption chromatography allows for both the isolation and concentration of a target protein, it is sometimes referred to as the "capture" phase of protein purification.

Ion Exchange Chromatography

In ion exchange chromatography, the support consists of tiny beads to which are attached chemicals possessing a charge. Before use, the beads are equilibrated in a solution containing an appropriate counter-ion to the charged molecule on the bead. Uses Ion exchange resins are useful for separating charged from uncharged, or oppositely charged, biomolecules in solution. The two types of ion chromatography are anion-exchange and cation-exchange.

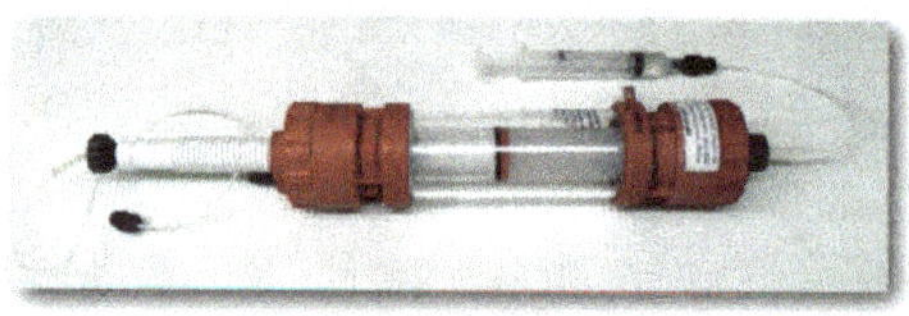

An ion exchange column apparatus

Ion exchange beads

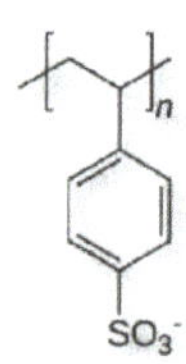

Polystyrolsulfonate, a cation exchange resin

Figure 14.18

Cation-exchange chromatography is used when the molecule of interest is positively charged. In this type of chromatography, the stationary phase is negatively charged, and positively charged molecules are loaded to be attracted to it.

Anion-exchange chromatography is when the stationary phase is positively charged, and negatively charged molecules are loaded to be attracted to it. The bound molecules then can be eluted and collected using an eluant which contains anions and cations by running higher concentration of ions through the column or changing pH of the column.

To remove the molecules "stuck" to a column, add a high concentration of counter-ions to release them.

The figure shows the repeating unit of polystyrolsulfonate, a compound used as a ***cation exchange resin***. The molecule is negatively charged, and thus the beads would be equilibrated in a buffer containing a positively charged ion, say sodium. Positively charged compounds from a cell lysate passed through the column will exchange with the counter-ions and "stick" to the negatively charged compounds covalently attached to the beads. Molecules in the sample that are neutral in charge or negatively charged will pass through the column. At this point, only positively charged molecules from the original sample would be bound to the column. These may then be washed off, or eluted, by using buffers containing high concentrations of salt. Under these conditions, the interaction between the positively charged molecules and the polystyrosulfonate would be disrupted, allowing the molecules that were bound to the column to be recovered.

Hydrophobic Interaction Chromatography (HIC)

Separation by HIC is based on the reversible interaction between a protein and a hydrophobic surface of chromatographic beads. This interaction is enhanced in the presence of a high ionic strength buffer solution, such as ammonium sulfate. Hydrophobic proteins bind to the column as they are loaded, and proteins with low affinity and other contaminants are washed away. Conditions are then altered so that the bound protein can be eluted differentially, usually by decreasing the salt concentration, or by the addition of chaotropic chemicals, or changes in pH.

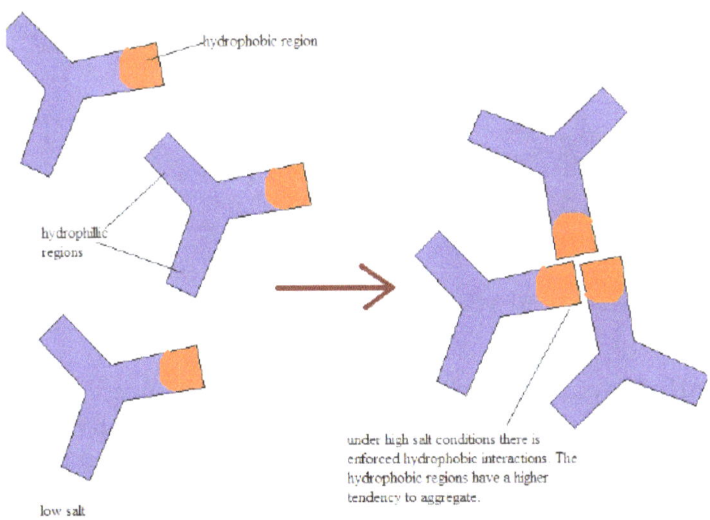

Figure 14.19 A cartoon image depicting hydrophobic interaction. Shown is 3, Y-shaped proteins, two prongs hydrophilic purple, one prong is hydrophobic and orange.

Affinity Chromatography

Affinity chromatography is a very powerful and selective technique that exploits the binding affinities of sample molecules (typically proteins) for molecules covalently linked to the support beads. In contrast to ion-exchange chromatography, where all molecules of a given charge would bind to the column, ***affinity chromatography exploits the specific binding of a protein or proteins to a ligand that is immobilized on the beads in the column.*** There are many high-resolution approaches and may include the use of antibody, nucleic acid, lectin, and other ligands.

Histidine Tagging

Histidine tagging (His-tagging) is a special kind of affinity chromatography and is a powerful tool for isolating a recombinant protein from a cell lysate. His-tagging relies on altering the DNA coding region for a protein to add a series of at least six histidine residues to the amino or carboxyl terminal of the encoded protein. This "His-Tag" is useful in purifying the tagged protein because histidine side chains are not found in nature and can readily bind to nickel or cobalt ions.

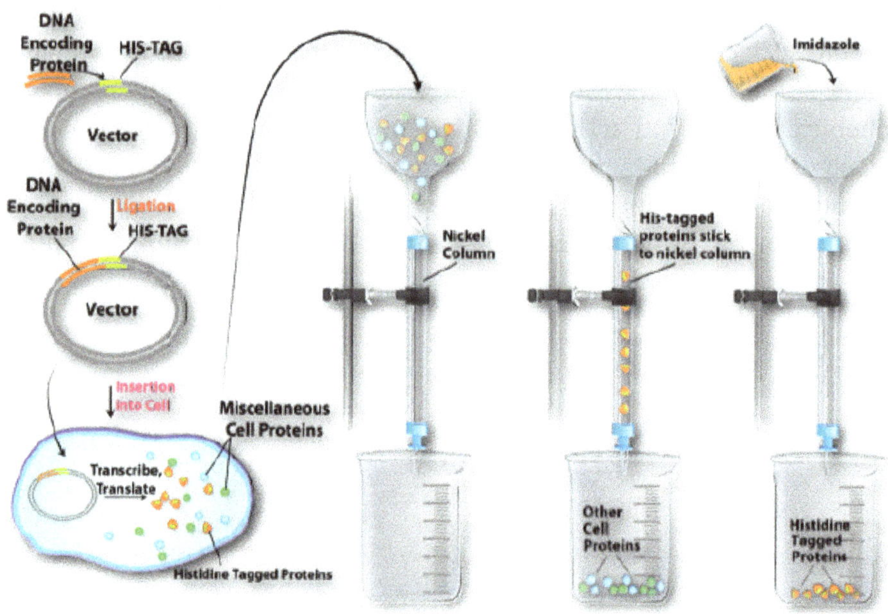

Figure 14.20 Affinity chromatographic purification of a protein by histidine tagging. A cartoon demonstrating how IMAC vector is made and the HIS-tag is used to purify target protein. Left is a vector diagram showing how a gene is inserted beside a HIS-tag.

Separation of His-tagged proteins from a cell lysate is relatively easy (Figure). Passing the crude cell lysate through a column with nickel or cobalt attached to beads allows the His-tagged proteins to "stick," while the remaining cell proteins all pass quickly through. The His-tagged proteins are then eluted by addition of imidazole to the column. Imidazole, which resembles the side chain of histidine, competes with the His-tagged proteins and displaces them from the column. Surprisingly, many His-tagged proteins appear to function normally despite the added histidine molecules, but if needed, the histidine tags may be cleaved from the purified protein by treatment with a protease that excises the added histidine's, allowing the recovery of the desired protein with its native sequence.

Fast Protein Liquid Chromatography (FPLC)

Fast protein liquid chromatography (FPLC), is a form of liquid column chromatography that is often used to analyze or purify mixtures of proteins. Separation of proteins is possible because the components of a protein mixture have different affinities for two materials: a moving mobile phase, and a porous solid stationary phase. ***In FPLC, the mobile phase is a mixture of aqueous buffers that moves through the column via a positive- displacement pump.*** The positive-displacement pump controls the mobile phase buffer flow at a constant rate while adjusting the composition of the buffer by drawing fluids in different proportions from two or more external reservoirs. The stationary phase is a resin composed of beads, usually of cross-linked agarose, packed into a cylindrical glass or plastic column. FPLC resins are available in a wide range of bead sizes and surface ligands depending on the application.

For example, in one type of FPLC strategy, ion exchange, the resin binds the protein of interest via a charge interaction while in buffer A (the running buffer) but become dissociated and return to solution in buffer B (the elution buffer). A mixture containing the target protein of interest is dissolved in 100% buffer A and pumped into the column. The proteins of interest bind to the resin while other contaminants are eluted out. The total flow rate of the buffer is kept constant; however, the proportion of Elution Buffer B is gradually increased from 0% to 100% according to a programmed change in concentration creating a "gradient" flow. The target protein dissociates and elutes; the effluent passes two detectors that measure conductivity and absorption of UV light at 280nm and collected in the fraction collector for further use.

The advantages to FPLC include a better resolution and much faster than gravity, it is automated in both the chromatography method and fraction collection and analysis, and it offers better separation through gradient elution techniques.

High-Performance Liquid Chromatography (HPLC)

HPLC High-performance liquid chromatography (HPLC) is a powerful tool for separating a variety of molecules based on their differential polarities. A more efficient form of column chromatography, it employs columns with tightly packed supports and very tiny beads such that flow of solvents/buffers through the columns requires high pressures. The supports used may be polar (normal phase separation) or non-polar (reverse phase separation). In normal phase separations, non-polar molecules elute first followed by the more polar compounds. This order is switched in reverse phase chromatography. Of the two, the reverse phase is commonly employed to do more reproducible chromatographic profiles (separations) that it typically produces.

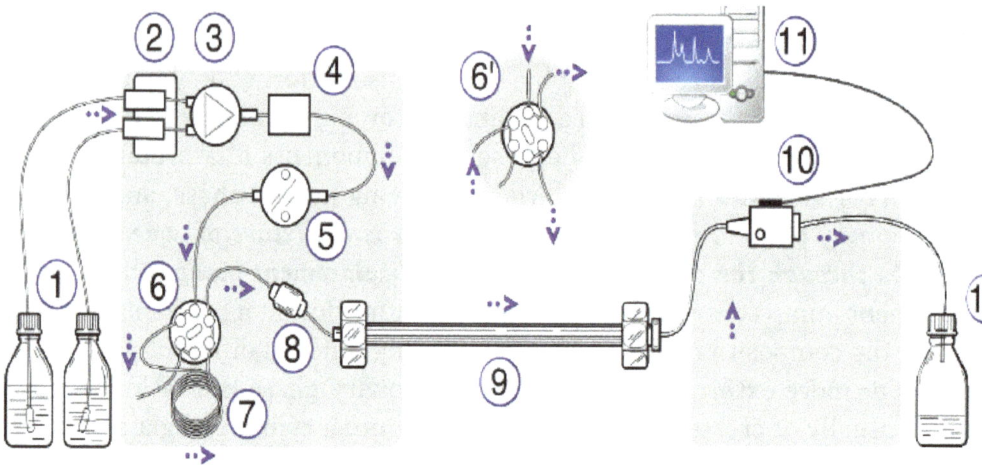

Figure 14.21 Schematic diagram of an HPLC system. (1) Solvent reservoirs, (2) Solvent degasser, (3) Gradient valve, (4) Mixing vessel for delivery of the mobile phase, (5) High-pressure pump, (6) Switching valve in "inject position", (6') Switching valve in "load position", (7) Sample injection loop, (8) Pre-column (guard column), (9) Analytical column, (10) Detector (i.e., IR, UV), (11) Data acquisition, (12) Waste or fraction collector. Image credit: https://en.wikipedia.org/ wiki/ High-performance_liquid_chromatography#/media/ File:HPLC_apparatus.svg (https://en.wikipedia.org/ wiki/High-performance_liquid_chromatography #/ media/File:HPLC_apparatus.svg)

ADVANCED CHROMATOGRAPHY APPLICATIONS

Accurate identification of bacterial isolates is essential in a clinical microbiology laboratory because the results often inform decisions about treatment that directly affect patient outcomes. For example, cases of food poisoning require accurate identification of the causative agent so that physicians can prescribe appropriate treatment. Likewise, it is important to accurately identify the causative pathogen during an outbreak of disease so that appropriate strategies can be employed to contain the epidemic.

There are many ways to detect, characterize, and identify microorganisms. Some methods rely on phenotypic biochemical characteristics, while others use genotypic identification. The biochemical characteristics of a bacterium provide many traits that are useful for classification and identification. Analyzing the nutritional and metabolic capabilities of the bacterial isolate is a common approach for determining the genus and the species of the bacterium. Some of the most important metabolic pathways that bacteria use to survive will be discussed in Microbial Metabolism . Some microorganisms store certain compounds as granules within their cytoplasm, and the contents of these granules can be used for identification purposes. For example, poly-β-hydroxybutyrate (PHB) is a carbon- and energy-storage compound found in some nonfluorescent bacteria of the genus *Pseudomonas*. Different species within this genus can be classified by the presence or the absence of PHB and fluorescent pigments. The human pathogen *P. aeruginosa* and the plant pathogen *P. syringae* are two examples of fluorescent *Pseudomonas* species that do not accumulate PHB granules.

Other systems rely on biochemical characteristics to identify microorganisms by their biochemical reactions, such as carbon utilization and other metabolic tests. In small laboratory settings or in teaching laboratories, those assays are carried out using a limited number of test tubes. However, more modern systems, such as the one developed by Biolog, Inc., are based on panels of biochemical reactions performed simultaneously and analyzed by software. Biolog's system identifies cells based on their ability to metabolize certain biochemicals and on their physiological properties, including pH and chemical sensitivity. It uses all major classes of biochemicals in its analysis. Identifications can be performed manually or with the semi- or fully automated instruments.

Another automated system identifies microorganisms by determining the specimen's mass spectrum and then comparing it to a database that contains known mass spectra for thousands of microorganisms. This method is based on matrix-assisted laser desorption/ionization time-of-flight mass spectrometry (MALDI-TOF) and uses disposable MALDI plates on which the microorganism is mixed with a specialized matrix reagent (Figure 14.22). The sample/reagent mixture is irradiated with a high-intensity pulsed ultraviolet laser, resulting in the ejection of gaseous ions generated from the various chemical constituents of the microorganism. These gaseous ions are collected and accelerated through the mass spectrometer, with ions travelling at a velocity determined by their mass-to-charge ratio (m/z), thus, reaching the detector at different times. A plot of detector signal versus m/z yields a mass spectrum for the organism that is uniquely related to its biochemical composition. Comparison of the mass spectrum to a library of reference spectra obtained from identical analyses of known microorganisms permits identification of the unknown microbe.

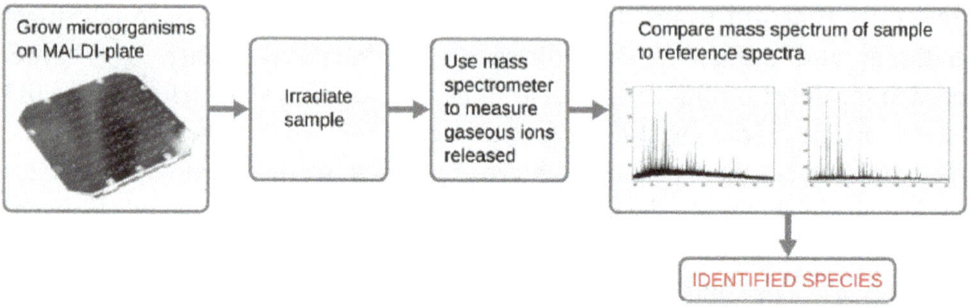

Figure 14.22 MALDI-TOF methods are now routinely used for diagnostic procedures in clinical microbiology laboratories. This technology is able to rapidly identify some microorganisms that cannot be readily identified by more traditional methods.

Microbes can also be identified by measuring their unique lipid profiles. As we have learned, fatty acids of lipids can vary in chain length, presence or absence of double bonds, and number of double bonds, hydroxyl groups, branches, and rings. To identify a microbe by its lipid composition, the fatty acids present in their membranes are analyzed. A common biochemical analysis used for this purpose is a technique used in clinical, public health, and food laboratories. It relies on detecting unique differences in fatty acids and is called **fatty acid methyl ester (FAME) analysis.** In a FAME analysis, fatty acids are extracted from the membranes of microorganisms, chemically altered to form volatile methyl esters, and analyzed by gas chromatography (GC). The resulting GC chromatogram is compared with reference chromatograms in a database containing data for thousands of bacterial isolates to identify the unknown microorganism (Figure 14.23).

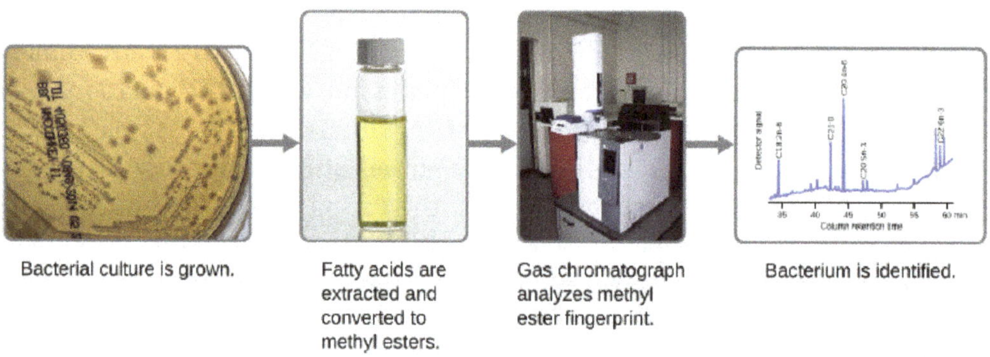

Figure 14.23 Fatty acid methyl ester (FAME) analysis in bacterial identification results in a chromatogram unique to each bacterium. Each peak in the gas chromatogram corresponds to a particular fatty acid methyl ester and its height is proportional to the amount present in the cell.

A related method for microorganism identification is called **phospholipid-derived fatty acids (PLFA) analysis.** Membranes are mostly composed of phospholipids, which can be saponified (hydrolyzed with alkali) to release the fatty acids. The resulting fatty acid mixture is then subjected to FAME analysis, and the measured lipid profiles can be compared with those of known microorganisms to identify the unknown microorganism.

Bacterial identification can also be based on the proteins produced under specific growth conditions within the human body. These types of identification procedures are called **proteomic analysis.** To perform proteomic analysis, proteins from the pathogen are first separated by high-pressure liquid chromatography (HPLC), and the collected fractions are then digested to yield smaller peptide fragments. These peptides are identified by mass spectrometry and compared with those of known microorganisms to identify the unknown microorganism in the original specimen.

Microorganisms can also be identified by the carbohydrates attached to proteins (glycoproteins) in the plasma membrane or cell wall. Antibodies and other carbohydrate-binding proteins can attach to specific carbohydrates on cell surfaces, causing the cells to clump together. Serological tests (e.g., the Lancefield groups tests, which are used for identification of *Streptococcus* species) are performed to detect the unique carbohydrates located on the surface of the cell.

VISUALIZING AND CHARACTERIZING DNA, RNA, AND PROTEIN

The sequence of a DNA molecule can help us identify an organism when compared to known sequences housed in a database. The sequence can also tell us something about the function of a particular part of the DNA, such as whether it encodes a particular protein. Comparing **protein signatures**—the expression levels of specific arrays of proteins—between samples is an important method for evaluating cellular responses to a multitude of environmental factors and stresses. Analysis of protein signatures can reveal the identity of an organism or how a cell is responding during disease.

The DNA and proteins of interest are microscopic and typically mixed in with many other molecules including

DNA or proteins irrelevant to our interests. Many techniques have been developed to isolate and characterize molecules of interest. These methods were originally developed for research purposes, but in many cases they have been simplified to the point that routine clinical use is possible. For example, many pathogens, such as the bacterium *Helicobacter pylori*, which causes stomach ulcers, can be detected using protein-based tests. In addition, an increasing number of highly specific and accurate DNA amplification-based identification assays can now detect pathogens such as antibiotic-resistant enteric bacteria, herpes simplex virus, varicella-zoster virus, and many others.

Molecular Analysis of DNA

In this subsection, we will outline some of the basic methods used for separating and visualizing specific fragments of DNA that are of interest to a scientist. Some of these methods do not require knowledge of the complete sequence of the DNA molecule. Before the advent of rapid DNA sequencing, these methods were the only ones available to work with DNA, but they still form the basic arsenal of tools used by molecular geneticists to study the body's responses to microbial and other diseases.

Nucleic Acid Probing

DNA molecules are small, and the information contained in their sequence is invisible. How does a researcher isolate a particular stretch of DNA, or having isolated it, determine what organism it is from, what its sequence is, or what its function is? One method to identify the presence of a certain DNA sequence uses artificially constructed pieces of DNA called probes. Probes can be used to identify different bacterial species in the environment and many DNA probes are now available to detect pathogens clinically. For example, DNA probes are used to detect the vaginal pathogens *Candida albicans*, *Gardnerella vaginalis*, and *Trichomonas vaginalis*.

To screen a genomic library for a particular gene or sequence of interest, researchers must know something about that gene. If researchers have a portion of the sequence of DNA for the gene of interest, they can design a **DNA probe**, a single-stranded DNA fragment that is complementary to part of the gene of interest and different from other DNA sequences in the sample. The DNA probe may be synthesized chemically by commercial laboratories, or it may be created by cloning, isolating, and denaturing a DNA fragment from a living organism. In either case, the DNA probe must be labelled with a molecular tag or beacon, such as a radioactive phosphorus atom (as is used for **autoradiography**) or a fluorescent dye (as is used in fluorescent *in situ* hybridization, or FISH), so that the probe and the DNA it binds to can be seen (Figure 14.25). The DNA sample being probed must also be denatured to make it single-stranded so that the single-stranded DNA probe can anneal to the single-stranded DNA sample at locations where their sequences are complementary. While these techniques are valuable for diagnosis, their direct use on sputum and other bodily samples may be problematic due to the complex nature of these samples. DNA often must first be isolated from bodily samples through chemical extraction methods before a DNA probe can be used to identify pathogens.

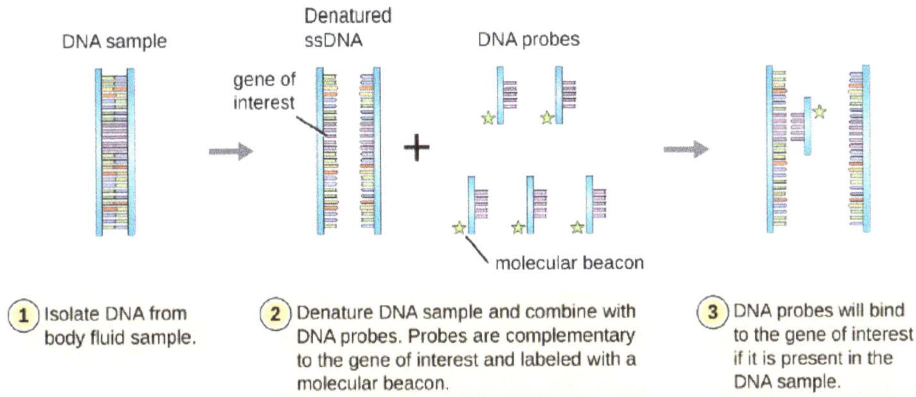

Figure 14.25 DNA probes can be used to confirm the presence of a suspected pathogen in patient samples. This diagram illustrates how a DNA probe can be used to search for a gene of interest associated with the suspected pathogen.

Agarose Gel Electrophoresis

There are a number of situations in which a researcher might want to physically separate a collection of DNA fragments of different sizes. A researcher may also digest a DNA sample with a restriction enzyme to form fragments. The resulting size and fragment distribution pattern can often yield useful information about the sequence of DNA bases that can be used, much like a bar-code scan, to identify the individual or species to which the DNA belongs.

Gel electrophoresis is a technique commonly used to separate biological molecules based on size and biochemical characteristics, such as charge and polarity. **Agarose gel electrophoresis** is widely used to separate DNA (or RNA) of varying sizes that may be generated by restriction enzyme digestion or by other means, such as the PCR (Figure 14.26).

Due to its negatively charged backbone, DNA is strongly attracted to a positive electrode. In agarose gel electrophoresis, the gel is oriented horizontally in a buffer solution. Samples are loaded into sample wells on the side of the gel closest to the negative electrode, then drawn through the molecular sieve of the agarose matrix toward the positive electrode. The agarose matrix impedes the movement of larger molecules through the gel, whereas smaller molecules pass through more readily. Thus, the distance of migration is inversely correlated to the size of the DNA fragment, with smaller fragments travelling a longer distance through the gel. Sizes of DNA fragments within a sample can be estimated by comparison to fragments of known size in a DNA ladder also run on the same gel. To separate very large DNA fragments, such as chromosomes or viral genomes, agarose gel electrophoresis can be modified by periodically alternating the orientation of the electric field during pulsed-field gel electrophoresis (PFGE). In PFGE, smaller fragments can reorient themselves and migrate slightly faster than larger fragments and this technique can thus serve to separate very large fragments that would otherwise travel together during standard agarose gel electrophoresis. In any of these electrophoresis techniques, the locations of the DNA or RNA fragments in the gel can be detected by various methods. One common method is adding ethidium bromide, a stain that inserts into the nucleic acids at non-specific locations and can be visualized when exposed to ultraviolet light. Other stains that are safer than ethidium bromide, a potential carcinogen, are now available.

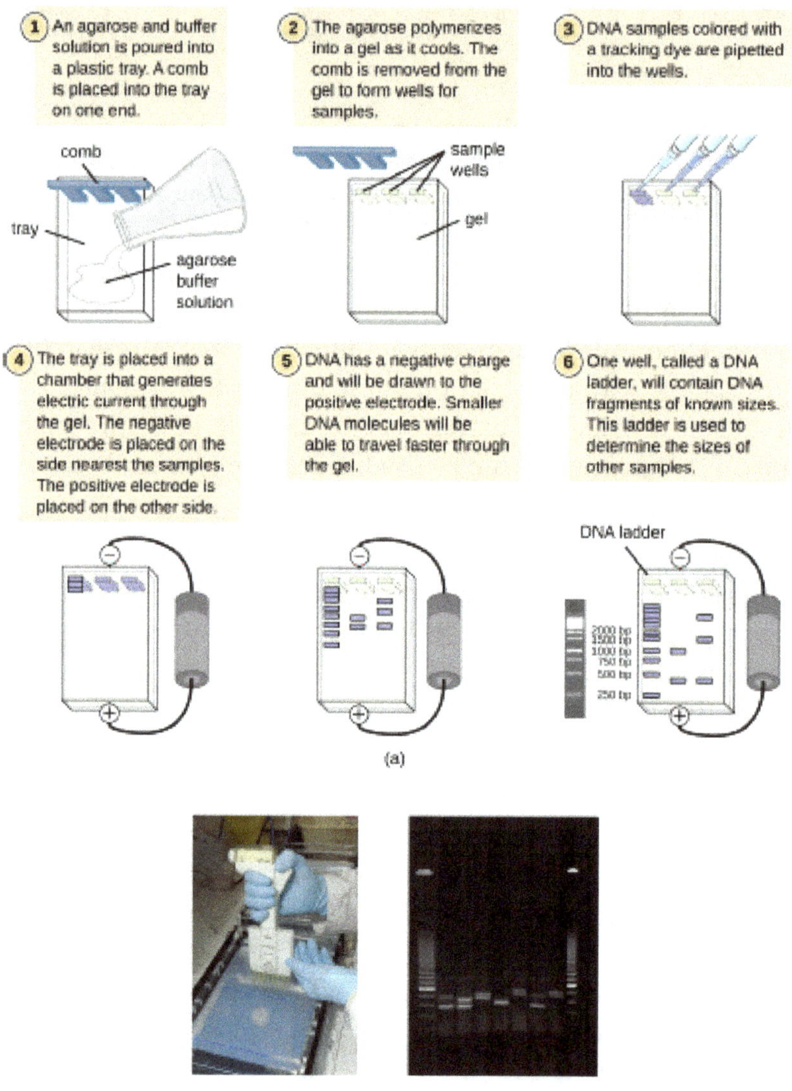

Figure 14.26 (a) The process of agarose gel electrophoresis. (b) A researcher loading samples into a gel. (c) This photograph shows a completed electrophoresis run on an agarose gel.

Restriction Fragment Length Polymorphism (RFLP) Analysis

Restriction enzyme recognition sites are short (only a few nucleotides long), sequence-specific palindromes, and may be found throughout the genome. Thus, differences in DNA sequences in the genomes of individuals will lead to differences in distribution of restriction-enzyme recognition sites that can be visualized as distinct banding patterns on a gel after agarose gel electrophoresis. **Restriction fragment length polymorphism (RFLP)** analysis compares DNA banding patterns of different DNA samples after restriction digestion (Figure 14.27).

RFLP analysis has many practical applications in both medicine and forensic science. For example, epidemiologists use RFLP analysis to track and identify the source of specific microorganisms implicated in outbreaks of food poisoning or certain infectious diseases. RFLP analysis can also be used on human DNA to determine inheritance patterns of chromosomes with variant genes, including those associated with heritable diseases or to establish paternity.

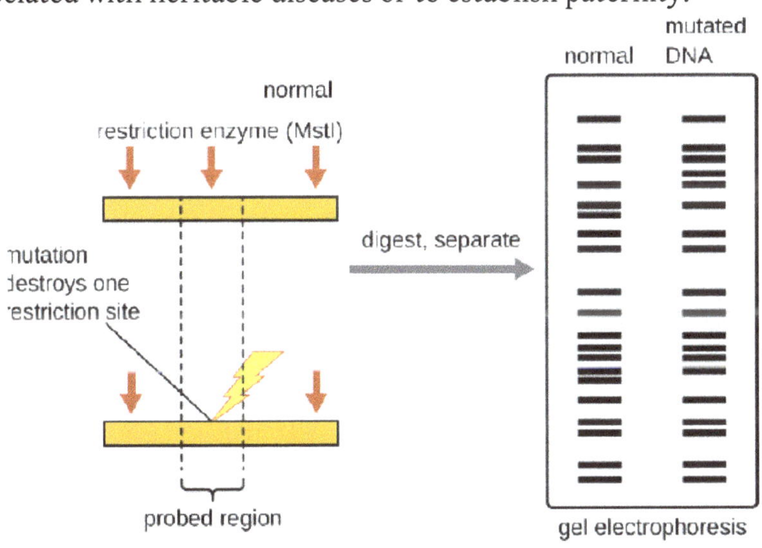

Figure 14.27 RFLP analysis can be used to differentiate DNA sequences. In this example, a normal chromosome is digested into two fragments, whereas digestion of a mutated chromosome produces only one fragment. The small red arrows pointing to the two different chromosome segments show the locations of the restriction enzyme recognition sites. After digestion and agarose gel electrophoresis, the banding patterns reflect the change by showing the loss of two shorter bands and the gain of a longer band. (credit: modification of work by National Center for Biotechnology Information)

Forensic scientists use RFLP analysis as a form of DNA fingerprinting, which is useful for analyzing DNA obtained from crime scenes, suspects, and victims. DNA samples are collected, the numbers of copies of the sample DNA molecules are increased using PCR, and then subjected to restriction enzyme digestion and agarose gel electrophoresis to generate specific banding patterns. By comparing the banding patterns of samples collected from the crime scene against those collected from suspects or victims, investigators can definitively determine whether DNA evidence collected at the scene was left behind by suspects or victims.

Southern Blots and Modifications

Several molecular techniques capitalize on sequence complementarity and hybridization between nucleic acids of a sample and DNA probes. Typically, probing nucleic-acid samples within a gel is unsuccessful because as the DNA probe soaks into a gel, the sample nucleic acids within the gel diffuse out. Thus, blotting techniques are commonly used to transfer nucleic acids to a thin, positively charged membrane made of nitrocellulose or nylon. In the **Southern blot** technique, developed by Sir Edwin Southern in 1975, DNA fragments within a sample are first separated by agarose gel electrophoresis and then transferred to a membrane through capillary action (Figure 14.28). The DNA fragments that bind to the surface of the membrane are then exposed to a specific single-stranded DNA probe labelled with a radioactive or fluorescent molecular beacon to aid in detection. Southern blots may be used to detect the presence of certain DNA sequences in a given DNA sample. Once the target DNA within the membrane is visualized, researchers can cut out the portion of the membrane containing the fragment to recover the DNA fragment of interest.

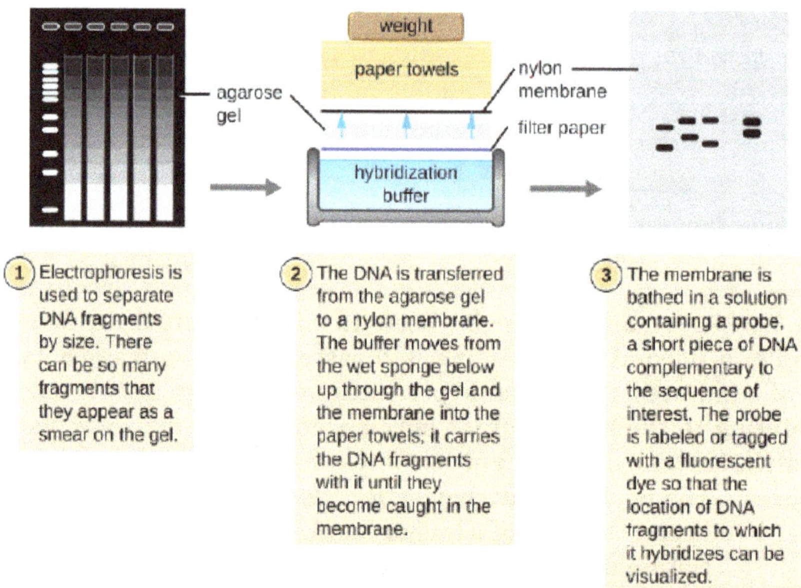

Figure 14.28 In the Southern blot technique, DNA fragments are first separated by agarose gel electrophoresis, then transferred by capillary action to a nylon membrane, which is then soaked with a DNA probe tagged with a molecular beacon for easy visualization.

Variations of the Southern blot—the dot blot, slot blot, and the spot blot—do not involve electrophoresis, but instead concentrate DNA from a sample into a small location on a membrane. After hybridization with a DNA probe, the signal intensity detected is measured, allowing the researcher to estimate the amount of target DNA present within the sample.

A colony blot is another variation of the Southern blot in which colonies representing different clones in a genomic library are transferred to a membrane by pressing the membrane onto the culture plate. The cells on the membrane are lysed and the membrane can then be probed to determine which colonies within a genomic library harbor the target gene. Because the colonies on the plate are still growing, the cells of interest can be isolated from the plate.

In the **northern blot**, another variation of the Southern blot, RNA (not DNA) is immobilized on the membrane and probed. Northern blots are typically used to detect the amount of mRNA made through gene expression within a tissue or organism sample.

Microarray Analysis

Another technique that capitalizes on the hybridization between complementary nucleic acid sequences is called **microarray analysis.** Microarray analysis is useful for the comparison of gene-expression patterns between different cell types—for example, cells infected with a virus versus uninfected cells, or cancerous cells versus healthy cells (Figure 14.29).

Typically, DNA or cDNA from an experimental sample is deposited on a glass slide alongside known DNA sequences. Each slide can hold more than 30,000 different DNA fragment types. Distinct DNA fragments (encompassing an organism's entire genomic library) or cDNA fragments (corresponding to an organism's full complement of expressed genes) can be individually spotted on a glass slide.

Once deposited on the slide, genomic DNA or mRNA can be isolated from the two samples for comparison. If mRNA is isolated, it is reverse-transcribed to cDNA using reverse transcriptase. Then the two samples of genomic DNA or cDNA are labelled with different fluorescent dyes (typically red and green). The labelled genomic DNA samples are then combined in equal amounts, added to the microarray chip, and allowed to hybridize to complementary spots on the microarray.

Hybridization of sample genomic DNA molecules can be monitored by measuring the intensity of fluorescence at particular spots on the microarray. Differences in the amount of hybridization between the samples can be readily observed. If only one sample's nucleic acids hybridize to a particular spot on the microarray, then that spot will appear either green or red. However, if both samples' nucleic acids hybridize, then the spot will appear yellow due to the combination of the red and green dyes.

Although microarray technology allows for a holistic comparison between two samples in a short time, it requires sophisticated (and expensive) detection equipment and analysis software. Because of the expense, this technology is typically limited to research settings. Researchers have used microarray analysis to study how gene expression is affected in organisms that are infected by bacteria or viruses or subjected to certain chemical treatments.

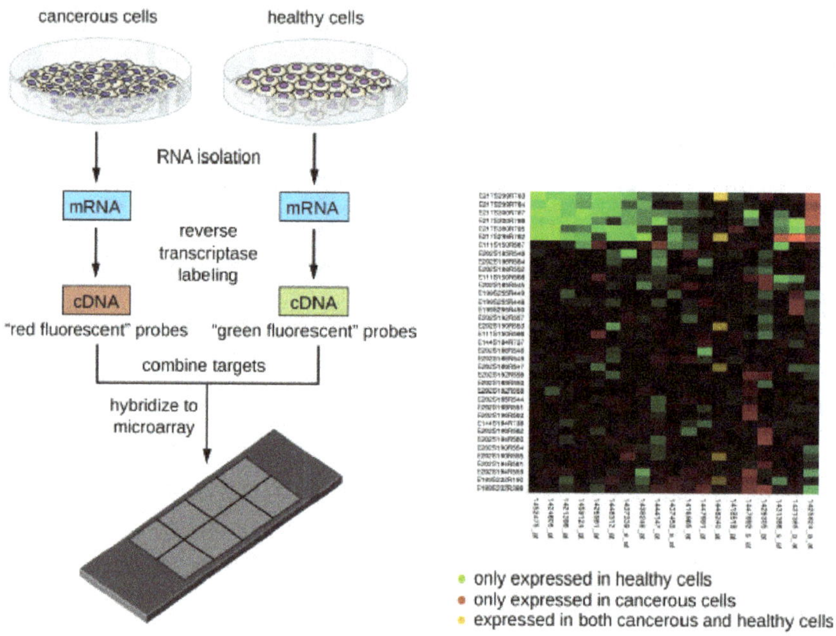

(a) (b)

Figure 14.29 (a) The steps in microarray analysis are illustrated. Here, gene expression patterns are compared between cancerous cells and healthy cells. (b) Microarray information can be expressed as a heat map. Genes are shown on the left side; different samples are shown across the bottom. Genes expressed only in cancer cells are shown in varying shades of red; genes expressed only in normal cells are shown in varying shades of green. Genes that are expressed in both cancerous and normal cells are shown in yellow.

Molecular Analysis of Proteins

In many cases it may not be desirable or possible to study DNA or RNA directly. Proteins can provide species- specific information for identification as well as important information about how and whether a cell or tissue is responding to the presence of a pathogenic microorganism. Various proteins require different methods for isolation and characterization.

Polyacrylamide Gel Electrophoresis

A variation of gel electrophoresis, called **polyacrylamide gel electrophoresis (PAGE)**, is commonly used for separating proteins. In PAGE, the gel matrix is finer and composed of polyacrylamide instead of agarose. Additionally, PAGE is typically performed using a vertical gel apparatus (Figure 14.30). Because of the varying charges associated with amino acid side chains, PAGE can be used to separate intact proteins based on their net charges. Alternatively, proteins can be denatured and coated with a negatively charged detergent called sodium dodecyl sulfate (SDS), masking the native charges and allowing separation based on size only. PAGE can be further modified to separate proteins based on two characteristics, such as their charges at various pHs as well as their size, through the use of two-dimensional PAGE. In any of these cases, following electrophoresis, proteins are visualized through staining, commonly with either Coomassie blue or a silver stain.

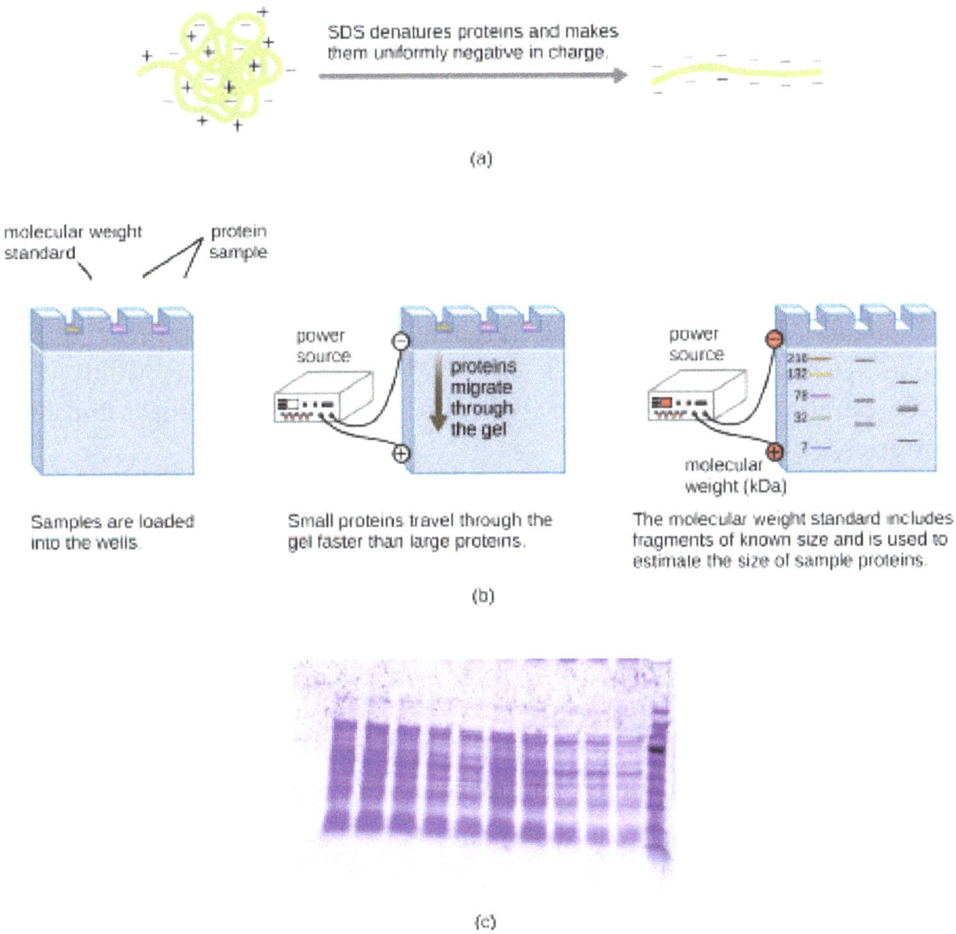

Figure 14.30 (a) SDS is a detergent that denatures proteins and masks their native charges, making them uniformly negatively charged. (b) The process of SDS-PAGE is illustrated in these steps. (c) A photograph of an SDS-PAGE gel shows Coomassie stained bands where proteins of different size have migrated along the gel in response to the applied voltage. A size standard lane is visible on the right side of the gel. (credit b: modification of work by "GeneEd"/YouTube)

Amplification-Based DNA Analysis Methods

Various methods can be used for obtaining sequences of DNA, which are useful for studying disease-causing organisms. With the advent of rapid sequencing technology, our knowledge base of the entire genomes of pathogenic organisms has grown phenomenally. We start with a description of the polymerase chain reaction, which is not a sequencing method but has allowed researchers and clinicians to obtain the large quantities of DNA needed for sequencing and other studies. The polymerase chain reaction eliminates the dependence we once had on cells to make multiple copies of DNA, achieving the same result through relatively simple reactions outside the cell.

Polymerase Chain Reaction (PCR)

Most methods of DNA analysis, such as restriction enzyme digestion and agarose gel electrophoresis, or DNA sequencing require large amounts of a specific DNA fragment. In the past, large amounts of DNA were produced by growing the host cells of a genomic library. However, libraries take time and effort to prepare and DNA samples of interest often come in minute quantities. The **polymerase chain reaction (PCR)** permits rapid amplification in the number of copies of specific DNA sequences for further analysis (**Figure 14.31**). One of the most powerful techniques in molecular biology, PCR was developed in 1983 by Kary Mullis while at Cetus Corporation. PCR has specific applications in research, forensic, and clinical laboratories, including:

- Determining the sequence of nucleotides in a specific region of DNA
- Amplifying a target region of DNA for cloning into a plasmid vector
- Identifying the source of a DNA sample left at a crime scene
- Analyzing samples to determine paternity
- Comparing samples of ancient DNA with modern organisms
- Determining the presence of difficult to culture, or unculturable, microorganisms in humans or environmental samples

PCR is an *in vitro* laboratory technique that takes advantage of the natural process of DNA replication. The heat-stable DNA polymerase enzymes used in PCR are derived from hyperthermophilic prokaryotes. *Taq DNA polymerase*, commonly used in PCR, is derived from the *Thermus aquaticus* bacterium isolated from a hot spring in Yellowstone National Park. DNA replication requires the use of primers for the initiation of replication to have free 3′-hydroxyl groups available for the addition of nucleotides by DNA polymerase. However, while primers composed of RNA are normally used in cells, DNA primers are used for PCR. **DNA primers** are preferable due to their stability, and DNA primers with known sequences targeting a specific DNA region can be chemically synthesized commercially. These DNA primers are functionally similar to the DNA probes used for the various hybridization techniques described earlier, binding to specific targets due to complementarity between the target DNA sequence and the primer.

PCR occurs over multiple cycles, each containing three steps: denaturation, annealing, and extension. Machines called thermal cyclers are used for PCR; these machines can be programmed to automatically cycle through the temperatures required at each step. First, double-stranded template DNA containing the target sequence is denatured at approximately 95°C. The high temperature required to physically (rather than enzymatically) separate the DNA strands is the reason the heat-stable DNA polymerase is required. Next, the temperature is lowered to approximately 50°C. This allows the DNA primers complementary to the ends of the target sequence to anneal (stick) to the template strands, with one primer annealing to each strand. Finally, the temperature is raised to 72°C, the optimal temperature for the activity of the heat-stable DNA polymerase, allowing for the addition of nucleotides to the primer using the single-stranded target as a template. Each cycle doubles the number of double-stranded target DNA copies. Typically, PCR protocols include 25–40 cycles, allowing for the amplification of a single target sequence by tens of millions to over a trillion.

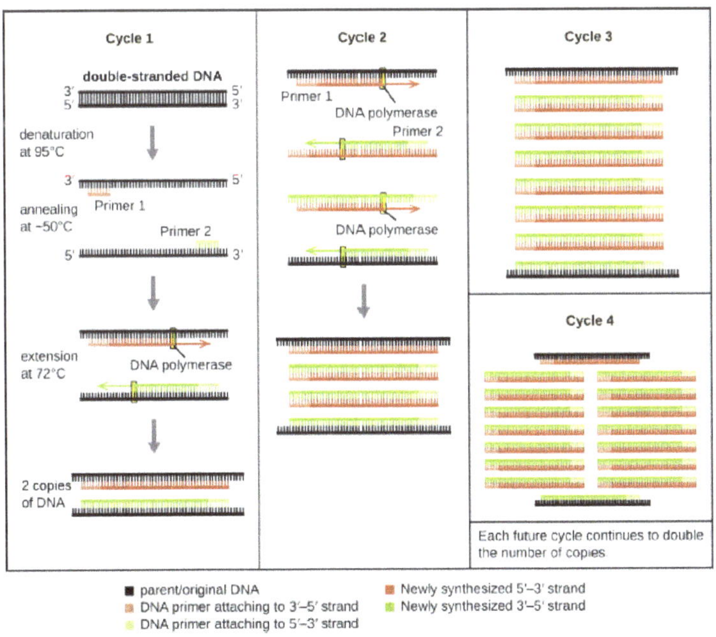

Figure 14.31 The polymerase chain reaction (PCR) is used to produce many copies of a specific sequence of DNA.

Natural DNA replication is designed to copy the entire genome, and initiates at one or more origin sites. Primers are constructed during replication, not before, and do not consist of a few specific sequences. PCR targets specific regions of a DNA sample using sequence-specific primers. In recent years, a variety of isothermal PCR amplification methods that circumvent the need for thermal cycling have been developed, taking advantage of accessory proteins that aid in the DNA replication process. As the development of these methods continues and their use becomes more widespread in research, forensic, and clinical labs, thermal cyclers may become obsolete.

PCR Variations

Several later modifications to PCR further increase the utility of this technique. **Reverse transcriptase PCR (RT-PCR)** is used for obtaining DNA copies of a specific mRNA molecule. RT-PCR begins with the use of the reverse transcriptase enzyme to convert mRNA molecules into cDNA. That cDNA is then used as a template for traditional PCR amplification. RT-PCR can detect whether a specific gene has been expressed in a sample. Another recent application of PCR is **real-time PCR**, also known as **quantitative PCR (qPCR).** Standard PCR and RT-PCR protocols are not quantitative because any one of the reagents may become limiting before all of the cycles within the protocol are complete, and samples are only analyzed at the end. Because it is not possible to determine when in the PCR or RT-PCR protocol a given reagent has become limiting, it is not possible to know how many cycles were completed prior to this point, and thus it is not possible to determine how many original template molecules were present in the sample at the start of PCR. In qPCR, however, the use of fluorescence allows one to monitor the increase in a double-stranded template during a PCR reaction as it occurs. These kinetics data can then be used to quantify the amount of the original target sequence. The use of qPCR in recent years has further expanded the capabilities of PCR, allowing researchers to determine the number of DNA copies, and sometimes organisms, present in a sample. In clinical settings, qRT-PCR is used to determine viral load in HIV-positive patients to evaluate the effectiveness of their therapy.

DNA Sequencing

A basic sequencing technique is the **chain termination method**, also known as the **dideoxy method** or the **Sanger DNA sequencing method**, developed by Frederick Sanger in 1972. The chain termination method involves DNA replication of a single-stranded template with the use of a DNA primer to initiate synthesis of a complementary strand, DNA polymerase, a mix of the four regular deoxynucleotide (dNTP) monomers, and a small proportion of dideoxynucleotides (ddNTPs), each labeled with a molecular beacon. The ddNTPs are monomers missing a hydroxyl group (–OH) at the site at which another nucleotide usually attaches to form a chain (Figure 14.32). Every time a ddNTP is randomly incorporated into the growing complementary strand, it terminates the process of DNA replication for that particular strand. This results in multiple short strands of replicated DNA that are each terminated at a different point during replication. When the reaction mixture is subjected to gel electrophoresis, the multiple newly replicated DNA strands form a ladder of differing sizes. Because the ddNTPs are labeled, each band on the gel reflects the size of the DNA strand when the ddNTP terminated the reaction.

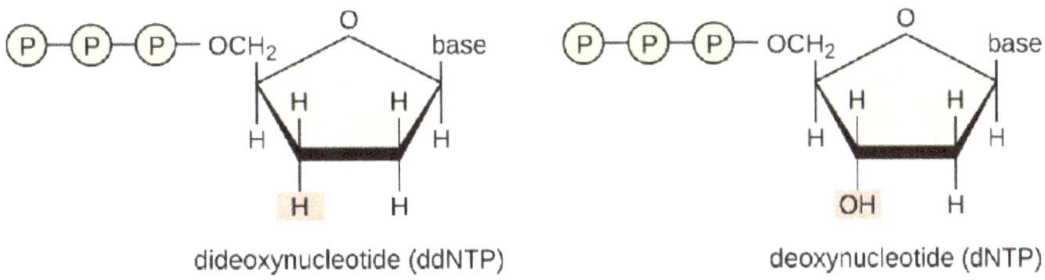

Figure 14.32 A dideoxynucleotide is similar in structure to a deoxynucleotide, but is missing the 3' hydroxyl group (indicated by the shaded box). When a dideoxynucleotide is incorporated into a DNA strand, DNA synthesis stops

In Sanger's day, four reactions were set up for each DNA molecule being sequenced, each reaction containing only one of the four possible ddNTPs. Each ddNTP was labeled with a radioactive phosphorus molecule. The products of the four reactions were then run in separate lanes side by side on long, narrow PAGE gels, and the bands of varying lengths were detected by autoradiography. Today, this process has been simplified with the use of ddNTPs, each labeled with a different colored fluorescent dye or fluorochrome (Figure 14.33), in one sequencing reaction containing all four possible ddNTPs for each DNA molecule being sequenced (Figure 14.34). These fluorochromes are detected by fluorescence spectroscopy. Determining the fluorescence color of each band as it passes by the detector produces the nucleotide sequence of the template strand.

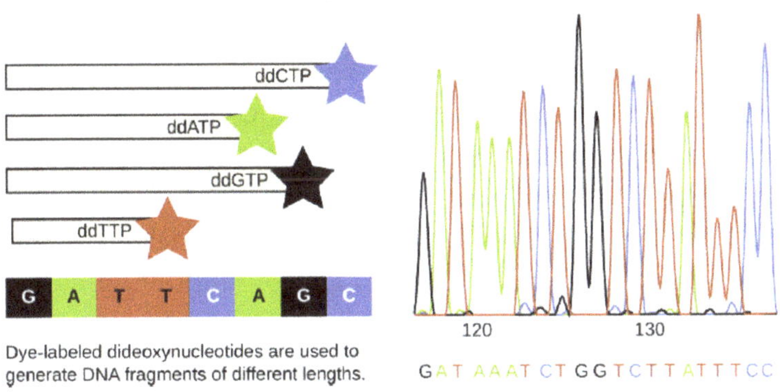

Figure 14.33 Frederick Sanger's dideoxy chain termination method is illustrated, using ddNTPs tagged with fluorochromes. Using ddNTPs, a mixture of DNA fragments of every possible size, varying in length by only one nucleotide, can be generated. The DNA is separated on the basis of size and each band can be detected with a fluorescence detector.

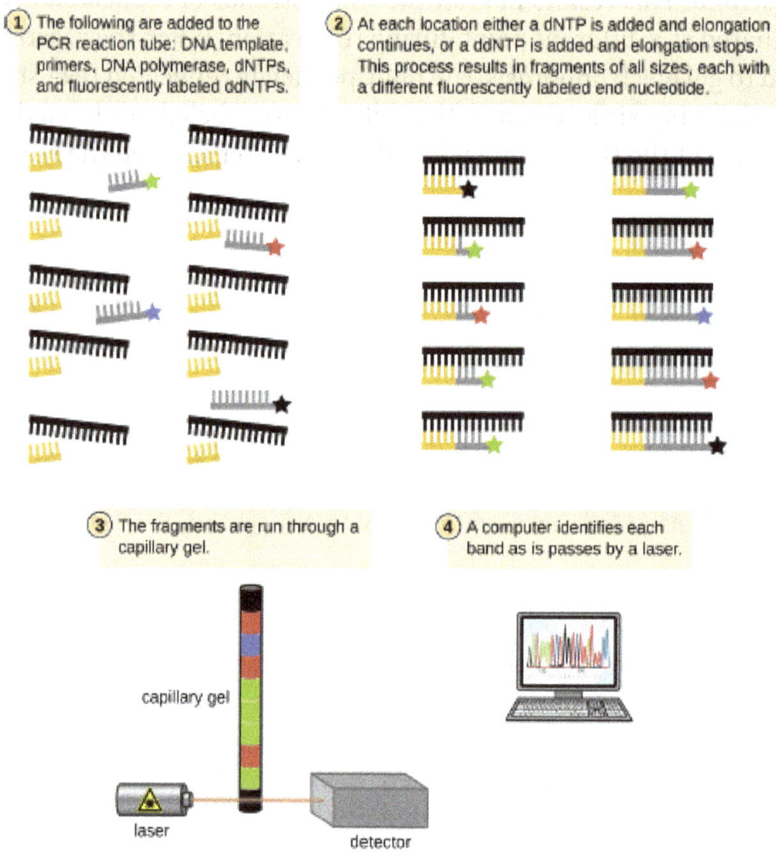

Figure 14.34 This diagram summarizes the Sanger sequencing method using fluorochrome-labeled ddNTPs and capillary gel electrophoresis.

Since 2005, automated sequencing techniques used by laboratories fall under the umbrella of **next generation sequencing**, which is a group of automated techniques used for rapid DNA sequencing. These methods have revolutionized the field of molecular genetics because the low-cost sequencers can generate sequences of hundreds of thousands or millions of short fragments (25 to 600 base pairs) just in one day. Although several variants of next generation sequencing technologies are made by different companies (for example, 454 Life Sciences' pyrosequencing and Illumina's Solexa technology), they all allow millions of bases to be sequenced quickly, making the sequencing of entire genomes relatively easy, inexpensive, and commonplace. In **454 sequencing (pyrosequencing)**, for example, a DNA sample is fragmented into 400–600-bp single-strand fragments, modified with the addition of DNA adapters to both ends of each fragment. Each DNA fragment is then immobilized on a bead and amplified by PCR, using primers designed to anneal to the adapters, creating a bead containing many copies of that DNA fragment. Each bead is then put into a separate well containing sequencing enzymes. To the well, each of the four nucleotides is added one after the other; when each one is incorporated, pyrophosphate is released as a byproduct of polymerization, emitting a small flash of light that is recorded by a detector. This provides the order of nucleotides incorporated as a new strand of DNA is made and is an example of synthesis sequencing. Next generation sequencers use sophisticated software to get through the cumbersome process of putting all the fragments in order. Overall, these technologies continue to advance rapidly, decreasing the cost of sequencing and increasing the availability of sequence data from a wide variety of organisms quickly.

GENOMICS AND PROTEOMICS

The study of nucleic acids began with the discovery of DNA, progressed to the study of genes and small fragments, and has now exploded to the field of **genomics.** Genomics is the study of entire genomes, including the complete set of genes, their nucleotide sequence and organization, and their interactions within a species and with other species. The advances in genomics have been made possible by DNA sequencing technology. Just as information technology has led to Google Maps that enable us to get detailed information about locations around the globe, genomic information is used to create similar maps of the DNA of different organisms.

Mapping Genomes

Genome mapping is the process of finding the location of genes on each chromosome. The maps that are created are comparable to the maps that we use to navigate streets. A **genetic map** is an illustration that lists genes and their location on a chromosome. Genetic maps provide the big picture (similar to a map of interstate highways) and use genetic markers (similar to landmarks). A genetic marker is a gene or sequence on a chromosome that shows genetic linkage with a trait of interest. The genetic marker tends to be inherited with the gene of interest, and one measure of distance between them is the recombination frequency during meiosis. Early geneticists called this linkage analysis.

Physical maps get into the intimate details of smaller regions of the chromosomes (similar to a detailed road map) (Figure 14.34). A physical map is a representation of the physical distance, in nucleotides, between genes or genetic markers. Both genetic linkage maps and physical maps are required to build a complete picture of the genome. Having a complete map of the genome makes it easier for researchers to study individual genes. Human genome maps help researchers in their efforts to identify human disease-causing genes related to illnesses such as cancer, heart disease, and cystic fibrosis, to name a few. In addition, genome mapping can be used to help identify organisms with beneficial traits, such as microbes with the ability to clean up pollutants or even prevent pollution. Research involving plant genome mapping may lead to methods that produce higher crop yields or to the development of plants that adapt better to climate change.

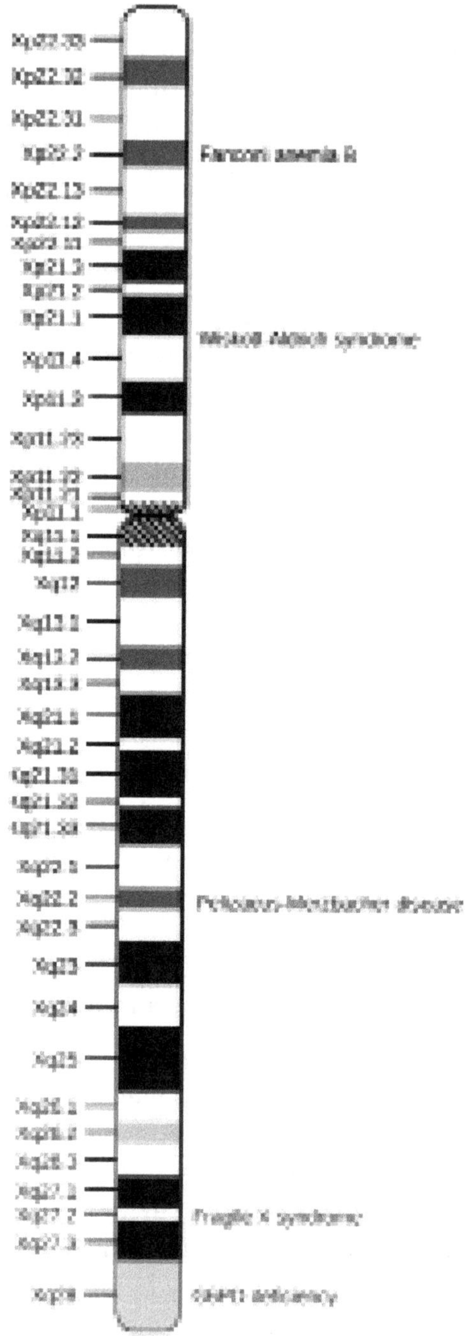

Figure 14.34 This is a physical map of the human X chromosome. (credit: modification of work by NCBI, NIH)

Genetic maps provide the outline, and physical maps provide the details. It is easy to understand why both types of genome-mapping techniques are important to show the big picture. Information obtained from each technique is used in combination to study the genome. Genomic mapping is used with different model organisms that are used for research. Genome mapping is still an ongoing process, and as more advanced techniques are developed, more advances are expected. Genome mapping is similar to completing a complicated puzzle using every piece of available data. Mapping information generated in laboratories all over the world is entered into central databases, such as the National Center for Biotechnology Information (NCBI). Efforts are made to make the information more easily accessible to researchers and the general public. Just as we use global positioning systems instead of paper maps to navigate through roadways, NCBI allows us to use a genome viewer tool to simplify the data mining process.

Whole Genome Sequencing

Although there have been significant advances in the medical sciences in recent years, doctors are still confounded by many diseases and researchers are using whole genome sequencing to get to the bottom of the problem. **Whole genome sequencing** is a process that determines the DNA sequence of an entire genome. Whole genome sequencing is a brute-force approach to problem solving when there is a genetic basis at the core of a disease. Several laboratories now provide services to sequence, analyze, and interpret entire genomes.

In 2010, whole genome sequencing was used to save a young boy whose intestines had multiple mysterious abscesses. The child had several colon operations with no relief. Finally, a whole genome sequence revealed a defect in a pathway that controls apoptosis (programmed cell death). A bone marrow transplant was used to overcome this genetic disorder, leading to a cure for the boy. He was the first person to be successfully diagnosed using whole genome sequencing.

The first genomes to be sequenced, such as those belonging to viruses, bacteria, and yeast, were smaller in terms of the number of nucleotides than the genomes of multicellular organisms. The genomes of other model organisms, such as the mouse (*Mus musculus*), the fruit fly (*Drosophila melanogaster*), and the nematode (*Caenorhabditis elegans*) are now known. A great deal of basic research is performed in **model organisms** because the information can be applied to other organisms. A model organism is a species that is studied as a model to understand the biological processes in other species that can be represented by the model organism. For example, fruit flies are able to metabolize alcohol like humans, so the genes affecting sensitivity to alcohol have been studied in fruit flies in an effort to understand the variation in sensitivity to alcohol in humans. Having entire genomes sequenced helps with the research efforts in these model organisms (Figure 14.35).

Figure 14.35 Much basic research is done with model organisms, such as the mouse, *Mus musculus*; the fruit fly, *Drosophila melanogaster*; the nematode *Caenorhabditis elegans*; the yeast *Saccharomyces cerevisiae*; and the common weed, *Arabidopsis thaliana*.

The first human genome sequence was published in 2003. The number of whole genomes that have been sequenced steadily increases and now includes hundreds of species and thousands of individual human genomes.

Applying Genomics

The introduction of DNA sequencing and whole genome sequencing projects, particularly the Human Genome Project, has expanded the applicability of DNA sequence information. Genomics is now being used in a wide variety of fields, such as metagenomics, pharmacogenomics, and mitochondrial genomics. The most commonly known application of genomics is to understand and find cures for diseases.

Predicting Disease Risk at the Individual Level

Predicting the risk of disease involves screening and identifying currently healthy individuals by genome analysis at the individual level. Intervention with lifestyle changes and drugs can be recommended before disease onset. However, this approach is most applicable when the problem arises from a single gene mutation. Such defects only account for about 5 percent of diseases found in developed countries. Most of the common diseases, such as heart disease, are multifactorial or polygenic, which refers to a phenotypic characteristic that is determined by two or more genes, and also environmental factors such as diet. In April 2010, scientists at Stanford University published the genome analysis of a healthy individual (Stephen Quake, a scientist at Stanford University, who had his genome sequenced); the analysis predicted his propensity to acquire various diseases. A risk assessment was done to analyze Quake's percentage of risk for 55 different medical conditions. A rare genetic mutation was found that showed him to be at risk for sudden heart attack. He was also predicted to have a 23 percent risk of developing prostate cancer and a 1.4 percent risk of developing Alzheimer's disease. The scientists used databases and several publications to analyze the genomic data. Even though genomic sequencing is becoming more affordable and analytical tools are becoming more reliable, ethical issues surrounding genomic analysis at a population level remain to be addressed. For example, could such data be legitimately used to charge more or less for insurance or to affect credit ratings?

Gene Editing

For thousands of years, humans have engaged in some level of control over genes and heredity regarding the plants and animals we rely on. The technology now exists to exert that control more directly by precisely altering the DNA of organisms. The technique is usually referred to as CRISPR, for the portions of DNA it targets: "Clustered Regularly Interspaced Short Palindromic Repeats." In essence, DNA contains repetitive sequences with "spacers" between them. CRISPR-associated nucleases (known as "Cas") are enzymes that can identify, attach to, and cut the strand at precise locations. In 2012, Jennifer Doudna and Emmanuelle Charpentier developed a method to combine the Cas nuclease with a synthetically produced "guide RNA" that leads the nuclease to selected locations on the DNA strand. The discovery revolutionized gene editing. Researchers around the world have used CRISPR to manipulate the actual DNA of plants, animals, laboratory cell lines, and (in trials) human patients. Doudna and Charpentier were awarded the Nobel Prize for their work.

Gene editing is so promising because it can be used experimentally to understand disease or organismal limitations, then be applied to overcome those issues. For example, in a human trial, cancerous cells were removed from a person, edited to remove their cancerous properties at the DNA level, and reintroduced into the patient so that those now edited cells could multiply and replace the cancerous ones. Using the person's own cells increases the likelihood of acceptance and success.

Like other applications of genomics, the prospect of directly editing genes brings up a number of ethical issues. Both Doudna and Chapentier, as well as many other genetic engineers, support only certain CRISPR applications. And many governments and other entities place strict guidelines on the uses of the powerful technology.

Genome-wide Association Studies

Since 2005, it has been possible to conduct a type of study called a genome-wide association study, or GWAS. A GWAS is a method that identifies differences between individuals in single nucleotide polymorphisms (SNPs) that may be involved in causing diseases. The method is particularly suited to diseases that may be affected by one or many genetic changes throughout the genome. It is very difficult to identify the genes involved in such a disease using family history information. The GWAS method relies on a genetic database that has been in development since 2002 called the International HapMap Project. The HapMap Project sequenced the genomes of several hundred individuals from around the world and identified groups of SNPs. The groups include SNPs that are located near to each other on chromosomes so they tend to stay together through recombination. The fact that the group stays together means that identifying one marker SNP is all that is needed to identify all the SNPs in the group. There are several million SNPs identified, but identifying them in other individuals who have not had their complete genome sequenced is much easier because only the marker SNPs need to be identified.

In a common design for a GWAS, two groups of individuals are chosen; one group has the disease, and the other group does not. The individuals in each group are matched in other characteristics to reduce the effect of confounding variables causing differences between the two groups. For example, the genotypes may differ because the two groups are mostly taken from different parts of the world. Once the individuals are chosen, and typically their numbers are a thousand or more for the study to work, samples of their DNA are obtained. The DNA is analyzed using automated systems to identify large differences in the percentage of particular SNPs between the two groups. Often the study examines a million or more SNPs in the DNA. The results of GWAS can be used in two ways: the genetic differences may be used as markers for susceptibility to the disease in undiagnosed individuals, and the particular genes identified can be targets for research into the molecular pathway of the disease and potential therapies. An offshoot of the discovery of gene associations with disease has been the formation of companies that provide so-called "personal genomics" that will identify risk levels for various diseases based on an individual's SNP complement. The science behind these services is controversial.

Because GWAS looks for associations between genes and disease, these studies provide data for other research into causes, rather than answering specific questions themselves. An association between a gene difference and a disease does not necessarily mean there is a cause-and-effect relationship. However, some studies have provided useful information about the genetic causes of diseases. For example, three different studies in 2005 identified a gene for a protein involved in regulating inflammation in the body that is associated with a disease-causing blindness called age-related macular degeneration. This opened up new possibilities for research into the cause of this disease. A large number of genes have been identified to be associated with Crohn's disease using GWAS, and some of these have suggested new hypothetical mechanisms for the cause of the disease.

Pharmacogenomics

Pharmacogenomics involves evaluating the effectiveness and safety of drugs on the basis of information from an individual's genomic sequence. Personal genome sequence information can be used to prescribe medications that will be most effective and least toxic on the basis of the individual patient's genotype. Studying changes in gene expression could provide information about the gene transcription profile in the presence of the drug, which can be used as an early indicator of the potential for toxic effects. For example, genes involved in cellular growth and controlled cell death, when disturbed, could lead to the growth of cancerous cells. Genome-wide studies can also help to find new genes involved in drug toxicity. The gene signatures may not be completely accurate, but can be tested further before pathologic symptoms arise.

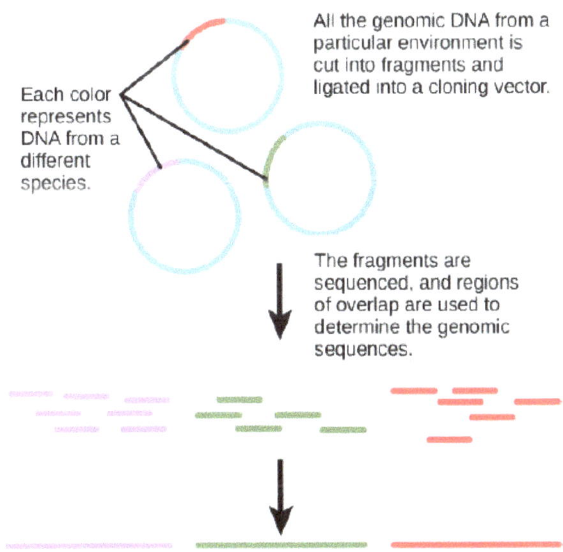

Figure 14.36 Metagenomics involves isolating DNA from multiple species within an environmental niche. The DNA is cut up and sequenced, allowing entire genome sequences of multiple species to be reconstructed from the sequences of overlapping pieces.

Metagenomics

Traditionally, microbiology has been taught with the view that microorganisms are best studied under pure culture conditions, which involves isolating a single type of cell and culturing it in the laboratory. Because microorganisms can go through several generations in a matter of hours, their gene expression profiles adapt to the new laboratory environment very quickly. On the other hand, many species resist being cultured in isolation. Most microorganisms do not live as isolated entities, but in microbial communities known as biofilms. For all of these reasons, pure culture is not always the best way to study microorganisms. **Metagenomics** is the study of the collective genomes of multiple species that grow and interact in an environmental niche. Metagenomics can be used to identify new species more rapidly and to analyze the effect of pollutants on the environment (Figure 14.36). Metagenomics techniques can now also be applied to communities of higher eukaryotes, such as fish.

Creation of New Biofuels

Knowledge of the genomics of microorganisms is being used to find better ways to harness biofuels from algae and cyanobacteria. The primary sources of fuel today are coal, oil, wood, and other plant products such as ethanol. Although plants are renewable resources, there is still a need to find more alternative renewable sources of energy to meet our population's energy demands. The microbial world is one of the largest resources for genes that encode new enzymes and produce new organic compounds, and it remains largely untapped. This vast genetic resource holds the potential to provide new sources of biofuels (Figure 14.37).

Figure 14.37 Renewable fuels were tested in Navy ships and aircraft at the first Naval Energy Forum. (credit: modification of work by John F. Williams, US Navy)

Mitochondrial Genomics

Mitochondria are intracellular organelles that contain their own DNA. Mitochondrial DNA mutates at a rapid rate and is often used to study evolutionary relationships. Another feature that makes studying the mitochondrial genome interesting is that in most multicellular organisms, the mitochondrial DNA is passed on from the mother during the process of fertilization. For this reason, mitochondrial genomics is often used to trace genealogy.

Genomics in Forensic Analysis

Information and clues obtained from DNA samples found at crime scenes have been used as evidence in court cases, and genetic markers have been used in forensic analysis. Genomic analysis has also become useful in this field. In 2001, the first use of genomics in forensics was published. It was a collaborative effort between academic research institutions and the FBI to solve the mysterious cases of anthrax (Figure 14.38) that was transported by the US Postal Service. Anthrax bacteria were made into an infectious powder and mailed to news media and two U.S. Senators. The powder infected the administrative staff and postal workers who opened or handled the letters. Five people died, and 17 were sickened from the bacteria. Using microbial genomics, researchers determined that a specific strain of anthrax was used in all the mailings; eventually, the source was traced to a scientist at a national biodefense laboratory in Maryland.

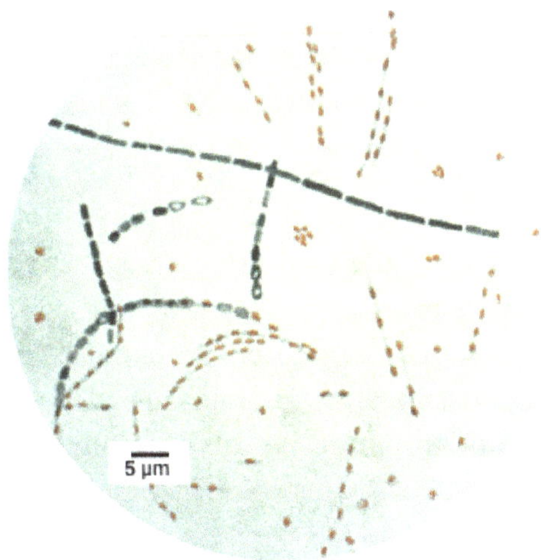

Figure 14.38 *Bacillus anthracis* is the organism that causes anthrax. (credit: modification of work by CDC; scale-bar data from Matt Russell)

Genomics in Agriculture

Genomics can reduce the trials and failures involved in scientific research to a certain extent, which could improve the quality and quantity of crop yields in agriculture (Figure 14.39). Linking traits to genes or gene signatures helps to improve crop breeding to generate hybrids with the most desirable qualities. Scientists use genomic data to identify desirable traits, and then transfer those traits to a different organism to create a new genetically modified organism, as described in the previous module. Scientists are discovering how genomics can improve the quality and quantity of agricultural production. For example, scientists could use desirable traits to create a useful product or enhance an existing product, such as making a drought-sensitive crop more tolerant of the dry season.

Figure 14.39 Transgenic agricultural plants can be made to resist disease. These transgenic plums are resistant to the plum pox virus. (credit: Scott Bauer, USDA ARS)

Proteomics

Proteins are the final products of genes that perform the function encoded by the gene. Proteins are composed of amino acids and play important roles in the cell. All enzymes (except ribozymes) are proteins and act as catalysts that affect the rate of reactions. Proteins are also regulatory molecules, and some are hormones. Transport proteins, such as hemoglobin, help transport oxygen to various organs. Antibodies that defend against foreign particles are also proteins. In the diseased state, protein function can be impaired because of changes at the genetic level or because of direct impact on a specific protein.

A proteome is the entire set of proteins produced by a cell type. Proteomes can be studied using the knowledge of genomes because genes code for mRNAs, and the mRNAs encode proteins. The study of the function of proteomes is called **proteomics**. Proteomics complements genomics and is useful when scientists want to test their hypotheses that were based on genes. Even though all cells in a multicellular organism have the same set of genes, the set of proteins produced in different tissues is different and dependent on gene expression. Thus, the genome is constant, but the proteome varies and is dynamic within an organism. In addition, RNAs can be alternatively spliced (cut and pasted to create novel combinations and novel proteins), and many proteins are modified after translation. Although the genome provides a blueprint, the final architecture depends on several factors that can change the progression of events that generate the proteome.

Genomes and proteomes of patients suffering from specific diseases are being studied to understand the genetic basis of the disease. The most prominent disease being studied with proteomic approaches is cancer (Figure 14.40). Proteomic approaches are being used to improve the screening and early detection of cancer; this is achieved by identifying proteins whose expression is affected by the disease process. An individual protein is called a **biomarker**, whereas a set of proteins with altered expression levels is called a **protein signature**. For a biomarker or protein signature to be useful as a candidate for early screening and detection of a cancer, it must be secreted in body fluids such as sweat, blood, or urine, so that large-scale screenings can be performed in a noninvasive fashion. The current problem with using biomarkers for the early detection of cancer is the high rate of false-negative results. A false-negative result is a negative test result that should have been positive. In other words, many cases of cancer go undetected, which makes biomarkers unreliable. Some examples of protein biomarkers used in cancer detection are CA-125 for ovarian cancer and PSA for prostate cancer. Protein signatures may be more reliable than biomarkers to detect cancer cells. Proteomics is also being used to develop individualized treatment plans, which involves the prediction of whether or not an individual will respond to specific drugs and the side effects that the individual may have. Proteomics is also being used to predict the possibility of disease recurrence.

Figure 14.40 This machine is preparing to do a proteomic pattern analysis to identify specific cancers so that an accurate cancer prognosis can be made. (credit: Dorie Hightower, NCI, NIH)

The National Cancer Institute has developed programs to improve the detection and treatment of cancer. The Clinical Proteomic Technologies for Cancer and the Early Detection Research Network are efforts to identify protein signatures specific to different types of cancers. The Biomedical Proteomics Program is designed to identify protein signatures and design effective therapies for cancer patients.

GENE THERAPY

Many types of genetic engineering have yielded clear benefits with few apparent risks. Few would question, for example, the value of our now abundant supply of human insulin produced by genetically engineered bacteria. However, many emerging applications of genetic engineering are much more controversial, often because their potential benefits are pitted against significant risks, real or perceived. This is certainly the case for **gene therapy**, a clinical application of genetic engineering that may one day provide a cure for many diseases but is still largely an experimental approach to treatment.

Mechanisms and Risks of Gene Therapy

Human diseases that result from genetic mutations are often difficult to treat with drugs or other traditional forms of therapy because the signs and symptoms of disease result from abnormalities in a patient's genome. For example, a patient may have a genetic mutation that prevents the expression of a specific protein required for the normal function of a particular cell type. This is the case in patients with Severe Combined Immunodeficiency (SCID), a genetic disease that impairs the function of certain white blood cells essential to the immune system.

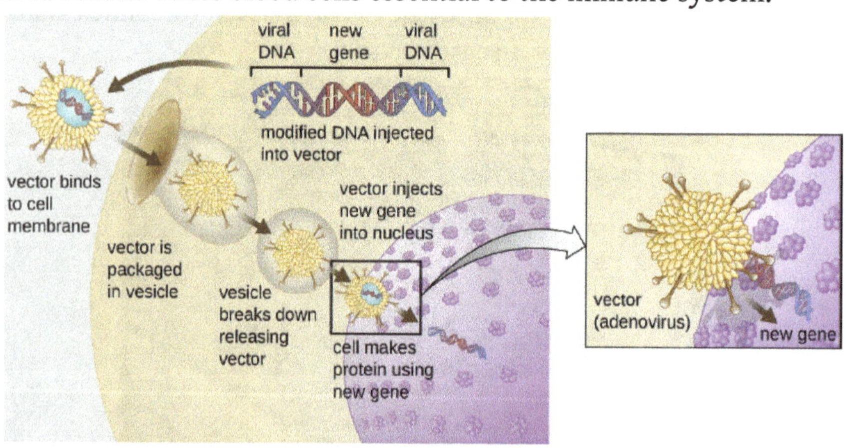

Figure 14.41 Gene therapy using an adenovirus vector can be used to treat or cure certain genetic diseases in which a patient has a defective gene. (credit: modification of work by National Institutes of Health)

Gene therapy attempts to correct genetic abnormalities by introducing a nonmutated, functional gene into the patient's genome. The nonmutated gene encodes a functional protein that the patient would otherwise be unable to produce. Viral vectors such as adenovirus are sometimes used to introduce the functional gene; part of the viral genome is removed and replaced with the desired gene (Figure 14.41). More advanced forms of gene therapy attempt to correct the mutation at the original site in the genome, such as is the case with treatment of SCID.

So far, gene therapies have proven relatively ineffective, with the possible exceptions of treatments for cystic fibrosis and adenosine deaminase deficiency, a type of SCID. Other trials have shown the clear hazards of attempting genetic manipulation in complex multicellular organisms like humans. In some patients, the use of an adenovirus vector can trigger an unanticipated inflammatory response from the immune system, which may lead to organ failure. Moreover, because viruses can often target multiple cell types, the virus vector may infect cells not targeted for the therapy, damaging these other cells and possibly leading to illnesses such as cancer. Another potential risk is that the modified virus could revert to being infectious and cause disease in the patient. Lastly, there is a risk that the inserted gene could unintentionally inactivate another important gene in the patient's genome, disrupting normal cell cycling and possibly leading to tumor formation and cancer. Because gene therapy involves so many risks, candidates for gene therapy need to be fully informed of these risks before providing informed consent to undergo the therapy.

Gene Therapy Gone Wrong

The risks of gene therapy were realized in the 1999 case of Jesse Gelsinger, an 18-year-old patient who received gene therapy as part of a clinical trial at the University of Pennsylvania. Jesse received gene therapy for a condition called ornithine transcarbamylase (OTC) deficiency, which leads to ammonia accumulation in the blood due to deficient ammonia processing. Four days after the treatment, Jesse died after a massive immune response to the adenovirus vector.

Until that point, researchers had not really considered an immune response to the vector to be a legitimate risk, but on investigation, it appears that the researchers had some evidence suggesting that this was a possible outcome. Prior to Jesse's treatment, several other human patients had suffered side effects of the treatment, and three monkeys used in a trial had died as a result of inflammation and clotting disorders. Despite this information, it appears that neither Jesse nor his family were made aware of these outcomes when they consented to the therapy. Jesse's death was the first patient death due to a gene therapy treatment and resulted in the immediate halting of the clinical trial in which he was involved, the subsequent halting of all other gene therapy trials at the University of Pennsylvania, and the investigation of all other gene therapy trials in the United States. As a result, the regulation and oversight of gene therapy overall was re-examined, resulting in new regulatory protocols that are still in place today.

Oversight of Gene Therapy

Presently, there is significant oversight of gene therapy clinical trials. At the federal level, three agencies regulate gene therapy in parallel: the Food and Drug Administration (FDA), the Office of Human Research Protection (OHRP), and the Recombinant DNA Advisory Committee (RAC) at the National Institutes of Health (NIH). Along with several local agencies, these federal agencies interact with the institutional review board to ensure that protocols are in place to protect patient safety during clinical trials. Compliance with these protocols is enforced mostly on the local level in cooperation with the federal agencies. Gene therapies are currently under the most extensive federal and local review compared to other types of therapies, which are more typically only under the review of the FDA. Some researchers believe that these extensive regulations actually inhibit progress in gene therapy research. In 2013, the Institute of Medicine (now the National Academy of Medicine) called upon the NIH to relax its review of gene therapy trials in most cases.However, ensuring patient safety continues to be of utmost concern.

Ethical Concerns

Beyond the health risks of gene therapy, the ability to genetically modify humans poses a number of ethical issues related to the limits of such "therapy." While current research is focused on gene therapy for genetic diseases, scientists might one day apply these methods to manipulate other genetic traits not perceived as desirable. This raises questions such as:

- Which genetic traits are worthy of being "corrected"?
- Should gene therapy be used for cosmetic reasons or to enhance human abilities?
- Should genetic manipulation be used to impart desirable traits to the unborn?
- Is everyone entitled to gene therapy, or could the cost of gene therapy create new forms of social inequality?

Who should be responsible for regulating and policing inappropriate use of gene therapies?

The ability to alter reproductive cells using gene therapy could also generate new ethical dilemmas. To date, the various types of gene therapies have been targeted to somatic cells, the non-reproductive cells within the body. Because somatic cell traits are not inherited, any genetic changes accomplished by somatic-cell gene therapy would not be passed on to offspring. However, should scientists successfully introduce new genes to germ cells (eggs or sperm), the resulting traits could be passed on to offspring. This approach, called germ- line gene therapy, could potentially be used to combat heritable diseases, but it could also lead to unintended consequences for future generations. Moreover, there is the question of informed consent, because those impacted by germ-line gene therapy are unborn and therefore unable to choose whether they receive the therapy. For these reasons, the U.S. government does not currently fund research projects investigating germ- line gene therapies in humans.

Risky Gene Therapies

While there are currently no gene therapies on the market in the United States, many are in the pipeline and it is likely that some will eventually be approved. With recent advances in gene therapies targeting p53, a gene whose somatic cell mutations have been implicated in over 50% of human cancers, cancer treatments through gene therapies could become much more widespread once they reach the commercial market.

Bringing any new therapy to market poses ethical questions that pit the expected benefits against the risks. How quickly should new therapies be brought to the market? How can we ensure that new therapies have been sufficiently tested for safety and effectiveness before they are marketed to the public? The process by which new therapies are developed and approved complicates such questions, as those involved in the approval process are often under significant pressure to get a new therapy approved even in the face of significant risks.

To receive FDA approval for a new therapy, researchers must collect significant laboratory data from animal trials and submit an Investigational New Drug (IND) application to the FDA's Center for Drug Evaluation and Research (CDER). Following a 30-day waiting period during which the FDA reviews the IND, clinical trials involving human subjects may begin. If the FDA perceives a problem prior to or during the clinical trial, the FDA can order a "clinical hold" until any problems are addressed. During clinical trials, researchers collect and analyze data on the therapy's effectiveness and safety, including any side effects observed. Once the therapy meets FDA standards for effectiveness and safety, the developers can submit a New Drug Application (NDA) that details how the therapy will be manufactured, packaged, monitored, and administered.

Because new gene therapies are frequently the result of many years (even decades) of laboratory and clinical research, they require a significant financial investment. By the time a therapy has reached the clinical trials stage, the financial stakes are high for pharmaceutical companies and their shareholders. This creates potential conflicts of interest that can sometimes affect the objective judgment of researchers, their funders, and even trial participants. The Jesse Gelsinger case (see Case in Point: Gene Therapy Gone Wrong) is a classic example. Faced with a life-threatening disease and no reasonable treatments available, it is easy to see why a patient might be eager to participate in a clinical trial no matter the risks. It is also easy to see how a researcher might view the short-term risks for a small group of study participants as a small price to pay for the potential benefits of a game-changing new treatment.

Gelsinger's death led to increased scrutiny of gene therapy, and subsequent negative outcomes of gene therapy have resulted in the temporary halting of clinical trials pending further investigation. For example, when children in France treated with gene therapy for SCID began to develop leukemia several years after treatment, the FDA temporarily stopped clinical trials of similar types of gene therapy occurring in the United States. Cases like these highlight the need for researchers and health professionals not only to value human well-being and patients' rights over profitability, but also to maintain scientific objectivity when evaluating the risks and benefits of new therapies.

CHAPTER-15

PATHOGENS, VACCINATION, AND IMMUNE RESPONSES

INTRODUCTION: AN OVERVIEW

In the intricate dance between the human body and the microscopic world of pathogens, the interplay of pathogens, vaccination, and immune responses forms the front line of defense against infectious diseases. Pathogens, ranging from viruses to bacteria, constantly challenge our well-being, prompting the immune system to mount a sophisticated defense. Vaccination, a powerful tool in the arsenal of preventive medicine, leverages the body's ability to remember and rapidly respond to invaders. This exploration delves into the complex dynamics of pathogens, the strategic deployment of vaccines, and the remarkable orchestration of immune responses, highlighting the intricate mechanisms that safeguard our health in the face of microbial threats.

Organisms have a wide array of adaptations for preventing attacks of parasites and diseases. The vertebrate defense systems, including those of humans, are complex and multilayered, with defenses unique to vertebrates. These unique vertebrate defenses interact with other defense systems inherited from ancestral lineages, and include complex and specific pathogen recognition and memory mechanisms. Research continues to unravel the complexities and vulnerabilities of the immune system.

Despite a poor understanding of the workings of the body in the early 18th century in Europe, the practice of inoculation as a method to prevent the often-deadly effects of smallpox was introduced from the courts of the Ottoman Empire. The method involved causing limited infection with the smallpox virus by introducing the pus of an affected individual to a scratch in an uninfected person. The resulting infection was milder than if it had been caught naturally and mortality rates were shown to be about two percent rather than 30 percent from natural infections. Moreover, the inoculation gave the individual immunity to the disease. It was from these early experiences with inoculation that the methods of vaccination were developed, in which a weakened or relatively harmless (killed) derivative of a pathogen is introduced into the individual. The vaccination induces immunity to the disease with few of the risks of being infected. A modern understanding of the causes of the infectious disease and the mechanisms of the immune system began in the late 19th century and continues to grow today.

VIRUSES

No one knows exactly when viruses emerged or from where they came, since viruses do not leave historical footprints such as fossils. Modern viruses are thought to be a mosaic of bits and pieces of nucleic acids picked up from various sources along their respective evolutionary paths. Viruses are **acellular**, parasitic entities that are not classified within any domain because they are not considered alive. They have no plasma membrane, internal organelles, or metabolic processes, and they do not divide. Instead, they infect a host cell and use the host's replication processes to produce progeny virus particles. Viruses infect all forms of organisms including bacteria, archaea, fungi, plants, and animals. Living things grow, metabolize, and reproduce. Viruses replicate, but to do so, they are entirely dependent on their host cells. They do not metabolize or grow, but are assembled in their mature form.

Viruses are diverse. They vary in their structure, their replication methods, and in their target hosts or even host cells. While most biological diversity can be understood through evolutionary history, such as how species have adapted to conditions and environments, much about virus origins and evolution remains unknown.

How Viruses Replicate

Viruses were first discovered after the development of a porcelain filter, called the Chamberland-Pasteur filter, which could remove all bacteria visible under the microscope from any liquid sample. In 1886, Adolph Meyer demonstrated that a disease of tobacco plants, tobacco mosaic disease, could be transferred from a diseased plant to a healthy one through liquid plant extracts. In 1892, Dmitri Ivanowski showed that this disease could be transmitted in this way even after the Chamberland-Pasteur filter had removed all viable bacteria from the extract. Still, it was many years before it was proven that these "filterable" infectious agents were not simply very small bacteria but were a new type of tiny, disease-causing particle.

Virions, single virus particles, are very small, about 20–250 nanometers (1 nanometer = 1/1,000,000 mm). These individual virus particles are the infectious form of a virus outside the host cell. Unlike bacteria (which are about 100 times larger), we cannot see viruses with a light microscope, with the exception of some large virions of the poxvirus family (Figure 15.1).

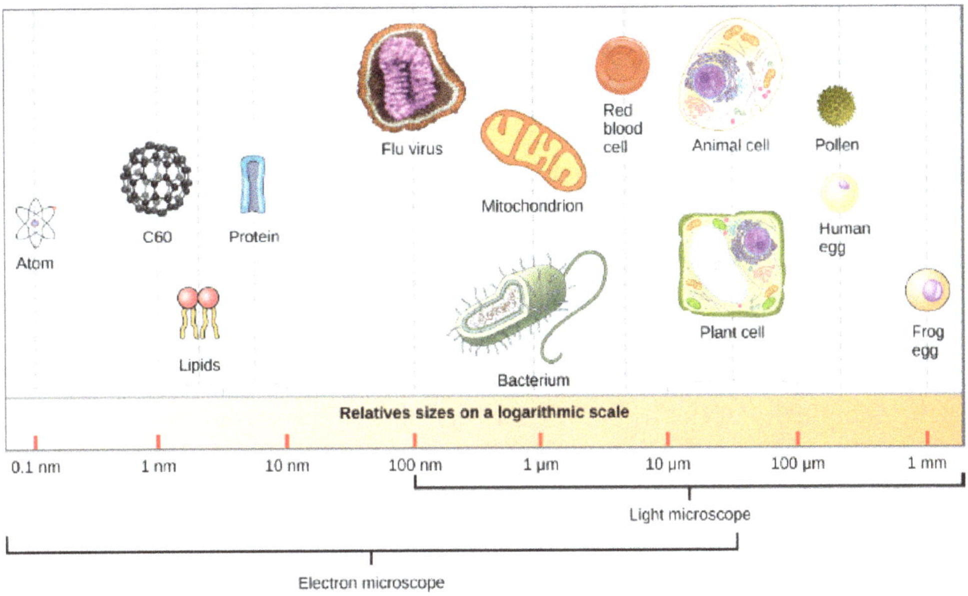

Figure 15.1 The size of a virus is very small relative to the size of cells and organelles.

It was not until the development of the electron microscope in the 1940s that scientists got their first good view of the structure of the tobacco mosaic virus and others. The surface structure of virions can be observed by both scanning and transmission electron microscopy, whereas the internal structures of the virus can only be observed in images from a transmission electron microscope (Figure 15.2).

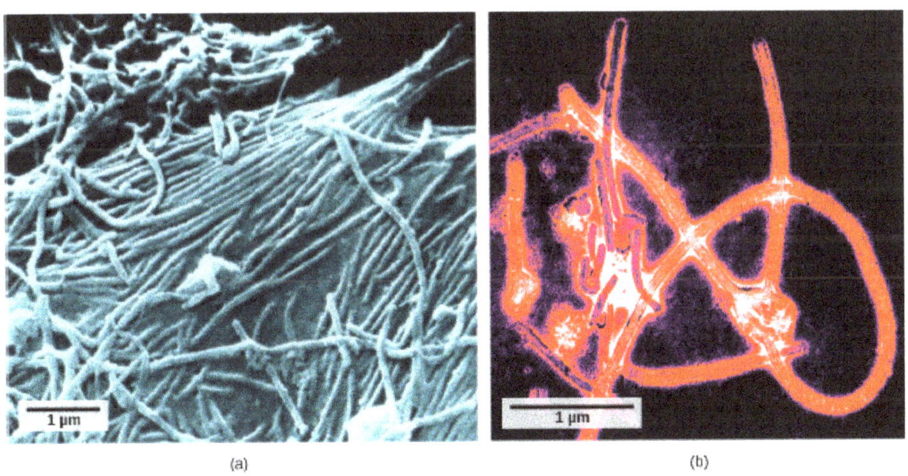

Figure 15.2 The ebola virus is shown here as visualized through (a) a scanning electron micrograph and (b) a transmission electron micrograph. (credit a: modification of work by Cynthia Goldsmith, CDC; credit b: modification of work by Thomas W. Geisbert, Boston University School of Medicine; scale-bar data from Matt Russell)

The use of this technology has allowed for the discovery of many viruses of all types of living organisms. They were initially grouped by shared morphology, meaning their size, shape, and distinguishing structures. Later, groups of viruses were classified by the type of nucleic acid they contained, DNA or RNA, and whether their nucleic acid was single- or double-stranded. More recently, molecular analysis of viral replication cycles has further refined their classification.

A **virion** consists of a nucleic-acid core, an outer protein coating, and sometimes an outer envelope made of protein and phospholipid membranes derived from the host cell. The most visible difference between members of viral families is their morphology, which is quite diverse. An interesting feature of viral complexity is that the complexity of the host does not correlate to the complexity of the virion. Some of the most complex virion structures are observed in bacteriophages, viruses that infect the simplest living organisms, bacteria.

Viruses come in many shapes and sizes, but these are consistent and distinct for each viral family (Figure 15.3). All virions have a nucleic-acid genome covered by a protective layer of protein, called a **capsid**. The capsid is made of protein subunits called capsomeres. Some viral capsids are simple polyhedral "spheres," whereas others are quite complex in structure. The outer structure surrounding the capsid of some viruses is called the **viral envelope**. All viruses use some sort of **glycoprotein** to attach to their host cells at molecules on the cell called viral receptors. The virus exploits these cell-surface molecules, which the cell uses for some other purpose, as a way to recognize and infect specific cell types. For example, the measles virus uses a cell-surface glycoprotein in humans that normally functions in immune reactions and possibly in the sperm-egg interaction at fertilization. Attachment is a requirement for viruses to later penetrate the cell membrane, inject the viral genome, and complete their replication inside the cell.

The T4 bacteriophage, which infects the *E. coli* bacterium, is among the most complex virion known; T4 has a protein tail structure that the virus uses to attach to the host cell and a head structure that houses its DNA.

Adenovirus, a nonenveloped animal virus that causes respiratory illnesses in humans, uses protein spikes protruding from its capsomeres to attach to the host cell. Nonenveloped viruses also include those that cause polio (poliovirus), plantar warts (papillomavirus), and hepatitis A (hepatitis A virus). Nonenveloped viruses tend to be more robust and more likely to survive under harsh conditions, such as the gut.

Enveloped virions like HIV (human immunodeficiency virus), the causative agent in AIDS (acquired immune deficiency syndrome), consist of nucleic acid (RNA in the case of HIV) and capsid proteins surrounded by a phospholipid bilayer envelope and its associated proteins (Figure 15.3). Chicken pox, influenza, and mumps are examples of diseases caused by viruses with envelopes. Because of the fragility of the envelope, nonenveloped viruses are more resistant to changes in temperature, pH, and some disinfectants than enveloped viruses.

Overall, the shape of the virion and the presence or absence of an envelope tells us little about what diseases the viruses may cause or what species they might infect, but is still a useful means to begin viral classification.

Visual Connection

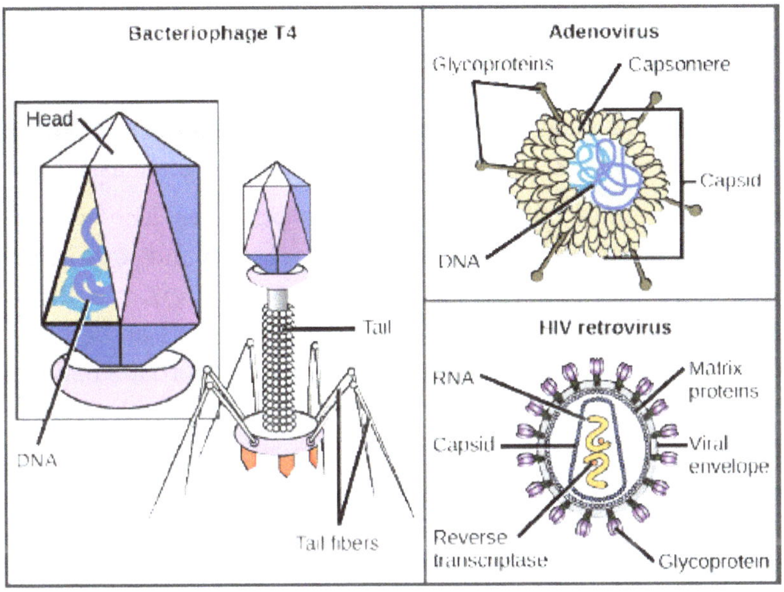

Figure 15.3 Viruses can be complex in shape or relatively simple. This figure shows three relatively complex virions: the bacteriophage T4, with its DNA-containing head group and tail fibers that attach to host cells; adenovirus, which uses spikes from its capsid to bind to the host cells; and HIV, which uses glycoproteins embedded in its envelope to do so. Notice that HIV has proteins called matrix proteins, internal to the envelope, which help stabilize virion shape. HIV is a retrovirus, which means it reverse transcribes its RNA genome into DNA, which is then spliced into the host's DNA. (credit "bacteriophage, adenovirus": modification of work by NCBI, NIH; credit "HIV retrovirus": modification of work by NIAID, NIH)

Unlike all living organisms that use DNA as their genetic material, viruses may use either DNA or RNA as theirs. The virus core contains the genome or total genetic content of the virus. Viral genomes tend to be small compared to bacteria or eukaryotes, containing only those genes that code for proteins the virus cannot get from the host cell. This genetic material may be single-stranded or double-stranded. It may also be linear or circular. While most viruses contain a single segment of nucleic acid, others have genomes that consist of several segments.

DNA viruses have a DNA core. The viral DNA directs the host cell's replication proteins to synthesize new copies of the viral genome and to transcribe and translate that genome into viral proteins. DNA viruses cause human diseases such as chickenpox, hepatitis B, and some venereal diseases like herpes and genital warts.

RNA viruses contain only RNA in their cores. To replicate their genomes in the host cell, the genomes of RNA viruses encode enzymes not found in host cells. RNA polymerase enzymes are not as stable as DNA polymerases and often make mistakes during transcription. For this reason, mutations, changes in the nucleotide sequence, in RNA viruses occur more frequently than in DNA viruses. This leads to more rapid evolution and change in RNA viruses. For example, the fact that influenza is an RNA virus is one reason a new flu vaccine is needed every year. Human diseases caused by RNA viruses include hepatitis C, measles, and rabies.

Viruses can be seen as obligate intracellular parasites. The virus must attach to a living cell, be taken inside, manufacture its proteins and copy its genome, and find a way to escape the cell so the virus can infect other cells and ultimately other individuals. Viruses can infect only certain species of hosts and only certain cells within that host. The molecular basis for this specificity is that a particular surface molecule, known as the viral receptor, must be found on the host cell surface for the virus to attach. Also, metabolic differences seen in different cell types based on differential gene expression are a likely factor in which cells a virus may use to replicate. The cell must be making the substances the virus needs, such as enzymes the virus genome itself does not have genes for, or the virus will not be able to replicate using that cell.

Steps of Virus Infections

A virus must "take over" a cell to replicate. The viral replication cycle can produce dramatic biochemical and structural changes in the host cell, which may cause cell damage. These changes, called **cytopathic** effects, can change cell functions or even destroy the cell. Some infected cells, such as those infected by the common cold virus (rhinovirus), die through lysis (bursting) or **apoptosis** (programmed cell death or "cell suicide"), releasing all the progeny virions at once. The symptoms of viral diseases result from the immune response to the virus, which attempts to control and eliminate the virus from the body, and from cell damage caused by the virus. Many animal viruses, such as HIV (human immunodeficiency virus), leave the infected cells of the immune system by a process known as budding, where virions leave the cell individually. During the budding process, the cell does not undergo lysis and is not immediately killed. However, the damage to the cells that HIV infects may make it impossible for the cells to function as mediators of immunity, even though the cells remain alive for a period of time. Most productive viral infections follow similar steps in the virus replication cycle: attachment, penetration, uncoating, replication, assembly, and release.

A virus attaches to a specific receptor site on the host-cell membrane through attachment proteins in the capsid or proteins embedded in its envelope. The attachment is specific, and typically a virus will only attach to cells of one or a few species and only certain cell types within those species with the appropriate receptors.

Unlike animal viruses, the nucleic acid of bacteriophages is injected into the host cell naked, leaving the capsid outside the cell. Plant and animal viruses can enter their cells through endocytosis, in which the cell membrane surrounds and engulfs the entire virus. Some enveloped viruses enter the cell when the viral envelope fuses directly with the cell membrane. Once inside the cell, the viral capsid is degraded and the viral nucleic acid is released, which then becomes available for replication and transcription.

The replication mechanism depends on the viral genome. DNA viruses usually use host cell proteins and enzymes to make additional DNA that is used to copy the genome or be transcribed to messenger RNA (mRNA), which is then used in protein synthesis. RNA viruses, such as the influenza virus, usually use the RNA core as a template for synthesis of viral genomic RNA and mRNA. The viral mRNA is translated into viral enzymes and capsid proteins to assemble new virions (Figure 15.4). Of course, there are exceptions to this pattern. If a host cell does not provide the enzymes necessary for viral replication, viral genes supply the information to direct synthesis of the missing proteins. Retroviruses, such as HIV, have an RNA genome that must be reverse transcribed to make DNA, which then is inserted into the host's DNA. To convert RNA into DNA, retroviruses contain genes that encode the virus-specific enzyme reverse transcriptase that transcribes an RNA template to DNA. The fact that HIV produces some of its own enzymes, which are not found in the host, has allowed researchers to develop drugs that inhibit these enzymes. These drugs, including the reverse transcriptase inhibitor AZT, inhibit HIV replication by reducing the activity of the enzyme without affecting the host's metabolism.

The last stage of viral replication is the release of the new virions into the host organism, where they are able to infect adjacent cells and repeat the replication cycle. Some viruses are released when the host cell dies and other viruses can leave infected cells by budding through the membrane without directly killing the cell.

Visual Connection

Influenza virus is packaged in a viral envelope, which fuses with the plasma membrane. This way, the virus can exit the host cell without killing it. What advantage does the virus gain by keeping the host cell alive?

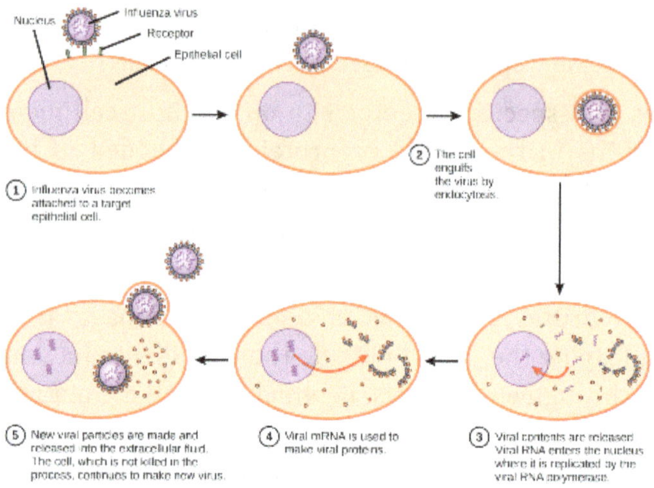

Figure 15.4 In influenza virus infection, glycoproteins attach to a host epithelial cell. As a result, the virus is engulfed. RNA and proteins are made and assembled into new virions.

Viruses and Disease

Viruses cause a variety of diseases in animals, including humans, ranging from the common cold to potentially fatal illnesses like meningitis (Figure 15.5). These diseases can be treated by antiviral drugs or by vaccines, but some viruses, such as HIV, are capable of avoiding the immune response and mutating so as to become resistant to antiviral drugs.

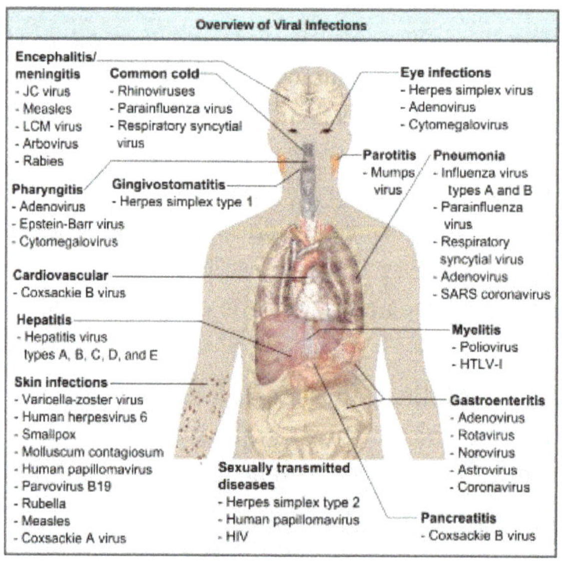

Figure 15.5 Viruses are the cause of dozens of ailments in humans, ranging from mild illnesses to serious diseases. (credit: modification of work by Mikael Häggström)

Vaccines for Prevention

While we do have limited numbers of effective antiviral drugs, such as those used to treat HIV and influenza, the primary method of controlling viral disease is by vaccination, which is intended to prevent outbreaks by building immunity to a virus or virus family. A **vaccine** may be prepared using weakened live viruses, killed viruses, or molecular subunits of the virus. In general, live viruses lead to better immunity, but have the possibility of causing disease at some low frequency. Killed viral vaccine and the subunit viruses are both incapable of causing disease, but in general lead to less effective or long-lasting immunity.

Weakened live viral vaccines are designed in the laboratory to cause few symptoms in recipients while giving them immunity against future infections. Polio was one disease that represented a milestone in the use of vaccines. Polio epidemics occurred with increasing frequency and impact as the twentieth century progressed, becoming a terrifying and tragic event each summer. Tens of thousands of people died and many more were paralyzed; children made up a large portion of the victims. Using killed virus tested on the HeLa cell line (originally obtained from Henrietta Lacks and then mass produced to meet the need), Jonas Salk developed a successful vaccine. Mass immunization campaigns in the U.S. in the 1950s (killed vaccine) and 1960s (live vaccine) essentially eradicated the disease. The success of the polio vaccine paved the way for the routine dispensation of childhood vaccines against measles, mumps, rubella, chickenpox, and other diseases.

Live vaccines are usually made by **attenuation** (weakening) of the "wild-type" (disease-causing) virus by growing it in the laboratory in tissues or at temperatures different from what the virus is accustomed to in the host. For example, the virus may be grown in cells in a test tube, in bird embryos, or in live animals. The adaptation to these new cells or temperature induces mutations in the virus' genomes, allowing them to grow better in the laboratory while inhibiting their ability to cause disease when reintroduced into the conditions found in the host. These attenuated viruses thus still cause an infection, but they do not grow very well, allowing the immune response to develop in time to prevent major disease. The danger of using live vaccines, which are usually more effective than killed vaccines, is the low but significant risk that these viruses will revert back to their disease- causing form by back mutations. Back mutations occur when the vaccine undergoes mutations in the host such that it readapts to the host and can again cause disease, which can then be spread to other humans in an epidemic. This happened as recently as 2007 in Nigeria where mutations in a polio vaccine led to an epidemic of polio in that country.

Some vaccines are in continuous development because certain viruses, such as influenza and HIV, have a high mutation rate compared to other viruses or host cells. With influenza, mutation in genes for the surface molecules helps the virus evade the protective immunity that may have been obtained in a previous influenza season, making it necessary for individuals to get vaccinated every year. Other viruses, such as those that cause the childhood diseases measles, mumps, and rubella, mutate so little that the same vaccine is used year after year.

Other vaccines are developed because of a new mutation of a well known type of virus. Corona viruses are very common in nature and various diseases, but a specific mutation of a particular corona virus led to the COVID-19 pandemic. The rapid development of successful vaccines was due in part to scientists who had been working with other corona viruses prior to the pandemic. Kizzmekia S. Corbett, a research fellow and scientific lead, had deep experience and knowledge of corona viruses, which was instrumental in developing one of the first vaccines (Moderna). She is now applying that experience to other respiratory diseases and vaccine development processes.

Vaccines and Antiviral Drugs for Treatment

In some cases, vaccines can be used to treat an active viral infection. In the case of rabies, a fatal neurological disease transmitted in the saliva of rabies virus-infected animals, the progression of the disease from the time of the animal bite to the time it enters the central nervous system may be two weeks or longer. This is enough time to vaccinate an individual who suspects being bitten by a rabid animal, and the boosted immune response from the vaccination is enough to prevent the virus from entering nervous tissue. Thus, the fatal neurological consequences of the disease are averted and the individual only has to recover from the infected bite. This approach is also being used for the treatment of Ebola, one of the fastest and most deadly viruses affecting humans, though usually infecting limited populations. Ebola is also a leading cause of death in gorillas. Transmitted by bats and great apes, this virus can cause death in 70–90 percent of the infected within two weeks. Using newly developed vaccines that boost the immune response, there is hope that immune systems of affected individuals will be better able to control the virus, potentially reducing mortality rates.

Another way of treating viral infections is the use of antiviral drugs. These drugs often have limited ability to cure viral disease but have been used to control and reduce symptoms for a wide variety of viral diseases. For most viruses, these drugs inhibit the virus by blocking the actions of one or more of its proteins. It is important that the targeted proteins be encoded for by viral genes and that these molecules are not present in a healthy host cell. In fact, for many years, scientists thought that drugs capable of impacting viruses would be too toxic for the body to endure. To meet this challenge, researcher Gertrude Elion sought to develop drugs that would target only the virus through processes such as inhibiting only viral DNA replication. For example, some of her medicines focused on purines, while others affected DNA polymerase. There are large numbers of antiviral drugs available to treat infections, some specific for a particular virus and others that can affect multiple viruses.

Antivirals have been developed to treat genital herpes (herpes simplex II) and influenza. For genital herpes, drugs such as acyclovir, developed by Elion, can reduce the number and duration of the episodes of active viral disease during which patients develop viral lesions in their skins cells. As the virus remains latent in nervous tissue of the body for life, this drug is not a cure but can make the symptoms of the disease more manageable. For influenza, drugs like Tamiflu can reduce the duration of "flu" symptoms by one or two days, but the drug does not prevent symptoms entirely. Other antiviral drugs, such as Ribavirin, have been used to treat a variety of viral infections.

By far the most successful use of antivirals has been in the treatment of the retrovirus HIV, which causes a disease that, if untreated, is usually fatal within 10–12 years after being infected. Anti-HIV drugs have been able to control viral replication to the point that individuals receiving these drugs survive for a significantly longer time than the untreated.

A particular challenge with HIV is its tendency to mutate quickly within the body of an individual patient. This leads to individual drug resistance, and requires a different treatment strategy than many other diseases. David Ho was among the first to propose and develop a method to treat multiple mutations of HIV at the same time. Ho's efforts were a turning point in fighting AIDS. Anti-HIV drugs inhibit viral replication at many different phases of the HIV replicative cycle. Drugs have been developed that inhibit the fusion of the HIV viral envelope with the plasma membrane of the host cell (fusion inhibitors), the conversion of its RNA genome to double-stranded DNA (reverse transcriptase inhibitors), the integration of the viral DNA into the host genome (integrase inhibitors), and the processing of viral proteins (protease inhibitors).

When any of these drugs are used individually, the virus' high mutation rate allows the virus to rapidly evolve resistance to the drug. The breakthrough in the treatment of HIV was the development of highly active anti- retroviral therapy (HAART), which involves a mixture of different drugs, sometimes called a drug "cocktail." By attacking the virus at different stages of its replication cycle, it is difficult for the virus to develop resistance to multiple drugs at the same time. Still, even with the use of combination HAART therapy, there is concern that, over time, the virus will evolve resistance to this therapy. Thus, new anti-HIV drugs are constantly being developed with the hope of continuing the battle against this highly fatal virus.

INNATE IMMUNITY

The vertebrate, including human, immune system is a complex multilayered system for defending against external and internal threats to the integrity of the body. The system can be divided into two types of defense systems: the innate immune system, which is nonspecific toward a particular kind of pathogen, and the adaptive immune system, which is specific (Figure 15.6). Innate **immunity** is not caused by an infection or vaccination and depends initially on physical and chemical barriers that work on all pathogens, sometimes called the first line of defense. The second line of defense of the innate system includes chemical signals that produce inflammation and fever responses as well as mobilizing protective cells and other chemical defenses. The adaptive immune system mounts a highly specific response to substances and organisms that do not belong in the body. The adaptive system takes longer to respond and has a memory system that allows it to respond with greater intensity should the body reencounter a pathogen even years later.

Vertebrate Immunity		
Innate Immune System		Adaptive Immune System
Physical Barriers	Internal Defenses	
• Skin, hair, cilia • Mucus membranes • Mucus and chemical secretions • Digestive enzymes in mouth • Stomach acid	• Inflammatory response • Complement proteins • Phagocytic cells • Natural killer (NK) cells	• Antibodies and the humoral immune response • Cell-mediated immune response • Memory response

Figure 15.6 There are two main parts to the vertebrate immune system. The innate immune system, which is made up of physical barriers and internal defenses, responds to all pathogens. The adaptive immune system is highly specific.

External and Chemical Barriers

The body has significant physical barriers to potential pathogens. The skin contains the protein keratin, which resists physical entry into cells. Other body surfaces, particularly those associated with body openings, are protected by the mucous membranes. The sticky mucus provides a physical trap for pathogens, preventing their movement deeper into the body. The openings of the body, such as the nose and ears, are protected by hairs that catch pathogens, and the mucous membranes of the upper respiratory tract have cilia that constantly move pathogens trapped in the mucus coat up to the mouth.

The skin and mucous membranes also create a chemical environment that is hostile to many microorganisms. The surface of the skin is acidic, which prevents bacterial growth. Saliva, mucus, and the tears of the eye contain an enzyme that breaks down bacterial cell walls. The stomach secretions create a highly acidic environment, which kills many pathogens entering the digestive system.

Finally, the surface of the body and the lower digestive system have a community of microorganisms such as bacteria, archaea, and fungi that coexist without harming the body. There is evidence that these organisms are highly beneficial to their host, combating disease-causing organisms and outcompeting them for nutritional resources provided by the host body. Despite these defenses, pathogens may enter the body through skin abrasions or punctures, or by collecting on mucosal surfaces in large numbers that overcome the protections of mucus or cilia.

Internal Defenses

When pathogens enter the body, the innate immune system responds with a variety of internal defenses. These include the inflammatory response, phagocytosis, natural killer cells, and the complement system. White blood cells in the blood and lymph recognize pathogens as foreign to the body. A **white blood cell** is larger than a red blood cell, is nucleated, and is typically able to move using amoeboid locomotion. Because they can move on their own, white blood cells can leave the blood to go to infected tissues. For example, a **monocyte** is a type of white blood cell that circulates in the blood and lymph and develops into a macrophage after it moves into infected tissue. A **macrophage** is a large cell that engulfs foreign particles and pathogens. **Mast cells** are produced in the same way as white blood cells, but unlike circulating white blood cells, mast cells take up residence in connective tissues and especially mucosal tissues. They are responsible for releasing chemicals in response to physical injury. They also play a role in the allergic response, which will be discussed later in the chapter.

When a pathogen is recognized as foreign, chemicals called cytokines are released. A **cytokine** is a chemical messenger that regulates cell differentiation (form and function), proliferation (production), and gene expression to produce a variety of immune responses. Approximately 40 types of cytokines exist in humans. In addition to being released from white blood cells after pathogen recognition, cytokines are also released by the infected cells and bind to nearby uninfected cells, inducing those cells to release cytokines. This positive feedback loop results in a burst of cytokine production.

One class of early-acting cytokines is the interferons, which are released by infected cells as a warning to nearby uninfected cells. An **interferon** is a small protein that signals a viral infection to other cells. The interferons stimulate uninfected cells to produce compounds that interfere with viral replication. Interferons also activate macrophages and other cells.

The Inflammatory Response and Phagocytosis

The first cytokines to be produced encourage **inflammation**, a localized redness, swelling, heat, and pain. Inflammation is a response to physical trauma, such as a cut or a blow, chemical irritation, and infection by pathogens (viruses, bacteria, or fungi). The chemical signals that trigger an inflammatory response enter the extracellular fluid and cause capillaries to dilate (expand) and capillary walls to become more permeable, or leaky. The serum and other compounds leaking from capillaries cause swelling of the area, which in turn causes pain. Various kinds of white blood cells are attracted to the area of inflammation. The types of white blood cells that arrive at an inflamed site depend on the nature of the injury or infecting pathogen. For example, a **neutrophil** is an early arriving white blood cell that engulfs and digests pathogens. Neutrophils are the most abundant white blood cells of the immune system (Figure 15.7). Macrophages follow neutrophils and take over the phagocytosis function and are involved in the resolution of an inflamed site, cleaning up cell debris and pathogens.

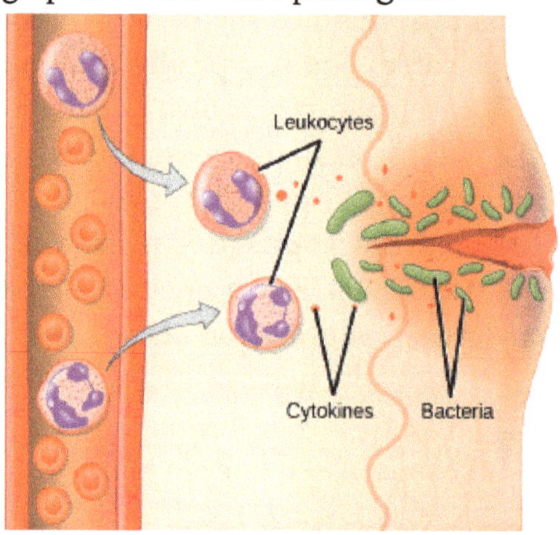

Figure 15.7 White blood cells (leukocytes) release chemicals to stimulate the inflammatory response following a cut in the skin.

Cytokines also send feedback to cells of the nervous system to bring about the overall symptoms of feeling sick, which include lethargy, muscle pain, and nausea. Cytokines also increase the core body temperature, causing a fever. The elevated temperatures of a fever inhibit the growth of pathogens and speed up cellular repair processes. For these reasons, suppression of fevers should be limited to those that are dangerously high.

Natural Killer Cells

A lymphocyte is a white blood cell that contains a large nucleus (Figure 15.8). Most lymphocytes are associated with the adaptive immune response, but infected cells are identified and destroyed by natural killer cells, the only lymphocytes of the innate immune system. A natural killer (NK) cell is a lymphocyte that can kill cells infected with viruses (or cancerous cells). NK cells identify intracellular infections, especially from viruses, by the altered expression of major histocompatibility complex (MHC) I molecules on the surface of infected cells. MHC class I molecules are proteins on the surfaces of all nucleated cells that provide a sample of the cell's internal environment at any given time. Unhealthy cells, whether infected or cancerous, display an altered MHC class I complement on their cell surfaces.

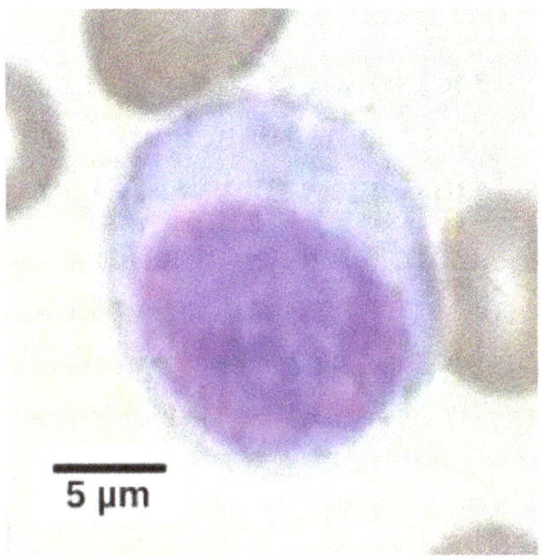

Figure 15.8 Lymphocytes, such as NK cells, are characterized by their large nuclei that actively absorb Wright stain and therefore appear dark colored under a microscope. (credit: scale-bar data from Matt Russell)

After the NK cell detects an infected or tumor cell, it induces programmed cell death, or apoptosis. Phagocytic cells then come along and digest the cell debris left behind. NK cells are constantly patrolling the body and are an effective mechanism for controlling potential infections and preventing cancer progression. The various types of immune cells are shown in Figure 15.9.

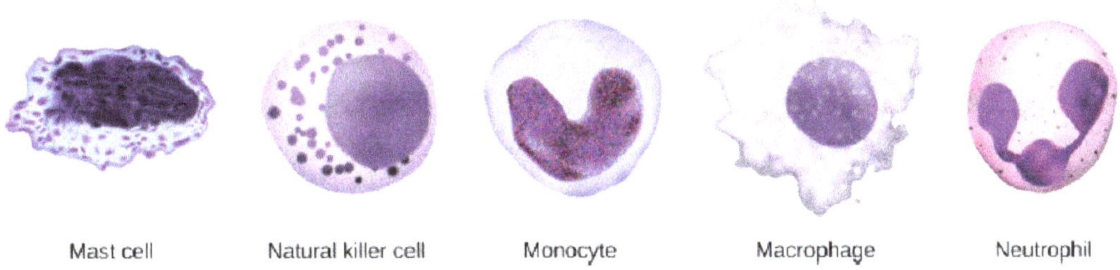

Figure 15.9 Cells involved in the innate immune response include mast cells, natural killer cells, and white blood cells, such as monocytes, macrophages and neutrophils.

Complement

An array of approximately 20 types of proteins, called a **complement system**, is also activated by infection or the activity of the cells of the adaptive immune system and functions to destroy extracellular pathogens. Liver cells and macrophages synthesize inactive forms of complement proteins continuously; these proteins are abundant in the blood serum and are capable of responding immediately to infecting microorganisms. The complement system is so named because it is complementary to the innate and adaptive immune system. Complement proteins bind to the surfaces of microorganisms and are particularly attracted to pathogens that are already tagged by the adaptive immune system. This "tagging" involves the attachment of specific proteins called antibodies (discussed in detail later) to the pathogen. When they attach, the antibodies change shape providing a binding site for one of the complement proteins. After the first few complement proteins bind, a cascade of binding in a specific sequence of proteins follows in which the pathogen rapidly becomes coated in complement proteins.

Complement proteins perform several functions, one of which is to serve as a marker to indicate the presence of a pathogen to phagocytic cells and enhance engulfment. Certain complement proteins can combine to open pores in microbial cell membranes and cause lysis of the cells.

ADAPTIVE IMMUNITY

The adaptive, or acquired, immune response takes days or even weeks to become established—much longer than the innate response; however, adaptive immunity is more specific to an invading pathogen. **Adaptive immunity** is an immunity that occurs after exposure to an antigen either from a pathogen or a vaccination. An **antigen** is a molecule that stimulates a response in the immune system. This part of the immune system is activated when the innate immune response is insufficient to control an infection. In fact, without information from the innate immune system, the adaptive response could not be mobilized. There are two types of adaptive responses: the **cell-mediated immune response**, which is controlled by activated **T cells**, and the **humoral immune response**, which is controlled by activated **B cells** and antibodies. Activated T and B cells whose surface binding sites are specific to the molecules on the pathogen greatly increase in numbers and attack the invading pathogen. Their attack can kill pathogens directly or they can secrete antibodies that enhance the phagocytosis of pathogens and disrupt the infection. Adaptive immunity also involves a memory to give the host long-term protection from reinfection with the same type of pathogen; on reexposure, this host memory will facilitate a rapid and powerful response.

B and T Cells

Lymphocytes, which are white blood cells, are formed with other blood cells in the red bone marrow found in many flat bones, such as the shoulder or pelvic bones. The two types of lymphocytes of the adaptive immune response are B and T cells (Figure 15.10). Whether an immature lymphocyte becomes a B cell or T cell depends on where in the body it matures. The B cells remain in the bone marrow to mature (hence the name "B" for "bone marrow"), while T cells migrate to the thymus, where they mature (hence the name "T" for "thymus").

Maturation of a B or T cell involves becoming immune-competent, meaning that it can recognize, by binding, a specific molecule or antigen (discussed below). During the maturation process, B and T cells that bind too strongly to the body's own cells are eliminated in order to minimize an immune response against the body's own tissues. Those cells that react weakly to the body's own cells, but have highly specific receptors on their cell surfaces that allow them to recognize a foreign molecule, or antigen, remain. This process occurs during fetal development and continues throughout life. The specificity of this receptor is determined by the genetics of the individual and is present before a foreign molecule is introduced to the body or encountered. Thus, it is genetics and not experience that initially provides a vast array of cells, each capable of binding to a different specific foreign molecule. Once they are immune-competent, the T and B cells will migrate to the spleen and lymph nodes where they will remain until they are called on during an infection. B cells are involved in the humoral immune response, which targets pathogens loose in blood and lymph, and T cells are involved in the cell-mediated immune response, which targets infected cells.

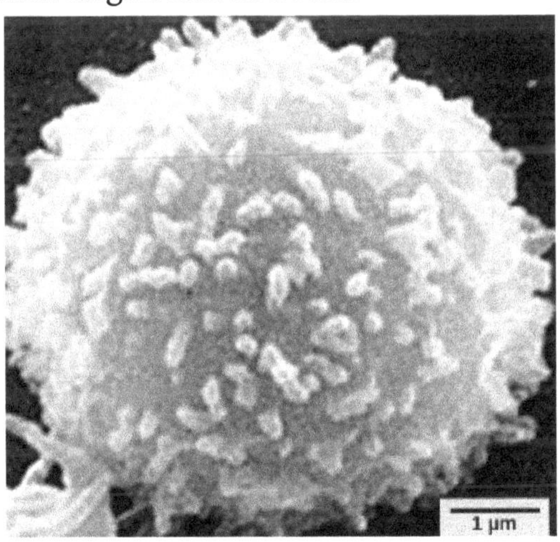

Figure 15.10 This scanning electron micrograph shows a T lymphocyte. T and B cells are indistinguishable by light microscopy but can be differentiated experimentally by probing their surface receptors. (credit: modification of work by NCI; scale-bar data from Matt Russell)

Humoral Immune Response

As mentioned, an antigen is a molecule that stimulates a response in the immune system. Not every molecule is antigenic. B cells participate in a chemical response to antigens present in the body by producing specific antibodies that circulate throughout the body and bind with the antigen whenever it is encountered. This is known as the humoral immune response. As discussed, during maturation of B cells, a set of highly specific B cells are produced that have many antigen receptor molecules in their membrane (Figure 15.11).

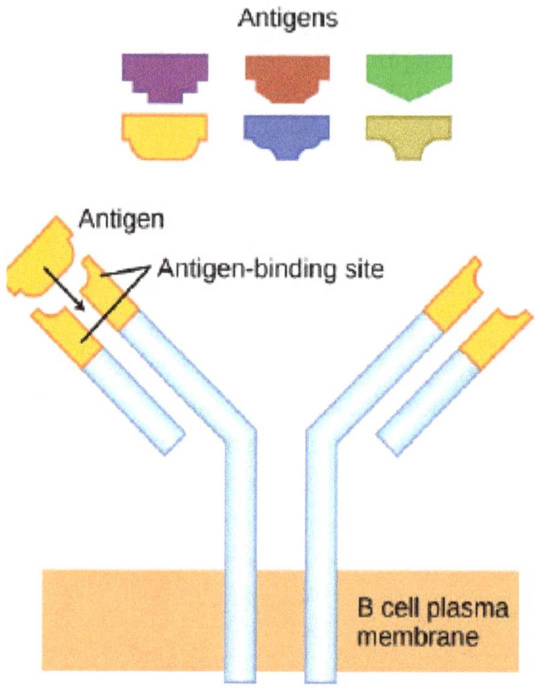

Figure 15.11 B cell receptors are embedded in the membranes of B cells and bind a variety of antigens through their variable regions.

Each B cell has only one kind of antigen receptor, which makes every B cell different. Once the B cells mature in the bone marrow, they migrate to lymph nodes or other lymphatic organs. When a B cell encounters the antigen that binds to its receptor, the antigen molecule is brought into the cell by endocytosis and reappears on the surface of the cell bound to an **MHC class II molecule**. When this process is complete, the B cell is sensitized. In most cases, the sensitized B cell must then encounter a specific kind of T cell, called a helper T cell, before it is activated. The helper T cell must already have been activated through an encounter with the antigen.

The helper T cell binds to the antigen-MHC class II complex and is induced to release cytokines that induce the B cell to divide rapidly, which makes thousands of identical (clonal) cells. These daughter cells become either plasma cells or memory B cells. The memory B cells remain inactive at this point, until another later encounter with the antigen, caused by a reinfection by the same bacteria or virus, results in them dividing into a new population of plasma cells. The plasma cells, on the other hand, produce and secrete large quantities, up to 100 million molecules per hour, of antibody molecules. An **antibody**, also known as an immunoglobulin (Ig), is a protein that is produced by plasma cells after stimulation by an antigen. Antibodies are the agents of humoral immunity. Antibodies occur in the blood, in gastric and mucus secretions, and in breast milk. Antibodies in these bodily fluids can bind pathogens and mark them for destruction by phagocytes before they can infect cells.

These antibodies circulate in the blood stream and lymphatic system and bind with the antigen whenever it is encountered. The binding can fight infection in several ways. Antibodies can bind to viruses or bacteria and interfere with the chemical interactions required for them to infect or bind to other cells. The antibodies may create bridges between different particles containing antigenic sites clumping them all together and preventing their proper functioning. The antigen-antibody complex stimulates the complement system described previously, destroying the cell bearing the antigen. Phagocytic cells, such as those already described, are attracted by the antigen-antibody complexes, and phagocytosis is enhanced when the complexes are present. Finally, antibodies stimulate inflammation, and their presence in mucus and on the skin prevents pathogen attack.

Antibodies coat extracellular pathogens and neutralize them by blocking key sites on the pathogen that enhance their infectivity (such as receptors that "dock" pathogens on host cells) (Figure 15.12). Antibody neutralization can prevent pathogens from entering and infecting host cells. The neutralized antibody-coated pathogens can then be filtered by the spleen and eliminated in urine or feces.

Antibodies also mark pathogens for destruction by phagocytic cells, such as macrophages or neutrophils, in a process called opsonization. In a process called complement fixation, some antibodies provide a place for complement proteins to bind. The combination of antibodies and complement promotes rapid clearing of pathogens.

The production of antibodies by plasma cells in response to an antigen is called **active immunity** and describes the host's active response of the immune system to an infection or to a vaccination. There is also a **passive immune** response where antibodies come from an outside source, instead of the individual's own plasma cells, and are introduced into the host. For example, antibodies circulating in a pregnant person's body move across the placenta into the developing fetus. The child benefits from the presence of these antibodies for up to several months after birth. In addition, a passive immune response is possible by injecting antibodies into an individual in the form of an antivenom to a snake-bite toxin or antibodies in blood serum to help fight a hepatitis infection. This gives immediate protection since the body does not need the time required to mount its own response.

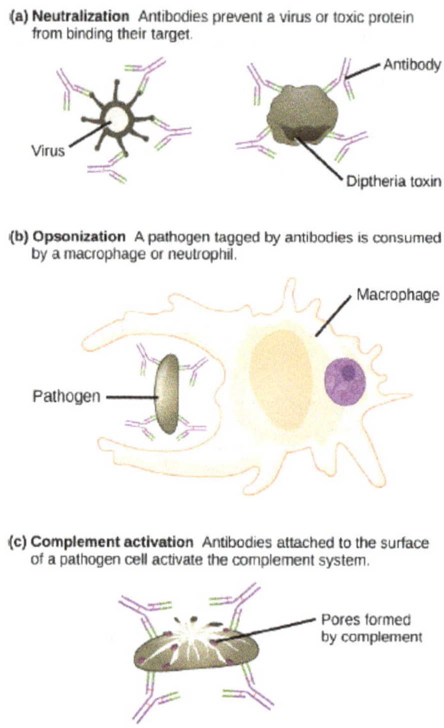

Figure 15.12 Antibodies may inhibit infection by (a) preventing the antigen from binding its target, (b) tagging a pathogen for destruction by macrophages or neutrophils, or (c) activating the complement cascade.

The availability and reliability of antibodies makes them ideally suited for use in medical tests and investigations. Radioimmunossays, for example, rely on the antigen-antibody interaction. Usually, a specific antigen is made radioactive, allowed to bind to its antibody, and then introduced into a sample substance, such as a patient's blood. By measuring eventual changes in the quantity of the bound and unbound antigen, analysts can detect the presence and/or concentration of certain substances. Developed by Rosalyn Sussman Yalow and Solomon Berson in the 1950s, the technique is known for extreme sensitivity, meaning that it can detect and measure very small quantities of a substance. It is used in narcotics detection, blood bank screening, early cancer screening, hormone measurement, and allergy diagnosis. Based on her significant contribution to the field, Yalow received a Nobel Prize, making her the second woman to be awarded the prize for medicine.

Cell-Mediated Immunity

Unlike B cells, T lymphocytes are unable to recognize pathogens without assistance. Instead, dendritic cells and macrophages first engulf and digest pathogens into hundreds or thousands of antigens. Then, an **antigen- presenting cell (APC)** detects, engulfs, and informs the adaptive immune response about an infection. When a pathogen is detected, these APCs will engulf and break it down through phagocytosis. Antigen fragments will then be transported to the surface of the APC, where they will serve as an indicator to other immune cells. A **dendritic cell** is an immune cell that mops up antigenic materials in its surroundings and presents them on its surface. Dendritic cells are located in the skin, the linings of the nose, lungs, stomach, and intestines. These positions are ideal locations to encounter invading pathogens. Once they are activated by pathogens and mature to become APCs they migrate to the spleen or a lymph node. Macrophages also function as APCs. After phagocytosis by a macrophage, the phagocytic vesicle fuses with an intracellular lysosome. Within the resulting phagolysosome, the components are broken down into fragments; the fragments are then loaded onto MHC class II molecules and are transported to the cell surface for antigen presentation (Figure 15.13). Helper T cells cannot properly respond to an antigen unless it is processed and embedded in an MHC class II molecule. The APCs express MHC class II on their surfaces, and when combined with a foreign antigen, these complexes signal an invader.

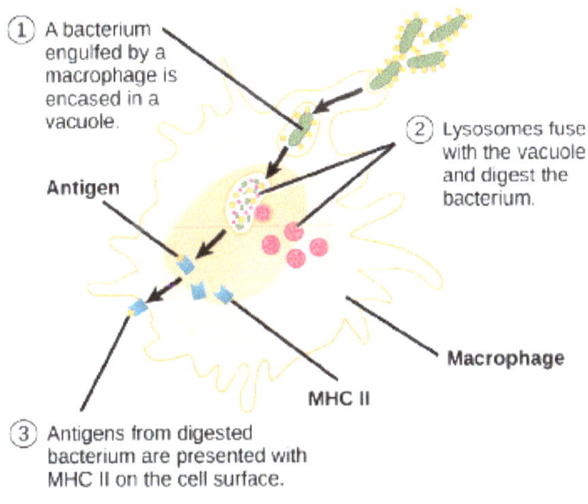

Figure 15.13 An antigen-presenting cell (APC), such as a macrophage, engulfs a foreign antigen, partially digests it in a lysosome, and then embeds it in an MHC class II molecule for presentation at the cell surface. Lymphocytes of the adaptive immune response must interact with antigen-embedded MHC class II molecules to mature into functional immune cells.

T cells have many functions. Some respond to APCs of the innate immune system and indirectly induce immune responses by releasing cytokines. Others stimulate B cells to start the humoral response as described previously. Another type of T cell detects APC signals and directly kills the infected cells, while some are involved in suppressing inappropriate immune reactions to harmless or "self" antigens.

There are two main types of T cells: helper T lymphocytes (T_H) and the cytotoxic T lymphocytes (T_C). The T_H lymphocytes function indirectly to tell other immune cells about potential pathogens. T_H lymphocytes recognize specific antigens presented by the MHC class II complexes of APCs. There are two populations of T_H cells: T_H1 and T_H2. T_H1 cells secrete cytokines to enhance the activities of macrophages and other T cells. T_H2 cells stimulate naïve B cells to secrete antibodies. Whether a T_H1 or a T_H2 immune response develops depends on the specific types of cytokines secreted by cells of the innate immune system, which in turn depends on the nature of the invading pathogen.

Cytotoxic T cells (T_C) are the key component of the cell-mediated part of the adaptive immune system and attack and destroy infected cells. T_C cells are particularly important in protecting against viral infections; this is because viruses replicate within cells where they are shielded from extracellular contact with circulating antibodies. Once activated, the T_C creates a large clone of cells with one specific set of cell-surface receptors, as in the case with proliferation of activated B cells. As with B cells, the clone includes active T_C cells and inactive memory T_C cells. The resulting active T_C cells then identify infected host cells. Because of the time required to generate a population of clonal T and B cells, there is a delay in the adaptive immune response compared to the innate immune response.

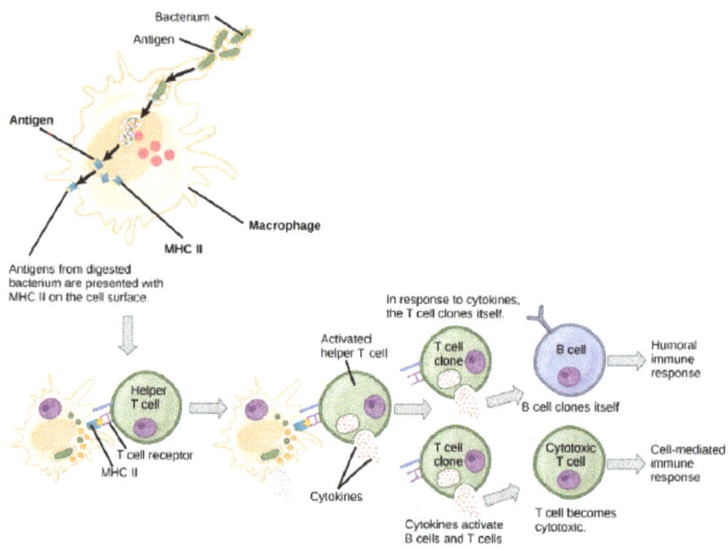

Figure 15.14 A helper T cell becomes activated by binding to an antigen presented by an APC via the MHCII receptor, causing it to release cytokines. Depending on the cytokines released, this activates either the humoral or the cell- mediated immune response.

T_C cells attempt to identify and destroy infected cells before the pathogen can replicate and escape, thereby halting the progression of intracellular infections. T_C

cells also support NK lymphocytes to destroy early cancers. Cytokines secreted by the T_H1 response that stimulates macrophages also stimulate T_C cells and enhance their ability to identify and destroy infected cells and tumors. A summary of how the humoral and cell-mediated immune responses are activated appears in Figure 15.14.

B plasma cells and T_C cells are collectively called **effector cells** because they are involved in "effecting" (bringing about) the immune response of killing pathogens and infected host cells.

Immunological Memory

The adaptive immune system has a memory component that allows for a rapid and large response upon reinvasion of the same pathogen. During the adaptive immune response to a pathogen that has not been encountered before, known as the **primary immune response**, plasma cells secreting antibodies and differentiated T cells increase, then plateau over time. As B and T cells mature into effector cells, a subset of the naïve populations differentiates into B and T memory cells with the same antigen specificities (Figure 15.15). A **memory cell** is an antigen-specific B or T lymphocyte that does not differentiate into an effector cell during the primary immune response, but that can immediately become an effector cell on reexposure to the same pathogen. As the infection is cleared and pathogenic stimuli subside, the effectors are no longer needed and they undergo apoptosis. In contrast, the memory cells persist in the circulation.

Visual Connection

The Rh antigen is found on Rh-positive red blood cells. An Rh-negative person can usually carry an Rh- positive fetus to term without difficulty. However, having a second Rh-positive fetus may launch an immune attack that causes hemolytic disease of the newborn. Why do you think hemolytic disease is only a problem during the second or subsequent pregnancies?

If the pathogen is never encountered again during the individual's lifetime, B and T memory cells will circulate for a few years or even several decades and will gradually die off, having never functioned as effector cells. However, if the host is re-exposed to the same pathogen type, circulating memory cells will immediately differentiate into plasma cells and T_C cells without input from APCs or T_H cells. This is known as the **secondary immune response**. One reason why the adaptive immune response is delayed is because it takes time for naïve B and T cells with the appropriate antigen specificities to be identified, activated, and proliferate. On reinfection, this step is skipped, and the result is a more rapid production of immune defenses. Memory B cells that differentiate into plasma cells output tens to hundreds-fold greater antibody amounts than were secreted during the primary response (Figure 15.16). This rapid and dramatic antibody response may stop the infection before it can even become established, and the individual may not realize they had been exposed.

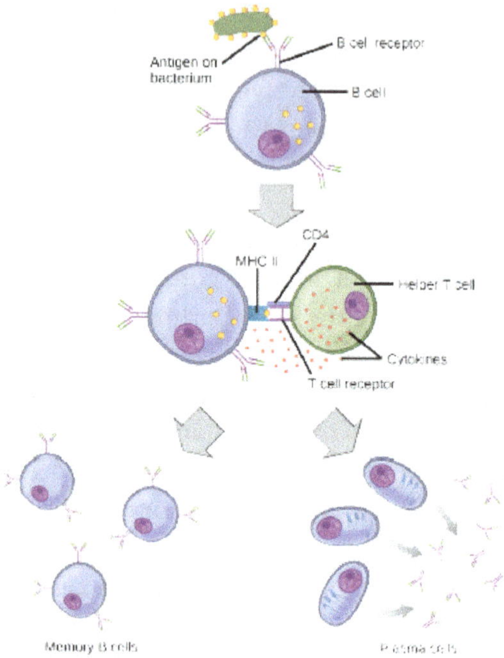

Figure 15.15 After initially binding an antigen to the B cell receptor, a B cell internalizes the antigen and presents it on MHC class II. A helper T cell recognizes the MHC class II- antigen complex and activates the B cell. As a result, memory B cells and plasma cells are made.

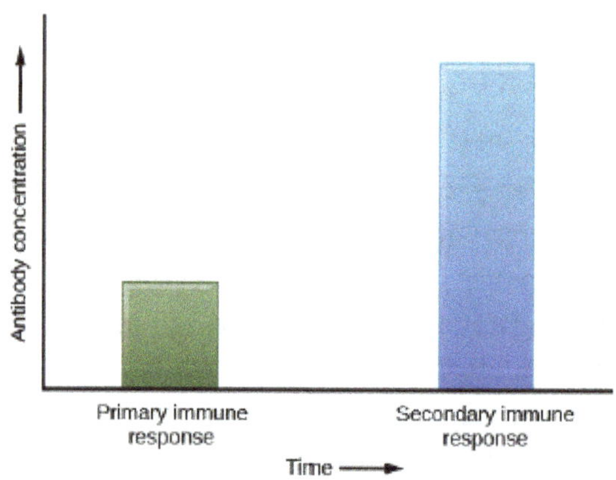

Figure 15.16 In the primary response to infection, antibodies are secreted first from plasma cells. Upon re-exposure to the same pathogen, memory cells differentiate into antibody-secreting plasma cells that output a greater amount of antibody for a longer period of time.

Vaccination is based on the knowledge that exposure to noninfectious antigens, derived from known pathogens, generates a mild primary immune response. The immune response to vaccination may not be perceived by the host as illness but still confers immune memory. When exposed to the corresponding pathogen to which an individual was vaccinated, the reaction is similar to a secondary exposure. Because each reinfection generates more memory cells and increased resistance to the pathogen, some vaccine courses involve one or more booster vaccinations to mimic repeat exposures.

The Lymphatic System

Lymph is the watery fluid that bathes tissues and organs and contains protective white blood cells but does not contain erythrocytes. Lymph moves about the body through the lymphatic system, which is made up of vessels, lymph ducts, lymph glands, and organs, such as tonsils, adenoids, thymus, and spleen.

Although the immune system is characterized by circulating cells throughout the body, the regulation, maturation, and intercommunication of immune factors occur at specific sites. The blood circulates immune cells, proteins, and other factors through the body. Approximately 0.1 percent of all cells in the blood are leukocytes, which include monocytes (the precursor of macrophages) and lymphocytes. Most cells in the blood are red blood cells. Cells of the immune system can travel between the distinct lymphatic and blood circulatory systems, which are separated by interstitial space, by a process called extravasation (passing through to surrounding tissue).

Recall that cells of the immune system originate from stem cells in the bone marrow. B cell maturation occurs in the bone marrow, whereas progenitor cells migrate from the bone marrow and develop and mature into naïve T cells in the organ called the thymus.

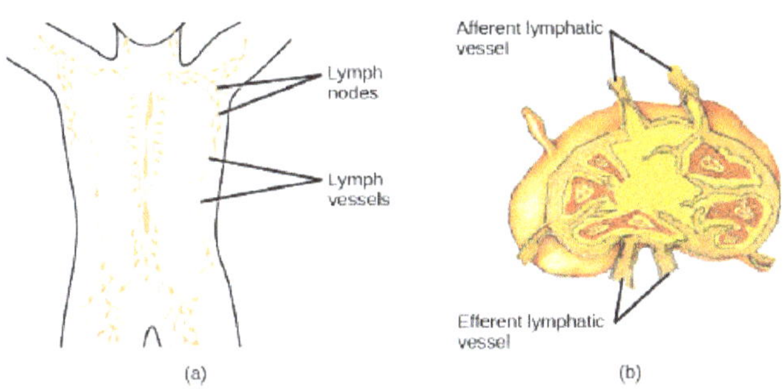

Figure 15.17 (a) Lymphatic vessels carry a clear fluid called lymph throughout the body. The liquid passes through (b) lymph nodes that filter the lymph that enters the node through afferent vessels and leaves through efferent vessels; lymph nodes are filled with lymphocytes that purge infecting cells. (credit a: modification of work by NIH; credit b: modification of work by NCI, NIH)

On maturation, T and B lymphocytes circulate to various destinations. Lymph nodes scattered throughout the body house large populations of T and B cells, dendritic cells, and macrophages (Figure 15.17). Lymph gathers antigens as it drains from tissues. These antigens then are filtered through lymph nodes before the lymph is returned to circulation. APCs in the lymph nodes capture and process antigens and inform nearby lymphocytes about potential pathogens.

The spleen houses B and T cells, macrophages, dendritic cells, and NK cells (Figure 15.18). The spleen is the site where APCs that have trapped foreign particles in the blood can communicate with lymphocytes. Antibodies are synthesized and secreted by activated plasma cells in the spleen, and the spleen filters foreign substances and antibody-complexed pathogens from the blood. Functionally, the spleen is to the blood as lymph nodes are to the lymph.

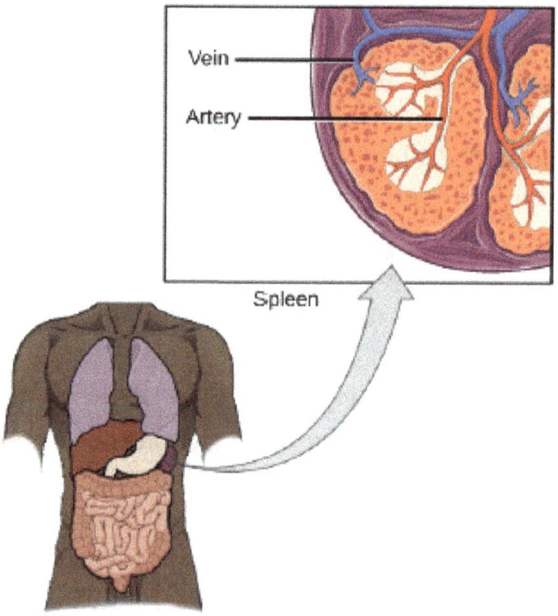

Figure 15.18 The spleen functions to immunologically filter the blood and allow for communication between cells corresponding to the innate and adaptive immune responses. (credit: modification of work by NCI, NIH)

Mucosal Immune System

The innate and adaptive immune responses compose the systemic immune system (affecting the whole body), which is distinct from the mucosal immune system. Mucosa associated lymphoid tissue (MALT) is a crucial component of a functional immune system because mucosal surfaces, such as the nasal passages, are the first tissues onto which inhaled or ingested pathogens are deposited. The mucosal tissue includes the mouth, pharynx, and esophagus, and the gastrointestinal, respiratory, and urogenital tracts.

Mucosal immunity is formed by MALT, which functions independently of the systemic immune system, and which has its own innate and adaptive components. MALT is a collection of lymphatic tissue that combines with epithelial tissue lining the mucosa throughout the body. This tissue functions as the immune barrier and response in areas of the body with direct contact to the external environment. The systemic and mucosal immune systems use many of the same cell types. Foreign particles that make their way to MALT are taken up by absorptive epithelial cells and delivered to APCs located directly below the mucosal tissue. APCs of the mucosal immune system are primarily dendritic cells, with B cells and macrophages having minor roles. Processed antigens displayed on APCs are detected by T cells in the MALT and at the tonsils, adenoids, appendix, or the mesenteric lymph nodes of the intestine. Activated T cells then migrate through the lymphatic system and into the circulatory system to mucosal sites of infection.

Immune Tolerance

The immune system has to be regulated to prevent wasteful, unnecessary responses to harmless substances, and more importantly, so that it does not attack "self." The acquired ability to prevent an unnecessary or harmful immune response to a detected foreign substance known not to cause disease, or self-antigens, is described as **immune tolerance**. The primary mechanism for developing immune tolerance to self-antigens occurs during the selection for weakly self-binding cells during T and B lymphocyte maturation. There are populations of T cells that suppress the immune response to self-antigens and that suppress the immune response after the infection has cleared to minimize host cell damage induced by inflammation and cell lysis. Immune tolerance is especially well developed in the mucosa of the upper digestive system because of the tremendous number of foreign substances (such as food proteins) that APCs of the oral cavity, pharynx, and gastrointestinal mucosa encounter. Immune tolerance is brought about by specialized APCs in the liver, lymph nodes, small intestine, and lung that present harmless antigens to a diverse population of regulatory T (T_{reg}) cells, specialized lymphocytes that suppress local inflammation and inhibit the secretion of stimulatory immune factors. The combined result of T_{reg} cells is to prevent immunologic activation and inflammation in undesired tissue compartments and to allow the immune system to focus on pathogens instead.

ISOLATION, CULTURE, AND IDENTIFICATION OF VIRUSES

At the beginning of this chapter, we described how porcelain Chamberland filters with pores small enough to allow viruses to pass through were used to discover TMV. Today, porcelain filters have been replaced with membrane filters and other devices used to isolate and identify viruses.

Isolation of Viruses

Unlike bacteria, many of which can be grown on an artificial nutrient medium, viruses require a living host cell for replication. Infected host cells (eukaryotic or prokaryotic) can be cultured and grown, and then the growth medium can be harvested as a source of virus. Virions in the liquid medium can be separated from the host cells by either centrifugation or filtration. Filters can physically remove anything present in the solution that is larger than the virions; the viruses can then be collected in the filtrate (see Figure 15.19).

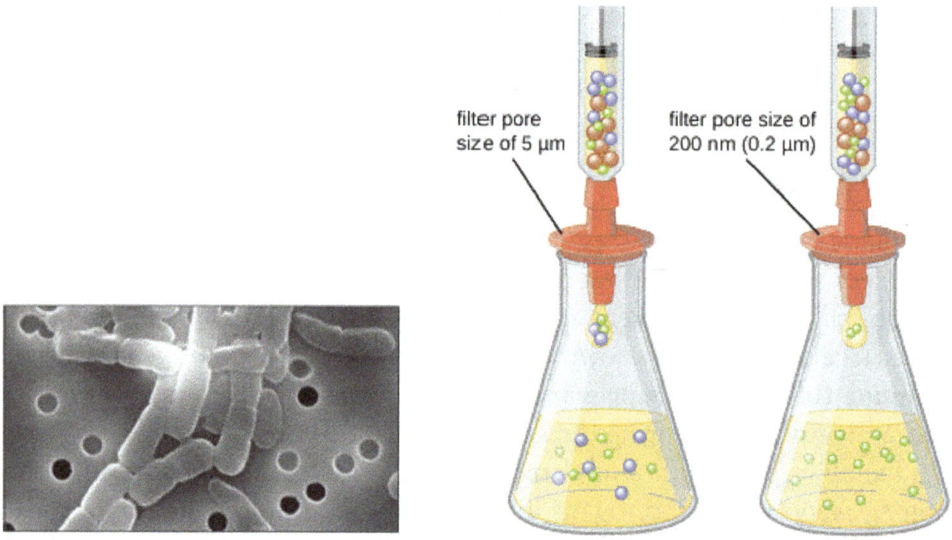

(a) (b)

Figure 15.19 Membrane filters can be used to remove cells or viruses from a solution. (a) This scanning electron micrograph shows rod-shaped bacterial cells captured on the surface of a membrane filter. Note differences in the comparative size of the membrane pores and bacteria. Viruses will pass through this filter. (b) The size of the pores in the filter determines what is captured on the surface of the filter (animal [red] and bacteria [blue]) and removed from liquid passing through. Note the viruses (green) pass through the finer filter. (credit a: modification of work by U.S. Department of Energy)

Cultivation of Viruses

Viruses can be grown in **vivo** (within a whole living organism, plant, or animal) or in **vitro** (outside a living organism in cells in an artificial environment. Flat horizontal cell culture flasks (see Figure 15.20(**a**)) are a common vessel used for in vitro work. Bacteriophages can be grown in the presence of a dense layer of bacteria (also called a **bacterial lawn**) grown in a 0.7 % soft agar in a Petri dish or flat (horizontal) flask (see Figure 15.20(**b**)). As the phage kills the bacteria, many plaques are observed among the cloudy bacterial lawn.

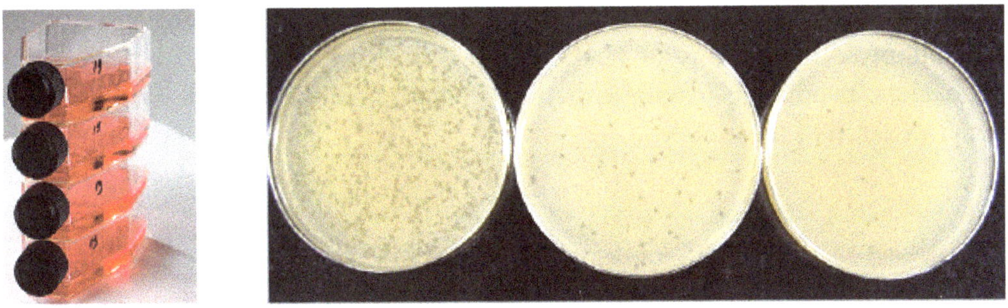

(a) (b)

Figure 15.20 (a) Flasks like this may be used to culture human or animal cells for viral culturing. (b) These plates contain bacteriophage T4 grown on an *Escherichia coli* lawn. Clear plaques are visible where host bacterial cells have been lysed. Viral titers increase on the plates to the left. (credit a: modification of work by National Institutes of Health; credit b: modification of work by American Society for Microbiology)

Animal viruses require cells within a host animal or tissue-culture cells derived from an animal. Animal virus cultivation is important for 1) identification and diagnosis of pathogenic viruses in clinical specimens, 2) production of vaccines, and 3) basic research studies. In vivo host sources can be a developing embryo in an embryonated bird's egg (e.g., chicken, turkey) or a whole animal. For example, most of the influenza vaccine manufactured for annual flu vaccination programs is cultured in hens' eggs.

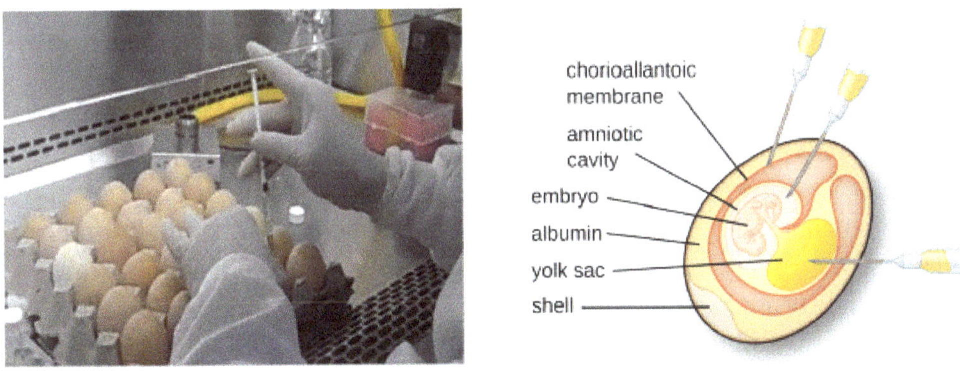

(a) (b)

Figure 15.21 (a) The cells within chicken eggs are used to culture different types of viruses. (b) Viruses can be replicated in various locations within the egg, including the chorioallantoic membrane, the amniotic cavity, and the yolk sac. (credit a: modification of work by "Chung Hoang"/YouTube)

The embryo or host animal serves as an incubator for viral replication (see Figure 15.21). Location within the embryo or host animal is important. Many viruses have a tissue tropism, and must therefore be introduced into a specific site for growth. Within an embryo, target sites include the amniotic cavity, the chorioallantoic membrane, or the yolk sac. Viral infection may damage tissue membranes, producing lesions called pox; disrupt embryonic development; or cause the death of the embryo.

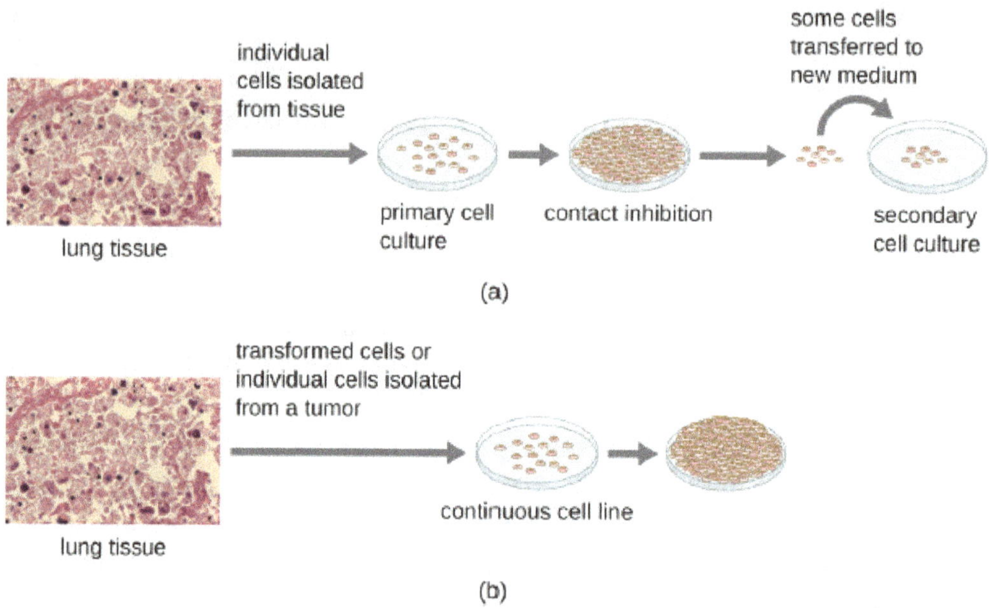

Figure 15.22 Cells for culture are prepared by separating them from their tissue matrix. (a) Primary cell cultures grow attached to the surface of the culture container. Contact inhibition slows the growth of the cells once they become too dense and begin touching each other. At this point, growth can only be sustained by making a secondary culture. (b) Continuous cell cultures are not affected by contact inhibition. They continue to grow regardless of cell density. (credit "micrographs": modification of work by Centers for Disease Control and Prevention)

For in vitro studies, various types of cells can be used to support the growth of viruses. A primary cell culture is freshly prepared from animal organs or tissues. Cells are extracted from tissues by mechanical scraping or mincing to release cells or by an enzymatic method using trypsin or collagenase to break up tissue and release single cells into suspension. Because of anchorage-dependence requirements, primary cell cultures require a liquid culture medium in a Petri dish or tissue-culture flask so cells have a solid surface such as glass or plastic for attachment and growth. Primary cultures usually have a limited life span. When cells in a primary culture undergo mitosis and a sufficient density of cells is produced, cells come in contact with other cells. When this cell-to-cell-contact occurs, mitosis is triggered to stop. This is called contact inhibition and it prevents the density of the cells from becoming too high. To prevent contact inhibition, cells from the primary cell culture must be transferred to another vessel with fresh growth medium. This is called a secondary cell culture. Periodically, cell density must be reduced by pouring off some cells and adding fresh medium to provide space and nutrients to maintain cell growth. In contrast to primary cell cultures, continuous cell lines, usually derived from transformed cells or tumors, are often able to be subcultured many times or even grown indefinitely (in which case they are called immortal). Continuous cell lines may not exhibit anchorage dependency (they will grow in suspension) and may have lost their contact inhibition. As a result, continuous cell lines can grow in piles or lumps resembling small tumor growths (see Figure 15.22).

An example of an immortal cell line is the HeLa cell line, which was originally cultivated from tumor cells obtained from Henrietta Lacks, a patient who died of cervical cancer in 1951. HeLa cells were the first continuous tissue- culture cell line and were used to establish tissue culture as an important technology for research in cell biology, virology, and medicine. Prior to the discovery of HeLa cells, scientists were not able to establish tissue cultures with any reliability or stability. More than six decades later, this cell line is still alive and being used for medical research. See Eye on Ethics: The Immortal Cell Line of Henrietta Lacks to read more about this important cell line and the controversial means by which it was obtained.

The Immortal Cell Line of Henrietta Lacks

In January 1951, Henrietta Lacks, a 30-year-old African American woman from Baltimore, was diagnosed with cervical cancer at Johns Hopkins Hospital. We now know her cancer was caused by the human papillomavirus (HPV). Cytopathic effects of the virus altered the characteristics of her cells in a process called transformation, which gives the cells the ability to divide continuously. This ability, of course, resulted in a cancerous tumor that eventually killed Mrs. Lacks in October at age 31. Before her death, samples of her cancerous cells were taken without her knowledge or permission. The samples eventually ended up in the possession of Dr. George Gey, a biomedical researcher at Johns Hopkins University. Gey was able to grow some of the cells from Lacks's sample, creating what is known today as the immortal HeLa cell line. These cells have the ability to live and grow indefinitely and, even today, are still widely used in many areas of research.

According to Lacks's husband, neither Henrietta nor the family gave the hospital permission to collect her tissue specimen. Indeed, the family was not aware until 20 years after Lacks's death that her cells were still alive and actively being used for commercial and research purposes. Yet HeLa cells have been pivotal in numerous research discoveries related to polio, cancer, and AIDS, among other diseases. The cells have also been commercialized, although they have never themselves been patented. Despite this, Henrietta Lacks's estate has never benefited from the use of the cells, although, in 2013, the Lacks family was given control over the publication of the genetic sequence of her cells.

This case raises several bioethical issues surrounding patients' informed consent and the right to know. At the time Lacks's tissues were taken, there were no laws or guidelines about informed consent. Does that mean she was treated fairly at the time? Certainly by today's standards, the answer would be no. Harvesting tissue or organs from a dying patient without consent is not only considered unethical but illegal, regardless of whether such an act could save other patients' lives. Is it ethical, then, for scientists to continue to use Lacks's tissues for research, even though they were obtained illegally by today's standards?

Ethical or not, Lacks's cells are widely used today for so many applications that it is impossible to list them all. Is this a case in which the ends justify the means? Would Lacks be pleased to know about her contribution to science and the millions of people who have benefited? Would she want her family to be compensated for the commercial products that have been developed using her cells? Or would she feel violated and exploited by the researchers who took part of her body without her consent? Because she was never asked, we will never know.

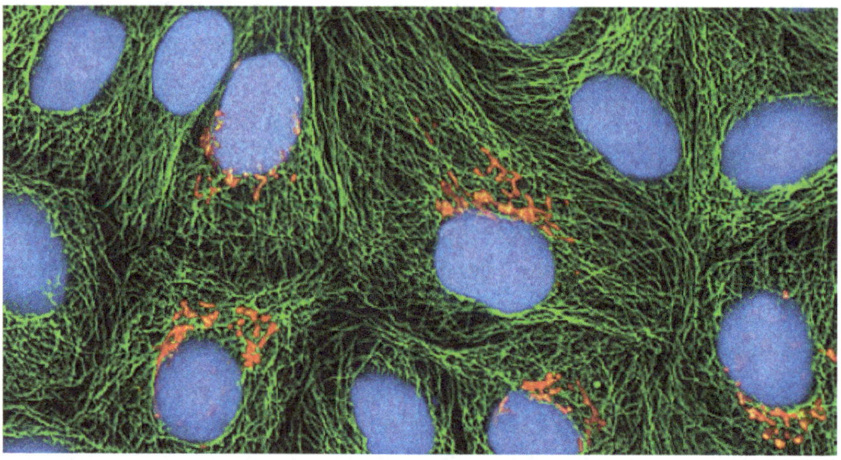

Figure 15.23 A multiphoton fluorescence image of HeLa cells in culture. Various fluorescent stains have been used to show the DNA (cyan), microtubules (green), and Golgi apparatus (orange). (credit: modification of work by National Institutes of Health)

Detection of a Virus

Regardless of the method of cultivation, once a virus has been introduced into a whole host organism, embryo, or tissue-culture cell, a sample can be prepared from the infected host, embryo, or cell line for further analysis under a brightfield, electron, or fluorescent microscope. **Cytopathic effects (CPEs)** are distinct observable cell abnormalities due to viral infection. CPEs can include loss of adherence to the surface of the container, changes in cell shape from flat to round, shrinkage of the nucleus, vacuoles in the cytoplasm, fusion of cytoplasmic membranes and the formation of multinucleated syncytia, inclusion bodies in the nucleus or cytoplasm, and complete cell lysis (see Figure 15.24).

Further pathological changes include viral disruption of the host genome and altering normal cells into transformed cells, which are the types of cells associated with carcinomas and sarcomas. The type or severity of the CPE depends on the type of virus involved. Figure 15.24 lists CPEs for specific viruses.

Cytopathic Effects of Specific Viruses		
Virus	**Cytopathic Effect**	**Example**
Paramyxovirus	Syncytium and faint basophilic cytoplasmic inclusion bodies	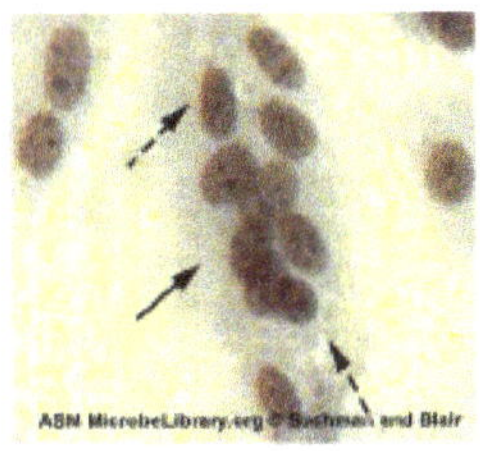
Poxvirus	Pink eosinophilic cytoplasmic inclusion bodies (arrows) and cell swelling	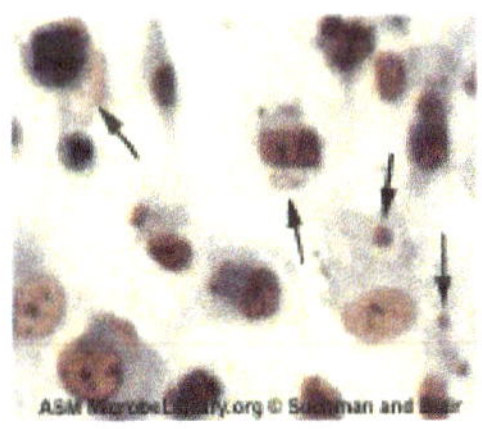
Herpesvirus	Cytoplasmic stranding (arrow) and nuclear inclusion bodies (dashed arrow)	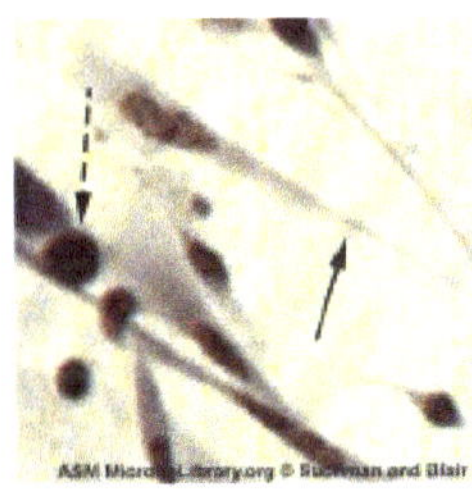
Adenovirus	Cell enlargement, rounding, and distinctive "grape-like" clusters	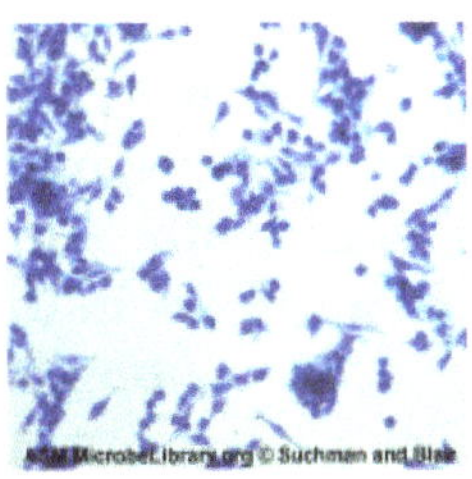

Figure 15.24 (credit "micrographs": modification of work by American Society for Microbiology)

Hemagglutination Assay

A serological assay is used to detect the presence of certain types of viruses in patient serum. Serum is the straw-colored liquid fraction of blood plasma from which clotting factors have been removed. Serum can be used in a direct assay called a hemagglutination assay to detect specific types of viruses in the patient's sample. Hemagglutination is the agglutination (clumping) together of erythrocytes (red blood cells). Many viruses produce surface proteins or spikes called hemagglutinins that can bind to receptors on the membranes of erythrocytes and cause the cells to agglutinate. Hemagglutination is observable without using the microscope, but this method does not always differentiate between infectious and non-infectious viral particles, since both can agglutinate erythrocytes.

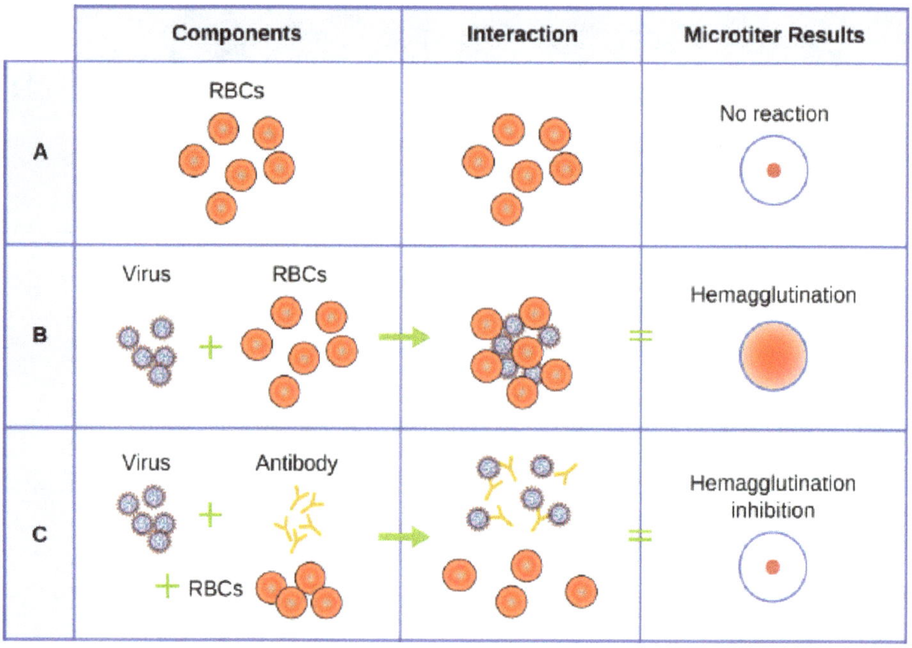

Figure 15.25 This chart shows the possible outcomes of a hemagglutination test. Row A: Erythrocytes do not bind together and will sink to the bottom of the well plate; this becomes visible as a red dot in the center of the well. Row B: Many viruses have hemagglutinins that causes agglutination of erythrocytes; the resulting hemagglutination forms a lattice structure that results in red colour throughout the well. Row C: Virus-specific antibody, the viruses, and the erythrocytes are added to the well plate. The virus-specific antibodies inhibit agglutination, as can be seen as a red dot in the bottom of the well. (credit: modification of work by Centers for Disease Control and Prevention)

To identify a specific pathogenic virus using hemagglutination, we must use an indirect approach. Proteins called antibodies, generated by the patient's immune system to fight a specific virus, can be used to bind to components such as hemagglutinins that are uniquely associated with specific types of viruses. The binding of the antibodies with the hemagglutinins found on the virus subsequently prevent erythrocytes from directly interacting with the virus. So when erythrocytes are added to the antibody-coated viruses, there is no appearance of agglutination; agglutination has been inhibited. We call these types of indirect assays for virus-specific antibodies hemagglutination inhibition (HAI) assays. HAI can be used to detect the presence of antibodies specific to many types of viruses that may be causing or have caused an infection in a patient even months or years after infection (see Figure 15.25). This assay is described in greater detail in Agglutination Assays.

Nucleic Acid Amplification Test

Nucleic acid amplification tests (NAAT) are used in molecular biology to detect unique nucleic acid sequences of viruses in patient samples. Polymerase chain reaction (PCR) is an NAAT used to detect the presence of viral DNA in a patient's tissue or body fluid sample. PCR is a technique that amplifies (i.e., synthesizes many copies) of a viral DNA segment of interest. Using PCR, short nucleotide sequences called primers bind to specific sequences of viral DNA, enabling identification of the virus.

Reverse transcriptase-PCR (RT-PCR) is an NAAT used to detect the presence of RNA viruses. RT-PCR differs from PCR in that the enzyme reverse transcriptase (RT) is used to make a cDNA from the small amount of viral RNA in the specimen. The cDNA can then be amplified by PCR. Both PCR and RT-PCR are used to detect and confirm the presence of the viral nucleic acid in patient specimens.

HPV Scare

Michelle, a 21-year-old nursing student, came to the university clinic worried that she might have been exposed to a sexually transmitted disease (STD). Her sexual partner had recently developed several bumps on the base of his penis. He had put off going to the doctor, but Michelle suspects they are genital warts caused by HPV. She is especially concerned because she knows that HPV not only causes warts but is a prominent cause of cervical cancer. She and her partner always use condoms for contraception, but she is not confident that this precaution will protect her from HPV.

Michelle's physician finds no physical signs of genital warts or any other STDs, but recommends that Michelle get a Pap smear along with an HPV test. The Pap smear will screen for abnormal cervical cells and the CPEs associated with HPV; the HPV test will test for the presence of the virus. If both tests are negative, Michelle can be more assured that she most likely has not become infected with HPV. However, her doctor suggests it might be wise for Michelle to get vaccinated against HPV to protect herself from possible future exposure.

Enzyme Immunoassay

Enzyme immunoassays (EIAs) rely on the ability of antibodies to detect and attach to specific biomolecules called antigens. The detecting antibody attaches to the target antigen with a high degree of specificity in what might be a complex mixture of biomolecules. Also included in this type of assay is a colourless enzyme attached to the detecting antibody.

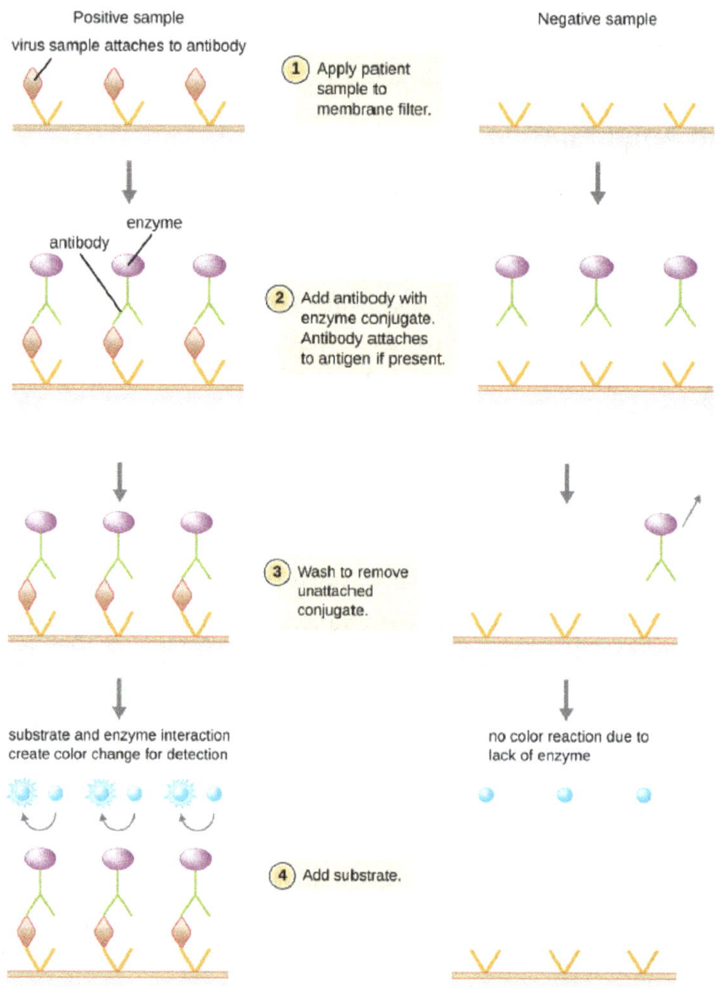

Figure 15.26 Similar to rapid, over-the-counter pregnancy tests, EIAs for viral antigens require a few drops of diluted patient serum or plasma applied to a membrane filter. The membrane filter has been previously modified and embedded with antibody to viral antigen and internal controls. Antibody conjugate is added to the filter, with the targeted antibody attached to the antigen (in the case of a positive test). Excess conjugate is washed off the filter. Substrate is added to activate the enzyme-mediated reaction to reveal the colour change of a positive test. (credit: modification of work by "Cavitri"/Wikimedia Commons)

The enzyme acts as a tag on the detecting antibody and can interact with a colourless substrate, leading to the production of a coloured end product. EIAs often rely on layers of antibodies to capture and react with antigens, all of which are attached to a membrane filter (see Figure 15.26). EIAs for viral antigens are often used as preliminary screening tests. If the results are positive, further confirmation will require tests with even greater sensitivity, such as a western blot or an NAAT. EIAs are discussed in more detail in EIAs and ELISAs.

Along with the RT/PCR analysis, David's saliva was also collected for viral cultivation. In general, no single

diagnostic test is sufficient for antemortem diagnosis, since the results will depend on the sensitivity of the assay, the quantity of virions present at the time of testing, and the timing of the assay, since release of virions in the saliva can vary. As it turns out, the result was negative for viral cultivation from the saliva. This is not surprising to David's doctor, because one negative result is not an absolute indication of the absence of infection. It may be that the number of virions in the saliva is low at the time of sampling. It is not unusual to repeat the test at intervals to enhance the chance of detecting higher virus loads.

VACCINES

For many diseases, prevention is the best form of treatment, and few strategies for disease prevention are as effective as vaccination. Vaccination is a form of artificial immunity. By artificially stimulating the adaptive immune defenses, a vaccine triggers memory cell production similar to that which would occur during a primary response. In so doing, the patient is able to mount a strong secondary response upon exposure to the pathogen—but without having to first suffer through an initial infection. In this section, we will explore several different kinds of artificial immunity along with various types of vaccines and the mechanisms by which they induce artificial immunity.

Classifications of Adaptive Immunity

All forms of adaptive immunity can be described as either active or passive. **Active immunity** refers to the activation of an individual's own adaptive immune defenses, whereas **passive immunity** refers to the transfer of adaptive immune defenses from another individual or animal. Active and passive immunity can be further subdivided based on whether the protection is acquired naturally or artificially.

Natural active immunity is adaptive immunity that develops after natural exposure to a pathogen (Figure 15.27). Examples would include the lifelong immunity that develops after recovery from a chickenpox or measles infection (although an acute infection is not always necessary to activate adaptive immunity).

The length of time that an individual is protected can vary substantially depending upon the pathogen and antigens involved. For example, activation of adaptive immunity by protein spike structures during an intracellular viral infection can activate lifelong immunity, whereas activation by carbohydrate capsule antigens during an extracellular bacterial infection may activate shorter-term immunity.

Natural passive immunity involves the natural passage of antibodies from a mother to her child before and after birth. IgG is the only antibody class that can cross the placenta from mother's blood to the fetal blood supply. Placental transfer of IgG is an important passive immune defense for the infant, lasting up to six months after birth. Secretory IgA can also be transferred from mother to infant through breast milk.

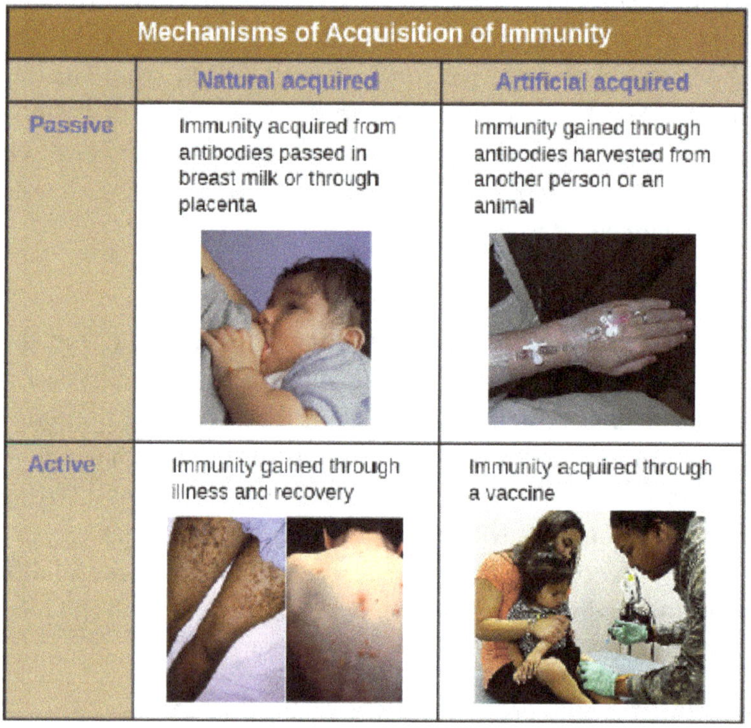

Figure 15.27 The four classifications of immunity. (credit top left photo: modification of work by USDA; credit top right photo: modification of work by "Michaelberry"/Wikimedia; credit bottom left photo: modification of work by Centers for Disease Control and Prevention; credit bottom right photo: Airman 1st Class Destinee Doughert / U.S. Air Force; Public Domain)

Artificial passive immunity refers to the transfer of antibodies produced by a donor (human or animal) to another individual. This transfer of antibodies may be done as a prophylactic measure (i.e., to prevent disease after exposure to a pathogen) or as a strategy for treating an active infection. For example, artificial passive immunity is commonly used for post-exposure prophylaxis against rabies, hepatitis A, hepatitis B, and chickenpox (in high risk individuals). Active infections treated by artificial passive immunity include cytomegalovirus infections in immune compromised patients and Ebola virus infections. In 1995, eight patients in the Democratic Republic of the Congo with active Ebola infections were treated with blood transfusions from patients who were recovering from Ebola. Only one of the eight patients died (a 12.5% mortality rate), which was much lower than the expected 80% mortality rate for Ebola in untreated patients. Artificial passive immunity is also used for the treatment of diseases caused by bacterial toxins, including tetanus, botulism, and diphtheria.

Artificial active immunity is the foundation for vaccination. It involves the activation of adaptive immunity through the deliberate exposure of an individual to weakened or inactivated pathogens, or preparations consisting of key pathogen antigens.

Herd Immunity

The four kinds of immunity just described result from an individual's adaptive immune system. For any given disease, an individual may be considered immune or susceptible depending on his or her ability to mount an effective immune response upon exposure. Thus, any given population is likely to have some individuals who are immune and other individuals who are susceptible. If a population has very few susceptible individuals, even those susceptible individuals will be protected by a phenomenon called **herd immunity**. Herd immunity has nothing to do with an individual's ability to mount an effective immune response; rather, it occurs because there are too few susceptible individuals in a population for the disease to spread effectively.

Vaccination programs create herd immunity by greatly reducing the number of susceptible individuals in a population. Even if some individuals in the population are not vaccinated, as long as a certain percentage is immune (either naturally or artificially), the few susceptible individuals are unlikely to be exposed to the pathogen. However, because new individuals are constantly entering populations (for example, through birth or relocation), vaccination programs are necessary to maintain herd immunity.

Vaccination: Obligation or Choice

A growing number of parents are choosing not to vaccinate their children. They are dubbed " antivaxxers," and the majority of them believe that vaccines are a cause of autism (or other disease conditions), a link that has now been thoroughly disproven. Others object to vaccines on religious or moral grounds (e.g., the argument that Gardasil vaccination against HPV may promote sexual promiscuity), on personal ethical grounds (e.g., a conscientious objection to any medical intervention), or on political grounds (e.g., the notion that mandatory vaccinations are a violation of individual liberties).

It is believed that this growing number of unvaccinated individuals has led to new outbreaks of whooping cough and measles. We would expect that herd immunity would protect those unvaccinated in our population, but herd immunity can only be maintained if enough individuals are being vaccinated.

Vaccination is clearly beneficial for public health. But from the individual parent's perspective the view can be murkier. Vaccines, like all medical interventions, have associated risks, and while the risks of vaccination may be extremely low compared to the risks of infection, parents may not always understand or accept the consensus of the medical community. Do such parents have a right to withhold vaccination from their children? Should they be allowed to put their children—and society at large—at risk?

Many governments insist on childhood vaccinations as a condition for entering public school, but it has become easy in most states to opt out of the requirement or to keep children out of the public system. Since the 1970s, West Virginia and Mississippi have had in place a stringent requirement for childhood vaccination, without exceptions, and neither state has had a case of measles since the early 1990s. California lawmakers recently passed a similar law in response to a measles outbreak in 2015, making it much more difficult for parents to opt out of vaccines if their children are attending public schools. Given this track record and renewed legislative efforts, should other states adopt similarly strict requirements?

What role should health-care providers play in promoting or enforcing universal vaccination? Studies have shown that many parents' minds can be changed in response to information delivered by health- care workers, but is it the place of health-care workers to try to persuade parents to have their children vaccinated? Some health-care providers are understandably reluctant to treat unvaccinated patients. Do they have the right to refuse service to patients who decline vaccines? Do insurance companies have the right to deny coverage to antivaxxers? These are all ethical questions that policymakers may be forced to address as more parents skirt vaccination norms.

Variolation and Vaccination

Thousands of years ago, it was first recognized that individuals who survived a smallpox infection were immune to subsequent infections. The practice of inoculating individuals to actively protect them from smallpox appears to have originated in the 10th century in China, when the practice of **variolation** was described (Figure 15.28). Variolation refers to the deliberate inoculation of individuals with infectious material from scabs or pustules of smallpox victims.

Infectious materials were either injected into the skin or introduced through the nasal route. The infection that developed was usually milder than naturally acquired smallpox, and recovery from the milder infection provided protection against the more serious disease.

Although the majority of individuals treated by variolation developed only mild infections, the practice was not without risks. More serious and sometimes fatal infections did occur, and because smallpox was contagious, infections resulting from variolation could lead to epidemics. Even so, the practice of variolation for smallpox prevention spread to other regions, including India, Africa, and Europe.

Figure 15.28 Variolation for smallpox originated in the Far East and the practice later spread to Europe and Africa. This Japanese relief depicts a patient receiving a smallpox variolation from the physician Ogata Shunsaku (1748–1810).

Although variolation had been practiced for centuries, the English physician Edward Jenner (1749–1823) is generally credited with developing the modern process of vaccination. Jenner observed that milkmaids who developed cowpox, a disease similar to smallpox but milder, were immune to the more serious smallpox. This led Jenner to hypothesize that exposure to a less virulent pathogen could provide immune protection against a more virulent pathogen, providing a safer alternative to variolation. In 1796, Jenner tested his hypothesis by obtaining infectious samples from a milkmaid's active cowpox lesion and injecting the materials into a young boy (Figure 15.29). The boy developed a mild infection that included a low-grade fever, discomfort in his axillae (armpit) and loss of appetite.

When the boy was later infected with infectious samples from smallpox lesions, he did not contract smallpox. This new approach was termed **vaccination**, a name deriving from the use of cowpox (Latin *vacca* meaning "cow") to protect against smallpox. Today, we know that Jenner's vaccine worked because the cowpox virus is genetically and antigenically related to the *Variola* viruses that caused smallpox. Exposure to cowpox antigens resulted in a primary response and the production of memory cells that identical or related epitopes of Variola virus upon a later exposure to smallpox.

The success of Jenner's smallpox vaccination led other scientists to develop vaccines for other diseases. Perhaps the most notable was Louis Pasteur, who developed vaccines for rabies, cholera, and anthrax. During the 20th and 21st centuries, effective vaccines were developed to prevent a wide range of diseases caused by viruses (e.g., chickenpox and shingles, hepatitis, measles, mumps, polio, and yellow fever) and bacteria (e.g., diphtheria, pneumococcal pneumonia, tetanus, and whooping cough,).

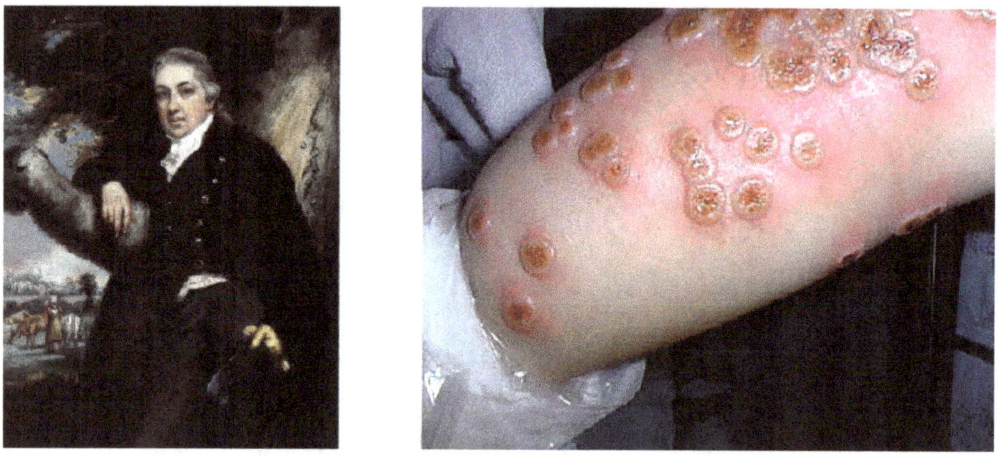

(a) (b)

Figure 15.29 (a) A painting of Edward Jenner depicts a cow and a milkmaid in the background. (b) Lesions on a patient infected with cowpox, a zoonotic disease caused by a virus closely related to the one that causes smallpox. (credit b: modification of work by the Centers for Disease Control and Prevention)

Classes of Vaccines

For a vaccine to provide protection against a disease, it must expose an individual to pathogen-specific antigens that will stimulate a protective adaptive immune response. By its very nature, this entails some risk. As with any pharmaceutical drug, vaccines have the potential to cause adverse effects. However, the ideal vaccine causes no severe adverse effects and poses no risk of contracting the disease that it is intended to prevent. Various types of vaccines have been developed with these goals in mind. These different classes of vaccines are described in the next section and summarized in Table 15.1.

Live Attenuated Vaccines

Live attenuated vaccines expose an individual to a weakened strain of a pathogen with the goal of establishing a subclinical infection that will activate the adaptive immune defenses. Pathogens are attenuated to decrease their virulence using methods such as genetic manipulation (to eliminate key virulence factors) or long-term culturing in an unnatural host or environment (to promote mutations and decrease virulence).

By establishing an active infection, live attenuated vaccines stimulate a more comprehensive immune response than some other types of vaccines. Live attenuated vaccines activate both cellular and humoral immunity and stimulate the development of memory for long-lasting immunity. In some cases, vaccination of one individual with a live attenuated pathogen can even lead to natural transmission of the attenuated pathogen to other individuals. This can cause the other individuals to also develop an active, subclinical infection that activates their adaptive immune defenses.

Disadvantages associated with live attenuated vaccines include the challenges associated with long-term storage and transport as well as the potential for a patient to develop signs and symptoms of disease during the active infection (particularly in immuno compromised patients). There is also a risk of the attenuated pathogen reverting back to full virulence. Table 15.1 lists examples live attenuated vaccines.

Inactivated Vaccines

Inactivated vaccines contain whole pathogens that have been killed or inactivated with heat, chemicals, or radiation. For inactivated vaccines to be effective, the inactivation process must not affect the structure of key antigens on the pathogen.

Because the pathogen is killed or inactive, inactivated vaccines do not produce an active infection, and the resulting immune response is weaker and less comprehensive than that provoked by a live attenuated vaccine. Typically the response involves only humoral immunity, and the pathogen cannot be transmitted to other individuals. In addition, inactivated vaccines usually require higher doses and multiple boosters, possibly causing inflammatory reactions at the site of injection.

Despite these disadvantages, inactivated vaccines do have the advantages of long-term storage stability and ease of transport. Also, there is no risk of causing severe active infections. However, inactivated vaccines are not without their side effects. Table 15.1 lists examples of inactivated vaccines.

Subunit Vaccines

Whereas live attenuated and inactive vaccines expose an individual to a weakened or dead pathogen, subunit vaccines only expose the patient to the key antigens of a pathogen—not whole cells or viruses. Subunit vaccines can be produced either by chemically degrading a pathogen and isolating its key antigens or by producing the antigens through genetic engineering. Because these vaccines contain only the essential antigens of a pathogen, the risk of side effects is relatively low. Table 15.1 lists examples of subunit vaccines.

Toxoid Vaccines

Like subunit vaccines, **toxoid vaccines** do not introduce a whole pathogen to the patient; they contain inactivated bacterial toxins, called toxoids. Toxoid vaccines are used to prevent diseases in which bacterial toxins play an important role in pathogenesis. These vaccines activate humoral immunity that neutralizes the toxins. Table 15.1 lists examples of toxoid vaccines.

Conjugate Vaccines

A **conjugate vaccine** is a type of subunit vaccine that consists of a protein conjugated to a capsule polysaccharide. Conjugate vaccines have been developed to enhance the efficacy of subunit vaccines against pathogens that have protective polysaccharide capsules that help them evade phagocytosis, causing invasive infections that can lead to meningitis and other serious conditions. The subunit vaccines against these pathogens introduce T-independent capsular polysaccharide antigens that result in the production of antibodies that can opsonize the capsule and thus combat the infection; however, children under the age of two years do not respond effectively to these vaccines. Children do respond effectively when vaccinated with the conjugate vaccine, in which a protein with T-dependent antigens is conjugated to the capsule polysaccharide. The conjugated protein-polysaccharide antigen stimulates production of antibodies against both the protein and the capsule polysaccharide. Table 15.1 lists examples of conjugate vaccines.

Table 15.1 Classes of Vaccines

Class	Description	Advantages	Disadvantages	Examples
Live attenuated	Weakened strain of whole pathogen	Cellular and humoral immunity	Difficult to store and transport	Chickenpox, German measles, measles, mumps, tuberculosis, typhoid fever, yellow fever
		Long-lasting immunity	Risk of infection in immune compromised patients	
		Transmission to contacts	Risk of reversion	

Class	Description	Advantages	Disadvantages	Examples
Inactivated	Whole pathogen killed or inactivated with heat, chemicals, or radiation	Ease of storage and transport	Weaker immunity (humoral only)	Cholera, hepatitis A, influenza, plague, rabies
		No risk of severe active infection	Higher doses and more boosters required	
Subunit	Immunogenic antigens	Lower risk of side effects	Limited longevity	Anthrax, hepatitis B, influenza, meningitis, papillomavirus, pneumococcal pneumonia, whooping cough
			Multiple doses required	
			No protection against antigenic variation	

Class	Description	Advantages	Disadvantages	Examples
Toxoid	Inactivated bacterial toxin	Humoral immunity to neutralize toxin	Does not prevent infection	Botulism, diphtheria, pertussis, tetanus
Conjugate	Capsule polysaccharide conjugated to protein	T-dependent response to capsule	Costly to produce	Meningitis (*Haemophilus influenzae, Streptococcus pneumoniae, Neisseria meningitides*)
			No protection against antigenic variation	
		Better response in young children	May interfere with other vaccines	

DNA Vaccines

DNA vaccines represent a relatively new and promising approach to vaccination. A DNA vaccine is produced by incorporating genes for antigens into a recombinant plasmid vaccine. Introduction of the DNA vaccine into a patient leads to uptake of the recombinant plasmid by some of the patient's cells, followed by transcription and translation of antigens and presentation of these antigens with MHC I to activate adaptive immunity. This results in the stimulation of both humoral and cellular immunity without the risk of active disease associated with live attenuated vaccines.

Although most DNA vaccines for humans are still in development, it is likely that they will become more prevalent in the near future as researchers are working on engineering DNA vaccines that will activate adaptive immunity against several different pathogens at once. First-generation DNA vaccines tested in the 1990s looked promising in animal models but were disappointing when tested in human subjects. Poor cellular uptake of the DNA plasmids was one of the major problems impacting their efficacy. Trials of second-generation DNA vaccines have been more promising thanks to new techniques for enhancing cellular uptake and optimizing antigens. DNA vaccines for various cancers and viral pathogens such as HIV, HPV, and hepatitis B and C are currently in development.

Some DNA vaccines are already in use. In 2005, a DNA vaccine against West Nile virus was approved for use in horses in the United States. Canada has also approved a DNA vaccine to protect fish from infectious hematopoietic necrosis virus. A DNA vaccine against Japanese encephalitis virus was approved for use in humans in 2010 in Australia.

Resolution

Based on Olivia's symptoms, her physician made a preliminary diagnosis of bacterial meningitis without waiting for positive identification from the blood and CSF samples sent to the lab. Olivia was admitted to the hospital and treated with intravenous broad-spectrum antibiotics and rehydration therapy. Over the next several days, her condition began to improve, and new blood samples and lumbar puncture samples showed an absence of microbes in the blood and CSF with levels of white blood cells returning to normal. During this time, the lab produced a positive identification of *Neisseria meningitidis*, the causative agent of meningococcal meningitis, in her original CSF sample.

N. meningitidis produces a polysaccharide capsule that serves as a virulence factor. *N. meningitidis* tends to affect infants after they begin to lose the natural passive immunity provided by maternal antibodies. At one year of age, Olivia's maternal IgG antibodies would have disappeared, and she would not have developed memory cells capable of recognizing antigens associated with the polysaccharide capsule of the *N. meningitidis.* As a result, her adaptive immune system was unable to produce protective antibodies to combat the infection, and without antibiotics she may not have survived. Olivia's infection likely would have been avoided altogether had she been vaccinated. A conjugate vaccine to prevent meningococcal meningitis is available and approved for infants as young as two months of age. However, current vaccination schedules in the United States recommend that the vaccine be administered at age 11–12 with a booster at age 16.

In countries with developed public health systems, many vaccines are routinely administered to children and adults. Vaccine schedules are changed periodically, based on new information and research results gathered by public health agencies.

INDEX

B

C

www.ingramcontent.com/pod-product-compliance
Lightning Source LLC
LaVergne TN
LVHW062212200926
848484LV00006B/1376
* 9 7 8 1 7 8 7 1 5 2 6 2 5 *